Dr. habil. Alexander Konyukhov

Geometrically Exact Theory for Contact Interactions

Geometrically Exact Theory for Contact Interactions

by
Dr. habil. Alexander Konyukhov

Habilitation, Karlsruher Institut für Technologie
Institut für Mechanik, 2010
Tag des Habilitationsvortrages: 16. April 2010
Tag des Habilitationskolloquiums: 1. Dezember 2010
Referenten: Prof. Dr.-Ing. Karl Schweizerhof
              Prof. Dr.-Ing. habil. Peter Wriggers

Impressum

Karlsruher Institut für Technologie (KIT)
KIT Scientific Publishing
Straße am Forum 2
D-76131 Karlsruhe
www.ksp.kit.edu

KIT – Universität des Landes Baden-Württemberg und nationales
Forschungszentrum in der Helmholtz-Gemeinschaft

KIT Scientific Publishing 2011
Print on Demand

ISBN: 978-3-86644-672-4

Chi conosce la geometria, può comprendere tutto in questo mondo.

(Galileo Galilei, 1564-1642)

من يفهم الهندسة يستطيع أن يفهم كل شيء في العالم.

Նա, ով հասկանում է երկրաչափություն, կարող է հասկանալ ամեն ինչ այս աշխարհում:

Който разбира геометрията, той разбира всичко на света.

他懂得幾何，也懂得世上的任何事情。

Tko razumije geometriju, može razumjeti sve na ovom svijetu.

Kdo rozumí geometrii, rozumí všemu ve vesmíru.

Who understands geometry understands anything in this world.

Joka geometrian ymmärtää, voi tässä maailmassa kaiken käsittää.

Celui qui comprend la géometrie, est capable de comprendre tout dans le monde.

Wer die Geometrie begreift, vermag in dieser Welt alles zu verstehen.

ვინც გაიგებს გეომეტრიას, იგი გაიგებს ამ სამყაროს

Όποιος καταλαβαίνει τη γεωμετρία, μπορεί να καταλάβει καθετί στον κόσμο.

Aki a geometriát érti, mindent ért a világon.

जो ज्यामिति को समझ लेता है , वह इस दुनिया में सब कुछ समझ सकता है।

Chi conosce la geometria, può comprendere tutto in questo mondo.

כל מי שמבין גיאומטריה מסוגל להבין כל דבר בעולם.

幾何学を理解する者は世界のすべてを理解し得る。

Qui geometriam intellegat, is universum totum intellegit.

Ten kto rozumie geometrię jest w stanie zrozumieć wszystko na tym świecie.

Quem compreende geometria é capaz de entender tudo neste mundo.

Тот, кто понимает геометрию, способен всё понять в этом мире.

Quien entiende la geometría puede entenderlo todo en este mundo.

Den som förstår geometri kan förstå allt i världen.

ใครผู้ใดรู้ลึกซึ้งในเรขาคณิต สามารถเข้าใจทุกสิ่งในโลกหล้า

O ki geometriyi anlayan bu dünyada herşeyi anlayabilecek olandır.

The author is very thankful to many international friends and colleagues who help this page to be written in their mother languages.
The list of languages is alphabetical according to the English names of the languages.

Acknowledgements

The current work is appearing as a summary of my research activity in the Institute of Mechanics of the University of Karlsruhe which nowdays changed the name into the Karlsruhe Institute of Technology.

First of all, I wish to express my great appreciation to Professor Dr.-Ing. Karl Schweizerhof for his huge ability to overview and to discuss many different topics in computational mechanics. This helped me a lot to run on research and to reconsider the computational contact mechanics from various points of view as well as to work productively together over long period.

Many thanks to Professor Dr.-Ing. habil. Peter Wriggers for his willingness to be a referent of this cumulative research work. His pioneering works on many subjects in contact became a lighthouse in the huge sea of computational contact mechanics.

I am also thankfull to all colleagues from the Institute of Mechanics for the fruitfull and creative atmosphere at the Institute.

I would like to thank my mother for encouraging me to stay on research for a long time far away from home.

Abstract

The intuitive understanding of contact between bodies is based on the geometry of adjoining bodies. A more sophisticated approach of an advanced analysis including the application of various numerical methods is to take advantage of the geometry of an analyzed object and describe the problem in the best coordinate system. The best coordinate system to describe contact interaction in all its geometrical details is a coordinate system attached to the geometrical features of contacting bodies. In the current thesis a systematical analysis of geometrical situations leading to contact pairs – surface-to-surface, line-to-surface, point-to-surface, line-to-line, point-to-line is presented. Each contact pair is inherited with a special coordinate system based on its geometrical properties such as a Gaussian surface coordinate system, or a Serret-Frenet curve coordinate system. Then standard methods well known in computational contact mechanics such as penalty methods, Lagrange multipliers methods and others are formulated in these coordinate systems. Such formulations require then the powerful apparatus of differential geometry of surfaces and curves and of convex analysis. The final goals of such transformations are then ready-for-implementation numerical algorithms within the finite element method which are most convenient for a certain geometrical situation because they contain their intrinsic geometrical properties. Among the advantages of this consideration are

- *the formulation of classical contact algorithms (penalty, Lagrange multipliers methods) in closed form independent on approximations of surfaces and curves;*

- *generalization of the well known Coulomb friction law into arbitrary surface interface laws with an example of a coupled anisotropic adhesion and friction law where the adhesion is defining the geometrical micro-structure of surfaces. This result includes even a proved experimental validation;*

- *development of a new description for curve-to-curve contact with applications to beam-to-beam and to edge-to-edge contact.*

In addition, a number of different numerical features appearing during this development are analyzed in several chapters of the current thesis.

Kurzfassung

Kontakt zwischen Körpern beruht intuitiv auf der Geometrie angrenzender Körper. Auch aktuelle Ansätze zur Lösung von Kontaktproblemen insbesonodere bei Anwendung diverser numerischer Methoden bedienen sich des Vorteils der bekannten Geometrie der gegebenen Strukturen. Zur Beschreibung des jeweiligen Kontakts zwischen den einzelnen Körpern ist dazu ein Bezugssystem zu wählen, das die geometrischen Eigenschaften der kontaktierenden Körper am besten wiedergibt. In der vorliegenden Arbeit wird dazu ein systematisches Vorgehen zur Ermittlung von geometrischen Konstellationen, Kontaktpaaren (surface-to-surface, line-to-surface, point-to-surface, line-to-line, point-to-line), aufgezeigt. Jedes dieser Kontaktpaare besitzt dabei ein spezielles Koordinatensystem. Es basiert je nach geometrischer Eigenschaft des Problems auf Koordinatensystemen wie dem auf einer Gaußschen Oberfläche oder dem an einer Serret-Frenet Kurve. Auf der Grundlage dieser Koordinatensysteme werden dann die in der computergestützten Kontaktmechanik bekannten Methoden wie das Penalty- und das Lagrange Multiplikatorenverfahren formuliert. Die Umsetzung dieser Formulierungen erfordert die Mittel der Differentialgeometrie für Flächen und Kurven sowie konvexe Analysis. Das Ziel dieser Umsetzung sind letztlich numerische Algorithmen, die sich einfach in die Finite Element Umgebung implementieren lassen und die aufgrund der geometrischen Eigenschaften der behandelten Problemstellung besonders geeignet sind.

Als Vorteile sind dabei u.a. aufzuführen:

- *die Unabhängigkeit der Formulierung von klassischen Kontaktalgorithmen in geschlossener Form (Penalty, Lagrange Multiplikatoren) von der Oberflächen- bzw. Kurvendiskretisierung;*

- *die Verallgemeinerung des bekannten Coulomb-Reibgesetzes in eine allgemeine Grenzflächenformulierung. Dies wird anhand eines Beispiels der Kopplung von anisotropem Haft- und Reibgesetz aufgezeigt und experimentell validiert, wobei die Haftung durch die geometrische Mikrostruktur der Oberfläche definiert wird;*

- *die Entwicklung einer neuen Beschreibung für curve-to-curve - Kontakt mit Anwendungen zum beam-to-beam und edge-to-edge - Kontakt.*

Außerdem werden unterschiedliche numerische Eigenschaften der Verfahren untersucht, die sich bei dieser Art von Kontaktbetrachtung ergeben.

I would like to dedicate this thesis to my children:

David and Zarina

Contents

CONTENTS

1

Introduction

1.1 Introduction

Computational *contact* mechanics has been developed into a separated branch of computational mechanics during the last decades. A fairly large number of publications on computational contact mechanics has been published since then. In order to locate the development in contact mechanics focusing on its computational issues and formulate the goals of the current work, first, we try to classify these developments.

1.1.1 Overview of approaches to model contact problems

The overview of the current state of the art in computational contact mechanics is arranged in such a way that the publications are classified by different approaches used for modeling, by different solution methods etc., therefore, a separated article can be mentioned in different subsections of the overview. Limits and drawbacks are then discussed in a special subsection.

In computational contact mechanics the modeling process depends generally on assumptions concerning the geometry of objects – 3D bodies, shells, beams, ropes and other combinations – leading to appropriate computational models. The robustness and tolerance of the computations depends on the numerical method involved in contact modeling.

In order to classify known approaches the modeling process can be split into several steps:

1. kinematics of contact interactions

2. constitutive relations for contact conditions

3. methods to enforce contact conditions

4. formulation of basic mechanical principles (conservation of momentum, equilibrium etc.) on contact boundaries

5. numerical solution and analysis of the proposed model.

Here we are discussing these steps classifying various numerical methods known in literature focusing mostly on the recent developments.

1.1.1.1 Step 1: Kinematics of contact interactions

In this step types of contact interactions together with measures of contact interaction are defined similar to deformation measures in continuum mechanics. Regarding the relative motion of contacting bodies it is important to distinguish the following interactions: *a)* **tied contact** and *b)* **arbitrary contact with** *b1)* **small** and with *b2)* **large relative displacements.**

a) **Tied contact** requires simple "sewing" of two contacting bodies, and, therefore, a pointwise enforcement of contact as

$$r = \left\| \mathbf{r}_A - \mathbf{r}_B \right\| = 0, \tag{1.1}$$

assuming that vectors from a body A and a body B describe the same geometrical point at the beginning of a deformation process, e.g. $\mathbf{r}_A - \mathbf{r}_B = 0$. Enforcement of a tied contact can be regarded as additional constraints on interface surfaces and has been introduced for the finite element method in earlier developments as mesh tying algorithms. This has been solved by the Lagrange multiplier method for dominantly straight contact boundaries already by Francavilla and Zienkiewicz [43] (1975). More recently mesh tying algorithms are developed for the case of curved contact interfaces. Tan [171] (2003) considered a special construction of contact matrices in 2D for mesh matching and for the contact patch test. Various techniques based on the construction of special volume elements after the projection procedure were developed by

Heinstein and Laursen [108] (2003), [66] (2003) for mesh tying of arbitrary curved interfaces. Puso and Laursen [149] (2003) used the Mortar method for mesh tying of curved interfaces in 3D.

b) Both normal and tangential displacements should be distinguished in the case of **arbitrary contact.** This leads to a splitting of displacements into a normal penetration and into tangential displacements. In early publications, see Kikuchi and Oden [84] (1988), *a normal penetration has been introduced as a result of the linearization of nonlinear non-penetration conditions eqn. (1.1) in the direction of the normal to the surface of a contacting body.* Hallquist et.al. [53] (1985) introduced this split, and therefore, a penetration via the projection operation. Thus, *a, so-called, "master-slave" algorithm* within an explicit time integration scheme for dynamic problems *has been introduced.* For applications with *implicit schemes the "master-slave" algorithm has then been modified* in Wriggers and Simo [192] (1985) and in Simo et.al. [162] (1986). The projection can be formulated in extremal form as

$$\|\mathbf{r}_s - \boldsymbol{\rho}\| \longrightarrow \min. \tag{1.2}$$

The "master-slave" approach is treating the normal contact such that a "slave" node $\mathbf{r}_s$ of one body penetrates into a "master" segment of another body parameterized with ρ. The value of penetration p is measured as a closest distance to the master body and, therefore, in the direction of the normal $\mathbf{n}$ of the "master" surface

$$p = (\mathbf{r}_s - \boldsymbol{\rho}) \cdot \mathbf{n}. \tag{1.3}$$

This measure can be defined differently depending either on the geometry of contacting bodies, or on further application of various numerical methods.

a) For contact between surfaces the penetration can be defined in the following ways:

   i in the direction of the normal $\mathbf{n}$ to a rigid undeformed body for Signorini problems as e.g. in Kikuchi and Oden [84] (1988);

   ii in the direction of the normal $\mathbf{n}$ to an initially undeformed surface for small displacement problems, see the "node-to-node" approach in Wriggers [188] (2002).

iii in the directions of normals to spheres around certain nodes representing the contacting body, see the "pinball algorithm" in Belytschko and Neal [10] (1991).

iv in the directions of the normal to a master segment with the assumption that the segment is updated during the deformation. This is called "node-to-segment" approach, first published in Hallquist et.al. [53] (1985) for explicit time integration and in Wriggers and Simo [192] (1985) for the implicit scheme. An alternative is the "segment-to-segment" approach with the projection of integration points from a slave segment, see Zavarise and Wriggers [199] (1998) and Harnau et.al. [57] (2005); for a Mortar approach with segment-wisely defined normals, see Puso and Laursen [150], [151] (2004); for a Mortar approach with point-wisely defined normals, see Fischer and Wriggers [41] (2006).

b) For contact between beams the penetration is defined in the direction of mutual normals assuming a shape of a cross sections which may vary.

c) For contact between particles for discrete models, see Cundall and Strack [170] (1979), the penetration is defined as distance between the centers of the particles subtracting the distance between corresponding centers and boundaries of the particles. Similar models for contact are used in the "Smoothed particle hydrodynamics" (SPH) method, see e.g. in Libersky and Petschek [110] (1991) and Campbell et.al. [25] (2000). For a specially shaped geometry, however, an exact measure is necessary, thus, a special projection procedure has been developed in Wellmann et.al. [178] (2008) in order to compute the distance between particles shaped as superellipsoids.

d) The measure of contact can be defined as a mutual volume of overlapping contacting bodies arising from inequality constraints. This approach is often used for contact between rigid bodies, see Kane et.al. [83] (1999) and Pandolfi et.al. [136] (2002).

e) A measure of the tangent interaction $\Delta\rho$ can be formally introduced via a projector operator constructed by subtracting a dyadic product of normals from the unit tensor:

$$\Delta\rho = (\mathbf{I} - \mathbf{n} \otimes \mathbf{n})(\mathbf{r}_A - \mathbf{r}_B). \tag{1.4}$$

This approach is widely used e.g. in Wriggers and Simo [192] (1985), Parisch and Lübbing [138], (1989) Peric and Owen [139] (1992) together with a penalty regularization, in Heegaard and Curnier [63] (1993), Heege and Alart [65] (1996) for Augmented Lagrangian methods and in McDevitt and Laursen [123] (2000), Puso [147] (2004), Puso and Laursen [150; 151] (2004) for the Mortar method.

The main difficulties of such a representation are the following:

- The absence of a geometrical interpretation leads to difficulties for the transformation of history variables.

- Some difficulties in the linearization – transformations are required to show that it would lead to a relative velocity vector. The linearizied results are very complex expressions for the tangent matrices, see the book of Wriggers [188] (2002) for surface-to-surface contact, Litewka and Wriggers [113; 114] (2002) for beam-to-beam contacts.

f) Another approach to describe a tangential interaction is to consider convective variables arising from the surface approximations: see Simo et.al. [162] (1986), Wriggers at.al. [194] (1990), Laursen and Simo [109] (1993), Laursen [105] (1994) and recently in Konyukhov and Schweizerhof [86] (2004), [89] (2005), [92] (2006) with

$$\Delta\rho = \Delta\xi^i \rho_i. \tag{1.5}$$

Then two convective variables $\Delta\xi^1, \Delta\xi^2$ in a surface covariant basis ρ_1, ρ_2 are introduced as tangential measure.

This approach has many advantages:

- objectivity is straightforwardly observed because the surface coordinates ξ^i are used.

- geometrical interpretation of a tangential contact measure is easily possible – line on a surface; geometrical interpretation of a linearized measure – relative tangent velocity of a contact point.

- the number of history variables is minimal (two for surface interaction).

- a complex constitutive law for the tangent interaction can be easily formulated in a robust form for computation.

- expressions for contact tangent matrices are by far less complex within the fully covariant approach than in schemes as in eqn. (1.4).

The last method, however, requires a lot of preliminary transformations based on the differential geometry of contacting objects (surfaces or even curves) and extensive application of the tensor analysis especially for differential operation and linearization. The complete transformation in a covariant form is, however, absolutely necessary – the lack of such transformation leads again to a very complex form of tangent matrices, see in Laursen [104] (1992), in Laursen and Simo [109] (1993) and in the monograph of Laursen [106]. Moreover, it can even lead to *the paradox result* such as "***the lack of symmetry in the stick linearization***" for 3D cases reported in Laursen and Simo [109] [P. 3468] (1993).

1.1.1.2 Step 2: Constitutive relations for contact conditions

A term "constitutive relations" here is understood in a *continuum mechanics* sense, however, being formulated as a constitutive law only between contacting surfaces. Thus, the relations are discussed separately for normal and tangential tractions. They can be formulated in several ways, but essentially two types can be distinguished: a) when a surface has randomly distributed asperities, and b) when asperities have algorithmic structure with prescribed geometrical and mechanical properties, e.g. the surface shows different macro properties in different directions. Statistical analysis is applied for the model of the first type in order to introduce and describe mechanical characteristics of rough surfaces observed in experiments. Various theoretical generalizations based on phenomenological observations are the basis for the model of rough surfaces of the second type. We outline here both approaches with the focus on computational analyses.

Statistical friction models. Statistical analysis of a real rough surface and experimental aspects of its measurements have been developed in a series of publications: Longuet-Higgins [116], [117] (1957), Greenwood and Williamson [50] (1966), Whitehouse and Archard [179] (1970) and more recently Whitehouse and Phillips [180] (1978), [181] (1982) and Greenwood [48] (1984). A comparative analysis of such surface models is presented in McCool and John [122] (1986). These statistical models formed a basis to construct a set of non-linear laws for normal compliance which has been then used for normal contact in the context of finite element methods: see Zavarise et.al. [202] (1992), Wriggers and Zavarise [196] (1993), [195] (1993) for the contact with asperities distributed by the Gaussian normal distribution law. Numerical tests for particular statistical models have been considered in Willner and Gaul [183] (1995), by Willner [182] (1997), [184] (1997) with experimental verification in Willner and Görke [185] (2006). Buczkowski and Kleiber [23] (1999) considered first an isotropical statistical distribution of asperities, and then in [24] (2000) an anisotropic statistical distribution of asperities. The novel combined approach is based on a multi-scale technique. In this case macro characteristics for friction are obtained via a micro model including randomly generated surfaces with certain statistical characteristics, see also in Bandeira, Wriggers and Pimenta [9] (2004) and in [8] (2005).

Phenomenological friction models. For this type of models the asperities possess an algorithmic structure and various generalizations of the isotropic macro characteristics have been proposed. A generalization can be derived separately for normal and tangential tractions.

Various interface non-linear constitutive relations *for normal traction* have been studied in Paggi et.al. [135] (2005). A generalization of a Coulomb friction law with regard to Maxwell and Kelvin viscoelastic models was considered in Araki and Hjelmstad [5] (2003). In the context of the discrete element method various models have been discussed in Luding [118] (2007) and models including normal adhesion can be found in Tomas [174] (2007).

A set of frictional models is obtained when various constitutive models are applied *for tangential traction*. One of the first anisotropic friction models has been proposed by Michalowski and Mroz [127] (1978) considering the sliding of a rigid block on an inclined surface. A model with periodically inclined asperities has been used in Mroz and Stupkiewicz [129] (1994). Zmitrowicz in [210] (1981) described various anisotropic structures of the friction tensor for the generalization of the Coulomb friction law and then developed a classification of anisotropic surfaces in [211] (1989). A rate independent theory of anisotropic friction is developed in Curnier [32] (1984) and thermodynamical restrictions for the friction tensor are discussed in He and Curnier [62] (1993). Finite element models of anisotropic friction are presented as follows: with an application to hot rolling processes these have been discussed in Montmitonnet and Hasquin [128] (1995) and also in Alart and Heege [4] (1995). Buczkowski and Kleiber [22] (1997), [23] (1999), [24] (2000) created an interface element containing an orthotropic sliding law. Hjiaj et.al. [68] (2004) formulated an anisotropic friction problem via the bi-potential and applied Lagrangian multiplier methods. Parametric quadratic programming was used in Zhang et.al. [203] (2004) to solve the almost identical problem. Jones and Papadopoulos [82] (2006) developed a finite element model for anisotropic friction, where the stick-slip condition is enforced via Lagrange multipliers for small relative sliding.

Adhesion models. In the earliest publications adhesion phenomena have been introduced in normal direction as force needed to detach one body from the other where lubrication plays an essential role, see Hardy and Nottage [55; 56] (1926). Experimental investigations of tangential adhesion for different pairs of materials are described in McFarlane and Tabor [124] (1950). The relation between friction force and normal adhesion is experimentally investigated in McFarlane and Tabor [125] (1950). The normal adhesion between clean metal surfaces is experimentally investigated in Bowden and Rowe [20] (1956). Here tangential adhesion forces have been observed in addition to the frictional force. Also a simplified theoretical model was proposed which describes this elastic force as a reaction of micro spherical asperities. Fuller and Tabor [44] (1975) made experiments with rubber surfaces with spherical asperities to show the influence of the surface roughness on normal adhesion. Surface adhesion on elastic spherical asperities has been introduced to describe global adhesion forces in Johnson, Kendall and Roberts [78] (1971). This model (so-called, JKR-model) has been developed further describing the coupling of normal adhesion and friction forces, see Johnson [76] (1997) and also Greenwood [49] (1997). The influence of the roughness of a different scale on normal adhesion is studied by Persson and Tosatti in [141] (2001). Concerning applications on different scales, a method of non-direct determination of adhesive and elastic properties of contacting materials is proposed for various adhesion models (JKR and others) for nano-indentation, Borodich and Keer [18] (2004), Borodich and Galanov [17] (2008).

Coupled adhesion-friction models. A constitutive model of tangential adhesion as a reversible elastic force has been discussed in Curnier [32] (1984). Complex models where adhesion and friction forces are uncoupled appear recently in computational mechanics. For statics Raous, Cangemi and Cocou in [153] (1999) introduce a model including a coupling of a regularization term called tangential adhesion and friction. Also an adhesion term representing elastic effects is considered in normal direction and introduced via additional internal variables. The frictional force and the normal adhesion are coupled via a yield friction function of Coulomb type. The model has been developed further in Cocou, Raous and Schryve [29] (2006) including the coupling of these adhesion terms and viscous forces in dynamics. A coupled model including sliding forces in accordance with the Coulomb friction model and viscoelastic tangential forces analogous to the tangential adhesion above in accordance with the Maxwell and Kelvin model was considered in Araki and Hjelmstad [5] (2003).

The discussed coupled models include only isotropic constitutive relations. Thus, a coupled interface contact model has been developed in Konyukhov and Schweizerhof [90], [91]. This model includes anisotropy for both the friction and the tangential adhesion region and is derived via the application of the covariant approach. A general approach for the combination of a complex friction law together with a complex adhesion law has been proposed. The approach necessarily leads to a computational model which allows a straightforward implementation into a finite element code. The important point is the correlation of such a model to experimentally observed phenomena. A first simple experimental test for a corrugated rubber mat has been reported in [98]. The importance of the coupled model has been shown experimentally in the so-called, "geometrically isotropic case", where the combination of both anisotropy for adhesion and anisotropy for friction can even lead to geometrically isotropic behavior of the sliding body. In [99] this effect was shown experimentally for a rubber surface possessing a periodical waviness.

Motivation of interface surface models from computational tribology. In experimental tribology The physical properties between contacting bodies are described via, a so-called, third body filling the space between the contacting bodies and possessing specific properties, though, differing from the properties of the contacting bodies. Recent developments in computational Molecular Dynamic (MD) simulations, see for 2D-simulations at the nano-scale [85] (2007), have shown that the behavior of such a third body (tribomaterial) during contact is equivalent to models known in computational mechanics, besides of course, the nano-effects. Another simplified MD model described as *movable cellular automata* is also used for simulations of third body behavior see Popov and Psakhie [146] (2007). A discussion of several models for interfaces, classical friction laws and dislocations can be found in Merkle and Marks [126] (2007).

Thus, the interface surface law can be also considered as an upper level in the chain of models: nanotribology, ... , micro-structural model, interface model. It also serves as a theoretical basis for the treatment of various interface surface laws such as a constitutive law reduced to the surface.

The advantages of the covariant approach is especially pronounced for complex contact interface laws. The coupled model including anisotropic friction and tangential adhesion, proposed in Konyukhov and Schweizerhof [90], [91] (2006), is directly formulated in a covariant form as a computational model for an iterative Newton type solver. The approach allows the straightforward coupling of various interface laws (plasticity, elasticity, viscosity etc.) formulated in an arbitrary curvilinear coordinate system given on the contact surface.

General discussions and reviews on constitutive models used in contact mechanics can be found in the monographs of Laursen [106] (2002), Persson [140] (2000), Sextro [157] (2002) and Wriggers [188] (2002).

1.1.1.3   Step 3: Methods to enforce contact conditions.

Several numerical methods are known in literature in order to enforce contact conditions. These method are used to resolve the contact conditions formulated as inequalities in the form of Kuhn-Tucker conditions. The terminology is initially found in the optimization methods, see e.g. the monographs of Bertsekas [13] (2003) and Borwein and Lewis [19] (2000). These methods can be classified by the structure of the functionals enforcing these conditions in a weak form: the Lagrange multiplier method, the Penalty method, the Nitsche method and the Augmented Lagrangian method.

Lagrange multipliers together with a trial-and-error procedure are introduced to describe additional contact nodal forces in the some first finite element models, see Francavilla and Zienkiewicz (1975) and Hughes at.al. [71] (1976). Cescotto and Charlier [27] (1993) considered four mixed variational formulations of frictional contact leading to contact elements with independent approximation of stress and displacement fields. Taylor and Papadopoulos in [173] (1991), Solberg and Papadopoulos in [163] (2005) studied this method in order to find out so-called Babuska-Brezzi stable cases of Lagrange multiplier schemes leading to the satisfaction of **the contact patch test**. For various orders of approximations, the patch test has been studied in Crisfield [31] (2000). A patch-test for the small displacement frictional case is studied in Jones and Papadopoulos in [79] (2000). Harnau et.al. [57] (2005) used various integration techniques for the contact integral within "a segment-to-segment" approach to improve the patch-test.

A special technique for the approximation of the Lagrange multipliers is the so-called *Mortar method* with dual multipliers, see Hüeber and Wohlmuth [70] (2005), though the Mortar method is not restricted to the enforcement via the Lagrange multipliers only, another application will be discussed later. For a small sliding problem, a perturbed Lagrangian method including both penalty and Lagrangian functionals was discussed in Wriggers and Simo [192] (1985).

Heintz and Hansbo [67] (2006) constructed a special functional including not only penalized displacements, but also penalized normal stresses from both contacting sides, the so-called *Nitsche method*, and obtained estimations for the penalty parameter necessary for convergence. The method is proved to be independent on the penalty parameter. However, due to the cumbersome formulation its finite element implementation have been considered only for 2D problems, see in Wriggers and Zavarise [198] (2008) and in Oliver et.al. (part 1) [132] and in Hartmann et.al. (part 2) [58] (2009).

The Penalty method theoretically discussed in detail in Kikuchi and Oden [84] (1988) is among the most popular methods for finite element implementations. Both normal and tangential displacements have to be penalized with a large penalty parameter in order to enforce contact conditions. This leads from a mechanical point of view to an "additional spring" acting at the contact point as proposed in Hallquist et.al. [53] (1985) for an explicit time integration scheme. The essential stage of the implicit solution scheme is an iterative solution of e.g. Newton's type, which requires, in due course, consistent linearization. This procedure has been introduced for the 2D non-frictional case in Wriggers and Simo [192] (1985). The penalty enforcement for non-frictional cases is discussed also in Parisch [137] (1989) within a covariant approach. A special linearization approach for the penalty term in the form of covariant derivatives in the local coordinate system has been introduced in and in Konyukhov and Schweizerhof [86] (2005). In order to compute correct tangent forces for frictional problems Wriggers et.al. [194] (1990) used the return-mapping algorithm to compute correct tangent forces for frictional problems. The penalty formulation in a global coordinate system is used in Peric and Owen [139] (1992) and Parisch and Lübbing [138] (1997). Starting with Laursen and Simo [109] (1993) (also in Laursen [104] (1992) and [105] (1994)) convective coordinates are directly used for regularization. The fully covariant description in the coordinate system defined by the closest point projection is developed in Konyukhov and Schweizerhof [89] (2005).

The penalty method is applied not only for contact between surfaces, but also for contact between other objects such as beams. Thus, contact between straight beams with circular cross sections is considered in Wriggers and Zavarise [197] (1997) for non-frictional contact and then in Zavarise and Wriggers [200] (2000) for frictional contact. Straight beams with rectangular cross-sections have been studied in Litewka and Wriggers [113] (2002) and in [114] (2002) for both frictional and non-frictional cases respectively. The Lagrange multiplier method for a beam-to-beam contact is discussed in Litewka and Wriggers [115] (2003). A special treatment (so-called "slide-line contact") also based on the penalty method for the straight rod/continuum contact is developed in Maker and Laursen [120] (1994).

The Mortar method together with penalty regularization with a specially defined Mortar projection is used in Puso and Laursen [150] (2004) for non-frictional and in [151] (2004) for frictional contact. The Augmented Lagrangian scheme is exploited to enforce frictional forces in [151]. Various approximations for contact surfaces concerning the order of the surface description within the Mortar method have been studied in Fischer and Wriggers [41] (2006).

An Augmented Lagrangian method allowing to satisfy contact constraints with prescribed tolerance, has been thoroughly studied in Pietrzak and Curnier [144] (1999). The problem has been formulated as a saddle point optimization. Convergence of various Augmented Lagrangian formulations in contact is studied by Stadler [167] (2007). A Mortar method together with Augmented Lagrangian enforcement of tangent forces is used in Hüeber et.al. [69] (2006).

A symmetrization procedure based on a nested Uzawa algorithm has been used in Simo and Laursen [161] (1992) in order to obtain symmetric tangent matrices for the frictional case. The Uzawa symmetrization technique has been extended also for a coupled anisotropic adhesion-friction model in Konyukhov and Schweizerhof [94] (2007).

Remark 1.

The application of various numerical methods requires a-priori knowledge about existence and uniqueness of the solutions of the contact mechanics problems. There is a large number of publications on this subject: more recent results are summarized in monographs of Han and Sofonea [54] (2002) and Shillor, Sofonea and Telega [158] (2004), see also references therein.

Remark 2.

In a geometrically exact theory the discussed solution methods depending on a geometrical feature (surface, line or point) should be formulated in the covariant form according to the corresponding geometrical feature (e.g. in a Frenet coordinate system for a beam-to-beam contact etc.).

1.1.1.4 Step 4: Formulation of basic principles (conservation of momentum, equilibrium etc.) on contact boundaries.

In this step fundamental equations such as balance of momentum etc. are formulated either in strong, or in weak forms. The form of differential equations (the strong form) is used directly in computations only in discrete models, see in Strack and Cundall [170] (1979), while the weak form is a key issue for the application of the finite element method as well as so-called various meshless methods. The current contribution is aimed at the development of computational algorithms in applications with the finite element method, therefore, we mention here only a specific problem in dynamic computations, namely, energy and momentum conserving algorithms for contact problems. Various integrations schemes on contact boundaries have been proposed in Laursen and Chawla [107] (1997) and in Armero and Petöcz [6] (1998) for non-frictional problems, and then, so-called, entropy dissipative schemes in Chawla and Laursen [28] (1998) and in Armero and Petöcz [6] (1999) for frictional problems. The discrete null space method developed earlier in Betsch [14] (2005) for continuum cases is extended for contact problems in Betsch and Hesch [15] (2007).

1.1.1.5   Step 5: Numerical solution and analysis of the proposed model

In this step the equations defined for the discretized model are solved numerically. Among numerous problems such as selection of methods, storage of data, parallel algorithms we consider only two particular problems typical for contact algorithms: the contact searching algorithm and the consistent linearization necessary for implicit solvers.

Global searching algorithms.   The computation of contact-impact problems for structures undergoing large deformations requires extensive searching of contact pairs, that is why robust searching algorithms are necessary. Benson and Hallquist [11] (1990) proposed a single surface contact algorithm effectively describing self-contact. Zhong and Nilsson [206] (1989), [207] (1996) developed a hierarchical algorithm carefully treating contact cases with edges, see also the book of Zhong [205] (1993). The Pinball algorithm proposed by Belytschko and Neal [10] (1991) is based on searching in spherical regions around certain nodes. Schweizerhof and Hallquist [155] (1992) described a searching strategy based on an additional subdivision of the searching space. Various searching algorithms based on hierarchical subdivisions are discussed also in Feng and Owen [39] (2002), Bruneel and Rycke [21] (2002). Especially for blow molding processes several algorithms have been developed: the inside-outside algorithm by Wang and Nakamachi [177] (1997); the moving box algorithm by Wang and Makinouchi [176] (2000); the Free-Form-Surface (FFS) algorithm for modeling the contact surface with C^1-continuity by Wang et.al. [175] (2001). Numerical aspects of searching especially for parallel computation are discussed in the thesis of Persson [142] (2000). Various global searching techniques in combination with both finite element and discrete element method are considered in monograph of Munjiza [130] (2004).

Consistent linearization.   Two approaches for linearization of the final functional representing the work of contact tractions can be distinguished in order to obtain consistent tangent matrices. *The direct approach* follows the following sequence: *functional – discretization – linearization* and *the covariant approach* follows the rule: *functional – lin-*

earization – discretization.

The direct approach assumes that the discretization is then involved in the process and the linearization is provided with regard to the displacement vector u and, therefore, of the discretized system. This leads to the final results containing a set of approximation matrices. Various aspects of this approach are considered in the following publications:

a) **for surface-to-surface contact** in Wriggers and Simo [192] (1985), Wriggers et.al. [194] (1990), Wriggers [187] (1995), Parisch [137] (1989), Parisch and Luebbing [138] (1997), Peric and Owen [139] (1992), Simo and Laursen [161] (1992), Laursen [104] (1992), Laursen and Simo [109] (1993);

b) **for anisotropic friction between surfaces** in Alart and Heege [4] (1995);

c) **for beam-to-beam contact** in Wriggers and Zavarise [197] (1997), Zavarise and Wriggers [200] (2000), Litewka and Wriggers [113] (2002), Litewka and Wriggers [114] (2002).

The complexity in the derivation for curved contact interfaces led to **the combination of a finite element code with a mathematical software for automatic derivations**, see **for surface-to-surface contact** in Heege and Alart [65] (1996), Stadler, Holzapfel and Korelc [169] (2003), Krstulovic-Opara and Wriggers [100] (2001), Krstulovic-Opara and Wriggers [101] (2002), Krstulovic-Opara, Wriggers and Korelc [102] (2002), **for anisotropic friction** in Montmitonnet and Hasquin [128] (1995) and **for beam-to-beam contact** in Litewka [112] (2007) and in [111] (2007).

The fully covariant approach, however, assumes only a local coordinate system associated with the deformed continuum (convective coordinates) and requires extensive application of covariant operations (derivatives etc.). Though the convective coordinates were partially involved in some publications mentioned above, the full description has been first derived in Konyukhov and Schweizerhof [86] (2004) for non-frictional; in [89] (2005) for frictional contact; a specific consideration for 2D cases in [92] (2006); for the coupled adhesion-friction anisotropic contact in [90] (2006) and in [91] (2006); in an application of the Augmented Lagrangian approach for the coupled adhesion-friction anisotropic contact in [94] (2007); for arbitrary contact between curves representing beam and cable type structures in [96] (2009). Concerning the final complexity of the derived algorithms the covariant approach leads to a simple structure of the contact matrices. Other advantages of the covariant approach are discussed in detail in the Section 1.4.2 describing the structure of the current work.

1.2   Discussion – why covariant approach?

Through all considered steps from 1 to 5 two general approaches can be distinguished: *a)* **covariant approach** when a contact description is fully based on local surface coordinate systems and *b)* **direct approach** (non-covariant one) when all parameters are described with the assumed finite element approximations in a global coordinate system. The latter is historically motivated by the development of the finite element method, while the fully covariant approach, though, is intended for the finite element method, but does not assume approximations from the beginning. The covariant approach serves to describe all parameters necessary for the solution based on the geometry of the contacting bodies in the local coordinate system.

Open questions and drawbacks of *the direct approach* can be summarized as follows:

- A closed form for tangent matrices is available only for linear approximations of surfaces. For curved interfaces, either a form depending on approximations (mathematical software), or a form of simplified matrices (taken for linear approximations) is reported.

- The structure of the derived matrices is very complicated and often intransparent. There is no clear interpretation of each part possible. Thus, simplifications are hardly possible.

- A specification of complex contact interface laws with properties explicitly depending on the surface geometry (e.g. **arbitrary** anisotropy) is not possible.

- A contact description of many geometrical features (curved line-to-curved line, curved line-to-surface) is almost not possible because of the necessity of convective surface coordinates.

- Geometrically motivated measures of contact interaction are coupled with convective variables in a specially defined coordinate system. This straightforwardly leads to a description via convective variables. However, in the direct approach, they are not defined separately for various geometrical objects such as surfaces, edges etc. and can be handled – mostly improperly – with great efforts.

1.3 On geometrical approaches in contact mechanics

Contact interaction from a geometrical point of view can be seen as an interaction between deformable surfaces and, therefore, geometrical approaches can be exploited. However, there are only a few publications uncovering geometrical issues to some extent. Gurtin et.al. [51] (1998) considered surface tractions on curvilinear interfaces describing them from a geometrical point of view. Jones and Papadopoulos [81] (2006) considered contact describing various mappings from the reference configuration employing the Lie derivative. Laursen and Simo [109] (1993) and Laursen [105] (1994) described some contact parameters via geometrical surface parameters. Heegaard and Curnier [64] (1996) considered geometrical properties of slip operators.

1.4 Goals and structure of the work

In order to formulate goals and describe the complete structure of the book we consider a model contact problem with two bodies possessing not only smooth surfaces, but also various geometrical features such as edges and vertexes – an example of this is a banana and a knife shown in Fig. 1.1. Considering all possible geometrical situations in which knife and banana can contact each other, the following hierarchical sequence of contact pairs is appearing:

Possible contact pairs:

1. Point-to-point contact pair, see Fig. 1.2

2. Point-to-line contact pair, see Fig. 1.7

3. Point-to-surface contact pair, Figs. 1.4 and 1.5.

4. Line-to-line contact pair, see Fig. 1.3

5. Line-to-surface contact pair, see Figs. 1.6 and 1.7

6. Surface-to-surface contact pair, see Figs. 1.4and 1.5.

1.4.1 Goals

Summarizing the literature review and the discussion in the previous sections the following open problems can be stated as goals for the current work:

1. Development of an unified geometrical formulation of contact conditions in a covariant form for various geometrical situations of contacting bodies leading to contact pairs: surface-to-surface, line-to-surface, point-to-surface, line-to-line, point-to-line, point-to-point. The description will be fully based on the differential geometry of specific features forming a continuum, because it is carried out in the local coordinate systems attached to this feature: in the case of a surface – in the Gaussian surface coordinate system; in the case of a curved line – in the Serret-Frenet basis; in the case of a point – in the coordinate system standard for rigid body rotation problems (e.g. via the Euler angles). **This general description is forming a geometrically exact theory for contact interaction.**

2. A full set of contact pairs requires various closest point projection (CPP) procedures. Thus, fundamental problems of existence and uniqueness of closest point projection routines corresponding to the following situations will be investigated: point-to-surface, point-to-line, line-to-line.

3. A solution of existence and uniqueness problems of closest point routines leads to "projection domains". Thus, the "maximum searching domain" allows to improve the searching routines including interactions between contact pairs.

4. Since contact interaction between arbitrary bodies is modeled via a corresponding set of contact pairs (surface-surface, surface-line etc.) the necessary transfer algorithm for history variables will be constructed.

5. Derivation of a unified covariant description of various applicable methods to enforce contact conditions: Lagrange multiplier methods, penalty methods, Augmented Lagrange multiplier method. Consistent tangent matrices are given in a closed covariant form possessing a clear geometrical structure.

6. Description of all geometrical situations in a covariant form which is a-priori independent of approximations of these geometrical features leads to straightforward recipes for the implementation with any order of approximation for finite elements. Application with bi-linear and bi-quadratic "solid-shell" finite elements will be considered in detail.

7. A special integration technique based on sub-domain integration is developed for "the segment-to-segment" approach (equivalent to the Mortar method).

8. Covariant contact description for high order approximations including exact representation of geometry for continua (iso-geometrical approach). Numerical tests show the efficiency for the classical Hertz problem.

9. Generalization of the classical Coulomb law into complex interface laws in covariant form for arbitrary geometries of the surfaces (e.g. coupled anisotropic friction and adhesion for surfaces). Development of an a-priori stable numerical algorithm for computations.

10. Experimental validation of the proposed anisotropic law for coupled tangential adhesion and friction.

11. Development of the curve-to-curve contact model allowing to consider the complete set of relative motions between curves including a rotational interaction (a novel in the current theory).

12. Development of the corresponding constitutive relations together with the corresponding numerical algorithm allowing an anisotropic behavior for curve-to-curve interaction (various relative adhesion and friction properties).

13. Application of the curve-to-curve contact algorithm to edge-to-edge contact as well as to beam-to-beam contact.

14. Curved beams possessing C^1-continuity allowing contact (a cable model). Application to the tying of knots (the theory of knots).

1.4.2 Structure of the current work

The current work is subjected to the rules for the submission the Habilitation thesis in a cumulative way at Karlsruhe University (TH) and, therefore, is organized as a set of research papers published within a research period of 2002-2009 years. The research has been supported by the following DFG grants:

1. **DFG SCHW 307/18-2** "Kovariante Kontaktformulierung von stark deformierbaren Köorpern mit FE Diskretisierung und separater Interpolation der Oberflächen bei statischer und dynamischer Belastung."

2. **DFG SCHW 307/22-1** "Geometrically exact theory of contact interaction of structures with curved beams, cables and surface edges – A covariant approach for all possible geometrical features of general bodies."

The publications are covering different parts of the goal stated in the previous section, therefore, the aim of the current section is to give to a reader the complete structure of particular details written in the following chapters.

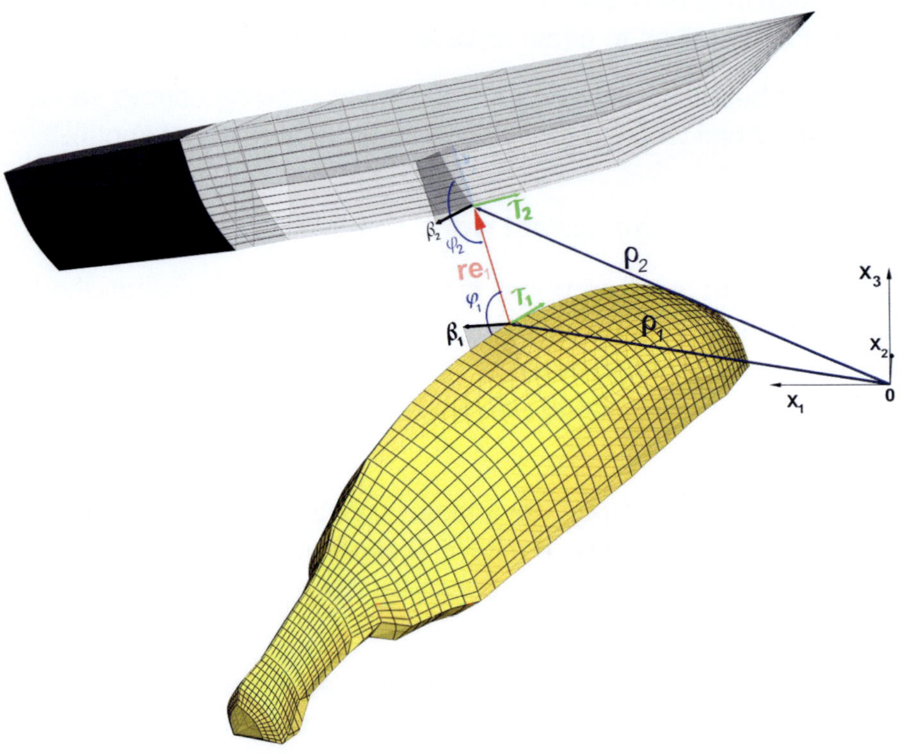

Figure 1.1: Various geometrical situations in contact lead to different contact algorithms: Surface-To-Surface, Line-To-Surface, Point-To-Line, Line-To-Line and Point-To-Point

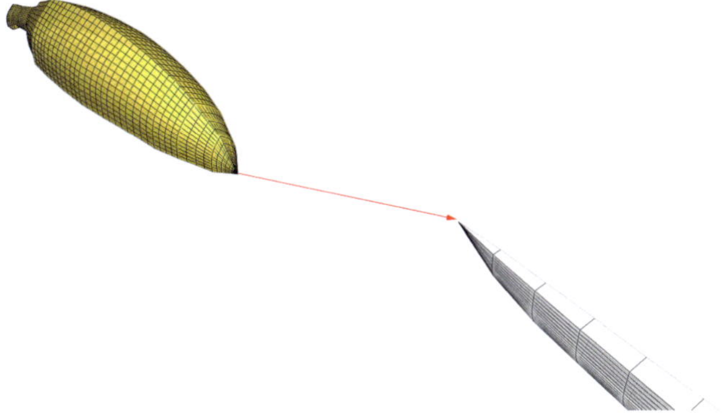

Figure 1.2: Point-To-Point (PTP) contact algorithm

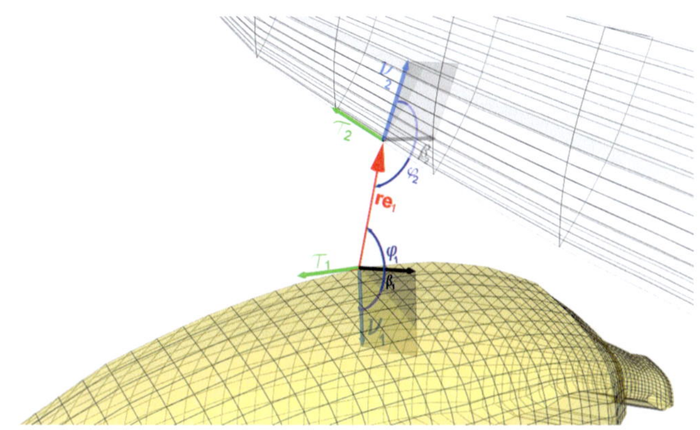

Figure 1.3: Line-To-Line (LTL) contact algorithm

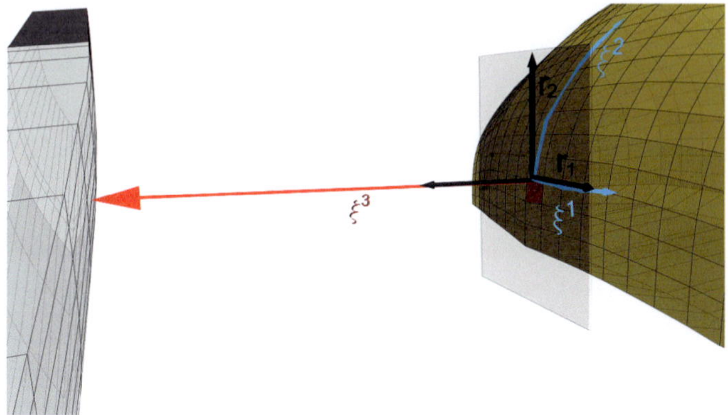

Figure 1.4: Surface-To-Surface (STS) contact algorithm

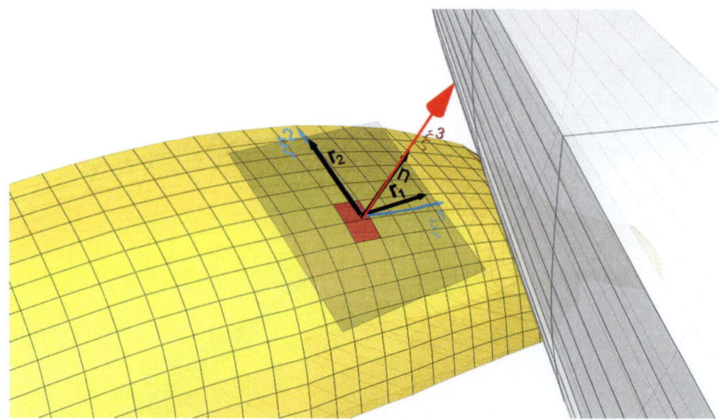

Figure 1.5: Surface-To-Surface (STS) contact algorithm:
transfer of variables between elements.

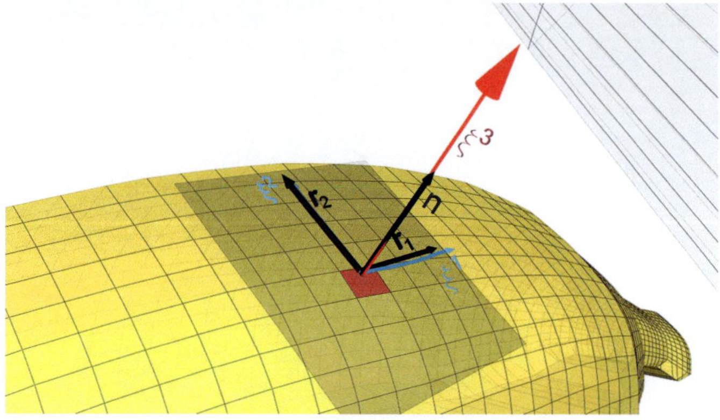

Figure 1.6: Line-To-Surface contact as Surface-To-Surface algorithm

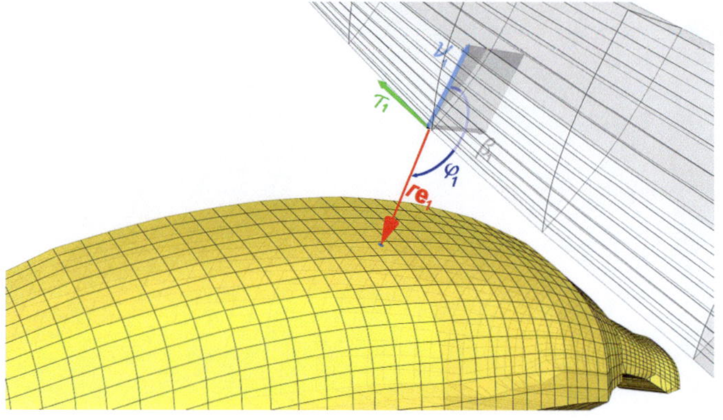

Figure 1.7: Line-To-Surface contact as Point-To-Line algorithm

The most powerful approach in the computational contact mechanics is to work in accordance with the geometry of contact bodies and construct all computational algorithms in a covariant form. This combination forms **a geometrically exact theory of contact interaction**.

As mentioned in the literature overview, the closest distance has become a natural measure of the contact interaction. The procedure is introduced via the closest point projection procedure (CPP), for which the solution requires the differentiability of the function representing the parameterization of the surface of contacting bodies. An analysis of the solvability for the CPP procedure allows then to classify all types of all possible contact pairs given in 1.4. **Chapter 2** starts, first, with a consideration of the CPP procedure for C^2-continuous surfaces. The concept of *the projection domain* is introduced as a domain from which any point can be uniquely projected, and therefore, the contact algorithm can be further constructed. This domain can be constructed for at the utmost C^1-continuous surfaces. If the surfaces contain edges and vertices then the CPP procedure should be generalized in order to include the projection onto edges and onto vertices. The criteria of uniqueness and existence of these projection routines and corresponding domains are studied in detail. Examples of the projection domain and a controversial for CPP procedure are shown.

The main idea for application for contact is then straightforward – the CPP procedure corresponding to a certain geometrical feature gives rise to a special, in general, curvilinear 3D coordinate system. This coordinate system is attached to a geometrical feature and its convective coordinates are directly used for further definition of the contact measures. Thus, all contact pairs listed in 1.4 can be described in the corresponding local coordinate system. The existence requirement for the generalized CPP procedure leads to the transformation rule between types of contact pairs according to which the corresponding coordinate system is chosen. Thus, all contact pairs can be uniquely described in most situations.

A surface-to-surface contact pair, see Fig. 1.4, is described via the well known "master-slave" contact algorithm based on the CPP procedure onto a surface. This projection allows to define a coordinate system as follows:

$$\mathbf{r}(\xi^1, \xi^2, \xi^3) = \boldsymbol{\rho}(\xi^1, \xi^2) + \xi^3 \mathbf{n}(\xi^1, \xi^2). \qquad (1.6)$$

The vector $\mathbf{r}$ is a vector for the "slave" point, ρ is a parameterization of the "master" surface, $\mathbf{n}$ is a normal to the surface. The slave point is taken to be a node from the finite element mesh for the well-known *Node-To-Segment* algorithm, however, the algorithm is applicable mainly to linear approximations due to failing the patch test. For arbitrary approximations *Segment-To-Segment*, or *Surface-To-Surface* algorithms are taken, in which an integration point is taken as a slave point. Such algorithms will be discussed later, however, it is obvious that the kinematics for both the surface-to-surface, and the point-to-surface contact algorithms are described in a coordinate system given in eqn. (1.6). Thus, point-to-surface and surface-to-surface contact pairs are equivalent. Eqn. (1.6) describes, in fact, a curvilinear coordinate transformation where convective coordinates are used for a measure of contact interaction: ξ^3 is a penetration, $\Delta\xi^1, \Delta\xi^2$ are measures for tangent interaction. The algorithm is applied only in the existence domain for *the surface CPP procedure*.

Consideration of existence of the CPP procedure for edges allows to define then the point-to-line contact algorithm used for the point-to-line contact pair directly as well as for the line-to-surface contact pair, see Fig. 1.7. The local coordinate system is constructed as follows:

$$\mathbf{r}(s, r, \varphi) = \boldsymbol{\rho}(s) + r\mathbf{e}(s, \varphi); \quad \mathbf{e} = \boldsymbol{\nu}\cos\varphi + \boldsymbol{\beta}\sin\varphi. \qquad (1.7)$$

For the line-to-surface algorithm, the vector $\mathbf{r}$ is describing a "slave" point from the surface, $\rho(s)$ is a parameterization of the "master" curve edge; $\mathbf{e}$ is a unit vector describing the shortest distance between the point and the curve written via the unit normal ν and bi-normal β of the curve $\rho(s)$. The convective coordinates are used as measures: r – for normal interaction; s – for tangential interaction; φ – for rotational interaction.

The Line-To-Surface contact pair, however, can be described in a dual fashion via the Surface-To-Surface contact algorithm if we consider a "slave" point on the edge and project it onto the "master" surface, see Fig. 1.6 *(Line-To-Surface contact as Surface-To-Surface algorithm)*. The contact is described then in the surface coordinate system (1.6).

The Line-To-Line contact pair requires the projection on both curves, therefore, there is no classical "master" and "slave" parts and both curves are equivalent. For the description both coordinate systems assigned to each contacting I-th curve are used:

$$\boldsymbol{\rho}_2(s_1, r, \varphi_1) = \boldsymbol{\rho}_1(s_1) + r\mathbf{e}_1(s_1, \varphi_1); \quad \mathbf{e}_1 = \boldsymbol{\nu}_1 \cos \varphi_1 + \boldsymbol{\beta}_1 \sin \varphi_1 \quad 1 \leftrightarrows 2.$$

$$(1.8)$$

Here, the vector $\boldsymbol{\rho}_2$ is a vector describing a contact point of the second curve, $\boldsymbol{\rho}_1(s_1)$ is a parameterization of the first curve; a unit vector describing the shortest distance $\mathbf{e}_1$ is written via the unit normal $\boldsymbol{\nu}_1$ and bi-normal $\boldsymbol{\beta}_1$ of the first curve. Eqn. (1.8) describes the motion of the second contact point in the coordinate system attached to the first curve. The description is symmetric with respect to the choice of the curve $1 \leftrightarrows 2$. Convective coordinates are used as measures: r – is mutual for both curve and a measure for normal interaction; s_I – for tangential interaction and φ_I – for rotational interaction for the I-th curve.

The Point-To-Point contact pair is described then in a coordinate system standard for rigid body rotation problems (e.g. via the Euler angles), however in contact situations it is a very rare case, and in computations it is rather improbable unless specially treated, and therfore, because of the numerical error would fall into other contact pair types.

Chapter 3 opens the development of the computational algorithm with non-frictional contact interaction for smooth surfaces. Here the description starts in the coordinate system given in eqn. (1.6), however, due to the small penetration $\xi^3 \approx 0$ it is mostly falling into a description in the Gaussian surface coordinate system arising from the surface parameterization $\boldsymbol{\rho}(\xi^1, \xi^2)$. All contact parameters such as sliding distance and tangent forces are described then on the tangent plane at $\xi^3 = 0$. The linearization procedure is given in a form of covariant derivatives. This leads to a closed form of the tangent matrix subdivided into a main, a rotational and a curvature parts. The influence of those parts on convergence is studied in numerical examples for the linear and quadratic finite elements.

Chapter 4 extends the problem into a Coulomb frictional contact. It is shown that for the correct regularization of tangential contact conditions the evolution equation for contact traction should be taken in a form of covariant derivatives on the tangent plane. The structures of all parts of tangent matrices are obtained due to covariant derivation in a compact tensor closed form. This makes it applicable for any surface approximation. It is shown that the tangent matrix in the sticking case is always symmetric for any kind of approximation. *A classification of parts of the tangent matrix* is given and their influence on convergence with regard to small and large sliding problems is considered. Small sliding problems are introduced as problems where the computation of the sticking-sliding zone is essential, while a sliding path is only of interest for large sliding problems. *An algorithm to transfer history variables* in contact problems overcoming the discontinuity of history variables on element boundaries, see illustration in Fig. 1.5, is created in a covariant form. Numerical examples illustrating the application of bilinear and biquadratic finite elements to frictional phenomena are shown.

Chapter 5 is devoted to the various computational aspects and implementation details of the developed theory. All known contact approaches such 'Node-To-Segment" (NTS) and, Mortar like "Segment-To-Segment" are reconsidered within the covariant description. All contact parameters are evaluated at integration points within the STS approach, therefore, **combinations of various adaptive integration techniques such as integration with sub-domains with independent application either Gauss-Legendre, or Gauss-Lobatto quadrature formulas are specially developed**. This allows to satisfy the "contact patch-test" on unstructured distorted meshes for arbitrarily chosen "master" or "slave" segments. Special attention is given to the implementation of the STS contact approach together with "solid-shell" elements. The influence of various integration techniques on computed results (especially on the force-displacement curves) is extensively studied in numerical examples. It is shown that geometrical contact conditions can be satisfied with high tolerance even for the linear finite elements by the application of adaptive integration techniques. A smoothing technique based on the application of NURBS surfaces is considered with quadrilateral meshes. A patch test is studied in a combination with smooth surfaces. A special development within the covariant approach is devoted to the contact with a rigid surface described analytically. **"The rigid surface is a slave"** and **"the rigid surface is a master"** approaches are analyzed. A reduced solution for the surface of revolution is studied and a set of closed form solutions for the penetration is derived for both strategies. The set contains the following geometrical figures: plane, sphere, cylinder, torus and cone. The cases requiring the application of special techniques such as the history transfer algorithm and higher then linear finite elements are specially studied for some deep drawing cases.

A general geometrical approach to treat contact kinematics in the 2D case either as a reduction of the 3D case, or as a development based on a plane curve geometry is described in **Chapter 6**. This leads to a more *simple kinematical interpretation of all parts of the tangent matrix* compared to the 3D case. A fast implementation of frictional contact in 2D is proposed. The algorithm to transfer the history variables, see Fig. 1.5, in contact problems overcoming the discontinuity of history variables on element boundaries and the algorithm to update the history variables in the case of reversible loading are studied and illustrated in detail. This part became a core for a course of *"Contact Mechanics"* at the University of Karlsruhe.

A special development of the covariant approach in combination with *high-order finite element methods* is given in **Chapter 7**. Both the penalty, and the Lagrange multiplier method are considered. Approximation and selection of the integration schemes for the Lagrange multipliers and for other terms are chosen in order to satisfy the discrete Babuska-Brezzi (BB) stability condition. The linearization procedure is enhanced for cases with the exact geometry of contact boundaries being represented by the blending function method. As a result a contact layer element allowing anisotropic $p-$refinement is created. The layer contact element is applied then to initially linear meshes. A good correlation with the analytical Hertz problem is achieved even within a single contact layer element.

Chapter 8 begins with *a systematic generalization of a contact interface law* from the Coulomb friction law into the anisotropic region in a covariant form including various known visco-elasto-plastic mechanical models. Thus, *a coupled model including anisotropy for tangential adhesion and for friction* is obtained. This model is formulated via the principle of maximum dissipation in a rate form. Finally, the computational model is derived via the application of the return-mapping scheme to the incremental form. As a result a frictional force is derived in a closed form including both, the adhesion and the friction tensors. The structure of the tensors is derived for various types of anisotropy: a uniform orthotropy of a plane given by the spectral decomposition, a nonuniform orthotropy of a plane inherited with the polar coordinate system and a spiral orthotropy of a cylindrical surface. A special update algorithm for history variables is developed for arbitrary coupled anisotropies. The geometrical interpretation of the return-mapping and the update algorithm is considered via an ellipse on the tangent plane.

Chapter 9 continues the development of the computational algorithm for the coupled anisotropic friction model. The linearization is obtained as a covariant derivation in the local surface coordinate system and, therefore, all tangent matrices possess the simple form similar to the isotropic friction models. The mechanical interpretation of the model as a two spring-slider-mass system is discussed. The behavior of contacting bodies for various types of anisotropy is numerically analyzed. The development of the sticking zone for the small displacement case, and the influence of orthotropic properties on a trajectory of the sliding block in the case of large displacements are analyzed for the uniform orthotropy of a plane given by the spectral decomposition and for the nonuniform orthotropy of a plane inherited with the polar coordinate system. As an interesting result, *a geometrically isotropic behavior* of the block has been found: in this case the combination of both, anisotropy for adhesion and anisotropy for friction leads to a trajectory which can be normally observed only for isotropic surfaces. It is shown that *the application of the spiral orthotropy on a cylindrical surface allows to simulate the kinematics of the bolt connection* on relatively rough mesh.

The tangent matrices for sliding in the case of the anisotropic coupled adhesion-friction model are non-symmetric, due to the non-associativity and due to the coupled anisotropy. Thus, *a symmetrized algorithm based on the Augmented Lagrangian method* for coupled anisotropic friction is developed in **Chapter 10**. It is shown that for small sliding problems both normal and tangential tractions should be augmented to enforce the non-penetration resp. sticking conditions, but for large sliding problems the augmentation of only the normal traction leads to a satisfactory tolerance for trajectories.

As a key for the practical application, the developed *model is experimentally investigated.* It is shown in **Chapter 11**, that the coupled anisotropic adhesion-friction model can *successfully describe a set of trajectories* of a block on a rubber mat with a periodical wavy profile, while the classical anisotropic friction model fails. Special attention is given to the analysis of *geometrically isotropic behavior.*

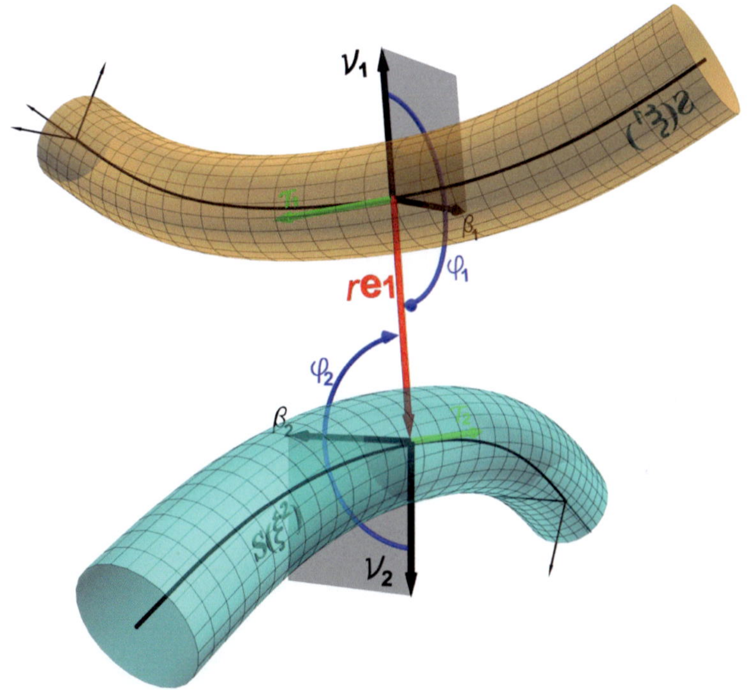

Figure 1.8: Contact between curvilinear beams with circular cross-sections

A geometrically exact description in a covariant form for curve-to-curve contact pairs shown in Fig. 1.3 is developed in **Chapter 12**. The development begins consistently with an analysis of the Closest Point Projection (CPP) procedure. This analysis leads to a special local coordinate system in which convective coordinates are used directly as measures of contact interaction between curves: normal, tangential and rotational. The existence and uniqueness of the CPP procedure

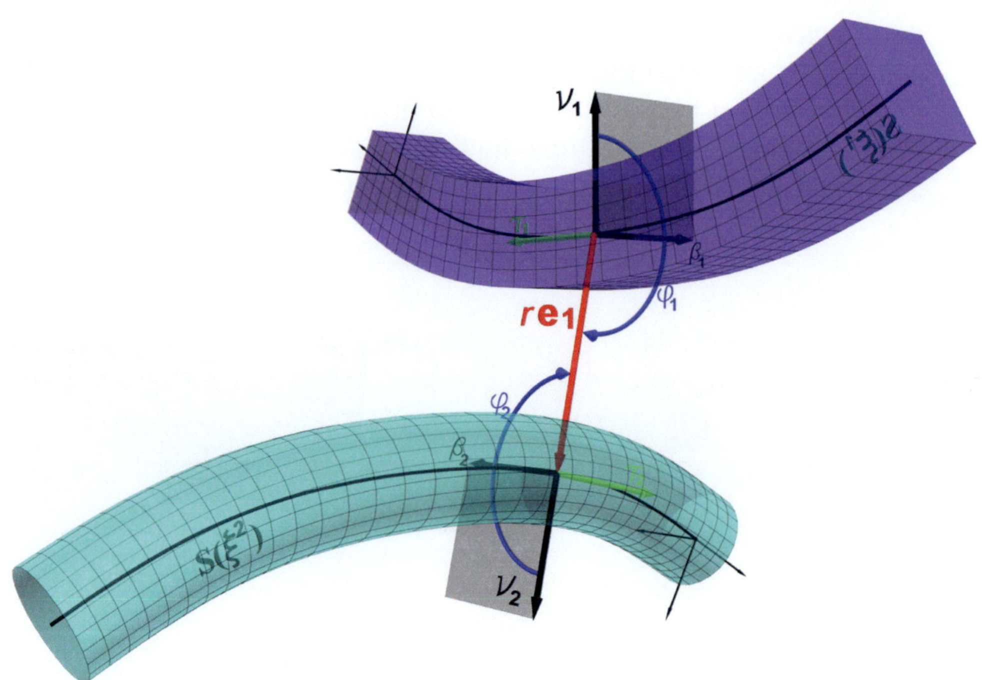

Figure 1.9: Contact between curvilinear beams: arbitrary cross-section can be defined in $(\nu_1 - \beta)$-plane

is studied in detail – projection domains with a-priori unique solutions are constructed in this coordinate system for curves with varying geometry. Several achievements appear to be novel for the line-to-line contact description: a) consideration of any relative motion separately for each curve is possible; b) rotational interactions including corresponding rotational moments between curves can be considered consistently.

The Coulomb friction law for tangential interaction and the Tresca friction law for rotational interaction are considered as examples for constitutive relations between curves. The curve-to-curve contact model is applied then for the beam-to-beam contact, see Fig. 1.8. In this case constitutive relations are formulated with regards to a cross section, while for the edge-to-edge contact they are supplied for the sector in the orthogonal plane $(\nu - \beta)$, see Fig. 1.9. However, not only circular cross sections are possible – any arbitrary cross section can be supplied in the orthogonal plane, see Fig. 1.9. All necessary linearizations for the iterative solution scheme are provided as covariant derivations in the introduced coordinate system for arbitrary large distances between curves. This leads to a closed form of tangent matrices independent of the approximation used for the finite elements.

The verification section contains the comparison between the beam-to-beam and the edge-to-edge finite element models as well as the verification with a famous "Equilibrium of Euler elastica problem" computed via a finite difference scheme. The further numerical examples are illustrating the ability to describe various kinematics for curve-to-curve contact situations e.g. partial sticking of a single curve. Special attention is devoted to the applications with curvilinear beams. C^1-continuous spline finite elements for the beam allowing large rotations are enriched with the possibility of contact. Thus, the isogeometrical approach for curvilinear beams together with contact is realized. A new application area for the tying of knots is considered in some numerical examples.

2

On the solvability of closest point projection procedures in contact analysis: analysis and solution strategy for surfaces of arbitrary geometry*

Abstract

The uniqueness and existence of the closest point projection procedure widely used in contact mechanics are analyzed in the current article. First, a projection domain for C^2-continuous surfaces is created based on the geometrical properties of surfaces. Then any point from the projection domain has a unique projection onto the given surface. It is shown that in order to construct a continuous projection domain for arbitrary globally C^1, or C^0–continuous surfaces, a projection routine should be generalized and also include a projection onto a curved edge and onto corner points. Criteria of uniqueness and existence of the corresponding projection routine are given and discussed from the geometrical point of view. Some examples showing the construction of the projection domain as well as the necessity of a generalized projection routine are given.

Keywords

closest point projection uniqueness and existence contact covariant description

*The chapter is published in [95]: A. Konyukhov and K. Schweizerhof, *On the solvability of closest point projection procedures in contact analysis: Analysis and solution strategy for surfaces of arbitrary geometry*, Computer Methods in Applied Mechanics and Engineering, **197** (33-40):3045–3056, 2008.

2.1 Introduction

The so-called closest point projection procedure is often introduced as a numerical scheme to compute coordinates of a point projected onto a surface. In variational formulations for contact problems it appears due to the split of contact displacements into a normal and a tangential part. In early publications for finite element models it was explained as a result of the linearization of non-penetration conditions, Kikuchi and Oden [84]. Hallquist et al. [53] considered the split into normal and tangential direction via the projection operation for a so-called "master-slave" approach within the finite element contact model. Such an approach is based on looking at normal contact where a "slave" node of one body penetrates into a "master" segment of another body. The value of penetration is measured as the closest distance to the master body. In a next step of the contact algorithm, the penalty method together with explicit time integration was used to enforce the contact conditions. For an implicit solution Wriggers and Simo [192] proposed a consistent linearization procedure of the penalty functional containing the penetration function and obtained the corresponding tangent stiffness matrices within the Newton iteration scheme. However, the penetration function based on the closest point projection is not only used for formulations in combination with a penalty functional, but also in other computational schemes to enforce the condition of non-penetration in contact. The concept of the closest distance is also important for contact searching routines in order to define potentially contacting points.

In order to show the importance of the projection operation, we briefly focus here on methods in contact mechanics, where the closest point procedure occurs. Contact elements for bilinear surface approximations based on the penalty method for non-frictional problems are discussed in Parisch [137]. Due to decoupling of normal and tangential forces the projection procedure necessarily appears for 2D frictional problems in Giannakopoulos [45] and Wriggers et al. [194], then also for 3D problems in Peric and Owen [139] and in Parisch and Lübbing [138]. All mentioned references combine the penalty method together with the regularization of the Coulomb friction model and a radial-return algorithm within the friction algorithm. This algorithm is predominantly used in computational plasticity, see Simo and Hughes [160], and also exploits the closest point procedure; however, then the projection is given onto the yield surface in the force space. Full Lagrange multipliers methods are standard for enforcing constraints in computational mechanics, but cannot be easily transported into frictional problems using the same penetration, or distance function as mentioned above, see the review in Zhong [205]. This is due to the nature of friction forces depending on velocity and, therefore, requiring an incremental update procedure. For a small sliding problem, a perturbed Lagrangian method including both penalty and Lagrangian functionals was discussed in Wriggers and Simo [192]. Then augmented Lagrangian methods appeared to be more robust. Thus, a 2D finite element algorithm based on the augmented Lagrangian method has been proposed in Wriggers et al. [194]. A mixed penalty-duality approach based on the augmented Lagrangian scheme is proposed in Alart and Curnier [3], where different distance functions have been discussed. Laursen and Simo [109] formulated contact conditions via convective coordinates on the contact surface within both, the penalty method and the augmented Lagrangian method. For various geometrical situations the method has been discussed in Heegaard and Curnier [63] for large-slip problems, in Heege and Alart [65] for contact with CAD surfaces and in Pietrzak and Curnier [144] for finite deformations and large sliding. The full Lagrangian approach for frictional problems has been discussed in Jones and Papadopoulos [79] utilizing stick-to-slide conditions in one step, and therefore, covering only small sliding problems. More recently the so-called mortar method, originally based on Lagrangian multipliers enforcing the non-penetration conditions is considered in Hüeber and Wohlmuth [70] for non-frictional problems, where a special discretization technique has been taken for the Lagrange multiplier functions. A mortar scheme based on a special integration technique for constraint equations is proposed in Puso and Laursen [150] for non-frictional problems and extended into the frictional case via the augmented Lagrangian method by the same authors [151].

The publications listed above cover the most applied methods in computational contact mechanics and all contain the closest point procedure as the first necessary step. Despite the enormous number of publications on contact mechanics, there are only a few publications covering to some completeness the problem of uniqueness and existence of the closest point procedure for arbitrary approximations of the contact surfaces as well as describing effective numerical algorithms to overcome problems. The problem of non-uniqueness and non-existence of the projection for e.g. bi-linear approximations of a surface by finite elements is known since the first publications, see Hallquist et al.[53], and mostly reported in theoretical manuals of popular commercial codes, see [52], [1]. Heegaard and Curnier in [64] mentioned that geometrical parameters of a surface, such as a focal point can be used to determine existence and uniqueness of the projection for smooth surfaces. However, for arbitrary surfaces described e.g. by CAD systems and containing combinations of smooth surfaces, curved edges and corner points, the situation can be far more complex. Some techniques dealing with non-existence of the projection in certain cases are well known, e.g. a situation when the slave point is passing over the edge of a locally concave part of a body can be overcome by taking an average vector from the normals of neighboring contact surface segments. A description of this rather heuristic approach can be found in the books of Wriggers [189] and Laursen [106] as well as in theory manuals [52], [1]. This technique is used in a similar manner in multi-surface plasticity when the C^1-continuity of the yield surface is broken, see Simo and Hughes [160]. Other techniques such as the projection onto the edge for such cases, see e.g. in Heegaard and Curnier [63] and in Zhong and Nilsson [207], are also reported to be implemented in commercial programs [52].

In the current contribution, we provide analytical tools allowing to create, a-priori, *proximity domains of contact surfaces* from which a given contact point is always uniquely projected. This approach is based on the geometrical properties of contact surfaces exploiting the covariant description for contact problems developed in Konyukhov and Schweizerhof [86], [89]. First, all C^2-continuous surfaces are classified according to their differential properties allowing a unique projection. Then, *proximity domains* are created for C^1-continuous surfaces. Finally, *proximity domains* are proposed for globally C^0-continuous surfaces covering all practical approximations. The projection scheme in the latter case is further generalized including the projection onto geometrical objects of lower order (curved edges, corner points). In such cases the corresponding proximity domains are created by a geometrical analysis of those objects. Also the reduction into the 2D plane case is discussed. The examples are chosen to show the necessity of the proposed generalized projection procedure in certain mechanical problems as well as the necessity of the proximity domain for searching techniques when a contact surface is given by arbitrary discretization, e.g. described by CAD-systems, or found in high order FEM analysis.

2.2 Formulation of the closest point projection procedure in geometrical terms

The most important operation for the data transfer between contacting bodies in contact mechanics is the closest point projection procedure, see Fig. 2.1, where one seeks the projection of a given contact point from one body $\mathbf{r}$, usually called a slave point, onto another contact body, usually called a master body, parameterized as

$$\rho = \rho(\xi^1, \xi^2). \tag{2.1}$$

The parameterization (12.1) by Gaussian coordinates is arbitrary and ξ^1, ξ^2 can be provided either by e.g. a finite element approximation, or by a spline approximation or by a NURBS approximation. The closest point procedure appears also in many other applications such as fluid-structure interactions, computational plasticity and others. The projection problem is formulated as an extremal problem

$$||\mathbf{r} - \rho(\xi^1, \xi^2)|| \to \min, \quad \longrightarrow (\mathbf{r} - \rho) \cdot (\mathbf{r} - \rho) \to \min, \tag{2.2}$$

which is solved then mostly numerically. However, then a fundamental problem arises: Does the solution of (2.2) exist? And, if it exists, then, is it unique for any arbitrary surface approximation?

The direct and strict answer is fully covered by the application of the famous theorem from the convex analysis to problem (2.2), see e.g. [13].

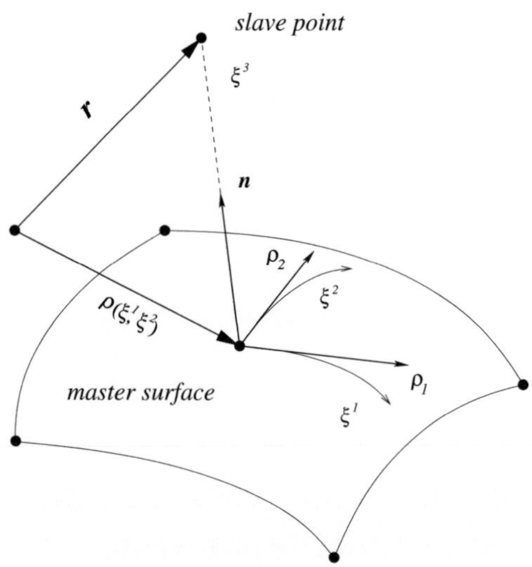

Figure 2.1: Closest point procedure and definition of the spatial coordinate system.

If the function

$$\mathbf{F}(\xi^1, \xi^2) = \frac{1}{2}(\mathbf{r} - \rho) \cdot (\mathbf{r} - \rho) \tag{2.3}$$

is convex in a domain $(\xi^1, \xi^2) \in D$, *then the solution of problem (2.2) exists and is unique in this domain.*

This leads to the fact that the solution can be obtained numerically by e.g. a Newton iteration procedure, which will converge from any initial point inside the domain $(\xi^1_{(0)}, \xi^2_{(0)}) \in D$.

Using this criterion we can focus on the geometrical properties of the surfaces. Then the goal of this contribution is to create a classification of surfaces from its differential geometry point of view onto which a point can be projected uniquely.

The analysis of the local geometrical structure of surfaces allows to create *a projection domain* Ω in 3D from which any point can be uniquely projected onto the given surface. Starting with the C^2-continuous case, we will consider also cases possessing parts with solely C^1- and C^0-continuity discussing the solvability of problem (2.2) and in addition the possible multiplicity of solutions.

Assuming that the function $\mathbf{F}$ is twice differentiable, i.e. for C^2-continuous surfaces, we can construct the Newton iterative process for the solution of the minimization problem (2.2) as follows:

$$\Delta\boldsymbol{\xi}_{(n)} = \begin{bmatrix} \Delta\xi^1_{(n)} \\ \\ \Delta\xi^2_{(n)} \end{bmatrix} = -(\mathbf{F}'')^{-1}_{(n)}\,\mathbf{F}'_{(n)} \tag{2.4}$$

$$\boldsymbol{\xi}_{(n+1)} = \boldsymbol{\xi}_{(n)} + \Delta\boldsymbol{\xi}_{(n)},$$

where the first derivative $\mathbf{F}'$ and the second derivative $\mathbf{F}''$ with respect to the surface coordinates ξ^1, ξ^2 are computed as:

$$\mathbf{F}' = \begin{bmatrix} \dfrac{\partial F}{\partial \xi^1} \\ \\ \dfrac{\partial F}{\partial \xi^2} \end{bmatrix} = -\begin{bmatrix} \boldsymbol{\rho}_1 \cdot (\mathbf{r} - \boldsymbol{\rho}) \\ \boldsymbol{\rho}_2 \cdot (\mathbf{r} - \boldsymbol{\rho}) \end{bmatrix} \tag{2.5}$$

$$\mathbf{F}'' = \begin{bmatrix} \boldsymbol{\rho}_1 \cdot \boldsymbol{\rho}_1 - \boldsymbol{\rho}_{11} \cdot (\mathbf{r} - \boldsymbol{\rho}) & \boldsymbol{\rho}_1 \cdot \boldsymbol{\rho}_2 - \boldsymbol{\rho}_{12} \cdot (\mathbf{r} - \boldsymbol{\rho}) \\ \boldsymbol{\rho}_2 \cdot \boldsymbol{\rho}_1 - \boldsymbol{\rho}_{21} \cdot (\mathbf{r} - \boldsymbol{\rho}) & \boldsymbol{\rho}_2 \cdot \boldsymbol{\rho}_2 - \boldsymbol{\rho}_{22} \cdot (\mathbf{r} - \boldsymbol{\rho}) \end{bmatrix}. \tag{2.6}$$

Here a short notation for the partial derivatives has been introduced as

$$\boldsymbol{\rho}_i = \frac{\partial \boldsymbol{\rho}}{\partial \xi^i}, \quad \boldsymbol{\rho}_{ij} = \frac{\partial^2 \boldsymbol{\rho}}{\partial \xi^i \partial \xi^j}. \tag{2.7}$$

Now, we introduce a 3D spatial coordinate system related to the surface coordinate system, see Fig. 2.1, as follows:

$$\mathbf{r}(\xi^1, \xi^2, \xi^3) = \boldsymbol{\rho} + \mathbf{n}\xi^3. \tag{2.8}$$

All necessary parameters are defined in the covariant basis of the surface tangent vectors $\boldsymbol{\rho}_1, \boldsymbol{\rho}_2$ and a unit normal vector $\mathbf{n} = \dfrac{\boldsymbol{\rho}_1 \times \boldsymbol{\rho}_2}{\|\boldsymbol{\rho}_1 \times \boldsymbol{\rho}_2\|}$. The solvability of the minimization problem (2.2) will be considered in this coordinate system. The introduction of such a coordinate system is the basis of the covariant description for contact problems, developed in Konyukhov and Schweizerhof [86], [89] for isotropical frictional problems, and then in [90], [91] for anisotropic frictional problems. The dominantly geometrical structure of all contact parameters is among the main advantages of this description, e.g. a value of penetration is simply the third convective coordinate ξ^3.

The following surface tensors are used to describe the metric and the curvature properties of surfaces in differential geometry [103]:

a) metric tensor with components:

$$a_{ij} = \boldsymbol{\rho}_i \cdot \boldsymbol{\rho}_j \qquad (2.9)$$

b) curvature tensor with components:

$$h_{ij} = \boldsymbol{\rho}_{ij} \cdot \mathbf{n} \qquad (2.10)$$

Inserting now eqn. (2.8) into eqn. (2.6) and using the notations given in eqns. (2.9) and (2.10) we obtain the expression $\mathbf{F}''$ in the introduced spatial coordinate system as:

$$\mathbf{F}'' = \begin{bmatrix} a_{11} - \xi^3 h_{11} & a_{12} - \xi^3 h_{12} \\ a_{21} - \xi^3 h_{21} & a_{22} - \xi^3 h_{22} \end{bmatrix}. \qquad (2.11)$$

Let us start with a case assuming a unique projection: In this case the function $\mathbf{F}$ must be convex. As is known for the convexity of $\mathbf{F}$, the second derivative $\mathbf{F}''$ must then be a positive definite matrix. Thus, the Sylvester criterion from basic algebra is exploited to check the positivity of the matrix given in eqn. (2.11):

$$\begin{aligned} (a_{11} - \xi^3 h_{11}) &> 0 \\ \det[(a_{ij} - \xi^3 h_{ij})] &> 0 \end{aligned} \qquad (2.12)$$

2.2.1 Necessary information about the surface structure

Surprisingly, the second equation in (12.18) is similar to that which is used in differential geometry for the analysis of the surface structure. To emphasize this, we present here main formulae necessary for the further developments. For more information including the specific derivations of the formulae see e.g. in [103].

The geometrical analysis of the surface structure is given by the generalized eigenvalue problem:

$$(h_{ij} - k a_{ij}) e_j = 0, \qquad (2.13)$$

which leads to the real roots k_1, k_2 of the equation

$$\det[(h_{ij} - ka_{ij})] = 0. \tag{2.14}$$

These roots are called *principal curvatures* and correspond to the orthogonal *principal directions* $\mathbf{e_1}$, $\mathbf{e_2}$. The principal curvature k_1 (resp. k_2) is the curvature of a line arising from the intersection of the surface with the plane containing both, normal vector $\mathbf{n}$ and principal vector $\mathbf{e_1}$ ($\mathbf{e_2}$). Their inverse values are principal radii $R_i = 1/k_i$, see Fig. 2.2. The local structure of the surface in the vicinity of the computed contact point can be then classified by the Gaussian curvature K

$$K = k_1 \cdot k_2 \tag{2.15}$$

into four cases as follows:

1) $K = k_1 \cdot k_2 > 0$ — an elliptic point, i.e. a surface in vicinity of the point *looks like an elliptic paraboloid*, see Fig. 2.2.

2) $K = k_1 \cdot k_2 < 0$ — a hyperbolic point, i.e. a surface *looks like a hyperbolic paraboloid*, see Fig. 2.3.

For the case with zero Gaussian curvature $K = 0$ a more careful analysis is required.

3) Either $k_1 = 0$, $k_2 \neq 0$, or $k_2 = 0$, $k_1 \neq 0$ — a parabolic point. A surface *looks then like a parabolic cylinder*, see Fig. 2.4. If these conditions are fulfilled for each point of the surface, then *a flat surface* is obtained. This surface (e.g. a cylinder, or a cone) can be unwrapped on a plane.

4) Both $k_1 = 0$ and $k_2 = 0$ — a planar point. The local structure can not be identified without the analysis of the higher order derivatives. Nevertheless, the case with $k_1 = 0$, $k_2 = 0$ for each point leads to a plane in 3D.

Eqn. (2.14) for a definition of the principal curvatures can be written in the following form:

$$k^2 - 2Hk + K = 0, \implies k_{1,2} = H \pm \sqrt{H^2 - K} \tag{2.16}$$

where the Gaussian curvature K is defined as

$$K = k_1 \cdot k_2 = \frac{\det[h_{ij}]}{\det[a_{ij}]} = \frac{h_{11}h_{22} - h_{12}^2}{a_{11}a_{22} - a_{12}^2}, \tag{2.17}$$

and the mean curvature H is defined as

$$H = \frac{1}{2}(k_1 + k_2) = \frac{1}{2}\frac{a_{22}h_{11} - 2a_{12}h_{12} + a_{11}h_{22}}{a_{11}a_{22} - a_{12}^2}. \tag{2.18}$$

Remark.
Any other structure of the surface in the vicinity of a point different from the described cases 1)-3) (e.g. edge point etc.) can arise either from a planar point (case 4), or from violation of the a-priori assumed C^2-continuity and a more advanced analysis would be necessary. Surfaces with more complicated local structure leading e.g. to a so-called "star"-looking domain, self-contacting and self-penetrating surfaces etc. are relatively seldom in practical applications and are out of the scope of the current contribution.

2.3   Proximity criteria for different surfaces

In this section we consider the structure of 3D domains (*proximity domains*) surrounding a given surface, from which a point can be a-priori uniquely projected onto the surface. This structure depends on the classification of surface points given in Sect. 2.2.1. Taking into account eqn. (2.16) for principal curvatures, the Sylvester criteria in eqn. (12.18) can be written in the following form:

$$(a_{11} - \xi^3 h_{11}) > 0$$

$$\left(\frac{1}{\xi^3}\right)^2 - 2H\frac{1}{\xi^3} + K > 0 \tag{2.19}$$

This system of inequalities can be reformulated in terms of principal curvatures k_1, k_2 as:

$$\left(\frac{1}{\xi^3} - k_1\right) > 0$$

$$\left(\frac{1}{\xi^3} - k_1\right)\left(\frac{1}{\xi^3} - k_2\right) > 0. \tag{2.20}$$

Within the last transformation, it was assumed that the coordinate ξ^3 is a positive value into the normal vector direction $\mathbf{n}$, as chosen in eqn. (2.8). Thus, zones with positive and negative coordinates ξ^3 should be distinguished.

2.3.1 Projection domains for an elliptic point

According to the geometry presented in Fig. 2.2, a normal $\mathbf{n}$ is pointing into a convex part, where $\xi^3 > 0$ and, both $k_1 > 0$ and $k_2 > 0$.

2.3.1.1 Domain for the convex part $\xi^3 > 0$

Using the geometrical interpretation of the principal curvatures as principal radii $R_i = 1/k_i$, eqn. (2.20) can be written as:

$$R_1 - \xi^3 > 0$$
$$(R_1 - \xi^3)(R_2 - \xi^3) > 0 \tag{2.21}$$

The solution allows to create a projection domain $\Omega(\xi^1, \xi^2, \xi^3)$ with a parameter $\hat{\xi}^3 = \min\{R_1, R_2\}$ with an a-priori unique solution of the projection problem, see Fig. 2.2:

$$\Omega(\xi^1, \xi^2, \xi^3) := \{\mathbf{r} = \boldsymbol{\rho} + \xi^3 \mathbf{n}, \ \ where \ \ 0 < \xi^3 < \hat{\xi}^3\}. \tag{2.22}$$

This domain is created in the local coordinate system and contains all points between the original surface and shifted by $\mathbf{n}\xi^3$ surface.

2.3.1.2 Domain for the concave part $\xi^3 < 0$

In this case, criterion (2.21) is automatically fulfilled leading to an infinite domain Ω above the surface:

$$\Omega(\xi^1, \xi^2, \xi^3) := \{\mathbf{r} = \boldsymbol{\rho} + \xi^3 \mathbf{n}, \quad with \quad -\infty < \xi^3 < 0\}. \qquad (2.23)$$

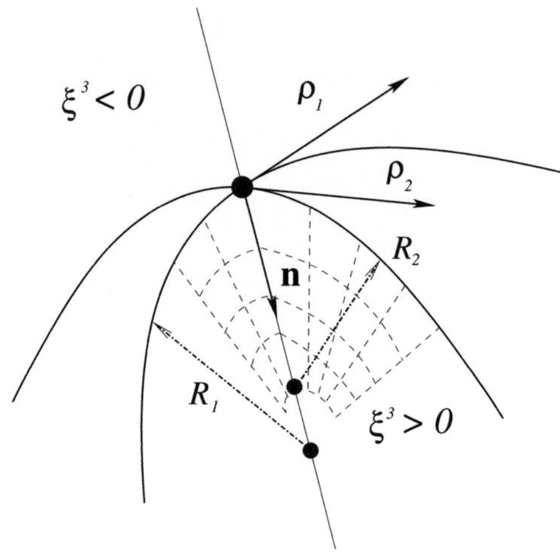

Figure 2.2: Elliptic point. Structure of the surface and projection domain assuming C^2-continuity.

2.3.2 Projection domain for a hyperbolic point

Considering a situation as presented in Fig. 2.3, in the case with $\xi^3 > 0$, where a projection domain is created as an overlapping of the semi-infinite domain due to the line with negative curvature k_1 and the finite domain due to the line with positive curvature k_2 similar to the domain as for a convex part for an elliptic point. The case with $\xi^3 < 0$ leads to a projection domain constructed in an identical fashion. Summarizing both cases, a finite projection domain $\Omega(\xi^1, \xi^2, \xi^3)$ is created as a domain between two shifted surfaces:

$$\Omega(\xi^1, \xi^2, \xi^3) := \{\mathbf{r} = \boldsymbol{\rho} + \xi^3 \mathbf{n}, \quad where \quad -R_1 < \xi^3 < R_2\}. \qquad (2.24)$$

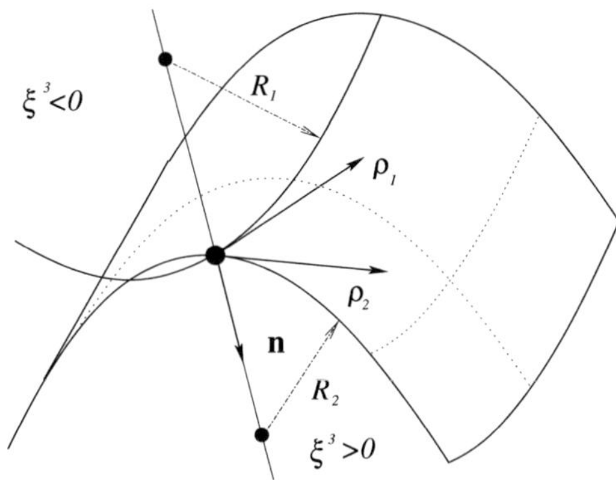

Figure 2.3: Hyperbolic point. Structure of the surface and projection domain assuming C^2-continuity.

2.3.3 Projection domain for a parabolic point

A projection domain for a parabolic point, see Fig. 2.4, consists of a finite domain for the convex part with $\xi^3 > 0$, and a semi-infinite domain for the concave part with $\xi^3 < 0$. Thus, the projection domain $\Omega(\xi^1, \xi^2, \xi^3)$ is described as follows:

$$\Omega(\xi^1, \xi^2, \xi^3) := \{\mathbf{r} = \boldsymbol{\rho} + \xi^3 \mathbf{n}, \quad where \quad -\infty < \xi^3 < R_2\}. \qquad (2.25)$$

2.3.4 Discussion about planar points – Required approximation for the plane

As mentioned in Sect. 2.2.1 case 4), the structure of a surface in the vicinity of the planar point requires an analysis of the higher order derivatives with respect to the surface coordinates. In this case, it is

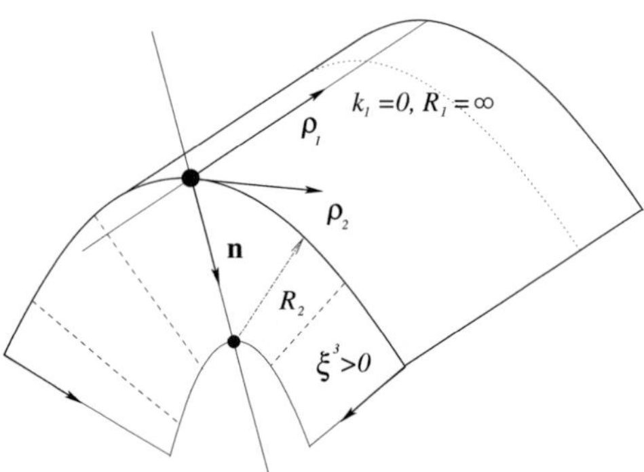

Figure 2.4: Parabolic point. Structure of the surface and projection domain assuming C^2-continuity.

also difficult to discuss the convergence of the Newton iterative process (2.4). However, for the plane with zero curvature tensor $h_{ij} = 0$ criterion (12.18) becomes

$$\det[a_{ij}] > 0. \tag{2.26}$$

Since the metric tensor a_{ij} is always positive, a trivial result is recovered: The projection onto a plane always exists and is unique. Nevertheless, from the numerical point of view it is necessary to avoid cases with $\det[a_{ij}] \approx 0$. In such cases the angle between the convective coordinate lines is close either to 0° or to 180°, which is normally avoided *ab initio* as e.g. in finite element approximations this would be a sign for inadmissibly distorted elements.

2.4 Solvability of the projection algorithm – allowable and non-allowable domains

In the previous section, the projection domains were constructed under the main assumption of sufficient smoothness of the corresponding surfaces, specifically C^2-continuity was necessary to derive all curvature parameters. In this section, we will obtain that C^1-continuity of a surface is sufficient for the construction of a continuous projection domain, while violation of C^1-continuity can cause either multiplicity of solutions, or non-existence of solutions at all. Starting with a 2D case as a preliminary case, we then develop "a remedy" for globally C^0-continuous surfaces by constructing a continuous projection domain allowing unique projections onto the surface. A generalized projection procedure includes then several projection procedures onto curved edges and corners.

2.4.1 Reduction to 2D plane geometry – Solvability criteria and uniqueness

A covariant approach for contact problems allows to look at a 2D case either as a special reduction from a cylindrical geometry in 2D space, or as a case based on a 2D plane curve geometry, see [92]. In the last case, the length of a curved line s is used as a convective coordinate. The closest point projection method (2.2) is then formulated as follows:

$$F = \frac{1}{2}||\mathbf{r} - \boldsymbol{\rho}(s)||^2 \rightarrow \min. \tag{2.27}$$

A curvilinear coordinate system in 2D is based on a flat curve geometry and is introduced as follows:

$$\mathbf{r}(s, \zeta) = \boldsymbol{\rho}(s) + \zeta \boldsymbol{\nu}. \tag{2.28}$$

A second derivative with respect to s in this coordinate system is computed as:

$$F'' = \frac{\partial \boldsymbol{\rho}}{\partial s} \cdot \frac{\partial \boldsymbol{\rho}}{\partial s} - \frac{\partial^2 \boldsymbol{\rho}}{\partial s^2} \cdot (\mathbf{r} - \boldsymbol{\rho}(s)) = \boldsymbol{\tau} \cdot \boldsymbol{\tau} - \frac{\partial \boldsymbol{\tau}}{\partial s} \cdot (\mathbf{r} - \boldsymbol{\rho}(s)) \tag{2.29}$$

For further transformation it is necessary to introduce the Serret-Frenet formulae

$$\boldsymbol{\tau} = \frac{\partial \boldsymbol{\rho}}{\partial s}, \quad \frac{\partial \boldsymbol{\tau}}{\partial s} = k \boldsymbol{\nu}, \tag{2.30}$$

where $\boldsymbol{\tau}$ is a unit tangent vector, $\boldsymbol{\nu}$ is a unit normal vector and k is a curvature of the curve (the radius of the curvature $R = \dfrac{1}{k}$ is also often used). Taking into account the Serret-Frenet formulae together with eqn. (2.28) the second derivative F'' is transformed as

$$F'' = 1 - k\zeta. \qquad (2.31)$$

Since the normal vector $\boldsymbol{\nu}$ is always pointing to the center of curvature C_1, see Fig. 2.5, we obtain from the condition $F'' = 1 - k\zeta > 0$ a finite projection domain $O_1\,O_2\,C_2\,C_1$ for the convex part as follows:

$$\Omega(s,\zeta) := \{\mathbf{r} = \boldsymbol{\rho}(s) + \zeta\boldsymbol{\nu}, \;\; with \;\; 0 < \zeta < \frac{1}{k} = R\}, \qquad (2.32)$$

and a semi-infinite projection domain for the concave part above the curve $O_1\,O_2$:

$$\Omega(s,\zeta) := \{\mathbf{r} = \boldsymbol{\rho}(s) + \zeta\boldsymbol{\nu}, \;\; with \;\; -\infty < \zeta < 0\}. \qquad (2.33)$$

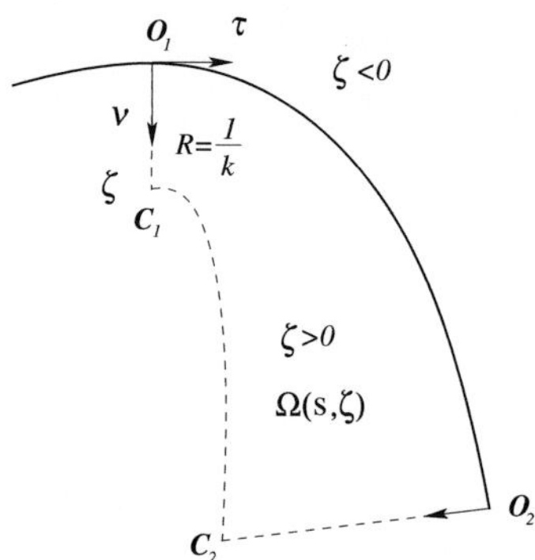

Figure 2.5: 2D case – plane curve. Projection domain.

The case with $F'' = 0$ can lead to a multiplicity of the solution. This is visible in Fig. 2.6, where an infinite number of projections is possible from a point O_1 with the coordinate $\zeta = R$ onto an arc $\mathbf{BC}$, and from a point O_2 with the coordinate $\zeta = -R$ onto another arc $\mathbf{CD}$.

2.4.1.1 Violation of C^2-continuity

The violation of C^2-continuity, but keeping C^1-continuity leads to a discontinuous projection domain. In this case, however, it is easily possible to construct a continuous projection domain, from where uniqueness of the projection operation is automatically fulfilled. The idea is presented in Table 2.1 and is illustrated in Fig. 2.6. Since discontinuity is resulting due to the only piecewise continuous parameter ζ, a continuous projection domain is constructed by taking the minimal parameter $\zeta = \min\{R, \infty\} = R$ along the curve.

Part of a curve	Curvature k	Computed ζ	Minimal ζ
A B	0	$+\infty$	R
B C	R	R	R
C D	-R	$+\infty$	R

Table 2.1: Violation of C^2-continuity of a curve leads to piecewise continuous curvature and consequently to a piecewise continuous parameter ζ. A continuous projection domain $\Omega(s, \zeta) := \{\mathbf{r} = \rho(s) + \zeta\nu\}$ can be constructed by setting ζ to a minimal value. An example is given for $\zeta > 0$ in Fig.2.6.

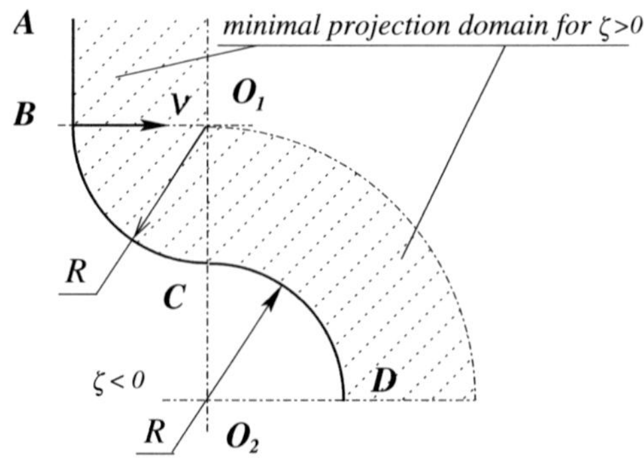

Figure 2.6: Violation of C^2-continuity results in changing the discontinuity of the projection domain. A continuous domain is constructed by setting ζ to a minimal value.

2.4.1.2 Violation of C^1-continuity

We consider here the most practical case which is standard in the triangulation of surfaces with low order finite elements: piecewise differentiable functions with finite jumps for the first derivative. From a geometrical point of view this situation leads to an angular point for curves or an edge for surfaces. The violation of C^1-continuity for a surface parameterization leads also to a discontinuity of the normal vector $\mathbf{n}$ causing either difficulty with definition of the local coordinate system, or multiplicity of projection. The situation in 2D is presented in Fig. 2.7. Now, we consider a point following the path S_1 S_2 S_3 S_4 S_5 in the local coordinate system (2.28). Since a normal vector ν has jumps (points **B** and **C**), there are portions of the trajectory which can not be described in this local coordinate system. They are located in the non-allowable domain with regard to the projection onto the curve: any point S_2 laying in the non-allowable domain can not be described in the local coordinate system given by eqn. (2.28). The term *non allowable domain with regard to the projection onto surface resp. curve resp. point* is then used for a domain where any point can not be described in the local coordinate system corresponding to the projection onto surface resp. curve resp. point.

However, in such a situation it is possible to create a continuous mapping of the path S_1 S_2 S_3 S_4 S_5 onto the curve by introducing a new projection operation in the non-allowable domain with regard to the projection onto the curve. This is a projection of the point into an angular point of curves (e.g. S_2 into B and S_4 into C in Fig. 2.7), or a closest point projection onto the edge of surfaces. This method is mentioned for 2D examples in Heegaard and Curnier [63] and in Zhong and Nilsson [207]; it is also reported to be done e.g. in LS-DYNA [52]. A projection procedure keeping a continuous mapping of any path on the curve will be called *the generalized closest point procedure*. In 2D cases this procedure includes both a projection onto the curve and onto the corner point.

Another situation arising from the violation of C^1-continuity is overlapping of two or more projection domains causing consequently the multiplicity of projection, see e.g. domain of multiple solutions in Fig. 2.7.

We already could observe the multiplicity of solutions in the case of the violation of C^2-continuity where the second derivative of the distance function is zero $F'' = 0$, see point O_1 in Fig. 2.6. In this case, a strong inequality in the formulation of the projection domains allows to eliminate the multiplicity. A case with violation of C^1-continuity (e.g. domain of multiple solutions in Fig. 2.7) can lead to a more severe situation, see point M passing over corner point C. A natural remedy in this case is to define a minimum distance in the sense of the generalized closest point procedure onto neighboring segments and onto corner point C. However, one can find a line with points which are equidistantly situated from both segments.

In this situation, though relatively seldom in numerical computations due to round-off errors, a choice either with a random selection of projection sides, or with an averaged normal from all projection sides can be applied, implicitly assuming that the distance is fairly small. Within a contact algorithm the storage of the "slave" point path as history variable would be then necessary.

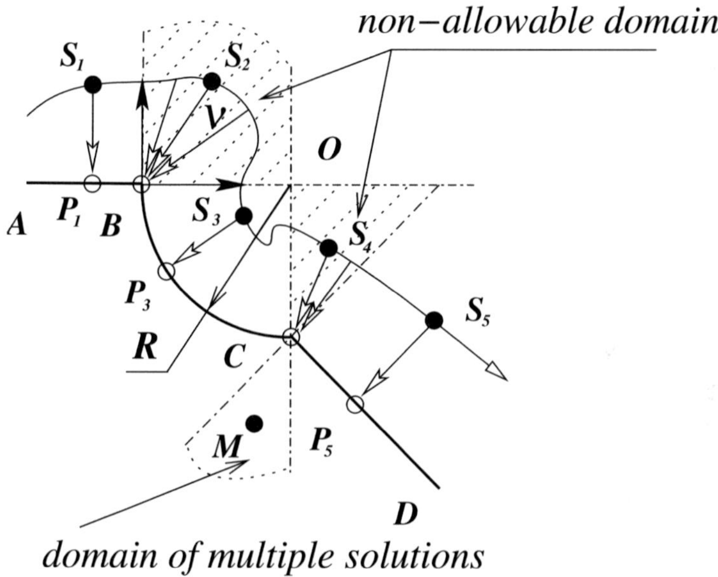

Figure 2.7: Violation of C^1-continuity leads to non-allowable domains with regard to the projection onto curve $ABCD$. A special treatment is necessary to preserve a continuous mapping.

2.4.2 Proximity domain for globally C^0-continuous surface in 3D

It is necessary to define additional projection procedures in order to create a proximity domain for the 3D space allowing a continuous mapping of any path laying inside, similar to that discussed for the 2D case in 2.4.1.2. These projections include a projection onto an edge and onto a corner. The combination of proximity domains leads finally to a continuous domain. The idea is illustrated in Fig. 2.8 for the contact surface of a hexaeder focusing on the surfaces containing the corner point O. The continuous proximity domain surrounding corner point O consists of

1. three domains for sides arising from the standard projection procedure:
$$S_1 = \{x, y, z \,|\, x > 0, y > 0, z < 0\} \text{ for side } BOCF,$$
$$S_2 = \{x, y, z \,|\, x > 0, y < 0, z > 0\} \text{ for side } AOBE,$$
$$S_3 = \{x, y, z \,|\, x < 0, y < 0, z < 0\} \text{ for side } AOC;$$

2. three domains for edges arising from the point-to-edge projection procedure:
$$E_1 = \{x, y, z \,|\, x < 0, y < 0, z > 0\} \text{ for edge } OA,$$
$$E_2 = \{x, y, z \,|\, x > 0, y > 0, z > 0\} \text{ for edge } OB,$$
$$E_3 = \{x, y, z \,|\, x < 0, y > 0, z < 0\} \text{ for edge } OC;$$

3. one domain arising from the point-to-corner point projection procedure:
$$P = \{x, y, z | x < 0, y > 0, z > 0\} \text{ for corner point } O.$$

The last domain is added to fulfill the continuity, because from this domain a projection onto the cube in general is possible only into a corner point. This projection is trivial, always exists and can be defined simply as difference between "slave" and corner point: $\mathbf{r}_s - \rho_O$.

The point-to-edge projection, in case of a curved edge, requires a numerical solution and again the problem of existence and uniqueness arises.

2.4.3 Point-to-edge closest point projection and corresponding projection domain

An arbitrary curved edge as a result of the intersection of two surfaces is represented by a curved line, see e.g. AOB in Fig. 2.9. This line can be parameterized by the arc-length parameter s as well as by an

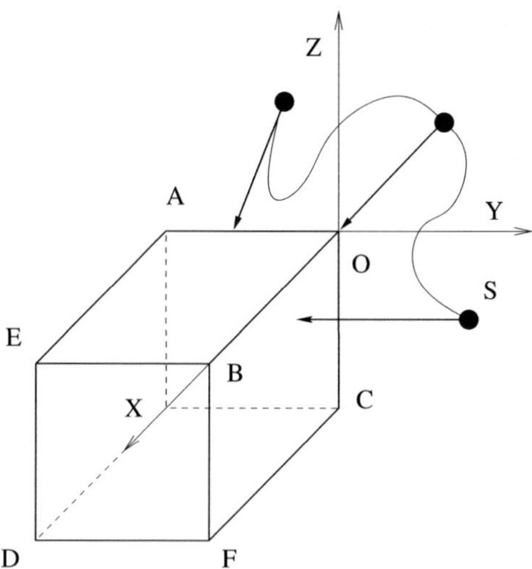

Figure 2.8: Hexaedral contact surface in the vicinity of O. A continuous proximity domain (i.e. allowing continuous projection onto the C^1-surface) consist of domains for sides, edges OA, OB, OC and, finally, of a domain for the corner point O.

arbitrary parameter ξ, e.g. a normalized coordinate in a finite element approximation:

$$\boldsymbol{\rho} = \boldsymbol{\rho}(s) = \boldsymbol{\rho}(s(\xi)). \tag{2.34}$$

The transformation of the parameters $s \leftarrow \xi$ is provided by the formula:

$$ds = \sqrt{\frac{\partial \boldsymbol{\rho}}{\partial \xi} \cdot \frac{\partial \boldsymbol{\rho}}{\partial \xi}} d\xi = \sqrt{\boldsymbol{\rho}_\xi \cdot \boldsymbol{\rho}_\xi} d\xi. \tag{2.35}$$

The projection routine, though it is looking similar to eqn. (2.27), is now formulated in 3D as follows:

$$F(s(\xi)) = \frac{1}{2}||\mathbf{r}_s - \boldsymbol{\rho}(s(\xi))||^2 \rightarrow \min. \tag{2.36}$$

The necessary optimum condition leads to the projection operation onto the curve

$$F' = -(\mathbf{r}_s - \boldsymbol{\rho}(s(\xi))) \cdot \frac{\partial \boldsymbol{\rho}}{\partial s} = -(\mathbf{r}_s - \boldsymbol{\rho}(s(\xi))) \cdot \boldsymbol{\tau} = 0, \tag{2.37}$$

showing the orthogonality of the vector $(\mathbf{r}_s - \boldsymbol{\rho})$ and the tangent vector $\boldsymbol{\tau}$. According to this, a new coordinate system $\boldsymbol{\tau}, \boldsymbol{\nu}, \boldsymbol{\beta}$ is introduced as a natural coordinate system of the curved line, see Fig. 2.9, where $\boldsymbol{\tau}$ is a unit tangent vector, $\boldsymbol{\nu}$ is a unit main normal vector and $\boldsymbol{\beta}$ is a unit binormal vector of the curve, see [103]. This results in orthogonal planes **I, II, III**. The given point S ("slave" point) is defined as follows:

$$\mathbf{r}_s = \boldsymbol{\rho}(s) + r\mathbf{e}, \tag{2.38}$$

where r is the shortest distance between the point S and the curve $\mathbf{AB}$. $\mathbf{e}$ is a vector giving a director of the shortest distance in the plane $\boldsymbol{\nu}0\boldsymbol{\beta}$ (plane **II**), defined via $\boldsymbol{\nu}$ and $\boldsymbol{\beta}$ as follows:

$$\mathbf{e} = \boldsymbol{\nu} \cos\varphi + \boldsymbol{\beta} \sin\varphi. \tag{2.39}$$

The second derivative F'' taking into account eqn. (2.38) is transformed in the curvilinear coordinate system as follows:

$$F'' = \boldsymbol{\tau} \cdot \boldsymbol{\tau} - (\mathbf{r} - \boldsymbol{\rho}(s)) \cdot k\boldsymbol{\nu} = 1 - kr(\mathbf{e} \cdot \boldsymbol{\nu}) \tag{2.40}$$

and finally using representation (2.39) for the vector e as

$$F'' = 1 - kr \cos\varphi. \tag{2.41}$$

Considering now $F'' > 0$, the projection domain is constructed from two parts **1** and **2**, see Fig. 2.9:

1: a semi-infinite domain with negative $\boldsymbol{\nu}$, or with $\varphi \in [\pi/2, 3\pi/2]$. Here the projection of the vector $r\mathbf{e}$ onto the normal $\boldsymbol{\nu}$ is negative leading to the automatic satisfaction of $F'' > 0$,

and,

2: a layer with positive $\boldsymbol{\nu}$, or with $\varphi \in (-\pi/2, \pi/2)$. Here the projection of the vector $r\mathbf{e}$ on the normal $\boldsymbol{\nu}$ must be less than the radius of the curvature R, i.e.

$$r \cos\varphi < \frac{1}{k} = R. \tag{2.42}$$

As one can see, two domains surround a curve densely, e.g. without any void. A problem can occur if vector $\boldsymbol{\rho}(s)$ looses C^1-continuity, however in this case we obtain a corner point as discussed above.

Remark 1.

If the edge is a straight line then the coordinate system τ, ν, β should not be derived via the curved properties of the line according to the Serret-Frenet formulae, however, it can be defined arbitrarily.

In the case of an arbitrary parameterization of a line, the projection routine is fulfilled via the Newton iterative process in eqn. (2.4) for the parameter ξ, where $\Delta\xi$ is computed as:

$$\Delta\xi = -\frac{F'}{F''} = -\frac{(\mathbf{r}_s - \boldsymbol{\rho}) \cdot \boldsymbol{\rho}_\xi}{(\boldsymbol{\rho}_\xi \cdot \boldsymbol{\rho}_\xi) - (\mathbf{r}_s - \boldsymbol{\rho}) \cdot \boldsymbol{\rho}_{\xi\xi}}, \quad \xi_{(n+1)} = \xi_{(n)} + \Delta\xi. \quad (2.43)$$

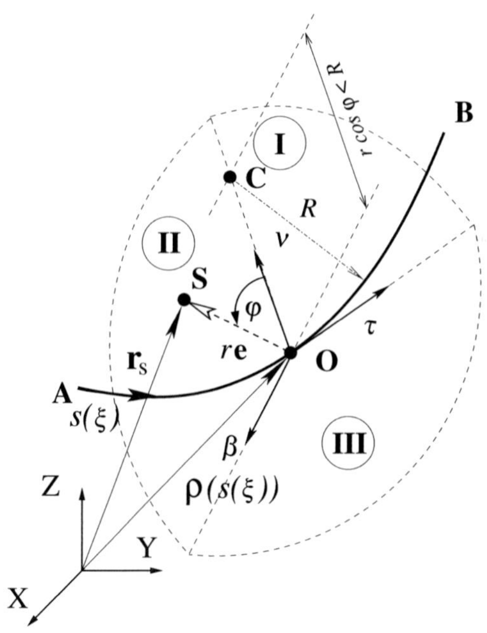

Figure 2.9: Point-to-edge projection. Proximity criteria and projection domain.

Summary: A continuous projection domain for a globally C^0-continuous surface allowing unique continuous projections of any path laying inside this domain can be constructed via *a generalized projection procedure* including three projection operations: a) onto a surface; b) onto an edge; c) onto a corner point.

Remark 2.

An algorithm for the continuous projection domain can be used also as a preparation stage for the global searching routine. Thus, in following computations within an analysis of a contact problem any possible "slave" point can be uniquely projected onto the "master" surface via the corresponding projection onto a surface, an edge or a point.

2.5   Kinematics of the point-to-edge contact element

The introduction of the projection procedure and a corresponding curvilinear coordinate system allows also to consider the kinematics of the contact, and therefore, to compute all necessary contact parameters such as forces, measures of displacements etc.

Consider a motion of the given point S ("a slave" point) in the moving coordinate system τ, ν, β. The velocity vector $\mathbf{v}_s$ is computed as a full time derivative:

$$\frac{d\mathbf{r}_s}{dt} = \frac{\partial \boldsymbol{\rho}}{\partial t} + \dot{s}\boldsymbol{\tau} + \dot{r}\mathbf{e} + r\frac{d\mathbf{e}}{dt}, \tag{2.44}$$

where the full time derivative $\dfrac{d\mathbf{e}}{dt} = \boldsymbol{\omega}$ describes a rigid body rotation of the coordinate system $\tau, \mathbf{e}, \mathbf{g}$ both, due to the changing curvature along a curve, and due to the rigid motion of a curve. For the further numerical model a value of r must lay within an allowable small distance surrounding a curve, see Fig. 2.10. This assumption necessarily will lead also to the corresponding small load or time increments to solve the contact problem. Altogether these assumptions allow to neglect the contribution of the last term $(r\frac{d\mathbf{e}}{dt})$ similar to the covariant description for the 3D-dimensional case [89], where a relative velocity vector has been considered on the tangent plane. With the dot product of the vectors τ and $\mathbf{e}$, the following components of the velocity vector are obtained:

$$\dot{s} = (\mathbf{v}_s - \mathbf{v}) \cdot \boldsymbol{\tau}, \tag{2.45}$$

$$\dot{r} = (\mathbf{v}_s - \mathbf{v}) \cdot \mathbf{e}, \tag{2.46}$$

where $\mathbf{v}_s = \dfrac{d\mathbf{r}_s}{dt}$ is absolute velocity of the slave points, and $\mathbf{v} = \dfrac{\partial \boldsymbol{\rho}}{\partial t}$ is the translational velocity of the projected point **O**.

Remark. According to the assumptions mentioned above, all variation parameters for the further weak form are created in a similar fashion as the kinematical eqns. (2.44), (2.45), (2.46), e.g.

$$\delta \mathbf{r}_s - \delta \boldsymbol{\rho} = \delta s \boldsymbol{\tau} + \delta r \mathbf{e}. \tag{2.47}$$

The convective velocities in eqns. (2.45), (2.46) define essential measures for the contact interaction as: r – for the normal interaction, Δs – for the tangential interaction. In addition for the normal interaction a penetration area can be defined in the plane $\nu O \beta$ (plane II in Fig. 2.9 as a curve in the polar coordinate system

$$r = r(\varphi), \quad \phi \in [\varphi_0, \varphi_n]. \tag{2.48}$$

The simplest case is a circular area with a radius R_ε, see Fig. 2.10. Thus, a measure can be taken as

$$\zeta = r - R_\varepsilon = \|\mathbf{r}_s - \boldsymbol{\rho}\| - R_\varepsilon. \tag{2.49}$$

Remark. A case with artificial R_ε is rather necessary for the point-to-curve algorithm applicable for edge-to-edge contact. In the case of globally C^0 surface, e.g. in the case of an edge, penetration is computed exactly at projection point: $\zeta = \|\mathbf{r}_s - \boldsymbol{\rho}\|$.

2.5.1 Weak formulation of contact equilibrium

The components of the contact force vector **F** in the curvilinear coordinate system are chosen to be conjugate variables with regard to the work W of the contact forces. Thus, vector **F** is decomposed as:

$$\mathbf{R} = N\boldsymbol{\nu} + T\boldsymbol{\tau}. \tag{2.50}$$

A pointwise equilibrium contact condition $\mathbf{F_S} + \mathbf{F_O} = 0$ is formulated in variational form as:

$$\delta W = \mathbf{F_S} \cdot \delta \mathbf{r}_s + \mathbf{F_S} \cdot \delta \boldsymbol{\rho}, \tag{2.51}$$

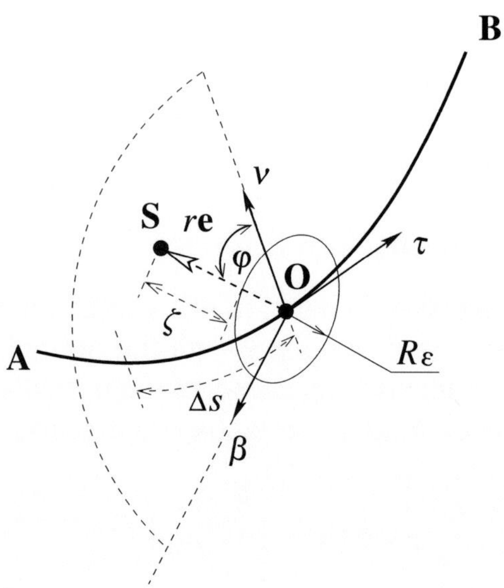

Figure 2.10: Point-to-edge contact element. Definition of measures of contact interactions: ζ – for normal interaction defined according to a circular area with radius R_ε; Δs – for tangential interaction.

and is transformed exploiting eqns. (2.50) and (2.47) into

$$\delta W = (\delta \mathbf{r}_s - \delta \boldsymbol{\rho}) \cdot \mathbf{F_S} = N \delta r + T \delta s. \tag{2.52}$$

2.5.2 Regularization of contact forces – Return-mapping scheme for the Coulomb friction model

The regularization of contact conditions is derived with regard to the decomposition into normal and tangential force as described in the covariant description, see [86], [89]. Thus, the normal component N is regularized in closed form as:

$$N = \epsilon_N \zeta, \tag{2.53}$$

and the tangent component T is regularized in rate form as:

$$\frac{dT}{dt} = -\epsilon_T \dot{s}. \tag{2.54}$$

A computational algorithm for the tangent traction can be constructed according to the standard return-mapping scheme for the Coulomb friction law, see e.g. in [189], [106]. First, the trial force T_{tr} is computed assuming to be an elastic predictor:

$$T_{tr}^{(n+1)} = T^{(n)} - \epsilon_T(s^{(n+1)} - s^{(n)}), \tag{2.55}$$

then the real force is computed as:

$$T^{(n+1)} = \begin{cases} T_{tr}^{(n+1)} & if \ \ |T_{tr}^{(n+1)}| < \mu|N| \ \ sticking \\[2ex] sign(T_{tr}^{(n+1)})\mu|N| & if \ \ |T_{tr}^{(n+1)}| \geq \mu|N| \ \ sliding \end{cases} . \tag{2.56}$$

2.5.3 Linearization of the weak form

The linearization of the weak form has to be done in the form of covariant derivatives in the curvilinear coordinate system in a similar fashion to the case of the point-to-surface contact algorithm, for the 3D case see [89], for the 2D case see [92]. Here we skip all mathematical details providing the final result only.

Linearized part for normal interaction:

$$L(\delta W_N) = L(N\delta\zeta) = L(\epsilon_N\zeta\delta\zeta) = (\delta\mathbf{r}_s - \delta\boldsymbol{\rho})\epsilon_N\boldsymbol{\nu} \otimes \boldsymbol{\nu}(\mathbf{v}_s - \mathbf{v}). \tag{2.57}$$

Linearized part for tangent interaction:
− sticking

$$L(\delta W_T^{stick}) = L(T_{tr}\delta s) = (\delta\mathbf{r}_s - \delta\boldsymbol{\rho})\epsilon_T\boldsymbol{\tau} \otimes \boldsymbol{\tau}(\mathbf{v}_s - \mathbf{v}). \tag{2.58}$$

− sliding

$$L(\delta W_T^{slide}) = L(T\delta s) = (\delta\mathbf{r}_s - \delta\boldsymbol{\rho})sign(T_{tr}^{(n+1)})\epsilon_N\boldsymbol{\tau} \otimes \boldsymbol{\nu}(\mathbf{v}_s - \mathbf{v}). \tag{2.59}$$

2.6 Numerical examples

In this section, we give first a reference example for the construction of the projection domain for a hyperbolical surface, and then discuss a situation where the generalized projection procedure including both, projection onto a segment and onto an edge is needed.

2.6.1 Reference example: projection domain for a hyperbolical surface

Consider a quadratic surface of the form: $z = (x^2 - y^2)/2,\ 0 \le x, y \le 1$ see Fig. 2.11. The upper projection domain is then a domain between the original surface $OABC$ and the surface $O'A'B'C'$ created by shifting in normal direction as follows:

$$\mathbf{r} = \boldsymbol{\rho}(x,y) + R(x,y)\mathbf{n}(x,y), \quad \boldsymbol{\rho}(x,y) = \{x, y, (x^2 - y^2)/2\}^T, \qquad (2.60)$$

where $R(x,y)$ is a function of the corresponding radius of curvature. Exemplarily, the position of point C' is computed as follows. A point on the surface C is defined by the vector $\boldsymbol{\rho} = \{1.000, 0.000, 0.500\}^T$, with the corresponding normal vector $\mathbf{n} = \{-0.707, 0.000, 0.707\}^T$. The main curvatures at point C are computed via eqn. (2.16), (2.17) and (2.18): $k_1 = 0.354,\ k_2 = -0.700$. A positive value here defines a curvature of a corresponding line which is locally convex with respect to the chosen direction of the normal $\mathbf{n}$, i.e. a normal is pointing into a center of curvature of the line. According to the rule for a hyperbolic point discussed in Sect. 2.3.2, the value k_1 is taken for the shift as $R_1 = 1/k_1 = 2.825$. Thus, the upper boundary of the projection domain at point C' is computed as $\mathbf{r} = \boldsymbol{\rho} + R_1\mathbf{n} = \{-0.997, 0.000, 2.497\}^T$.

Remark

It is obvious, that the complete algorithm for the projection for an arbitrarily composed CAD surface requires the application of some Boolean operations with domains known from CAD applications. Thus, in general, a fairly large number of cases has to be analyzed within a contact search routine taking advantage of the derived proximity domains.

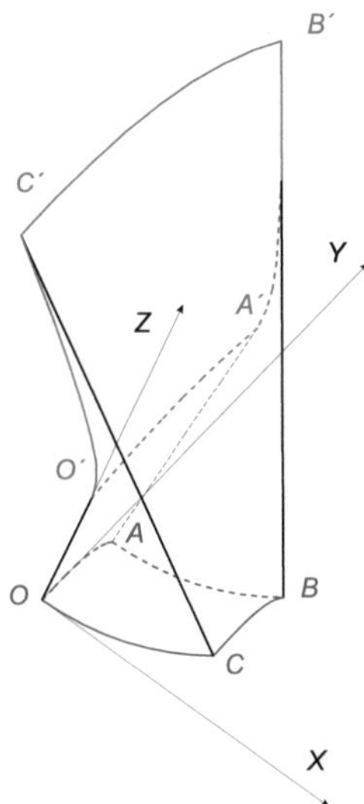

Figure 2.11: Structure of the upper proximity domain for a given surface $OABC$ with $z = (x^2 - y^2)/2, 0 \leq x, y \leq 1$.

2.6.2 Sliding block

This example is intentionally chosen as simple as possible in order to discuss the influence of various contact algorithms on kinematical effects only. A heavy rigid block $ABCD$ with mass m, see Fig. 2.12 starts to slide (position **I**) without friction on an inclined surface OE till it is impacting an edge 0 in position **II**. Assuming frictional contact at the horizontal surface OF, we seek two parameters: the hight H and the friction coefficient μ such that after the impact the block would turn over the edge O in position **III** without jumping on it until the second impact with a surface OF in position **IV**. The simplicity of the case allows us to obtain an analytical solution, which we will try to represent numerically. The equation of motion immediately after impacting the edge O (contact only at edge):

$$m\mathbf{a} = \mathbf{R} + m\mathbf{g} \tag{2.61}$$

leads to the result that a coefficient of friction $\mu > \tan \alpha$ would prevent the block from sliding on the surface OF. An equation for the angular momentum is involved in a polar coordinate system in order to get the information about the possibility of jumping. The full set of equations is then written as:

$$J_O \ddot{\varphi} = mga \frac{\sqrt{2}}{2} \sin \varphi$$

$$ma \frac{\sqrt{2}}{2} \ddot{\varphi} = mg \sin \varphi + R_\tau$$

$$ma \frac{\sqrt{2}}{2} \dot{\varphi}^2 = mg \cos \varphi + R_\nu, \tag{2.62}$$

where a is an edge of the cube, J_O is the moment of inertia about the edge O. The solution leads to the reaction force $R_\nu = 5mg \cos \varphi / 2 - 3mg/2 - ma\sqrt{2}/2 \dot{\varphi}_0^2$. Defining via the energy theorem the initial angular velocity $\dot{\varphi}_0$ after the impact, we finally obtain $R_\nu = 5mg \cos \varphi / 2 - mgL\sqrt{2}/a - 3mg/2$. If the reaction is positive $R_\nu > 0$, the block will not jump before the second contact along the side CD occurs (i.e. $\cos \varphi = 45^o$). The sliding length L satisfies the following inequality:

$$L < a(5\sqrt{2} - 6). \tag{2.63}$$

Now, we model this problem via the node-to-segment approach (NTS), see the FE algorithms in [189] and [106]. The full problem is modeled with only three finite elements, see Fig. 2.12 b. A penalty approach to enforce contact conditions is applied and a Newmark time integration scheme is used. Inclined and horizontal segments are chosen to be "master" segments, then nodes from the cube are "slaves". It is obvious to see, that during the rotation of the block over the edge the slave nodes are running through a non-allowable domain for the segment projection procedure. This leads to the impossibility to describe the rotation. This artefact causes the block to jump on the second segment, see screen-shots in Fig. 2.13.

Several possibilities to get a correct solution exist, e.g. Heege and Alart [65] mentioned that correct forces can be recovered as a superposition of forces from neighboring segments within a full Lagrange multiplier method. Another of the possible remedies (also reported in [52]) to recover a correct force as a superposition is to allow an overlapping of segments and, therefore, double projections on both segments at the same time. However, this does not solve the problem completely and also leads to jumping. Thus, an additional projection onto the edge

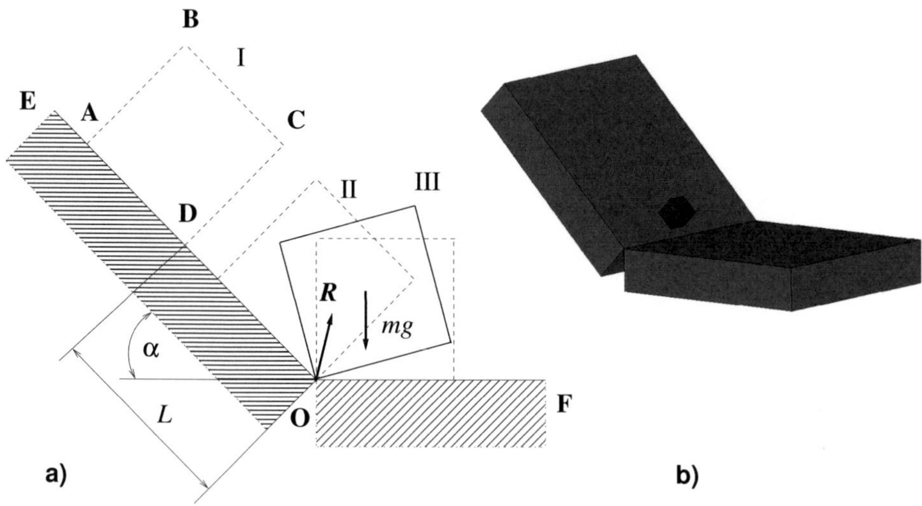

Figure 2.12: a) Process of sliding cube b) and its finite element model.

needs to be included into the contact algorithm within a penalty approach. But since, the penalty parameter plays the role of an additional spring hanging on the edge, this in general leads to unnecessary vibrations. Then only a careful selection of the penalty parameters leads to acceptable results leading to a rotational motion. This shows that correct kinematics are hard to achieve within a penalty approach. A full Lagrange multiplier method together with a generalized projection procedure would cover the kinematics of this specific example more exactly, allowing e.g. sliding along the edge.

Remark

The finite element model described above requires also additional contact elements such as a node-to-edge element. The simple version with a linear edge has been implemented so-far for the current example, see also Wriggers [189]. For the fully nonlinear element for curvilinear edges also the application the covariant approach for kinematics as well as for linearization is needed, which is out of scope of the current contribution.

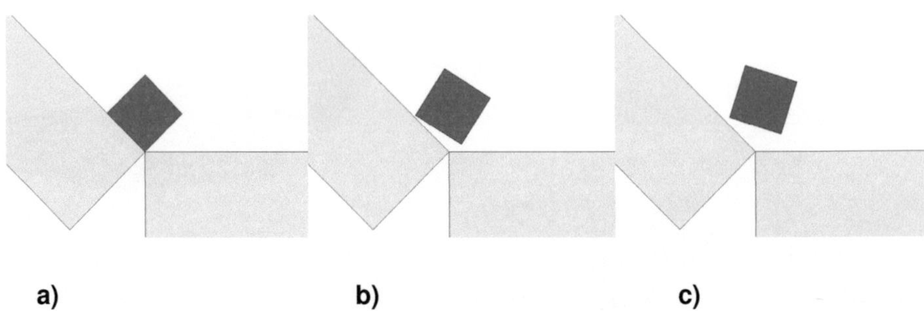

Figure 2.13: Absence of the projection of the slave node onto the edge in the contact algorithm (only Node-to-Segment approach) leads to the impossibility to capture the rotational part of the motion. a) cube impacting the edge, b)-c) results of artefact: block jumps on the second segment

2.7 Conclusion

In this contribution fundamental problems of existence and uniqueness of the closest point projection procedure are investigated. The analysis is given in a surface coordinate system, which has also been a basis for a covariant description of the contact. The consideration of the differential properties of smooth surfaces allows to create "projection domains" from which a projection of e.g. a slave node is uniquely defined. For arbitrary C^0 continuous surfaces a projection routine should be generalized to include projections not only onto surfaces, but also onto objects of lower geometrical dimension, such as curved lines and points. The corresponding criteria of existence and uniqueness and, therefore, projection domains are also given in the contribution. The general results are illustrated within a simple example, where the lack of a fully generalized projection scheme may result in a completely different kinematical behavior.

3

Contact formulation via a velocity description allowing efficiency improvements in frictionless contact analysis[*]

Abstract

A velocity description, based on the consideration of contact from the surface geometry point of view, is used for a consistent formulation of contact conditions and for the derivation of the corresponding tangent matrix. Within this approach differential operations are treated as covariant derivatives in the local surface coordinate system. The main advantage is a more algorithmic and geometrical structure of the tangent matrix, which consists of a "main", a "rotational" and a pure "curvature" term. Each part of the tangent matrix contains the information either about the internal geometry of the contact surface or about the change of the geometry during incremental loading and can be estimated in a norm during the analysis. Representative examples with contact and bending of shells modelled with linear and quadratic elements over some classical second order geometrical figures serve to show situations where keeping all parts of the tangent matrix is not necessary.

Keywords

contact problem   velocity description   covariant differential operations

---

[*]The chapter is published in [86]: A. Konyukhov and K. Schweizerhof, *Contact formulation via a velocity description allowing efficiency improvements in frictionless contact analysis*, Computational Mechanics, **33**:165-173, 2004.

tangent matrix, penalty method.

3.1 Introduction

From the variety of methods which are mainly used for the solution of contact problems, the "master-slave" concept is one of the most robust methods. This concept is based on the determination of the penetration of the "slave" surface, represented by "slave" nodes, resp. points, into a "master" surface. The penetration can be used for regularization methods like the penalty method and the Augmented Lagrange multiplier method. The penalty method see e.g. Wriggers and Simo [192], Laursen [104] and an extensive theoretical discussion in Kikuchi and Oden [84] leads to the exact solution in the limit when the penalty approaches infinity. As an improvement, concerning the satisfaction of impenetrability, the Augmented Lagrange method has been proposed and developed for contact problems by Wriggers, Simo and Taylor [193], Simo and Laursen [161], Pietrzak and Curnier [144]. A contemporary and comprehensive review about contact problems in general can be found in the books of Wriggers [188] and Laursen [106].

In nonlinear contact problems the penetration is a function of the current geometry and it is used for the "constitutive" model of the contact tractions. For the solution of the nonlinear equilibrium equations by a Newton method the corresponding equations have to be linearized. The correct linearization, taking, in particular, also algorithmic aspects into account, is called consistent linearization. This procedure was considered by Wriggers and Simo [192] for the 2D case, where the penalty functional has been linearized in the global coordinate system directly. Parisch [137] developed the consistent linearization for the 3D case also in global coordinates and used equations based on an orthogonal projection of the "slave" node onto the "master" surface to get convective coordinate increments during linearization. This approach was generalized for the linearization procedure in the local surface coordinate system by Simo and Laursen [161]. In both contributions, the solution of the projection problem for every "slave" node and the consistent linearization of the global equations were considered together in one step making it difficult to distinguish the different contributions to e.g. the stiffness matrix.

In the alternative approach considered in the following, the coordinate increment vector can be treated as a velocity vector during linearization, see Bonet and Wood [16], and the linearization itself can be treated as a covariant differential operation in the local surface coordinate system, see Marsden and Hughes [121]. The main approach of the proposed velocity description is to consider the global linearization separately from the local "slave" node searching procedure and derive linearized equations from kinematic equations in the local surface coordinate system. Focusing on frictionless contact, it leads to a very simple structure of the tangent matrix for the contact element, which is naturally divided into three parts. The first "main" part, or the constitutive part, consists of the tensor product of surface normal vectors only, while the second "rotational" part contains information about rotations of the contact element during the iteration process and the third "pure curvature" part contains the curvature tensor of the "master" surface.

For an extensive test of the proposed technique numerical examples with curved surfaces are presented. These tests serve to check the influence of different parts of the contact matrix on convergence within a nonlinear solution process. Two surfaces of second order (cylinder and sphere) have been chosen for this purpose. So-called "solid-shell"elements with various orders of approximation are used to model the shell structures, see [60], [61] and [59].

3.2 Covariant formulation of contact conditions and linearization

We introduce two coordinate systems: a reference global coordinate system for the finite element discretization only and a spatial local surface coordinate system in the contact consideration. All geometric properties of the element as well as the differential operations will be described in the local coordinate system.

3.2.1 Geometry of the contact condition

A surface 2D coordinate system is usually defined to describe the surface geometry. In addition, a special 3D local coordinate system which is related to this surface can be constructed to describe any spatial object. This system will be used to define any characteristics that belongs to the surface as well as to transfer the result of the linearization into the global coordinate system for the purpose of a finite element implementation. All geometric and kinematic characteristics of the contact are investigated in the local coordinate system. First, all necessary operations in the surface coordinate system are described. For this a surface element (Fig.1), the so-called "master" element, is considered, which is parameterized by local coordinates ξ^1, ξ^2. ρ is a vector, describing an arbitrary point on the surface. It has the following form

$$\rho = \sum_k N_k(\xi^1, \xi^2)\mathbf{x}^{(k)}. \tag{3.1}$$

where $N_k(\xi^1, \xi^2)$, $k = 1, 2, .., n$ are later the shape functions of e.g. finite elements.

Though a rather general description is given, we present for implementation purposes the expression for a 4-node bilinear surface element in detail. The approximation for this element can be written as

$$\rho = \sum_{k=1}^{4} N_k(\xi^1, \xi^2)\mathbf{x}^{(k)} = \frac{1}{4}\sum_{k=1}^{4}(1 + \xi^1\xi^{(k)})(1 + \xi^2\xi^{(k)})\mathbf{x}^{(k)}. \tag{3.2}$$

In order to describe the geometry of the surface element from the internal differential geometry point of view, surface tangent vectors ρ_i, $i = 1, 2$ have to be specified

$$\rho_1 = \frac{\partial \rho}{\partial \xi^1}, \quad \rho_2 = \frac{\partial \rho}{\partial \xi^2}. \tag{3.3}$$

The normal surface vector is computed as a cross product of the tangent vectors

$$\mathbf{n} = \frac{\rho_1 \times \rho_2}{|\rho_1 \times \rho_2|}. \tag{3.4}$$

The surface vectors ρ_1, ρ_2 define a surface coordinate system, while the normal $\mathbf{n}$ is used to describe geometrical properties of the surface and to construct a local 3D coordinate system as well. The coordinate vectors serve to obtain two fundamental tensors of the surface: the first (also called a metric tensor) and the second fundamental tensors [47] (also called a curvature tensor). The covariant components of the metric tensor are defined by the dot product of the base surface vectors

$$a_{ij} = \boldsymbol{\rho}_i \cdot \boldsymbol{\rho}_j, \quad i,j = 1,2. \tag{3.5}$$

Assuming invertibility of the metric tensor (3.5), the contravariant components of the metric tensor a^{ij} can be defined as

$$a^{ij} : \quad \frac{1}{a}\begin{bmatrix} a_{22} & -a_{12} \\ -a_{12} & a_{11} \end{bmatrix}, \quad a = det(a_{ij}) = a_{11}a_{22} - (a_{12})^2 \tag{3.6}$$

Covariant components of the second fundamental tensor are given as a dot product of the second derivative of the vector ρ and the normal $\mathbf{n}$

$$h_{ij} = \boldsymbol{\rho}_{ij} \cdot \mathbf{n}, \tag{3.7}$$

and contravariant components are defined as a double summation with the contravariant components of the metric tensor given as

$$h^{ij} = h_{kl}a^{ik}a^{jl}. \tag{3.8}$$

The formulae of partial derivatives of the base vectors are necessary to describe any differential operation in the local surface coordinate system. The Weingarten formula [47] for the derivative of the normal vector $\mathbf{n}$

$$\mathbf{n}_i = -h_{ij}a^{jk}\boldsymbol{\rho}_k = -h_i^k\boldsymbol{\rho}_k \tag{3.9}$$

and the Gauss-Kodazzi formula [47], for the derivatives of the surface vectors $\boldsymbol{\rho}_{,i}$,

$$\boldsymbol{\rho}_{ij} = \Gamma_{ij}^k\boldsymbol{\rho}_k + h_{ij}\mathbf{n} \tag{3.10}$$

are among them. In the last equation (3.10) Γ_{ij}^k are the Christoffel symbols, defined as

$$\Gamma_{ij}^k = \boldsymbol{\rho}_{ij} \cdot \boldsymbol{\rho}^k = \boldsymbol{\rho}_{ij} \cdot \boldsymbol{\rho}_m a^{mk}. \tag{3.11}$$

3.2.1.1 Projection of the contact node vector onto the master surface.

The penetration is computed by a projection procedure, see [188], [106]. Let $\mathbf{r}_s$ be a position vector of a "slave" node in the 3D space and $\boldsymbol{\rho}$ its projection onto the "master" surface. The standard closest point projection procedure leads then to the following extremal problem

$$||(\mathbf{r}_s - \boldsymbol{\rho})|| \to \min \implies (\mathbf{r}_s - \boldsymbol{\rho}) \cdot (\mathbf{r}_s - \boldsymbol{\rho}) \to \min . \qquad (3.12)$$

As is well known, the solution of this problem can be achieved by the application of a Newton procedure for the function

$$F(\xi^1, \xi^2) = (\mathbf{r}_s - \boldsymbol{\rho})^2 \qquad (3.13)$$

$$\Delta \xi_{n+1} = \begin{pmatrix} \Delta \xi^1_{n+1} \\ \Delta \xi^2_{n+1} \end{pmatrix} = -(F'')_n^{-1} F'_n \qquad (3.14)$$

$$\xi_{n+1} = \xi_n + \Delta \xi_{n+1}$$

The first derivative with respect to the surface coordinates in the form of

$$F' = \begin{pmatrix} \dfrac{\partial F}{\partial \xi^1} \\ \dfrac{\partial F}{\partial \xi^2} \end{pmatrix} = -2 \cdot \begin{pmatrix} \boldsymbol{\rho}_1 \cdot (\mathbf{r}_s - \boldsymbol{\rho}) \\ \boldsymbol{\rho}_2 \cdot (\mathbf{r}_s - \boldsymbol{\rho}) \end{pmatrix} \qquad (3.15)$$

must be finally zero. The second derivative has the form

$$
F'' = \begin{bmatrix} \dfrac{\partial^2 F}{\partial \xi^1 \partial \xi^1} & \dfrac{\partial^2 F}{\partial \xi^1 \partial \xi^2} \\ \dfrac{\partial^2 F}{\partial \xi^2 \partial \xi^1} & \dfrac{\partial^2 F}{\partial \xi^2 \partial \xi^2} \end{bmatrix} =
$$

$$
= 2 \cdot \begin{bmatrix} \boldsymbol{\rho}_1 \cdot \boldsymbol{\rho}_1 - \boldsymbol{\rho}_{11}(\mathbf{r}_s - \boldsymbol{\rho}) & \boldsymbol{\rho}_1 \cdot \boldsymbol{\rho}_2 - \boldsymbol{\rho}_{12}(\mathbf{r}_s - \boldsymbol{\rho}) \\ \boldsymbol{\rho}_2 \cdot \boldsymbol{\rho}_1 - \boldsymbol{\rho}_{12}(\mathbf{r}_s - \boldsymbol{\rho}) & \boldsymbol{\rho}_2 \cdot \boldsymbol{\rho}_2 - \boldsymbol{\rho}_{22}(\mathbf{r}_s - \boldsymbol{\rho}) \end{bmatrix} = \qquad (3.16)
$$

$$
= 2 \cdot \begin{bmatrix} a_{11} - \boldsymbol{\rho}_{11}(\mathbf{r}_s - \boldsymbol{\rho}) & a_{12} - \boldsymbol{\rho}_{12}(\mathbf{r}_s - \boldsymbol{\rho}) \\ a_{21} - \boldsymbol{\rho}_{22}(\mathbf{r}_s - \boldsymbol{\rho}) & a_{22} - \boldsymbol{\rho}_{22}(\mathbf{r}_s - \boldsymbol{\rho}) \end{bmatrix} .
$$

3.2.1.2 Spatial local coordinate system and internal geometry of the element

Now we construct a special local coordinate system, introducing the third coordinate ξ^3 in the direction of the surface normal $\mathbf{n}$, and keeping a surface point $\boldsymbol{\rho}(\xi^1, \xi^2)$ as a projection of the "slave" point. Any spatial vector $\mathbf{r}_s$ in this system can be written as

$$\mathbf{r}_s(\xi^1, \xi^2, \xi^3) = \boldsymbol{\rho}(\xi^1, \xi^2) + \xi^3 \mathbf{n}. \tag{3.17}$$

One should notice that the projection procedure is taken into account within our local coordinate system. The Lie type derivative in the form of a covariant derivative [121] is used for any differential operation on the surface. If e.g. $\mathbf{a}$ is a vector which is defined in the local coordinate system, then its material time derivative is defined as

$$\frac{d}{dt}\mathbf{a} = \left(\frac{\partial a^i}{\partial t} + \nabla_j a^i \dot{\xi}^j\right)\boldsymbol{\rho}_i, \tag{3.18}$$

where the term $\nabla_j a^i$ is a covariant derivative of the contravariant component a^i

$$\nabla_j a^i = \frac{\partial a^i}{\partial \xi^j} + a^k \Gamma^i_{jk}. \tag{3.19}$$

The "slave" point in the local coordinate system (3.17) has the local coordinate ξ^3 (Fig. 3.1). We now consider the motion of the "slave" point S in the local coordinate system, assuming that the "master" surface is moving, i.e. the surface vector $\boldsymbol{\rho}(t, \xi^1, \xi^2, \xi^3)$ as well as the normal $\mathbf{n}(t, \xi^1, \xi^2, \xi^3)$ are time dependent. Within a static process, the time t is an incremental load parameter. Then the full time derivative becomes

$$\frac{d}{dt}\mathbf{r}_s(t, \xi^1, \xi^2, \xi^3) = \frac{d}{dt}\boldsymbol{\rho} + \frac{d}{dt}(\mathbf{n}\xi^3) = \tag{3.20}$$

$$= \frac{\partial \boldsymbol{\rho}}{\partial t} + \frac{\partial \boldsymbol{\rho}}{\partial \xi^j}\dot{\xi}^j + \frac{\partial \mathbf{n}}{\partial t}\xi^3 + \mathbf{n}\dot{\xi}^3 + \frac{\partial \mathbf{n}}{\partial \xi^j}\xi^3\dot{\xi}^j, \quad j = 1, 2.$$

Let a point C be a projection point of the "slave" node onto the master surface. Denote the translation velocity of the point C as $\mathbf{v} = \dfrac{\partial \boldsymbol{\rho}}{\partial t}$ and the velocity of the "slave" point as $\mathbf{v}_s = \dfrac{d}{dt}\mathbf{r}_s(t, \xi^1, \xi^2, \xi^3)$. Then equation

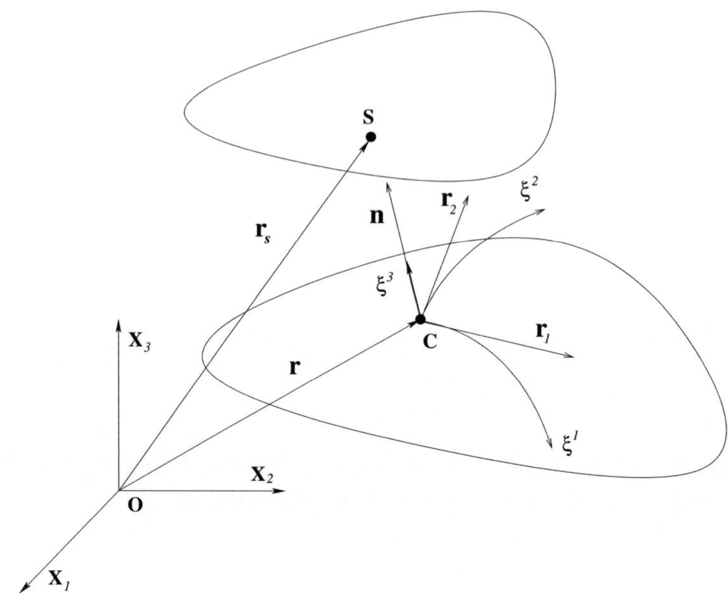

Figure 3.1: Definition of coordinate systems

(3.20) has the following form, using the Weingarten formula (3.9),

$$\mathbf{v}_s = \mathbf{v} + \xi^3\frac{\partial \mathbf{n}}{\partial t} + \mathbf{n}\dot{\xi}^3 + (\boldsymbol{\rho}_j - \xi^3 h^i_j \boldsymbol{\rho}_i)\dot{\xi}^j, \quad i,j = 1,2, \tag{3.21}$$

where h^i_j are mixed components of the curvature tensor.

The difference between the velocity $\mathbf{v}_s$ of "slave" point S and the velocity of point C is a relative velocity $\mathbf{v}^{rel}$ of the "slave" point, or in other words, the velocity of point S as it can be seen from point C. In order to define its projection in the local coordinate system, the dot product of the relative velocity $\mathbf{v}^{rel}$ and the coordinate vector $\boldsymbol{\rho}_i$ is taken

$$(\mathbf{v}_s - \mathbf{v}) \cdot \boldsymbol{\rho}_i = (a_{ij} - \xi^3 h_{ij})\dot{\xi}^j + \xi^3(\frac{\partial \mathbf{n}}{\partial t} \cdot \boldsymbol{\rho}_i), \tag{3.22}$$

where a_{ij} are the components of the metric tensor. Therefore, the convective velocity is defined as

$$\dot{\xi}^j = \hat{a}^{ij}[(\mathbf{v}_s - \mathbf{v}) \cdot \boldsymbol{\rho}_i - \xi^3(\frac{\partial \mathbf{n}}{\partial t} \cdot \boldsymbol{\rho}_i)] \tag{3.23}$$

where $\hat{a}^{ij}$ are contravariant components of the tensor with components $a_{ij} - \xi^3 h_{ij}$.

The third coordinate ξ^3 is a penetration

$$\xi^3 = g = (\mathbf{r}_s - \boldsymbol{\rho}) \cdot \mathbf{n}. \tag{3.24}$$

The scalar product of the normal $\mathbf{n}$ and the equation (3.21) gives the time derivative of the penetration

$$\dot{\xi}^3 = (\mathbf{v}_s - \mathbf{v}) \cdot \mathbf{n}. \tag{3.25}$$

All further considerations are based on the following assumption: Only the contact problem is considered, but not the motion and deformation of the two body system connected by means of the normal vector with coordinate ξ^3. The penetration is assumed to be very small, as usual during the solution of contact problems. Further, the global iteration procedure for the solution leads to a decreasing value of the penetration g. Thus, the convective velocity, with the additional assumption $\xi^3 = 0$, can be defined in the form

$$\dot{\xi}^j = a^{ij}(\mathbf{v}_s - \mathbf{v}) \cdot \boldsymbol{\rho}_i. \tag{3.26}$$

This definition of the convective velocity (3.26) on the tangent plane allows to consider the contact kinematics on the master surface only and to exploit the differential geometry of the surface during further considerations. The formula in eq. (3.26) was mentioned in Wriggers [187] as a possible simplifiction as well. Here we show that from a mathematical point of view this simplifiction leads to a consistent expression of the contact integral and, as it can be seen from numerical results, also leads to high numerical efficiency.

In order to estimate the difference between the exact definition in equation (3.23) and the proposed form (3.26), a series expansion of A^{ij} assuming ξ^3 as a small parameter is performed. Suppose A is a metric tensor with components a_{ij}, H is a curvature tensor with components h_{ij} and tensor $\hat{A}$ with components $a_{ij} - \xi^3 h_{ij}$

$$\hat{A}^{-1} = (A - \xi^3 H)^{-1} = A^{-1} + \xi^3 A^{-1} H A^{-1} + \mathcal{O}((\xi^3)^2) \tag{3.27}$$

where A^{-1} is the contravariant metric tensor with components a^{ij}. Then the convective velocity has the following form:

$$\dot{\xi}^j = a^{ij}(\mathbf{v}_s - \mathbf{v}) \cdot \boldsymbol{\rho}_i + \xi^3 \left[a^{ik} a^{mj} h_{km}(\mathbf{v}_s - \mathbf{v}) \cdot \boldsymbol{\rho}_i - a^{ij}\dot{\mathbf{n}} \cdot \boldsymbol{\rho}_i \right] + \mathcal{O}((\xi^3)^2) \quad (3.28)$$

It is obvious that eqn. (3.26) describes the constant main part of the full equation.

3.3   Weak formulation of the contact conditions

For the complete description contact tractions $\mathbf{T}_1$, $\mathbf{T}_2$ on the surfaces s_1 and s_2 in the current configuration have to be considered, $\delta \mathbf{u}$ is then the variation of the displacement field on the surface. The virtual work δW_c of the contact tractions is obtained by the following surface integral

$$\delta W_c = \int_{s_1} \mathbf{T}_1 \cdot \delta \mathbf{u}_1 ds_1 + \int_{s_2} \mathbf{T}_2 \cdot \delta \mathbf{u}_2 ds_2 \quad (3.29)$$

which must be added to the global work of the internal and external forces. Due to the equilibrium equation at the contact boundary, $\mathbf{T}_1 ds_1 = -\mathbf{T}_2 ds_2$, equation (3.29) can be written as

$$\delta W_c = \int_{s_1} \mathbf{T}_1 \cdot (\delta \mathbf{u}_1 - \delta \mathbf{u}_2) ds_1 \quad (3.30)$$

Up to now, one surface has to be specified as the "master" and the other one as the "slave" surface. The contact integral is computed over the "slave" surface. It can be computed using quadrature formulae [106] or, in simple cases, nodal quadrature [194]. With s_1 as the "slave" surface, the redefined previous notation leads to $\delta \mathbf{u}_1 = \delta \boldsymbol{\rho}_s$ as the variation of the "slave" point and $\delta \mathbf{u}_2 = \delta \boldsymbol{\rho}$ as a projection of the the variation of "slave" point onto the "master" surface.

The traction vector in the local coordinate system can be split into a normal and into a tangential part

$$\mathbf{T}_1 \equiv \mathbf{T} = N\mathbf{n} + T^i \boldsymbol{\rho}_i. \quad (3.31)$$

Here the traction vector is defined as a contravariant vector. An equation for the variation is derived following the kinematic equation (3.21):

$$\delta \mathbf{r}_s - \delta \boldsymbol{\rho} = (\boldsymbol{\rho}_j - \xi^3 h^i_j \boldsymbol{\rho}_i)\delta \xi^j + \mathbf{n}\delta \xi^3 + \xi^3 \delta \mathbf{n} \qquad (3.32)$$

It should be mentioned, that the variation itself is time independent. Then the contact integral (3.30) can be written in the following form:

$$\delta W_c = \int_s N \delta \xi^3 ds + \int_s [a_{ij} T^i \delta \xi^j + \xi^3 T^i (\delta \mathbf{n} \cdot \boldsymbol{\rho}_i - h^k_j a_{ik} \delta \xi^j)] ds \qquad (3.33)$$

with

$$\delta \xi^3 = \delta g = (\delta \mathbf{r}_s - \delta \boldsymbol{\rho}) \cdot \mathbf{n}. \qquad (3.34)$$

In most formulations (see Wriggers [188] and Laursen [106]), it is assumed that the virtual work is considered on the surface, i.e. $\xi^3 = 0$. Therefore, the contact integral (3.33) can be reduced to the following form:

$$\delta W_c = \int_s N \delta g ds + \int_s a_{ij} T^i \delta \xi^j ds = \qquad (3.35)$$

$$= \int_s N(\delta \mathbf{r}_s - \delta \boldsymbol{\rho}) \cdot \mathbf{n} ds + \int_s T^i (\delta \mathbf{r}_s - \delta \boldsymbol{\rho}) \cdot \boldsymbol{\rho}_i ds$$

This form (3.35) is mostly used in contact mechanics. Therefore it appears contradictory to use equations for the convective velocity in the full form of eqn. (3.23) with the reduced form of the contact integral (3.35). One can show that the contact integral in the form (3.35) is the main part of the full form (3.33) after consistent expansion into a Taylor series with the small parameter ξ^3, i.e. with taking into account the expansion for the convective velocity (3.28). However, if the problem of two bodies with large arbitrary penetration is considered, then the contact integral in the full form (3.33) together with the full convective velocity equations (3.23) must be used.

In the current contribution the further discussion is restricted to the non-frictional case, i.e. $T^i = 0$. The case with friction is considered in a following paper.

3.3.1 Penalty regularization

The penalty regularization of the contact condition, see [188], [106], leads to the following regularized functional:

$$\delta W_c = \int_S \epsilon_N \langle g \rangle \delta g \, ds \tag{3.36}$$

where ϵ_N is a penalty parameter and $\langle \rangle$ are Macauley brackets, which means that the integral is computed only if the value of penetration g is nonpositive

$$\langle g \rangle = \begin{cases} 0, & if \ \ g > 0 \\ g, & if \ \ g \le 0 \end{cases}.$$

The contact integral (3.36) is computed over the master surface. Within the "node-to-surface" approach the value of penetration is taken from the node and, in fact, there is no computation of the surface integral over the master surface. Following the velocity description, we take the full time derivative in order to linearize it as well as to develop the variation

$$D_v(\delta W_c) = \int_S \epsilon_N H(-g)(\dot{g}\,\delta g + g\,\delta\dot{g})ds \tag{3.37}$$

where $H(-g)$ is the Heaviside function, replacing the Macauley brackets. An expression for the linearized penetration follows from the kinematical equation (3.25)

$$\dot{g} = \dot{\xi}^3 = (\mathbf{v}_s - \mathbf{v}) \cdot \mathbf{n}. \tag{3.38}$$

The full time derivative (see eqn. 3.18) is used to linearize the variation of the penetration (3.34)

$$D_v(\delta g) = \delta\dot{g} = -\delta\boldsymbol{\rho}_i\dot{\xi}^i \cdot \mathbf{n} + (\delta\mathbf{r}_s - \delta\boldsymbol{\rho}) \cdot \left(\frac{\partial \mathbf{n}}{\partial t} + \mathbf{n}_i\dot{\xi}^i\right) =$$

$$= -(\delta\boldsymbol{\rho}_i \cdot \mathbf{n})a^{ij}(\mathbf{v}_s - \mathbf{v}) \cdot \boldsymbol{\rho}_j - (\delta\mathbf{r}_s - \delta\boldsymbol{\rho}) \cdot \boldsymbol{\rho}_i h^{ij}(\mathbf{v}_s - \mathbf{v}) \cdot \boldsymbol{\rho}_j - (\delta\mathbf{r}_s - \delta\boldsymbol{\rho}) \cdot \boldsymbol{\rho}_i a^{ij}(\mathbf{v}_i \cdot \mathbf{n}) \tag{3.39}$$

For this expression the Weingarten formula (3.9), the equation for the convective velocities (3.26) and the orthogonality condition $\mathbf{n} \cdot \boldsymbol{\rho}_i = 0$ are taken into account. The complete linearization of the contact integral (3.36) leads to the following result

$$
D(\delta W_c^N) \;=\; \int_S \epsilon_N \, H(-g) \, (\delta \mathbf{r}_s - \delta\boldsymbol{\rho}) \cdot (\mathbf{n} \otimes \mathbf{n})(\mathbf{v}_s - \mathbf{v}) ds - \quad (3.40)
$$

$$
-\int_S \epsilon_N \, H(-g) \, g \, \delta\boldsymbol{\rho}_j \cdot a^{ij}(\mathbf{n} \otimes \boldsymbol{\rho}_i)(\mathbf{v}_s - \mathbf{v}) ds - \quad (3.41)
$$

$$
-\int_S \epsilon_N \, H(-g) \, g \, (\delta \mathbf{r}_s - \delta\boldsymbol{\rho}) \cdot a^{ij}(\boldsymbol{\rho}_j \otimes \mathbf{n})\mathbf{v}_{,i} ds - \quad (3.42)
$$

$$
-\int_S \epsilon_N \, H(-g) \, g \, (\delta \mathbf{r}_s - \delta\boldsymbol{\rho}) \cdot h^{ij}(\boldsymbol{\rho}_i \otimes \boldsymbol{\rho}_j)(\mathbf{v}_s - \mathbf{v}) ds \quad (3.43)
$$

The full contact tangent matrix is then directly subdivided into the "main" part eq. (3.40) and the "curvature" part (3.41, 3.42, 3.43) which is small due to the small penetration g. The "curvature" part itself consist of a "rotational" part (eq. 3.41 and 3.42) and a "pure curvature" part (eq. 3.43). The "rotational" part contains derivatives of $\delta\rho$ and $\mathbf{v}$ with respect to the convective coordinates ξ^j and, therefore, is responsible for the rotation of a contact surface during the incremental solution procedure. The pure curvature part contains components of the curvature tensor h^{ij} and, therefore, is responsible for the change of the master surface curvature.

3.4   Finite element discretization

Though, all derivations and later numerical tests are provided also for elements based on higher order shape functions, in this section we consider only details of the finite element implementation for the bilinear element. All equations for the tangent matrix have an algorithmic structure and, therefore, the procedure of the tangent matrix derivation can be easily extended into any other case. The variables of the displacement field of the bilinear "contact element" are described by the vector

$$
\mathbf{u}^T = \{u_1^{(1)}, u_2^{(1)}, u_3^{(1)}, u_1^{(2)}, u_2^{(2)}, u_3^{(2)}, u_1^{(3)}, u_2^{(3)}, u_3^{(3)}, u_1^{(4)}, u_2^{(4)}, u_3^{(4)}, u_1^{(5)}, u_2^{(5)}, u_3^{(5)}\}^T, \quad (3.44)
$$

where the first 4 nodes belong to the master surface, while the 5'th node is the "slave" node.

Position matrices $\mathbf{A}^k$ of dimension 3×15, where the unit matrix 3×3

is on k'th position,

$$\mathbf{A}^k = \begin{bmatrix} 0 & 0 & 0 & \cdot & \cdot & \cdot & 1 & 0 & 0 & \cdot & \cdot & \cdot \\ 0 & 0 & 0 & \cdot & \cdot & \cdot & 0 & 1 & 0 & \cdot & \cdot & \cdot \\ 0 & 0 & 0 & \cdot & \cdot & \cdot & 0 & 0 & 1 & \cdot & \cdot & \cdot \end{bmatrix} \tag{3.45}$$

serve to define the variation of nodal displacements for the "master" surface as

$$\delta \mathbf{u}_k = \mathbf{A}^k \delta \mathbf{u} \tag{3.46}$$

and the variation of the "slave" node S as

$$\delta \mathbf{r}_s = \mathbf{A}^5 \delta \mathbf{u}, \tag{3.47}$$

therefore, the variation of the projection point C is defined as

$$\delta \boldsymbol{\rho} = \sum_{k=1}^{4} N_k \mathbf{A}^k \delta \mathbf{u} = \mathbf{A}^c \delta \mathbf{u}, \tag{3.48}$$

where $\mathbf{A}^c = \sum_{k=1}^{4} N_k \mathbf{A}^k$. Similar expressions can be given for the velocity vector. Using this notation, we will have

$$\delta \mathbf{r}_s - \delta \boldsymbol{\rho} = (\mathbf{A}^5 - \mathbf{A}^c)\delta \mathbf{u}. \tag{3.49}$$

The surface tangent vectors $\boldsymbol{\rho}_i$ are defined by differentiation of the shape functions

$$\boldsymbol{\rho}_i = \sum_{k=1}^{4} \frac{\partial N_k(\xi^1, \xi^2)}{\partial \xi^i} \mathbf{x}^{(k)}. \tag{3.50}$$

The normal (3.4), the first (3.5) and the second fundamental tensor (3.7) can be then computed according to their definition. After introducing a new matrix $\mathbf{A} = \mathbf{A}^5 - \mathbf{A}^c$

$$\mathbf{A} = \begin{bmatrix} -N_1 & 0 & 0 & -N_2 & 0 & 0 & -N_3 & 0 & 0 & -N_4 & 0 & 0 & 1 & 0 & 0 \\ 0 & -N_1 & 0 & 0 & -N_2 & 0 & 0 & -N_3 & 0 & 0 & -N_4 & 0 & 0 & 1 & 0 \\ 0 & 0 & -N_1 & 0 & 0 & -N_2 & 0 & 0 & -N_3 & 0 & 0 & -N_4 & 0 & 0 & 1 \end{bmatrix}, \tag{3.51}$$

the relative velocity vector $(\mathbf{v}_s - \mathbf{v})$ has the following form:

$$\mathbf{v}_s - \mathbf{v} = (\mathbf{A}^{(5)} - \mathbf{A}^c)\tilde{\mathbf{v}} = \mathbf{A}\tilde{\mathbf{v}}, \qquad (3.52)$$

where $\tilde{\mathbf{v}}$ is a nodal velocity vector of the contact element, similar to the nodal displacement vector $\mathbf{u}$ (3.44). It has to be noted that the velocity $\mathbf{v}$ introduced for the tangent matrix derivation has to be treated as an incremental displacement $\Delta\mathbf{u}$ within the computation.

With the matrix of the shape function derivative $\mathbf{A}_{,j}$

$$\mathbf{A}_{,j} = \begin{bmatrix} -N_{1,j} & 0 & 0 & -N_{2,j} & 0 & 0 \\ 0 & -N_{1,j} & 0 & 0 & -N_{2,j} & 0 \\ 0 & 0 & -N_{1,j} & 0 & 0 & -N_{2,j} \end{bmatrix}$$

$$\begin{bmatrix} -N_{3,j} & 0 & 0 & -N_{4,j} & 0 & 0 & 0 & 0 & 0 \\ 0 & -N_{3,j} & 0 & 0 & -N_{4,j} & 0 & 0 & 0 & 0 \\ 0 & 0 & -N_{3,j} & 0 & 0 & -N_{4,j} & 0 & 0 & 0 \end{bmatrix}, \qquad (3.53)$$

the vectors $\delta\boldsymbol{\rho}_j$ and $\mathbf{v}_{,j}$ are written as

$$\delta\boldsymbol{\rho}_j = -\mathbf{A}_{,j}\delta\mathbf{u}, \quad \mathbf{v}_{,j} = -\mathbf{A}_{,j}\tilde{\mathbf{v}}. \qquad (3.54)$$

The "main" part, often also called "constitutive" part, of the normal tangent matrix (3.40) has then the following form

$$\mathbf{K}^{(m)} = H(-g)\,\epsilon_N \int_S \mathbf{A}^T(\mathbf{n}\otimes\mathbf{n})\mathbf{A}ds \qquad (3.55)$$

The "curvature" part of the tangent matrix, in the general case, consists of three matrices. The first two matrices (3.41) and (3.42) build the rotation matrix

$$\mathbf{K}_r^{(1)} = -H(-g)\,\epsilon_N \int_S g\,\mathbf{A}_{,j}^T a^{ij}(\mathbf{n}\otimes\boldsymbol{\rho}_i)\mathbf{A}ds \qquad (3.56)$$

$$\mathbf{K}_r^{(2)} = -H(-g)\,\epsilon_N \int_S g\,\mathbf{A}^T a^{ij}(\boldsymbol{\rho}_i\otimes\mathbf{n})\mathbf{A}_{,j}ds \qquad (3.57)$$

and the third matrix (3.43) is the "pure curvature" matrix

$$\mathbf{K}^{(h)} = -H(-g)\,\epsilon_N \int_S g\,\mathbf{A}^T h^{ij}(\boldsymbol{\rho}_i\otimes\boldsymbol{\rho}_j)\mathbf{A}ds \qquad (3.58)$$

Thus, the full normal tangent matrix is set up as

$$\mathbf{K} = \mathbf{K}^{(m)} + \mathbf{K}^{(curv)} = \mathbf{K}^{(m)} + \mathbf{K}_r^{(1)} + \mathbf{K}_r^{(2)} + \mathbf{K}^{(h)} \qquad (3.59)$$

which is symmetric due to $\mathbf{K}_r^{(1)^T} = \mathbf{K}_r^{(2)}$, and due to the symmetry of $h^{ij}(\boldsymbol{\rho}_i \otimes \boldsymbol{\rho}_j)$.

The proposed procedure has been implemented into the finite element code FEAP-MeKA documented in [172] and [209].

3.5   Numerical examples

In this section a series of numerical examples of contact problems between a flexible structure and rigid surfaces of second order (cylinder and sphere) are investigated. Bilinear and biquadratic contact elements are used for parameterization of both contact surfaces. The aim is to estimate the influence of rotational and curvature parts of the contact tangent matrix on the convergence of the iterative algorithm. In order to investigate the corresponding contribution of each part of the tangent matrix, the following three alternatives are considered

a. Use of the full tangent matrix $\mathbf{K} = \mathbf{K}^{(m)} + \mathbf{K}^{(curv)}$.

b. Use the main part $\mathbf{K}^{(m)}$ and the rotational part $\mathbf{K}_r$, i.e. the contact matrix $\mathbf{K} = \mathbf{K}^{(m)} + \mathbf{K}_r^{(1)} + \mathbf{K}_r^{(2)}$.

c. Use the main part $\mathbf{K}^{(m)}$ and the pure curvature part $\mathbf{K}^h$, i.e. the contact matrix $\mathbf{K} = \mathbf{K}^{(m)} + \mathbf{K}^{(h)}$.

d. Use only the main part $\mathbf{K}^{(m)}$ as a contact matrix, i.e. $\mathbf{K} = \mathbf{K}^{(m)}$

The main contribution into the full contact tangent matrix comes usually from the main matrix $\mathbf{K}^{(m)}$, therefore a relative measure ε is chosen as an estimate

$$\varepsilon = \frac{||\mathbf{K} - \mathbf{K}^{(m)}||}{||\mathbf{K}^{(m)}||} \cdot 100\% \qquad (3.60)$$

with the following matrix norm:

$$||K|| = \max_i \sum_{j=1}^{n} |k_{ij}| \qquad (3.61)$$

The influence on the convergence rate is given by the number of equilibrium iterations at each load step, while the value ε, computed at each load step after the equilibrium iteration, is given in order to estimate the contribution of the "curvature" part.

3.5.1 Bending of a beam over a rigid cylinder

3.5.1.1 Linear approximation of the contact surfaces

A clamped elastic beam is loaded by prescribed displacements at the free end in vertical direction, that leads to bending over a rigid cylinder. The parameters of the beam are chosen as: length l=24 cm, height h=0.25 cm, width b=1 cm; the material model is St. Venant elastic material with an elasticity modulus of $1.0 \cdot 10^4$; the Poisson ratio is 0.3; for contact a penalty factor of $1.0 \cdot 10^4$ MPa/cm is chosen. The beam is modeled with 24 "solid-shell" elements [61] with linear shape functions plus some added shell specific enhancements. The rigid cylinder with radius R=2 cm is modelled with 49 contact elements in the circumferential direction. The central axis of the cylinder is positioned at 12 cm from the clamping. For the contact elements a linear approximation for the contact surfaces is used; the beam surface was treated as a "master" surface. The prescribed final displacement $u_{ext} = 9$ cm is subdivided into 90 identical load steps. Fig. 2 shows a sequence of the deformed beam during loading at the 0, 20, 40, 60, 80 and 90'th load step respectively. The value of penetration does not exceed 0.18 % of the beam thickness.

The results of the computation are presented in table 3.1. The convergence rate in each load step is estimated by the number of equilibrium iterations (column *No. it. / l.s.* in table 3.1), which has been changing over load steps, e.g. the computation shows 3 equilibrium iterations per load step within the first 15 load steps. The contribution of various parts of the tangent matrix is estimated by the norm in eq. (3.60).

We obtain 319 total iterations for the cases **a** and **b** and 345 iterations for the cases **c** and **d**. As expected, there is no difference between the results if the "pure curvature" part is taken into account or not, because the curvature tensor is zero. It is also obvious that the rotational part is much more important.

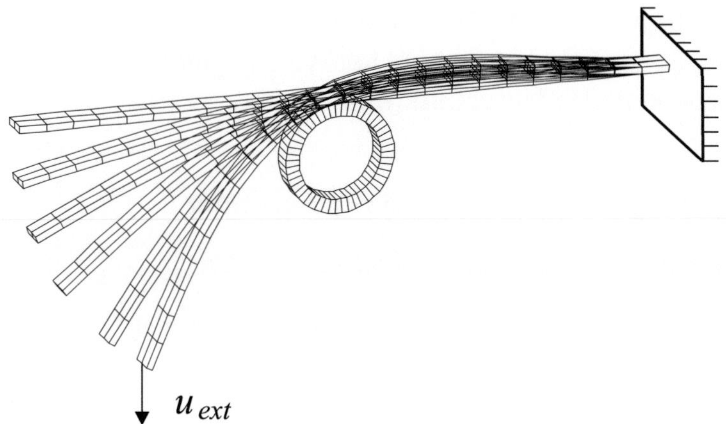

Figure 3.2: Bending of a clamped beam over a rigid cylinder. Sequence of deformations.

Case a/b				Case c/d		
No. l.s.	it./l.s.	Cum. it.	$\varepsilon \cdot 10^{-2}$ % eq. (3.60)	No. l.s.	it./l.s.	Cum. it.
1-15	3	45	0.18	1-15	3	45
16	5	50	0.19	16	5	50
17-45	3	137	0.55	17-26	3	80
46-73	4	249	0.45	27-73	4	268
74	5	254	0.62	74	5	273
75	4	258	1.04	75	4	277
76	5	263	0.62	76	5	282
77-90	4	319	1.26	77-83	4	310
				84-90	5	345

Table 3.1: Bending over a rigid cylinder. Bilinear elements for the beam. Node-to-surface contact elements. Influence of various contact stiffness parts on convergence. Case a: full matrix, case b: without curvature part; case c: without rotational part; case d: only main matrix. Comparison of no. of iterations in each load step (l.s.) and accumulated over several load steps.

3.5.1.2 Quadratic approximation of the contact surfaces

Now the beam from the previous example is modeled with 12 quadratic "solid-shell" elements [59]. The penalty factor is chosen as $5.0 \cdot 10^4$ MPa/cm, which leads to a maximum penetration of 0.048 % of the beam thickness.

The total number of iterations for case **a** with full matrix and case **b**, when the rotational matrix is taken into account, is identical, therefore, table 3.2 presents the influence of the "curvature" matrix on convergence in case of **a**, **c** and **d**. Though, elements with quadratic shape functions are used, the influence of the "pure curvature" part is still negligible. It is even possible to obtain the result with the main matrix only with a loss of 7 % of the total number of iterations.

3.5.2 Bending of a beam over a rigid sphere

A more general case to examine the influence of all parts of the contact matrix is to consider the contact with a body with a curvature in both directions.

The clamped elastic beam of the previous example, but with a width of $b = 4$ cm, is now bending over a rigid sphere. The radius of the sphere is 4 cm. The center of the sphere is positioned at 0.5 cm from the edge of the beam and at 12 cm along the beam measured from the clamping. The rigid sphere was modeled with 512 bilinear contact elements. The contact elements of the beam inherit the geometry of the beam and, therefore, have biquadratic approximations; the beam surface is represented as a "master" surface. Two asymmetric forces $F_1 = 17.5$ N and $F_2 = 70$ N are applied in the nodes at the free end as presented in Fig. 3.3. In the computation they were applied incrementally with 100 identical load steps. Fig. 3.3 shows the evolution diagram of the deformation for the 0, 20, 40, 60, 80, 100'th load step respectively. In this example we have bending and twisting of the beam as well.

Again the composition of the tangent matrix is varied. The value of the contact penalty is taken as $0.5 \cdot 10^4$ MPa/cm for the cases **c** and **d**, but $0.5 \cdot 10^3$ MPa/cm for the cases **a** and **b** due to convergence problems. The last value leads to a maximum penetration of 0.75 % of the beam thickness. Table 3.5.2 shows the result of the analysis with biquadratic

Case a/b				Case c			
No. l.s.	No. it./l.s.	Cum. No. it.	$\varepsilon \cdot 10^{-2}\%$	No. l.s.	No. it./l.s.	Cum. No. it.	$\varepsilon \cdot 10^{-4}\,\%$
1-16	3	48	0.19	1-16	3	48	0.09
17-18	4	56	0.19	17-18	4	56	0.12
19-46	3	140	0.55	19-30	3	92	0.35
47	4	144	0.45	31-47	4	160	0.49
48-50	3	153	0.62	48	3	163	1.03
51-69	4	229	1.04	49-86	4	315	4.81
70	5	234	0.63	87-90	5	335	9.76
71-86	4	298	1.27				
87	5	303	2.00				
88-90	4	315	2.37				

Case d		
No. l.s.	No. it./l.s.	Cum. No. it.
1-16	3	48
17-18	4	56
19-30	3	92
31-47	4	160
48	3	163
49-69	4	247
70	5	252
71-86	4	316
87	6	322
88-90	5	337

Table 3.2: Bending over a rigid cylinder. Biquadratic elements for the beam. Node-to-surface contact elements. Influence of various contact stiffness parts on convergence. Case a: full matrix; case b: without curvature part; case c: without rotational part; case d: only main matrix. Comparison of no. of iterations in each load step (l.s.) and accumulated over several load steps.

"solid-shell" elements and biquadratic contact elements.

In this case the curvature is changing in both directions and, as it can be seen from the result for case **b**, the influence of the "rotational" and the "pure curvature" part is larger. The higher the curvature, the more equilibrium iterations are necessary. One can see that the influence of the "rotational" part in this example is crucial, because the exclusion of the "rotational" part leads to an increase of the total number of iterations of more than 40 % , while the influence of the "pure curvature" part is still small. For the case **c**, which is not presented in the table 3.5.2, we obtain

Case a				Case b			Case d		
No. l.s.	No. it./l.s.	Cum. No. it.	$\varepsilon \cdot 10^{-2}$ %	No. l.s.	No. it./l.s.	Cum. No. it.	No. l.s.	No. it./l.s.	Cum. No. it.
1	27	27	0.731	1	27	27	1	20	20
2-23	4	115	7.382	2-22	4	111	2-8	4	48
24	6	121	18.08	23	6	117	9-23	5	123
25-33	4	157	16.26	24-33	4	157	24	6	129
34	5	162	12.20	34	6	163	25-33	5	174
35-100	4	426	17.32	35-100	4	427	34-48	6	264
							49-58	5	314
							59-68	6	374
							69-85	7	493
							86-100	8	613

Table 3.3: Bending over a rigid sphere. Biquadratic elements for the beam. Node-to-surface contact elements. Influence of various contact stiffness parts on convergence. Case a: full matrix; case b: without curvature part; case d: only main matrix. Comparison of no. of iterations in all load steps (l.s.)

609 iterations. For the full matrix (case **a**), however, the computation of the curvature part is rather costly in comparison with the "main" and with the "rotational" part, due to the necessity to compute the second derivative and the double summation for the contravariant components of the curvature tensor h^{ij}, see eqn. (3.8). Thus it appears to be the most efficient choice to consider the analysis without "pure curvature" matrix.

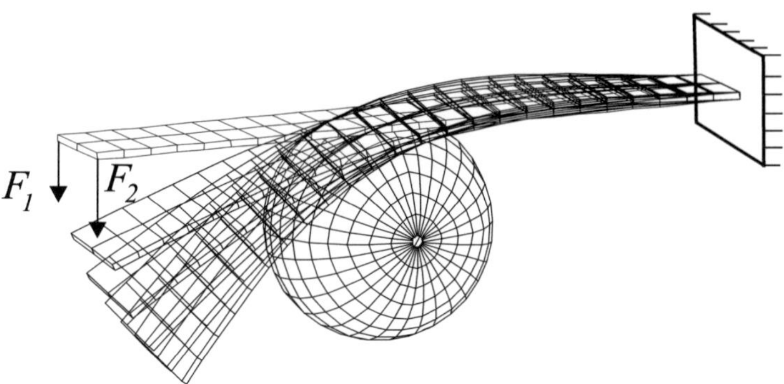

Figure 3.3: Bending of a beam over a rigid sphere. Sequence of deformations.

3.6 Conclusions

In this contribution a velocity description for the development of a consistent contact tangent matrix has been proposed. It allows to distinguish between three parts of a tangent matrix, namely the "main" part, the "rotational" part and the "pure curvature" part.

The numerical examples show that in the case of linear approximations and aligned contact elements keeping of the "pure curvature" part is meaningless. Then, it even appears sufficient to keep only the main part as a contact tangent matrix.

If elements with higher order approximations are used, the influence of the "rotational" part is larger, but the influence of the "pure curvature" part remains still small. Therefore, the last part, which is computationally more expensive then the others, can be eliminated from the complete tangent matrix without loss of efficiency.

4

Covariant description for frictional contact problems*

Abstract
A fully covariant description, based on the consideration of contact from the surface geometry point of view, is used for a consistent formulation of frictional contact conditions. All necessary operations for the description of the contact problems: kinematics, all differential operations etc. are defined in the covariant form in the local coordinate system which corresponds to the closest point procedure. The main advantage is a geometrical structure of the full tangent matrix, which is is subdivided into main, rotational and curvature parts. The consistent linearization of the penalty regularized contact integral leads to a symmetrical tangent matrix in the case of sticking. Representative examples show the effectiveness of the approach for problems where the definition of sticking-sliding zones is necessary as well as for the case of fully developed sliding zones.

Keywords
frictional contact problem covariant description tangent matrix sticking sliding evolution equations

4.1 Introduction

With frictional contact a specific interaction between bodies contacting each other along surfaces of those bodies is described. Differential ge-

---

*The chapter is published in [89]: A. Konyukhov and K. Schweizerhof. *Covariant description for frictional contact problems*, Computational Mechanics, **35**:190–213, 2005.

ometry provides a powerful mathematical tool to capture the change of these surfaces in the covariant form. Another essential feature to model frictional contact problems is the formulation of the contact conditions as kinematicalconstraints which leads to a nonlinear problem and, therefore, in the correct description of the solution process, to a consistent linearization problem. The Lagrange multiplier method as well as various regularization techniques are among the solutions schemes available to satisfy the contact conditions. E. g. for 2D frictional problems Wriggers et. al. [194] used the elasto-plastic analogy and the penalty regularization of contact conditions. By then the return mapping algorithm developed for the plasticity problem was linearized in the global coordinate system. Peric and Owen [139] used the penalty method for 3D frictional contact problems with small deformations. Laursen and Simo [109], however, formulated the penalty based contact conditions and the return mapping algorithm via convective surface coordinates, but the following linearization performed in the global coordinate system led to an artificial non-symmetry of the tangent matrix in the case of sticking. The symmetrization based on the nested Augmented Lagrangian algorithm was proposed in Simo and Laursen [161] to gain back the symmetry of the tangent matrices, but this is not a consistent procedure. Pietrzak and Curnier [32] worked extensively with the Augmented Lagrangian formulation, which was still formulated in global coordinates though with an usage of the convective coordinates. Parish and Lübbing [138] used also the convective conditions together with the penalty regularization for sticking and sliding, but still obtain a non-symmetric stick tangent matrix. Wriggers in [188] mentions the regularization of the stick conditions based on a functional used in mesh tying procedures which consequently leads to a symmetric tangent matrix. An alternative approach preserving symmetry in 2D for sticking, based on the so-called moving cone was proposed in Krstulovic-Opara and Wriggers [101]. Another problems arises from the artificial non-smoothness of the contact surfaces modeled by low-order polynomial functions leading to oscillations of the major characteristics of the solution. Various techniques based on smooth approximations of contact surfaces can be found in [134], [37], [2], [32]. Wriggers et. al. [190] mentioned e.g. a problem concerning the discontinuity of the history variables at element boundaries for smooth surfaces and proposed to use the path length in the 2D case. Various techniques based on geometrical forms in global coordinates were later considered for 3D problems in Krstulovic-Opara et. al. [102] and in Puso and Laursen [148].

Despite the large amount of contributions the fully covariant description of contact is still not available in literature. In this contribution we employ the highly developed "apparatus" of differential geometry (see e.g. Gray [47]) to reconsider the contact conditions in a specially defined spatial local coordinate system which corresponds to the well-known closest point procedure. All differential operations necessary for kinematics and linearization are considered as covariant derivatives (see Marsden and Hughes [121]). Special attention is on the consideration of the operations and the weak form on the tangent plane. The constitutive equations for the tangential tractions within the penalty regularization, or, so called, the evolution equations, are considered in the covariant description as a parallel translation on the contact surface. It is important to use this form of the constitutive equations, because the consistent linearization of the contact integral together with these equations leads to a symmetrical tangent matrix on the tangent plane in the case of sticking. Each part of the full tangent matrix, such as the normal tangent matrix, the tangent matrix in the case of sticking and the tangent matrix in the case of sliding has a geometrical structure, and, in due course, is subdivided into main, rotational and curvature parts. In addition the geometrical interpretation of the parallel translation allows to develop an integration scheme for the tangential tractions and to overcome the problem of the discontinuity of the history variables at element boundaries. The frictional contact problem can be subdivided for numerical solutions into two types depending on the necessity to capture the stick-slip behavior precisely by considering the numerical integration of the evolution equations. The "segment-to-segment", the "node-to-segement" and the "segment-to-analytical surface" finite element approaches are considered and discussed for different types of contact problems.

The article is organized as follows. In the first section of the part "Geometry and Kinematics of Contact" we recall all the operations necessary for our development, known from differential geometry. The core of the contribution is the second section where a spatial coordinate system corresponding to the closest point projection procedure is built. Kinematics of contact and differential operations are revisited in this coordinate system. In the third section the numerical algorithms to compute the characteristics from the geometrical point of view are presented. In particular the weak form, the penalty regularization and the return-mapping algorithm are considered with a special attention on the construction of the evolution equations for the tangent tractions. The developed equations are combined during the linearization in the fourth section. The fifth section contains a summary of the results which are necessary for finite element implementation. A series of the numerical examples shows the effectiveness of the proposed technique in the sixth section.

4.2 Geometry and Kinematics of Contact

We consider two interacting bodies (Figure 4.1). One of them is chosen as the contact body: its surface is called "master" surface. On the surface of the second body, we consider a "slave" point S, which is e.g. an integration or a nodal point. Two bodies are coming into contact, if a slave point of the second surface penetrates into the master surface, where penetration is defined as the shortest distance between the two surfaces of the contacting bodies.

As contact between two bodies is dominantly an interaction between these two surfaces, the main aim of the following consideration is to take advantage of the differential geometry of the contact surfaces in order to describe the kinematics of the contact conditions. First, we consider the geometry of the master surface and its characteristics and then define a special spatial coordinate system attached to this surface.

4.2.1 Local surface coordinate system and its geometrical characteristics

The "master" surface of the body (Fig. 4.1), is a 2D manifold, and therefore, can be parameterized by the surface coordinates ξ^1, ξ^2. Let ρ be a surface vector, describing any point on the surface. In a finite element

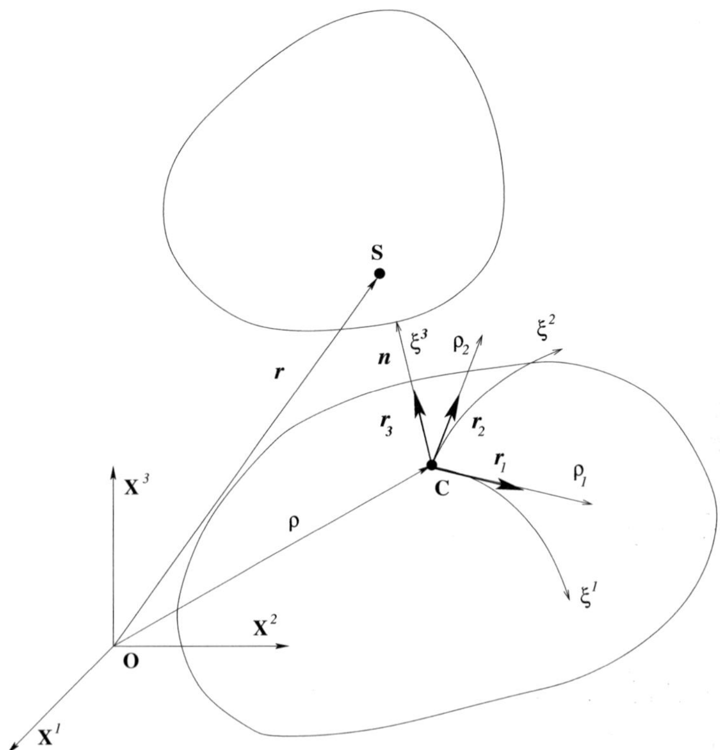

Figure 4.1: Two body contact. Local surface coordinate system on master surface.

discretization this can be done in the following form

$$\rho = \sum_k N_k(\xi^1, \xi^2)\mathbf{x}^{(k)}, \tag{4.1}$$

where $N_k(\xi^1, \xi^2)$ are shape functions and $\mathbf{x}^{(k)}$ are nodal coordinates. The set of shape functions can be either of the same order as for the finite discretization of the contact body, or it can be constructed differently as for the case of the smooth approximation of the contact surfaces. It must be noted that the parameterization (4.1) is locally defined on the surface element. Therefore, the internal variables ξ^i are not continuous between the boundaries of adjacent surface elements. This leads to a problem for the kinematical characteristics while crossing the element boundaries.

We consider here quasi-statical problems, therefore, we assume that $\mathbf{x}^{(k)} = \mathbf{x}^{(k)}(t)$, where time is treated as a load increment parameter. In general, we consider the geometry of moving surfaces. A specific focus is on the solution scheme, for which the nonlinear equations have to be linearized. Within the velocity description the increment vector is then treated as a velocity vector.

4.2.1.1 The fundamental tensors and property of the contact surface

Two fundamental tensors of the surface: the metrics tensor, or the first fundamental tensor, and the curvature tensor, or the second fundamental tensor, fully contain the properties of a surface. The metrics tensor is responsible for any metric operation on the surface (length, area or angle). The curvature tensor is responsible for the inclusion of a surface as a 2D manifold into the 3D space $\mathbf{R}^3$ (Cartesian space) respectively for the local structure of the surface in the 3D space.

First, two surface base vectors $\boldsymbol{\rho}_i$, $i = 1, 2$ in the tangent plane of a surface are introduced

$$\boldsymbol{\rho}_1 = \frac{\partial \boldsymbol{\rho}}{\partial \xi^1}, \quad \boldsymbol{\rho}_2 = \frac{\partial \boldsymbol{\rho}}{\partial \xi^2}, \tag{4.2}$$

then the normal unit surface vector is given as the cross product of the basis vectors

$$\mathbf{n} = \frac{\boldsymbol{\rho}_1 \times \boldsymbol{\rho}_2}{|\boldsymbol{\rho}_1 \times \boldsymbol{\rho}_2|}. \tag{4.3}$$

These three vectors $\boldsymbol{\rho}_1$, $\boldsymbol{\rho}_2, \mathbf{n}$ define a local surface coordinate system and they are used to obtain the two fundamental tensors of the surface [47], [154].

The metric tensor. The covariant components of the metric tensor on the surface are defined as the dot product of the base surface vectors (4.2)

$$a_{ij} = \boldsymbol{\rho}_i \cdot \boldsymbol{\rho}_j, \quad i, j = 1, 2 \tag{4.4}$$

The contravariant components of the metric tensor a^{ij} are obtained via the equation

$$a_{ik}a^{kj} = a_i^{\,j} = \delta_i^j, \tag{4.5}$$

i. e. as the inverse matrix is given in the following form:

$$a^{ij} : \quad \frac{1}{a}\begin{bmatrix} a_{22} & -a_{12} \\ -a_{12} & a_{11} \end{bmatrix}, \quad a = \det(a_{ij}) = a_{11}a_{22} - (a_{12})^2, \qquad (4.6)$$

the mixed components $a_{i.}^{j}$ are in fact identical to the Kronecker delta δ_i^j.

An adjacent basis of the surface is defined by the contravariant base vectors $\boldsymbol{\rho}^i$, which are obtained via a linear form of the covariant base vectors:

$$\boldsymbol{\rho}^i = a^{ij}\boldsymbol{\rho}_j. \qquad (4.7)$$

Thus, the metric tensor can be defined either in the covariant, or the contravariant basis, or the mixed basis

$$\mathbf{A} = a^{ij}\boldsymbol{\rho}_i \otimes \boldsymbol{\rho}_j = a_{ij}\boldsymbol{\rho}^i \otimes \boldsymbol{\rho}^j = a_{i.}^{j}\boldsymbol{\rho}^i \otimes \boldsymbol{\rho}_j \qquad (4.8)$$

The metrical characteristics, which are necessary for the further description, are length and area. The differential dl of the length is obtained as

$$dl = \sqrt{(\boldsymbol{\rho}_i \cdot \boldsymbol{\rho}_j)d\xi^i d\xi^j} = \sqrt{a_{ij}d\xi^i d\xi^j}. \qquad (4.9)$$

The differential ds of the area can be obtained either via the determinant a of the matrix of the metric tensor eqn. (4.4), or via the absolute value of the cross product of the surface vectors eqn. (4.3):

$$ds = \sqrt{|\det(a_{ij})|}d\xi^1 d\xi^2 = |\boldsymbol{\rho}_1 \times \boldsymbol{\rho}_2|d\xi^1 d\xi^2. \qquad (4.10)$$

The curvature tensor. In differential geometry the curvature tensor is used to describe a local surface structure via e.g. main curvatures, for more information see [47], [154]. The tensor is defined by its covariant components h_{ij}, which are computed as the dot product of the second derivative of the vector ρ and the normal $\mathbf{n}$

$$h_{ij} = \boldsymbol{\rho}_{ij} \cdot \mathbf{n}. \qquad (4.11)$$

The contravariant components are obtained as a bilinear combination of the covariant components with the contravariant metrics components

$$h^{ij} = h_{kn}a^{ik}a^{nj}. \qquad (4.12)$$

Equation (4.12) gives also a general rule how to compute contravariant components of any second order tensor via covariant components.

4.2.1.2 Differential operations in the surface coordinate system

For any further derivations the most important mathematical operations in the surface coordinate system are frame independent differential operations. They are defined in order to describe the kinematics from the local surface coordinate system point of view. For this the derivatives of base vectors have to be determined. The Weingarten formula and the Gauss-Codazzi formula [154] give us a complete set for derivatives of base vectors and are used to define covariant derivatives on the surface.

The Weingarten formula gives directly derivatives of the unit normal – prove see Appendix **A**:

$$\mathbf{n}_i = -h_{ij}a^{jk}\boldsymbol{\rho}_k = -h_i^k\boldsymbol{\rho}_k. \tag{4.13}$$

The Gauss-Codazzi formula allows directly the computation of derivatives of the basis vectors $\boldsymbol{\rho}_i$.

$$\boldsymbol{\rho}_{ij} = \Gamma_{ij}^k\boldsymbol{\rho}_k + h_{ij}\mathbf{n}, \tag{4.14}$$

where Γ_{ij}^k are Christoffel symbols [154], [121], defined on the surface as follows

$$\Gamma_{ij}^k = \boldsymbol{\rho}_{ij} \cdot \boldsymbol{\rho}^k = \boldsymbol{\rho}_{ij} \cdot \boldsymbol{\rho}_n a^{nk}. \tag{4.15}$$

For the prove see Appendix **B**.

Indifferent covariant derivative. The derivatives of the base vectors allow to evaluate a frame indifferent derivative of any object defined on the surface. We now consider a vector $\mathbf{T}$, defined by its local contravariant coordinates T^i in the surface coordinate system

$$\mathbf{T} = T^i\boldsymbol{\rho}_i. \tag{4.16}$$

The full time derivative of $\mathbf{T}$ – with the assumption that the vector $\boldsymbol{\rho}_i$ is implicitly time dependent via the coordinates ξ^i – gives

$$\frac{d}{dt}\mathbf{T} = \frac{\partial T^i}{\partial t}\boldsymbol{\rho}_i + \frac{\partial T^i}{\partial \xi^j}\dot{\xi}^j\boldsymbol{\rho}_i + T^i\dot{\xi}^j\boldsymbol{\rho}_{ij}. \qquad (4.17)$$

Further applying the Gauss-Codazzi formula, we get

$$\frac{d}{dt}\mathbf{T} = \frac{\partial T^i}{\partial t}\boldsymbol{\rho}_i + (\frac{\partial T^i}{\partial \xi^j} + T^k\Gamma^i_{jk})\dot{\xi}^j\boldsymbol{\rho}_i.$$

Finally, the full material time derivative of the vector has the following form

$$\frac{d}{dt}\mathbf{T} = (\frac{\partial T^i}{\partial t} + \nabla_j T^i\dot{\xi}^j)\boldsymbol{\rho}_i. \qquad (4.18)$$

The term $\nabla_j T^i$ is a covariant derivative of the contravariant component T^i

$$\nabla_j T^i = \frac{\partial T^i}{\partial \xi^j} + T^k\Gamma^i_{jk}. \qquad (4.19)$$

A similar expression can be found for the covariant derivative of the covariant components T_i, see prove in Appendix **C**:

$$\nabla_j T_i = \frac{\partial T_i}{\partial \xi^j} - T_k\Gamma^k_{ij}. \qquad (4.20)$$

4.2.2 Spatial coordinate system and its characteristics

As discussed above two bodies come into contact if a slave point penetrates at least at the closest distance into the master surface. This point is computed via the well known closest point procedure, see details for the finite element implementation in Wriggers [188], Laursen [106]. This procedure can be included in the variational formulation, see also the theoretical details in Kikuchi and Oden [84]. One of the important aspects in the current contribution is to construct a special spatial coordinate system on the master surface corresponding to the projection procedure and to consider then the contact integral as well as a linearization procedure in this system.

4.2.2.1 Projection of the contact point vector onto the master surface

We recall here the projection procedure with specific attention on the definition of all necessary parameters via the surface characteristics. At the location C on the surface described by the vector $\rho(t, \xi_1, \xi_2)$ (see Fig. 4.1), the value of the penetration of a surface into another one is defined as the minimal distance between these surfaces, see Kikuchi and Oden [84], Wriggers [188], Laursen [106]. This leads to the following extremal problem:

$$||(\mathbf{r}_s - \rho)|| \to \min, \quad \longrightarrow (\mathbf{r}_s - \rho) \cdot (\mathbf{r}_s - \rho) \to \min . \tag{4.21}$$

As is well known, the solution of eqn. (4.21) can be achieved by the application of a Newton procedure for the function

$$\mathbf{F}(\xi^1, \xi^2) = (\mathbf{r}_s - \rho)^2. \tag{4.22}$$

The convective coordinates ξ_{n+1}^i at the penetration location C are computed with the Newton scheme for the iteration $n + 1$

$$\Delta \boldsymbol{\xi}_n = \begin{bmatrix} \Delta \xi_{n+1}^1 \\ \Delta \xi_{n+1}^2 \end{bmatrix} = -(\mathbf{F}'')_n^{-1} \, \mathbf{F}'_n \tag{4.23}$$

$$\boldsymbol{\xi}_{n+1} = \boldsymbol{\xi}_n + \Delta \boldsymbol{\xi}_n,$$

where the first derivative $\mathbf{F}'$ and the second derivative $\mathbf{F}''$ with respect to the surface coordinates are described via the surface characteristics as:

$$\mathbf{F}' = \begin{bmatrix} \dfrac{\partial F}{\partial \xi^1} \\ \dfrac{\partial F}{\partial \xi^2} \end{bmatrix} = -2 \cdot \begin{bmatrix} \rho_1 \cdot (\mathbf{r}_s - \rho) \\ \rho_2 \cdot (\mathbf{r}_s - \rho) \end{bmatrix} \tag{4.24}$$

$$\mathbf{F}'' = 2 \cdot \begin{bmatrix} a_{11} - \rho_{11}(\mathbf{r}_s - \rho) & a_{12} - \rho_{12}(\mathbf{r}_s - \rho) \\ a_{21} - \rho_{22}(\mathbf{r}_s - \rho) & a_{22} - \rho_{22}(\mathbf{r}_s - \rho) \end{bmatrix}, \tag{4.25}$$

with the components of $\mathbf{F}'_n$ and $\mathbf{F}''_n$ evaluated at state n.

4.2.2.2 Spatial local coordinate system. Geometrical characteristics on the tangent plane.

Now we define a special local coordinate system related to the master surface at the penetration point C. Any spatial vector in space can be defined as

$$\mathbf{r}(\xi^1, \xi^2, \xi^3) = \boldsymbol{\rho} + \mathbf{n}\xi^3 \tag{4.26}$$

By assuming the normal vector to be known, the projection procedure has already been taken into account into this consideration. The equilibrium equations for contact will now be formulated in the defined local coordinate system, but since contact is an interaction between surfaces then each necessary equation especially for the linearization will be considered on the tangent plane, i.e. at $\xi^3 = 0$. For this, we define all the geometrical and differential characteristics with special attention on their values on the tangent plane.

The penetration. A value of the penetration g, essential for formulation of the non-penetration conditions in the contact mechanics, see [188], [106], [84], is exactly the third coordinate in our surface coordinate system:

$$\xi^3 = g = (\mathbf{r}_s - \boldsymbol{\rho}) \cdot \mathbf{n}. \tag{4.27}$$

In the spatial curvilinear coordinate system all the characteristics as metrics, covariant derivative etc. considered before can be defined. We consider only those which are necessary for the further development. The base vectors of the system are given as

$$\mathbf{r}_i = \frac{\partial \mathbf{r}}{\partial \xi^i} = \boldsymbol{\rho}_i + \mathbf{n}_i\xi^3 = (a_i^k - h_i^k\xi^3)\boldsymbol{\rho}_k, \quad i = 1, 2, \quad \mathbf{r}_3 = \mathbf{n}, \tag{4.28}$$

where the Weingarten formula (4.13) and the first fundamental tensor in the mixed formulation (4.5) have been used to obtain a more compact formula. The covariant components of the metric tensor of the spatial coordinate system are defined via the dot product of vectors eqn. (4.28).

$$g_{ij} = (\mathbf{r}_i \cdot \mathbf{r}_j) = a_{ij} - 2\,\xi^3 h_{ij} + h_{ik}h_j^k(\xi^3)^2, \quad i = 1, 2 \quad g_{i3} = 0, \quad g_{33} = 1. \tag{4.29}$$

Contravariant metric components g^{ij}, as well as contravariant base vectors $\mathbf{r}^i$ are defined in a similar fashion, eqn. (4.5), (4.7).

Time derivative of the covariant metrics components a_{ij}. During the forthcoming linearization it is essential to consider this procedure as a 3D process in the spatial coordinate system (4.26), therefore, in general, derivatives also with respect to the third coordinate ξ^3 should be considered. Thus, time derivatives of the surface metric components a_{ij} are calculated as values of the spatial metric components g_{ij} on the tangent plane at $\xi^3 = 0$, namely

$$\frac{da_{mn}}{dt} = \frac{dg_{mn}}{dt}\bigg|_{\xi^3=0} =$$

$$\left[\frac{\partial}{\partial t} + \dot{\xi}^j \frac{\partial}{\partial \xi^j} + \dot{\xi}^3 \frac{\partial}{\partial \xi^3}\right] \left(a_{mn} - 2\xi^3 h_{mn} + h_{mk}h_n^k(\xi^3)^2\right)\bigg|_{\xi^3=0} =$$

$$= \left(\frac{\partial}{\partial t} + \dot{\xi}^j \frac{\partial}{\partial \xi^j}\right)(\boldsymbol{\rho}_m \cdot \boldsymbol{\rho}_n) - 2h_{mn}\dot{\xi}^3 =$$

$$= (\mathbf{v}_m \cdot \boldsymbol{\rho}_n) + (\boldsymbol{\rho}_m \cdot \mathbf{v}_n) + \left(\Gamma^l_{mj}(\boldsymbol{\rho}_l \cdot \boldsymbol{\rho}_n) + \Gamma^l_{nj}(\boldsymbol{\rho}_m \cdot \boldsymbol{\rho}_l)\right)\dot{\xi}^j - 2h_{mn}\dot{\xi}^3. \quad (4.30)$$

The Christoffel symbols appear in eqn. (4.30) due to the usage of the Codazzi formula. All indices are running from 1 to 2.

Time derivative of the contravariant metrics components a^{ij}. The time derivative of the contravariant component of the metric tensor a^{ij} is obtained from the derivation of eqn. (4.5):

$$\frac{d}{dt}(a^{ik}a_{kj}) = 0 \quad \longrightarrow \quad a_{kj}\frac{da^{ik}}{dt} + a^{ik}\frac{da_{kj}}{dt} = 0 \quad \longrightarrow \quad \frac{da^{ik}}{dt} = -a^{im}a^{nk}\frac{da_{mn}}{dt}.$$

$$(4.31)$$

Spatial Christoffel symbols. Covariant derivative on the tangent plane. In order to distinguish in the summation agreement a spatial object from the surface one, we will use capital letters, i.e. $I, J, ... = \{1, 2, 3\}$. Covariant derivatives in the spatial coordinate system require the spatial Christoffel symbols $\hat{\Gamma}_{IJ}^{K}$. They are defined, similar to eqn. (4.15) but with the spatial base vectors $\mathbf{r}_I$, as $\tilde{\Gamma}_{IJ}^{K} = (\mathbf{r}^K \cdot \mathbf{r}_{IJ})$. The full time derivative in the spatial coordinate system in the form of eqn. (4.18), computed in convective coordinates ξ^I via covariant derivatives for contravariant components in eqn. (4.19) or for covariant components in eqn. (4.20), is a frame indifferent derivative and coincides with the Lie time derivative definition $\mathcal{L}_t$ in the form

$$\mathcal{L}_t \mathbf{T} := \mathbf{F}\frac{d}{dt}(\mathbf{F}^{-1}\mathbf{T}) = \frac{d}{dt}\mathbf{T} \tag{4.32}$$

where $\mathbf{F}$ is a push-forward and $\mathbf{F}^{-1}$ is a pull-back operator, see more in Bonet and Wood [16], Marsden and Hughes [121]. For the prove of formula (4.32) see Appendix **D**. The Lie time derivative is usually exploited for the linearization, therefore, the computation of the covariant derivatives will be employed for further linearization. In the further considerations we concentrate on the full time derivative on the tangent plane. Then values of the spatial Christoffel symbols on the surfaces $\hat{\Gamma}_{IJ}^{K}|_{\xi^3=0}$, i.e. if $\xi^3 = 0$, define a value of covariant derivatives for any spatial object on the tangent plane. It can be easily seen from their definition and the Weingarten formula that the following relations between the spatial and surface terms hold:

$$\hat{\Gamma}_{ij}^{k}|_{\xi^3=0} = \Gamma_{ij}^{k}, \quad i,j,k = 1,2$$

$$\hat{\Gamma}_{ij}^{3}|_{\xi^3=0} = 0 \tag{4.33}$$

$$\hat{\Gamma}_{3j}^{k}|_{\xi^3=0} = -h_j^k,$$

where Γ_{ij}^{k} are the surface Christoffel symbols (4.15) and h_j^k are mixed components of the curvature tensor.

With the vector $\mathbf{T}$ in the tangent plane in covariant components, i.e.

$$\mathbf{T} = T_i \mathbf{r}^i|_{\xi^3=0} = T_i \boldsymbol{\rho}^i; \tag{4.34}$$

its full time derivative is computed employing the rules given in (4.17) and (4.20)

$$\frac{dT_i}{dt} = \frac{\partial T_i}{\partial t} + \left(\frac{\partial T_i}{\partial \xi^J} - \hat{\Gamma}_{IJ}^{K} T_K\right)_{\xi^3=0} \dot{\xi}^J \longrightarrow$$

$$= \frac{\partial T_i}{\partial t} + \left(\frac{\partial T_i}{\partial \xi^j} - \Gamma^k_{ij} T_k \right) \dot{\xi}^j + h^k_i T_k \dot{\xi}^3 \qquad (4.35)$$

One should distinguish that the full time derivative with the surface Christoffel symbols in the form eqn. given in (4.17) and (4.20) can be applied to an object that belongs to the internal geometry of the surface, e.g. for $\ddot{\xi}^i$; for the full time derivative of an arbitrary spatial object, positioned in the tangent plane, the form in eqn. (4.35) must be used.

4.2.2.3 Motion of a slave point. Convective velocity on the tangent plane

During the quasi-statical loading the contact surfaces are moving and may change. This process can be observed in the local coordinate system of the surface as a motion of a slave point S, defined in eqn. (4.26). As mentioned before for the quasi-statical problems, all parameters are time dependent, where time is seen as a load parameter. Thus the "master" surface is moving and the surface vector $\rho(t, \xi^1, \xi^2)$ as well as the normal $\mathbf{n}(t, \xi^1, \xi^2)$ are time dependent. Taking a full time derivative we obtain:

$$\frac{d}{dt} \mathbf{r}_s(t, \xi^1, \xi^2, \xi^3) = \frac{d}{dt} \rho + \frac{d}{dt} (\mathbf{n} \xi^3) = \qquad (4.36)$$

$$= \frac{\partial \rho}{\partial t} + \frac{\partial \rho}{\partial \xi^j} \dot{\xi}^j + \frac{\partial \mathbf{n}}{\partial t} \xi^3 + \mathbf{n} \dot{\xi}^3 + \xi^3 \frac{\partial \mathbf{n}}{\partial \xi^j} \xi^3 \dot{\xi}^j.$$

With the translation velocity of the penetration point C as $\mathbf{v} = \dfrac{\partial \rho}{\partial t}$ and the velocity of the slave point as $\mathbf{v}_s = \dfrac{d}{dt} \mathbf{r}_s(t, \xi^1, \xi^2, \xi^3)$, the latter can be written using the Weingarten formula

$$\mathbf{v}_s = \mathbf{v} + \xi^3 \frac{\partial \mathbf{n}}{\partial t} + \mathbf{n} \dot{\xi}^3 + (\rho_j - \xi^3 h^i_j \rho_i) \dot{\xi}^j. \qquad (4.37)$$

The convective velocities $\dot{\xi}^i$ and the rate of penetration $\dot{g} = \dot{\xi}^3$ are obtained from eqn. (4.37) as a projection in the local coordinate system by evaluating the dot product with the base vectors defined in eqn. (4.28).

The vector $\dfrac{\partial \mathbf{n}}{\partial t}$ is orthogonal to $\mathbf{n}$ due to the fact that $\mathbf{n}$ is a unit vector:

$$\mathbf{n} \cdot \mathbf{n} = 1 \longrightarrow \frac{\partial \mathbf{n}}{\partial t} \cdot \mathbf{n} = 0 \tag{4.38}$$

Evaluating then the dot product of eqn. (4.37) with $\mathbf{r}_3 = \mathbf{n}$, and using the last expression (4.38), we obtain the projection of the relative velocity on the normal, or the full time derivative of the penetration:

$$\dot{\xi}^3 = \dot{g} = (\mathbf{v}_s - \mathbf{v}) \cdot \mathbf{n}. \tag{4.39}$$

A dot product of eqn. (4.37) with the base vectors $\mathbf{r}_i$ gives the following expression:

$$(\mathbf{v}_s - \mathbf{v}) \cdot (\boldsymbol{\rho}_i - \xi^3 h_i^k \boldsymbol{\rho}_k) = \xi^3 \frac{\partial \mathbf{n}}{\partial t} \cdot (\boldsymbol{\rho}_i - \xi^3 h_i^k \boldsymbol{\rho}_k) + (a_{ij} - 2\xi^3 h_{ij} + (\xi^3)^2 h_i^k h_{jk})\dot{\xi}^j, \tag{4.40}$$

from which an expression for the first two convective velocities is obtained:

$$\dot{\xi}^j = \hat{a}^{ij} \left[(\mathbf{v}_s - \mathbf{v}) \cdot \boldsymbol{\rho}_i - \xi^3 \left(\frac{\partial \mathbf{n}}{\partial t} \cdot \boldsymbol{\rho}_i + h_i^k (\mathbf{v}_s - \mathbf{v}) \cdot \boldsymbol{\rho}_k \right) \right], \tag{4.41}$$

where $\hat{a}^{ij}$ are components of the inverse matrix $(a_{ij} - 2\xi^3 h_{ij} + (\xi^3)^2 h_i^k h_{jk})$. Having taken $\xi^3 = 0$, we obtain the values of the convective velocities (4.41) on the tangent plane as

$$\dot{\xi}^j = a^{ij} (\mathbf{v}_s - \mathbf{v}) \cdot \boldsymbol{\rho}_i. \tag{4.42}$$

Again the assumption of a small value of the penetration g allows to consider each characteristics on the tangent plane. This is a main feature of the velocity description which leads to simplification of the tangent matrix and an efficient application to non-frictional problems, see Konyukhov and Schweizerhof [86].

4.2.3 Geometrical interpretation of covariant derivative and numerical realization

The covariant derivatives require C^1 continuity of the surface. Lack of the surface continuity leads to oscillations in the characteristics, e.g. at the crossing of element boundaries. Therefore, various approaches based on the usage of a C^1 approximation of the surface with Hermite splines, NURBS etc. were developed e.g. in the following articles [190], [134], [37], [2], [148], [169]. Wriggers et. al. [190] shown that for C^1 continuous contact surfaces a continuity problem of internal parameters on the element boundary arises and proposed an algorithm for the 2D case, based on the usage of the path length of the projection point. Puso and Laursen [148] proposed to determine increments of convective coordinates in the geometric form for the 3D case. Here we construct a numerical algorithm based on a geometrical interpretation of the covariant derivative as a parallel translation, see Marsden and Hughes [121]. The result of this section will be used for the computation of the contact tractions within the return-mapping algorithm.

4.2.3.1 Continuous numerical integration algorithm for a relative motion vector $\Delta\rho$

Consider a relative motion of the projection point C on the master surface. The relative velocity vector of this motion is laying in the tangent plane, i.e.

$$\mathbf{v}_r = \dot{\xi}^i \boldsymbol{\rho}_i.\qquad(4.43)$$

We are interested in the relative distance $\Delta\rho$ which was passed by point C from step (n) to step $(n+1)$. For the C^1-continuous surface and continuous convective coordinates we can write the following

$$\boldsymbol{\rho}^{(n+1)} - \boldsymbol{\rho}^{(n)} = \boldsymbol{\rho}(\xi^i + \Delta\xi^i) - \boldsymbol{\rho}(\xi^i) = \boldsymbol{\rho}_i\Delta\xi^i + \mathcal{O}((\Delta\xi^i)^2)\qquad(4.44)$$

We define the incremental vector $\Delta\rho$ at step $(n+1)$ as

$$\Delta\boldsymbol{\rho} = \boldsymbol{\rho}_i^{(n+1)}\Delta\xi^i,\qquad(4.45)$$

from which the incremental components $\Delta\xi^i$ are derived as

$$\Delta\xi^i = (\Delta\boldsymbol{\rho}\cdot\boldsymbol{\rho}_j)\,a^{ij}_{(n+1)}.\qquad(4.46)$$

If the convective coordinates are no longer continuous then the incremental vector $\Delta\rho$ can not be derived via eqn. (4.45), but it can be derived directly in the 3D space. For illustration, see Fig. 4.2, we consider at step (n) two adjacent patches $A^{(n)}B^{(n)}D^{(n)}G^{(n)}$ and $G^{(n)}D^{(n)}E^{(n)}F^{(n)}$

for a C^1-continuous surface, i.e. a surface normal $\mathbf{n}$ being continuous while crossing the line $D^{(n)}G^{(n)}$, but with independently defined convective coordinates of the patches. Let $S^{(n)}$ be a slave point and $C^{(n)}$ its projection onto the patch $A^{(n)}B^{(n)}D^{(n)}G^{(n)}$ at step (n). A pair of points $S^{(n)}$ and $C^{(n)}$ defines then a spatial coordinate system, eqn. (4.26). Now we consider a case, when at the next step $(n+1)$ the same pair is shifted into a position $S^{(n+1)}$ and $C^{(n+1)}$ with the slave point projected onto the adjacent patch $G^{(n+1)}D^{(n+1)}E^{(n+1)}F^{(n+1)}$ to obtain $C^{(n+1)}$. On the surface it can be interpreted as a motion of the projection point from position $\tilde{C}^{(n)}$ to position $C^{(n+1)}$, where the projection point has been crossing the line $D^{(n+1)}G^{(n+1)}$ (see a vector $\Delta\rho = \tilde{\mathbf{C}}^{(n)}\mathbf{C}^{(n+1)}$ in Fig. 4.2). Since a moving surface is considered, point $C^{(n)}$ is shifted in the 3D space to the position $\tilde{C}^{(n)}$ by the vector $\mathbf{u}$. Thus, the increment vector $\Delta\rho$ is obtained in the global reference Cartesian system as

$$\Delta\boldsymbol{\rho} = \boldsymbol{\rho}_{C^{(n+1)}}\big|_{\xi^1_{(n+1)},\ \xi^2_{(n+1)}} - \left(\boldsymbol{\rho}_{C^{(n)}} + \mathbf{u}_{C^{(n)}}\right)\big|_{\xi^1_{(n)},\ \xi^2_{(n)}} \tag{4.47}$$

The computation in the global reference Cartesian system clearly defines the increment vector and, therefore, allows to avoid jumps which would occur with the local convective coordinates ξ^i. It should be noted that vectors $\boldsymbol{\rho}_{C^{(n)}}$ and $\boldsymbol{\rho}_{C^{(n+1)}}$ are defined after the closest point projection procedure, therefore the information about internal variables $\xi_1^{(n)}$, $\xi_2^{(n)}$ must be stored. However, within the "segment-to-analytical surface"–approach the value of penetration is computed at the same integration points, i.e. $\boldsymbol{\rho}_{C^{(n)}} \equiv \boldsymbol{\rho}_{C^{(n+1)}}$. Then it is only necessary to keep the information about the increment vector $\mathbf{u}$ from the last load step in the global coordinate system. Eqn. (4.46) is then reduced to

$$\Delta\xi^i = -(\mathbf{u}\cdot\boldsymbol{\rho}_j)a^{ij}_{(n+1)}. \tag{4.48}$$

Summarizing the result we obtain the rule for the continuous numerical algorithm to compute the increment vector:
The increment vector $\Delta\rho$ is defined in the spatial coordinate system at step $(n+1)$ by its projection in eqn. (4.45), where the increments $\Delta\xi^i$ are computed via eqn. (4.46) and (4.47), or in the case of the "segment-to-analytical surface"–approach via (4.48).

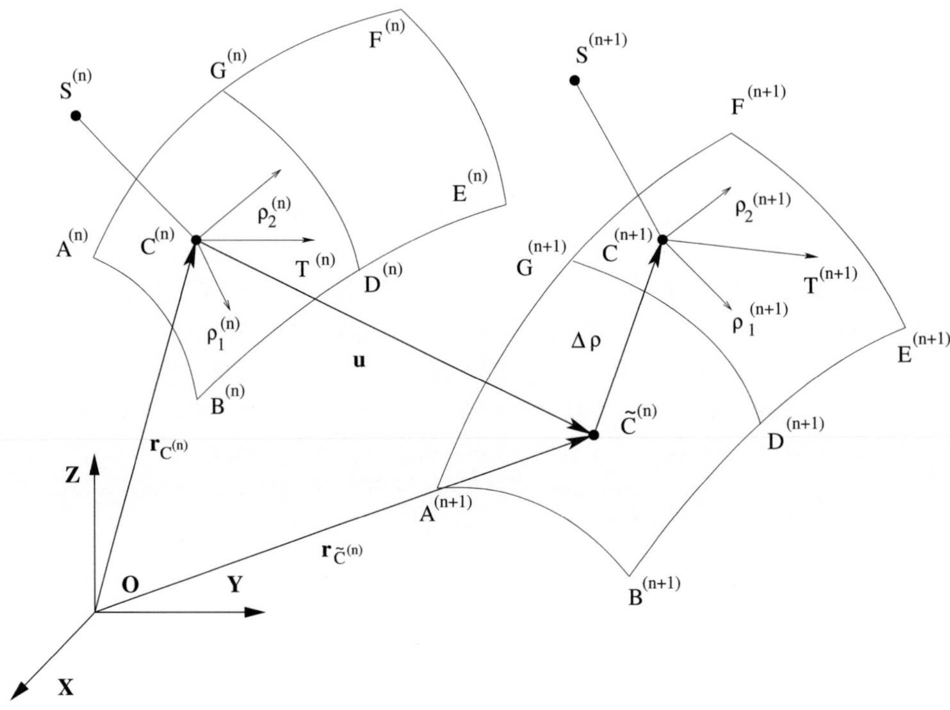

Figure 4.2: Contact point moving across element boundaries. Covariant derivatives. Sketch of integration scheme.

4.2.3.2 Parallel translation of a vector $\mathbf{T}$ on the tangent plane

The full time derivative of a vector $\mathbf{T}$ in the covariant form in eqn. (4.35) describes its change along the tangent plane. The geometrical interpretation of the numerical increment analogy is to consider the evolution of the vector $\mathbf{T}$ by enforcing its position on the tangent plane. This operation is called "parallel translation" in differential geometry terminology, see e.g. Gray [47], Schoen [154] and application in mechanics in Marsden and Hughes [121]. This interpretation also allows to overcome a variation in the representation of the vector $\mathbf{T}$ due to different local element coordinate systems.

If $\mathbf{T}^{(n)}$ is defined at the step (n), see Fig. 4.2, and $\mathbf{e}_K$ are basis vectors of the global Cartesian coordinate system, then the vector $\mathbf{T}$

can be written in both local and global coordinate systems as

$$\mathbf{T} = T_i^{(n)} a_{(n)}^{ij} \boldsymbol{\rho}_j^{(n)} = T_i^{(n)} a_{(n)}^{ij} \frac{\partial x_{(n)}^K}{\partial \xi^j} \mathbf{e}_K. \tag{4.49}$$

Projections of this vector to the new basis at state $(n+1)$ gives us the vector $\mathbf{T}^{(n+1)}$ translated in parallel. This operation in the Cartesian coordinate system leads to:

$$T_l^{(n+1)} = \mathbf{T}^{(n)} \cdot \boldsymbol{\rho}_l^{(n+1)} = T_i^{(n)} a_{(n)}^{ij} \frac{\partial x_{(n)}^K}{\partial \xi^j} \mathbf{e}_K \cdot \frac{\partial x_{(n+1)}^M}{\partial \xi^l} \mathbf{e}_M = T_i^{(n)} a_{(n)}^{ij} \frac{\partial x_{(n)}^K}{\partial \xi^j} \frac{\partial x_{(n+1)}^K}{\partial \xi^l}, \tag{4.50}$$

or in compact form

$$T_l^{(n+1)} = T_i^{(n)} a_{(n)}^{ij} \left(\boldsymbol{\rho}_j^{(n)} \cdot \boldsymbol{\rho}_l^{(n+1)} \right). \tag{4.51}$$

In other words, this operation can be seen as a pull-back from the current configuration at time (n) into the reference configuration and then a push-forward into the current configuration at time $(n+1)$. This procedure allows to keep continuity due to the use of the same reference configuration.

Remark

In the case of translation in a plane, the metric tensor is constant and eqn. (4.51) defines a standard parallel shifting

$$T_l^{(n+1)} = T_i^{(n)} a_{(n)}^{ij} \left(\boldsymbol{\rho}_j^{(n)} \cdot \boldsymbol{\rho}_l^{(n+1)} \right) = T_i^{(n)} a^{ij} a_{jl} = T_i^{(n)} \delta_l^i = T_l^{(n)}. \tag{4.52}$$

4.3 Weak form for finite element formulation and regularized contact conditions

The previous parts give us all the necessary operations to build a weak formulation. Due to the a-priori small value of the penetration the weak form in the spatial coordinate system is considered on the tangent plane. A penalty method for a simple Coloumb friction law is now used as a regularization within the contact algorithm.

4.3.1 Weak formulation in the spatial coordinate system

Now we consider the contact tractions $\mathbf{T}_1$ and $\mathbf{T}_2$ on both contact surfaces s_1 and s_2 in the current configuration. Let $\delta \mathbf{u}_i$ be a variation of the displacement field on the surface s_i. Then the work of the contact forces is determined in the following integral

$$\delta W_c = \int_{s_1} \mathbf{T}_1 \cdot \delta \mathbf{u}_1 ds_1 + \int_{s_2} \mathbf{T}_2 \cdot \delta \mathbf{u}_2 ds_2 \qquad (4.53)$$

which must be added to the global work of the internal and external forces. Due to equilibrium at the contact boundary $\mathbf{T}_1 ds_1 = -\mathbf{T}_2 ds_2$, equation (4.53) can be also written as

$$\delta W_c = \int_{s_1} \mathbf{T}_1 \cdot (\delta \mathbf{u}_1 - \delta \mathbf{u}_2) ds_1. \qquad (4.54)$$

The integral in (4.54) is considered in the local coordinate system, therefore, since this point one surface must be specified as master surface and the other as slave surface. With s_1 as slave surface, the previous notation is now slightly redefined:

$\mathbf{u}_1 = \mathbf{r}_s$ is a slave point; $\mathbf{u}_2 = \boldsymbol{\rho}$ is a projection of the slave point onto the master surface; the traction vector in the local coordinate system becomes then:

$$\mathbf{T}_1 = \mathbf{T} = N\mathbf{n} + T_i \boldsymbol{\rho}^i. \qquad (4.55)$$

Here the traction vector is defined as a covariant vector. The variation of $(\mathbf{u}_1 - \mathbf{u}_2)$ is directly obtained from the kinematic equation (4.37):

$$\delta \mathbf{r}_s - \delta \boldsymbol{\rho} = (\boldsymbol{\rho}_j - \xi^3 h_j^i \boldsymbol{\rho}_i)\delta\xi^j + \mathbf{n}\delta\xi^3 + \xi^3 \delta\mathbf{n}. \qquad (4.56)$$

It should be mentioned, that the variations themselves are time independent. Now the contact integral (4.54) can be written as:

$$\delta W_c = \int_s N\delta\xi^3 ds + \int_s [T_i\delta\xi^i + \xi^3 T_i(\delta\mathbf{n} \cdot \boldsymbol{\rho}^i - h_j^i\delta\xi^j)]ds. \qquad (4.57)$$

The full integral must be considered with the variation of the convective coordinates which are obtained from eqn. (4.39) for the penetration as the third coordinate $g = \xi^3$ in the form

$$\delta\xi^3 = \delta g = (\delta\mathbf{r}_s - \delta\boldsymbol{\rho}) \cdot \mathbf{n}, \tag{4.58}$$

and from eqn. (4.41) for the convective coordinate ξ^j in the form

$$\delta\xi^j = \hat{a}^{ij} \left[(\delta\mathbf{r}_s - \delta\boldsymbol{\rho}) \cdot \boldsymbol{\rho}_i - \xi^3 \left(\delta\mathbf{n} \cdot \boldsymbol{\rho}_i + h_i^k(\delta\mathbf{r}_s - \delta\boldsymbol{\rho}) \cdot \boldsymbol{\rho}_k \right) \right]. \tag{4.59}$$

The full formulation with eqns. (4.57), (4.58) and (4.59) in the local coordinate system is very cumbersome. However, as the value of penetration g must be small during the solution, which is an important feature of the current covariant description, we consider the full contact integral only on the tangent plane, i.e. $\xi^3 = 0$. Thus, we obtain the following form:

$$\delta W_c = \int_s N\delta g\, ds + \int_s T_j \delta\xi^j ds = \tag{4.60}$$

$$= \int_s N(\delta\mathbf{r}_s - \delta\boldsymbol{\rho}) \cdot \mathbf{n}\, ds + \int_s T_j a^{ij}(\delta\mathbf{r}_s - \delta\boldsymbol{\rho}) \cdot \boldsymbol{\rho}_i ds,$$

which is accompanied with the variation of the convective coordinates on the tangent plane in the form:

$$\delta\xi^j = a^{ij}(\delta\mathbf{r}_s - \delta\boldsymbol{\rho}) \cdot \boldsymbol{\rho}_i \tag{4.61}$$

The formulation of the contact integral in the form presented in (4.60) is mostly used in contact mechanics (see Wriggers [188] and Laursen [106]).

4.3.2 Regularization by the penalty method

The contact tractions N and T_j are additional unknowns in the contact integral (4.60). If they are treated as independent variables, the Lagrangian multiplier method is used. If they are treated as dependent variables, additional assumptions are necessary to define the contact tractions, leading to regularization schemes. Here we follow the regularization technique as described e.g. in Kikuchi [84], Wriggers [188], [187], Laursen [106] and Zhong [205]. This regularization is based on an elasto-plastic analogy to model the Coulomb friction. Other types of regularization based on elasto-visco-plastic models of the Maxwell type and the Kelvin type are considered in Araki and Hjelmstad [5].

4.3.2.1 Normal contact conditions

We describe contact conditions in terms of the spatial coordinate system. For normal contact they can be formulated as the Kuhn-Tucker complementary conditions for the variational problem.

1. Contact occurs when a slave point penetrates into the tangent plane: $\xi^3 = g \leq 0$.

2. At the penetration point the normal nonnegative traction appears: $N \geq 0$.

3. The contact traction N exists only, if the slave point is on the tangent plane, i.e. when $\xi^3 = g = 0$: $N \cdot g = 0$.

The penalty method, allowing a small penetration, is often used to overcome numerical difficulties in satisfying conditions 1-3. These three conditions can be accomplished by the following regularization:

$$N = \epsilon_N \langle g \rangle, \tag{4.62}$$

where ϵ_N is a penalty parameter and $\langle \rangle$ are Macauley brackets in the form

$$\langle g \rangle = \left\{ \begin{array}{l} 0, \; if \; g > 0 \\ g, \; if \; g \leq 0 \end{array} \right. .$$

4.3.2.2 Tangential contact conditions. Evolution equations

Additional constitutive equations are necessary for the tangential contact tractions T_j. Frictional problems in the finite element formulation are considered as quasi-statical ones with the loading from zero up to a certain value. This kinematical approach allows to describe stick and slide conditions in our spatial coordinate system, see Fig. 4.1.

a) The slave point S sticks, if its projection point C is not moving on the tangent plane, i.e. has zero relative velocity $\mathbf{v}_r = 0$.

b) The slave point S slides, if during quasi-statical loading there is a relative motion of its projection point C, i.e. $\mathbf{v}_r \neq 0$.

These conditions for the simplest case as a model of Coulomb dry friction can be specified as follows:

1. The slave point sticks as long as the Coulomb dry friction inequality holds

$$\mathbf{v}_r = 0 \quad if \quad \Phi := \|\mathbf{T}\| - \mu N \leq 0 \tag{4.63}$$

where μ is a friction coefficient, and $\|\mathbf{T}\|$ is the absolute value of the tangential traction $\mathbf{T}$, which is computed as

$$\|\mathbf{T}\| = \sqrt{T_i T_j a^{ij}}. \tag{4.64}$$

2. Beyond the threshold defined by the friction condition (4.63) the slave point starts to slide in the direction of the relative velocity vector; the tangential tractions are then acting in the opposite direction.

$$if \quad \Phi > 0 \quad then \quad \exists \zeta > 0 \quad \mathbf{v}_r = -\zeta \frac{\mathbf{T}}{\|\mathbf{T}\|}, \tag{4.65}$$

where ζ is a consistency parameter.

3. Sliding happens only if $\Phi = 0$, thus

$$\zeta \Phi = 0. \tag{4.66}$$

Again the contact conditions lead to a lack of differentiability and, therefore, numerical problems. In order to overcome this Kikuchi [84] considered a penalty regularization for the contact functional assuming a small tangential motion in the case of sticking; a penalty regularization based on the elasto-plastic analogy was developed then in Wriggers et. al. [194], Laursen and Simo [161]. In the last article the following regularization was proposed in convective coordinates for the trial tractions:

$$a_{ij}\dot{\xi}^j - \zeta \frac{T_i}{\|\mathbf{T}\|} = -\frac{1}{\epsilon_T} \frac{\partial T_i}{\partial t}, \tag{4.67}$$

where ϵ_T is a penalty parameter. Then a return-mapping algorithm known from plasticity can be used to satisfy the stick-slide condition.

From a mathematical point of view (see Marsden and Hughes [121]), it appears to be more correct to consider **a parallel translation of the vector field** $T_i(\xi^i(t))$ **on the master surface**. In this situation the relative velocity vector $\mathbf{v}_r$ of the projection point C on the master surface, see eqn. (4.43) must be equal to the full time derivative in the covariant form (4.18) of the vector $\mathbf{T}$ defined on the tangent plane in the spatial coordinate system. Thus, for the corresponding regularization we propose the following form

$$\mathbf{v}_r - \zeta \frac{\mathbf{T}}{\|\mathbf{T}\|} = -\frac{1}{\epsilon_T} \frac{d\mathbf{T}}{dt}, \tag{4.68}$$

or employing the covariant derivative of $\mathbf{T}$ on the tangent plane in eqn. (4.35), we obtain the following expression for the components

$$a_{ij}\dot{\xi}^j - \zeta \frac{T_i}{\|\mathbf{T}\|} = -\frac{1}{\epsilon_T}\left(\frac{\partial T_i}{\partial t} + \left(\frac{\partial T_i}{\partial \xi^j} - \Gamma_{ij}^k T_k\right)\dot{\xi}^j + h_i^k T_k \dot{\xi}^3\right) \tag{4.69}$$

or finally, having denoted the time derivative of T_i as

$$\frac{dT_i}{dt} = \frac{\partial T_i}{\partial t} + \frac{\partial T_i}{\partial \xi^j}\dot{\xi}^j, \tag{4.70}$$

we obtain

$$a_{ij}\dot{\xi}^j - \zeta \frac{T_i}{\|\mathbf{T}\|} = -\frac{1}{\epsilon_T}\left(\frac{dT_i}{dt} - \Gamma_{ij}^k T_k \dot{\xi}^j + h_i^k T_k \dot{\xi}^3\right). \tag{4.71}$$

In order to integrate the differential equation (4.71) we employ a return-mapping algorithm based on the backward Euler implicit scheme for the ordinary differential equations, see e.g. Simo and Hughes [160]. The trial step is assumed to be with sticking, therefore $\zeta = 0$. The consistent backward Euler scheme for eq. (4.71) has the following form

$$\left(\delta_k^i - \Gamma_{ij}^k\big|_{(n+1)}\xi_{(n+1)}^j + h_i^k\big|_{(n+1)}\xi_{(n+1)}^3\right)(T^{trial})_k^{(n+1)} =$$

$$\tag{4.72}$$

$$= (T^{trial})_i^{(n)} - \epsilon_T\left(a_{ij}^{(n+1)}\xi_{(n+1)}^j - a_{ij}^{(n)}\xi_{(n)}^j\right) - \Gamma_{ij}^k\big|_{(n)}T_k^{(n)}\xi_{(n)}^j + h_i^k\big|_{(n)}\xi_{(n)}^3 T_k^{(n)},$$

which can be seen as a backward scheme for the following ordinary differential equations

$$\frac{dT_i}{dt} = (-\epsilon_T a_{ij} + \Gamma_{ij}^k T_k)\dot{\xi}^j - h_i^k T_k \dot{\xi}^3 \tag{4.73}$$

The system of ordinary differential equations for the computation of the tangential traction (4.73) is called *the evolution equations*. They are important for the linearization process. Keeping the form with the covariant derivatives (4.73) instead of the form in eqn. (4.67) leads to a symmetrical tangent matrix for sticking, while as used in Laursen and Simo [109], [104], the form (4.67) leads to a non-symmetrical tangent matrix for the arbitrary 3D case.

Remark 1. Consider the backward Euler scheme (4.72) in the case with a linear approximation of the master surface. Then, having taken all Christoffel symbols and components of the curvature tensor as zero, we obtain the following equations:

$$(T^{tr})_i^{(n+1)} = T_i^{(n)} - \epsilon_T a_{ij}(\xi^j_{(n+1)} - \xi^j_{(n)}). \tag{4.74}$$

This algorithm can be found in Laursen [106] for the trial step solution of equation (4.67).

Any analysis based on equation (4.72) becomes computationally rather expensive, because a full matrix appears on the left side and additional history variables $\Gamma^k_{ij}|_{(n)}, a^{(n)}_{ij}$ have to be used. Moreover, the integration scheme (4.72) as well as (4.74) suffers from jumps occurring at element boundaries due to the different internal coordinates ξ^i. Thus, we propose a discrete analog of the evolution equations (4.73) for the numerical computation

$$\Delta \mathbf{T} = -\epsilon_T \Delta \boldsymbol{\rho}. \tag{4.75}$$

The application of the results of section 4.2.3 to eqn. (4.75) together with the sliding condition leads to the following return-mapping scheme:

Trial step.

$$\left. \begin{aligned} &N^{(n+1)} = \epsilon_N \langle g^{(n+1)} \rangle \\[2ex] &(T^{tr})_i^{(n+1)} = T_k^{(n)} a_{(n)}^{kj} \left(\boldsymbol{\rho}_j^{(n)} \cdot \boldsymbol{\rho}_i^{(n+1)} \right) - \epsilon_T \Delta \xi^j a_{ij}^{(n+1)} \\[2ex] &\Phi^{tr}_{(n+1)} := \|\mathbf{T}^{tr}_{(n+1)}\| - \mu N^{(n+1)} \\[2ex] &\|\mathbf{T}^{tr}_{(n+1)}\| = \sqrt{(T^{tr})_i^{(n+1)}(T^{tr})_j^{(n+1)} a^{ij}_{(n+1)}} \end{aligned} \right\}, \tag{4.76}$$

where $\Delta\xi^j$ is obtained as

$$\Delta\xi^j = \begin{cases} (\Delta\boldsymbol{\rho}\cdot\boldsymbol{\rho}_k)\,a^{jk}_{(n+1)} & \text{for node-to-surface (NTS) and} \\ & \text{surface-to-surface (STS) approaches, where} \\[2mm] & \Delta\boldsymbol{\rho} = \boldsymbol{\rho}_{C(n+1)}\big|_{\xi^1_{(n+1)},\,\xi^2_{(n+1)}} - \left(\boldsymbol{\rho}_{C(n)} + \mathbf{u}_{C(n)}\right)\big|_{\xi^1_{(n)},\,\xi^2_{(n)}} \\[2mm] -(\mathbf{u}\cdot\boldsymbol{\rho}_k)a^{kj}_{(n+1)} & \text{for segment-to-analytical surface (STAS) approach} \end{cases}$$

$$(4.77)$$

Return mapping. The stick-slip condition is checked within the return mapping process:

$$T_i^{(n+1)} = \begin{cases} (T^{tr})_i^{(n+1)} & if \ \Phi^{tr}_{(n+1)} \leq 0 \ (stick) \\[4mm] \mu N^{(n+1)}\dfrac{(T^{tr})_i^{(n+1)}}{\|\mathbf{T}^{tr}_{(n+1)}\|} & if \ \Phi^{tr}_{(n+1)} > 0 \ (slide) \end{cases} \qquad (4.78)$$

Remark 2. The regularized frictional problem is strictly path-dependent: it follows from the fact that the contact tractions T_i in the contact functional in eqn. (4.60) must satisfy the evolution equations (4.73). The return-mapping algorithm for the incremental solution, as is known, is unconditionally stable, but a problem of choosing the displacement increments arises due to the correct definition of sticking and sliding zones. A simple a-priori estimation will be proposed further for the numerical example.

Remark 3. For 2D problems Krstulovic-Opara and Wriggers [101] proposed the so-called moving cone description. Under the assumption of **Remark 1**, now a point of the cone axis on the tangent plane with coordinates ξ^1_0, ξ^2_0 is considered. One can show that the friction condition (4.76. 3) together with eqn. (4.74) defines an ellipse on the tangent plane, which can be obtained by projection of the frictional cone onto the tangent plane. For a stick case the initial frictional forces T_i are zero at the initial point $\xi^1_{(0)}, \xi^2_{(0)}$ in algorithm (4.74). Considering the absolute value $\|\mathbf{T}\|$ in eqn. (4.64) at step (n) we obtain

$$\begin{aligned}
\|\mathbf{T}^{(n)}\|^2 &= T_i^{(n)} T_j^{(n)} a^{ij} = \\
&= \epsilon_T a_{ik}(\xi_{(n)}^k - \xi_{(0)}^k)\epsilon_T a_{jl}(\xi_{(n)}^l - \xi_{(0)}^l)a^{ij} = \\
&= \epsilon_T^2 a_{kl}(\xi_{(n)}^k - \xi_{(0)}^k)(\xi_{(n)}^l - \xi_{(0)}^l).
\end{aligned} \tag{4.79}$$

Having taken an incremental analog of the differential of length in eqn. (4.9) together with eqn. (4.76. 3), we can find that

$$\Delta l^2 = \epsilon_T^2 a_{kl}(\xi_{(n)}^k - \xi_{(0)}^k)(\xi_{(n)}^l - \xi_{(0)}^l) \leq (\mu N)^2. \tag{4.80}$$

Eqn. (4.80) defines an ellipse as allowable domain inside which the projection point C can move in the case of sticking leading to a symmetric tangent matrix finally.

4.4 Consistent linearization

The idea behind the consistent linearization for a Newton type solution process is to exploit the full material time derivative in the form of the co-variant derivative in the spatial coordinate system, see sections 4.2.1.2 and 4.2.2.2, together with the evolution equations for the contact tangent frictional forces (4.73).

4.4.1 Linearization of the normal contact expression

The contact integral, see e.g. eqn. (3.30) is computed over the "slave" surface, which is defined by a set of "slave" points. Each parameter in the contact integral is considered in the spatial local coordinate system of the "master" surface, (i.e. as a function of the convective coordinates ξ^i), therefore, linearization of the "slave" surface element ds will not be included in process. Thus ds is assumed to remain constant within linearization. Further it must be noted that the use of different quadrature schemes for the computation of the contact integral may lead to different contact elements.

The normal part of the contact integral (4.60) has the following form:

$$\delta W_c^N = \int_s \epsilon_N \langle g \rangle \delta g \, ds = \int_s \epsilon_N \langle (\mathbf{r}_s - \boldsymbol{\rho}) \cdot \mathbf{n} \rangle \, (\delta \mathbf{r}_s - \delta \boldsymbol{\rho}) \cdot \mathbf{n} \, ds \qquad (4.81)$$

The details of the linearization of the normal part δW_c^N and the application to the non-frictional problems are outlined in Konyukhov and Schweizerhof [86]. Here we only include the result for the full normal tangent matrix:

$$D(\delta W_c^N) =$$

$$= \int_S \epsilon_N \, H(-g) \, (\delta \mathbf{r}_s - \delta \boldsymbol{\rho}) \cdot (\mathbf{n} \otimes \mathbf{n})(\mathbf{v}_s - \mathbf{v}) dS - \qquad (4.82a)$$

$$- \int_S \epsilon_N \, H(-g) \, g \, \left(\delta \boldsymbol{\rho}_{,j} \cdot a^{ij}(\mathbf{n} \otimes \boldsymbol{\rho}_i)(\mathbf{v}_s - \mathbf{v}) + (\delta \mathbf{r}_s - \delta \boldsymbol{\rho}) \cdot a^{ij}(\boldsymbol{\rho}_j \otimes \mathbf{n})\mathbf{v}_{,i} \right) dS - \quad (4.82b)$$

$$- \int_S \epsilon_N \, H(-g) \, g \, (\delta \mathbf{r}_s - \delta \boldsymbol{\rho}) \cdot h^{ij}(\boldsymbol{\rho}_i \otimes \boldsymbol{\rho}_j)(\mathbf{v}_s - \mathbf{v}) dS. \qquad (4.82c)$$

The full contact tangent matrix is subdivided into the main part eqn. (4.82a), the "rotational" part (4.82b) and the "curvature" part (4.82c). The last two terms are small due to the small value of the penetration g. The "rotational" part contains derivatives of $\delta \boldsymbol{\rho}$ and $\mathbf{v}$ with respect to the convective coordinates ξ^j and, therefore, represents the rotation of a contact surface during the incremental solution procedure. The "curvature" part contains components of the curvature tensor h^{ij} and, therefore, represents the change of the curvature of the master surface.

4.4.2 Linearization of the tangential contact expression

The tangential part of the contact integral (4.60)

$$\delta W_c^T = \int_S T_i \delta \xi^i ds \qquad (4.83)$$

has to be considered together with the evolution equations (4.73) and the return mapping algorithm eqn. (4.76), (4.77), (4.78). The cases of sticking and sliding have to be treated separately.

For the linearization either a covariant or a contravariant component, two operators, based on the covariant derivative are necessary. The operator for the linearization of the contravariant component has the form

$$L(x^i) \equiv \left(\frac{\partial}{\partial t} + \dot{\xi}^j \nabla_j \right) (x^i) = \frac{\partial x^i}{\partial t} + \left(\frac{\partial x^i}{\partial \xi^j} + \Gamma^i_{kj} x^k \right) \dot{\xi}^j \tag{4.84}$$

and the linearization operator for the covariant component has the form

$$L(x_i) \equiv \left(\frac{\partial}{\partial t} + \dot{\xi}^j \nabla_j \right) (x_i) = \frac{\partial x_i}{\partial t} + \left(\frac{\partial x_i}{\partial \xi^j} - \Gamma^k_{ij} x_i \right) \dot{\xi}^j. \tag{4.85}$$

It is obvious that the Christoffel symbols disappear in the final result after the linearization of the scalar, i.e. the full time derivative of the scalar is the covariant derivative of the scalar

$$L(x^i v_i) = \left\{ \frac{\partial x^i}{\partial t} + \left(\frac{\partial x^i}{\partial \xi^j} + \Gamma^i_{kj} x^k \right) \dot{\xi}^j \right\} v_i + \left\{ \frac{\partial v_i}{\partial t} + \left(\frac{\partial v_i}{\partial \xi^j} - \Gamma^k_{ij} v_i \right) \dot{\xi}^j \right\} x^i =$$

$$= \left\{ \frac{\partial x^i}{\partial t} + \frac{\partial x^i}{\partial \xi^j} \dot{\xi}^j \right\} v_i + \left\{ \frac{\partial v_i}{\partial t} + \frac{\partial v_i}{\partial \xi^j} \dot{\xi}^j \right\} x^i = v_i \frac{dx^i}{dt} + x^i \frac{dv_i}{dt} \tag{4.86}$$

Therefore, the linearization leads to the following expression

$$D_v(\delta W^T_c) = \int_s \left(\delta\xi^i \frac{dT_i}{dt} + T_i \frac{d\delta\xi^i}{dt} \right) ds. \tag{4.87}$$

As the handling of the complete expression is rather complex, we focus on each term separately in the following.

4.4.2.1 Linearization of $\delta\xi^i$

The linearization of the variation of the convective coordinates $\delta\xi^i$ is one of the important parts which requires the results about differential operations in the spatial coordinate system from section 4.2.2.2 together with the tensor algebra operations on the tangent plane. The full time derivative gives

$$L(\delta\xi^i) = \left\{ \frac{\partial}{\partial t} + \frac{\partial}{\partial \xi^j} \dot{\xi}^j \right\} (\delta\xi^i) = \frac{da^{ik}}{dt} (\delta\mathbf{r}_s - \delta\boldsymbol{\rho}) \cdot \boldsymbol{\rho}_k + a^{ik} \frac{d}{dt} [(\delta\mathbf{r}_s - \delta\boldsymbol{\rho}) \cdot \boldsymbol{\rho}_k]. \tag{4.88}$$

Linearization of $(\delta \mathbf{r}_s - \delta \boldsymbol{\rho}) \cdot \boldsymbol{\rho}_k$ requires the application of the Gauss-Codazzi formula (4.14).

$$\frac{d}{dt}\left[(\delta \mathbf{r}_s - \delta \boldsymbol{\rho}) \cdot \boldsymbol{\rho}_k)\right] = \tag{4.89}$$

$$= ((\delta \mathbf{r}_s - \delta \boldsymbol{\rho})_{,j} \cdot \boldsymbol{\rho}_k)\dot{\xi}^j + (\delta \mathbf{r}_s - \delta \boldsymbol{\rho}) \cdot \mathbf{v}_k) +$$
$$+ \Gamma^l_{kj}((\delta \mathbf{r}_s - \delta \boldsymbol{\rho}) \cdot \boldsymbol{\rho}_l)\dot{\xi}^j + h_{kj}((\delta \mathbf{r}_s - \delta \boldsymbol{\rho}) \cdot \mathbf{n})\dot{\xi}^j.$$

Linearization of the contravariant components a^{ij} was already given in the section 4.2.2.2.

Simplification of $\frac{d}{dt}\delta\xi^i$. The final formula is long, but can be simplified. In addition, the following transformations are cumbersome but necessary to show the symmetry of the tangent matrix in the case of sticking. Summarizing the results in one formula, we obtain

$$\frac{d}{dt}(\delta\xi^i) =$$

$$= -a^{im}a^{nk}(\mathbf{v}_m \cdot \boldsymbol{\rho}_n)((\delta \mathbf{r}_s - \delta \boldsymbol{\rho}) \cdot \boldsymbol{\rho}_k) \tag{4.90a}$$

$$-a^{im}a^{nk}(\boldsymbol{\rho}_m \cdot \mathbf{v}_n)((\delta \mathbf{r}_s - \delta \boldsymbol{\rho}) \cdot \boldsymbol{\rho}_k) \tag{4.90b}$$

$$-a^{im}a^{nk}\Gamma^l_{mj}(\boldsymbol{\rho}_l \cdot \boldsymbol{\rho}_n)\dot{\xi}^j((\delta \mathbf{r}_s - \delta \boldsymbol{\rho}) \cdot \boldsymbol{\rho}_k) \tag{4.90c}$$

$$-a^{im}a^{nk}\Gamma^l_{nj}(\boldsymbol{\rho}_m \cdot \boldsymbol{\rho}_l)\dot{\xi}^j((\delta \mathbf{r}_s - \delta \boldsymbol{\rho}) \cdot \boldsymbol{\rho}_k) \tag{4.90d}$$

$$+2a^{im}a^{nk}h_{mn}\dot{\xi}^3((\delta \mathbf{r}_s - \delta \boldsymbol{\rho}) \cdot \boldsymbol{\rho}_k) \tag{4.90e}$$

$$+a^{ik}((\delta \mathbf{r}_s - \delta \boldsymbol{\rho}))_{,j} \cdot \boldsymbol{\rho}_k)\dot{\xi}^j \tag{4.90f}$$

$$+a^{ik}(\delta\mathbf{r}_s - \delta\boldsymbol{\rho}) \cdot \mathbf{v}_k \tag{4.90g}$$

$$+a^{ik}\Gamma^l_{kj}((\delta\mathbf{r}_s - \delta\boldsymbol{\rho}) \cdot \boldsymbol{\rho}_l)\dot{\xi}^j \tag{4.90h}$$

$$+a^{ik}h_{kj}((\delta\mathbf{r}_s - \delta\boldsymbol{\rho}) \cdot \mathbf{n})\dot{\xi}^j \tag{4.90i}$$

The nine parts in eqns. (4.90a–4.90i) will be tremendously simplified, if we take into account the expression for the convective velocities (4.42) and consider tensor operations on the tangent plane. The following five transformations will lead to a simple structure:

a.

The sum of the terms (4.90a) and (4.90g) becomes zero on the surface:

$$-a^{im}a^{nk}(\mathbf{v}_m \cdot \boldsymbol{\rho}_n)((\delta\mathbf{r}_s - \delta\boldsymbol{\rho}) \cdot \boldsymbol{\rho}_k) + a^{ik}(\delta\mathbf{r}_s - \delta\boldsymbol{\rho}) \cdot \mathbf{v}_k = 0. \tag{4.91}$$

In order to show this the dot product in the second term is expressed on the tangent plane, i.e. as double sum with the surface metric tensor components a^{ij}:

$$a^{im}(\delta\mathbf{r}_s - \delta\boldsymbol{\rho}) \cdot \mathbf{v}_m = a^{im}((\delta\mathbf{r}_s - \delta\boldsymbol{\rho}) \cdot \boldsymbol{\rho}_k)\boldsymbol{\rho}^{\,k} \cdot (\mathbf{v}_m \cdot \boldsymbol{\rho}_n)\boldsymbol{\rho}^{\,n} =$$

$$= a^{im}a^{kn}((\delta\mathbf{r}_s - \delta\boldsymbol{\rho}) \cdot \boldsymbol{\rho}_k)(\mathbf{v}_m \cdot \boldsymbol{\rho}_n),$$

from which (4.91) is obtained.

b.

The sum of the terms (4.90c) and (4.90h) becomes zero on the surface:

$$-a^{im}a^{nk}\Gamma^l_{mj}(\boldsymbol{\rho}_l \cdot \boldsymbol{\rho}_n)\dot{\xi}^j((\delta\mathbf{r}_s - \delta\boldsymbol{\rho}) \cdot \boldsymbol{\rho}_k) + a^{ik}\Gamma^l_{kj}((\delta\mathbf{r}_s - \delta\boldsymbol{\rho}) \cdot \boldsymbol{\rho}_l)\dot{\xi}^j =$$

$$= \left(-a^{im}a^{nk}a_{ln}\Gamma^l_{mj}((\delta\mathbf{r}_s - \delta\boldsymbol{\rho}) \cdot \boldsymbol{\rho}_k) + a^{ik}\Gamma^l_{kj}((\delta\mathbf{r}_s - \delta\boldsymbol{\rho}) \cdot \boldsymbol{\rho}_l)\right)\dot{\xi}^j =$$

$$= \left(-a^{im}a^k_l\Gamma^l_{mj}((\delta\mathbf{r}_s - \delta\boldsymbol{\rho}) \cdot \boldsymbol{\rho}_k) + a^{ik}\Gamma^l_{kj}((\delta\mathbf{r}_s - \delta\boldsymbol{\rho}) \cdot \boldsymbol{\rho}_l)\right)\dot{\xi}^j =$$

$$= \left(-a^{im}\Gamma^k_{mj}((\delta\mathbf{r}_s - \delta\boldsymbol{\rho}) \cdot \boldsymbol{\rho}_k) + a^{ik}\Gamma^l_{kj}((\delta\mathbf{r}_s - \delta\boldsymbol{\rho}) \cdot \boldsymbol{\rho}_l)\right)\dot{\xi}^j = 0.$$

Here the properties of the covariant and contravariant components (4.4) and (4.5) have been used.

c.

The sum of (4.90b) and (4.90f) leads to a symmetrical rotational part.

We start with using the expression for the convective velocities (4.42):

$$-a^{im}a^{nk}(\boldsymbol{\rho}_m \cdot \mathbf{v}_n)((\delta \mathbf{r}_s - \delta\boldsymbol{\rho}) \cdot \boldsymbol{\rho}_k) + a^{ik}((\delta \mathbf{r}_s - \delta\boldsymbol{\rho})_{,j} \cdot \boldsymbol{\rho}_k)\dot{\xi}^j =$$

$$= -a^{il}a^{jk}(\boldsymbol{\rho}_l \cdot \mathbf{v}_j)((\delta \mathbf{r}_s - \delta\boldsymbol{\rho}) \cdot \boldsymbol{\rho}_k) - a^{ik}a^{jl}(\delta\boldsymbol{\rho}_{,j} \cdot \boldsymbol{\rho}_k)((\mathbf{v}_s - \mathbf{v}) \cdot \boldsymbol{\rho}_l) =$$

$$= -(\delta \mathbf{r}_s - \delta\boldsymbol{\rho})\, a^{il}a^{jk}\, \boldsymbol{\rho}_k \otimes \boldsymbol{\rho}_l\, \mathbf{v}_j - \delta\boldsymbol{\rho}_{,j}\, a^{ik}a^{jl}\, \boldsymbol{\rho}_k \otimes \boldsymbol{\rho}_l\, (\mathbf{v}_s - \mathbf{v}). \quad (4.92)$$

The final expression is found via the tensor product.

d.

After grouping (4.90e) with (4.90i), we obtain

$$2a^{im}a^{nk}h_{mn}\dot{\xi}^3((\delta \mathbf{r}_s - \delta\boldsymbol{\rho}) \cdot \boldsymbol{\rho}_k) + a^{ik}h_{kj}((\delta \mathbf{r}_s - \delta\boldsymbol{\rho}) \cdot \mathbf{n})\dot{\xi}^j =$$

$$= a^{im}a^{nk}h_{mn}\dot{\xi}^3((\delta \mathbf{r}_s - \delta\boldsymbol{\rho}) \cdot \boldsymbol{\rho}_k) + \quad (4.93a)$$

$$a^{im}a^{nk}h_{mn}(\mathbf{v}_s - \mathbf{v}) \cdot \mathbf{n})((\delta \mathbf{r}_s - \delta\boldsymbol{\rho}) \cdot \boldsymbol{\rho}_k) + a^{ik}a^{jm}h_{kj}((\delta \mathbf{r}_s - \delta\boldsymbol{\rho}) \cdot \mathbf{n})((\mathbf{v}_s - \mathbf{v}) \cdot \boldsymbol{\rho}_m) = \quad (4.93b)$$

In order to show the symmetry of the part in eqn. (4.93b), the tensor product and contravariant components of the curvature tensor eqn. (4.12) are used. For a reduction of eqn. (4.93a) the equation for the variation of the convective velocity (4.42) and mixed components of the curvature tensor are taken, leading finally to

$$= h^i_n\dot{\xi}^3\delta\xi^n + \quad (4.94a)$$

$$+h^{ij}(\delta \mathbf{r}_s - \delta\boldsymbol{\rho}) \cdot \left(\boldsymbol{\rho}_j \otimes \mathbf{n} + \mathbf{n} \otimes \boldsymbol{\rho}_j\right)(\mathbf{v}_s - \mathbf{v}) \quad (4.94b)$$

The last part (4.94b) defines the curvature part of the tangent matrix.

e.

The equation for the variation of the convective velocity (4.42) is used to simplify (4.90d):

$$-a^{im}a^{nk}\Gamma^l_{nj}(\boldsymbol{\rho}_m \cdot \boldsymbol{\rho}_l)\dot{\xi}^j((\delta \mathbf{r}_s - \delta\boldsymbol{\rho}) \cdot \boldsymbol{\rho}_k)$$

$$= -\Gamma^i_{kj}\dot{\xi}^j \delta\xi^k \tag{4.95}$$

The resulting parts in eqn. (4.94a) and (4.95) remain untransformed, however they will disappear in both sticking and sliding cases, after taking into account the fully linearized contact integral together with the evolution equations (4.73) as shown in the next section.

Summarizing the result of the complete transformation, we obtain

$$\frac{d}{dt}(\delta\xi^i) =$$

$$= -(\delta\mathbf{r}_s - \delta\boldsymbol{\rho})\, a^{il} a^{jk}\, \boldsymbol{\rho}_k \otimes \boldsymbol{\rho}_l\, \mathbf{v}_j - \delta\boldsymbol{\rho}_{,j}\, a^{ik} a^{jl}\, \boldsymbol{\rho}_k \otimes \boldsymbol{\rho}_l\, (\mathbf{v}_s - \mathbf{v}) \tag{4.96a}$$

$$+h^{ij}(\delta\mathbf{r}_s - \delta\boldsymbol{\rho})\cdot\left(\boldsymbol{\rho}_j \otimes \mathbf{n} + \mathbf{n} \otimes \boldsymbol{\rho}_j\right)(\mathbf{v}_s - \mathbf{v})+ \tag{4.96b}$$

$$+h^i_n \dot{\xi}^3 \delta\xi^n - \Gamma^i_{kj}\dot{\xi}^j \delta\xi^k \tag{4.96c}$$

Thus, the full time derivative consists of a symmetrical rotational part (4.96a), a symmetrical pure curvature part (4.96b) and a connection part with the Christoffel symbols (4.96c), describing the connection properties.

4.4.2.2 Sticking

In the sticking case, the trial tangential traction terms T_i are identical with the real traction, therefore, the linearized traction terms are obtained from the evolution equation in (4.73) directly. Starting with eqn. (4.87) and taking into account **Remark 1** in section 4.3.2 together with the evolution equation (4.73), and eqn. (4.96a), (4.96b), (4.96c) we finally obtain

$$D_v(\delta W^T_c) = \tag{4.97}$$

$$= \int_s \left((-\epsilon_T a_{ij} + \Gamma^k_{ij} T_k)\dot{\xi}^j - h^k_i T_k \dot{\xi}^3\right)\delta\xi^i ds+$$

$$+ \int_s T_i \Big[- \big((\delta \mathbf{r}_s - \delta \boldsymbol{\rho}) \, a^{il} a^{jk} \, \boldsymbol{\rho}_k \otimes \boldsymbol{\rho}_l \, \mathbf{v}_j + \delta \boldsymbol{\rho}_{,j} \, a^{ik} a^{jl} \, \boldsymbol{\rho}_k \otimes \boldsymbol{\rho}_l \, (\mathbf{v}_s - \mathbf{v}) \big) +$$

$$+ h^{ij} (\delta \mathbf{r}_s - \delta \boldsymbol{\rho}) \cdot \big(\boldsymbol{\rho}_j \otimes \mathbf{n} + \mathbf{n} \otimes \boldsymbol{\rho}_j \big) (\mathbf{v}_s - \mathbf{v}) +$$

$$+ \underbrace{h^i_n \dot{\xi}^3 \delta \xi^n - \Gamma^i_{kj} \dot{\xi}^j \delta \xi^k} \Big] \, ds.$$

Using the tensor notation and the equation for convective velocities (4.42) for the main part $a_{ij} \dot{\xi}^j \delta \xi^i$, we obtain the following form for the tangential tangent matrix in the case of sticking.

$$D_v (\delta W_c^T) =$$

$$-\varepsilon_T \int_s (\delta \mathbf{r}_s - \delta \boldsymbol{\rho}) a^{ij} \boldsymbol{\rho}_i \otimes \boldsymbol{\rho}_j (\mathbf{v}_s - \mathbf{v}) ds \qquad (4.98\text{a})$$

$$- \int_s T_i \big((\delta \mathbf{r}_s - \delta \boldsymbol{\rho}) \, a^{il} a^{jk} \, \boldsymbol{\rho}_k \otimes \boldsymbol{\rho}_l \, \mathbf{v}_j + \delta \boldsymbol{\rho}_{,j} \, a^{ik} a^{jl} \, \boldsymbol{\rho}_k \otimes \boldsymbol{\rho}_l \, (\mathbf{v}_s - \mathbf{v}) \big) \, ds \qquad (4.98\text{b})$$

$$+ \int_s T_i h^{ij} (\delta \mathbf{r}_s - \delta \boldsymbol{\rho}) \cdot \big(\boldsymbol{\rho}_j \otimes \mathbf{n} + \mathbf{n} \otimes \boldsymbol{\rho}_j \big) (\mathbf{v}_s - \mathbf{v}) ds. \qquad (4.98\text{c})$$

As we have a conservative problem for sticking it is obvious that the symmetric form is correct.

Similar to the normal tangent matrix, the tangential tangent matrix can be subdivided into a main (4.98a), a rotational (4.98b) and a pure curvature part (4.98c).

Remark 1. The artificial non-symmetry of the tangent matrix for the stick condition, based on the evolution equation (4.67) was mentioned by Laursen and Simo in [109], [106] and [104]. As an appropriate alternative within a solution scheme, a symmetrization based on a split technique with the Augmented Lagrangian method was proposed by Laursen in [161]. Wriggers [188], suggested to use the consistent linearization of the sticking conditions in the form $\| \mathbf{r}_s - \boldsymbol{\rho} \|^2$ directly, which then leads to the correct symmetric matrix. Here, it becomes obvious, that it is particularly important to use the evolution equation in the form of the covariant derivatives (see eqn. 4.73) together with the linearization of the metric components a_{ij} as 3D metric components g_{ij} (see eqn. 4.30). This allows to avoid the artificial non-symmetry and obtain the correct symmetric tangent matrix for sticking.

4.4.2.3 Sliding

The expressions for the linearized variation of the convective velocity eqn. (4.96a), (4.96b) and (4.96c) are also used in this case. In addition, the tangential force in the case of sliding, see eqn. (4.78) of the return-mapping algorithm, has to be linearized.

$$\frac{dT_i^{(n+1)}}{dt} = \frac{d}{dt}\left(\mu N^{(n+1)}\frac{(T^{tr})_i^{(n+1)}}{\|\mathbf{T}_{(n+1)}^{tr}\|}\right) = \tag{4.99}$$

$$= \mu\frac{dN^{(n+1)}}{dt}\frac{(T^{tr})_i^{(n+1)}}{\|\mathbf{T}_{(n+1)}^{tr}\|} + \mu N^{(n+1)}\frac{d}{dt}\left(\frac{(T^{tr})_i^{(n+1)}}{\|\mathbf{T}_{(n+1)}^{tr}\|}\right)$$

For the derivative of the unit vector on the tangent plane

$$\mathbf{e} = \frac{(T^{trial})_i^{(n+1)}}{\|\mathbf{T}_{(n+1)}^{trial}\|}a^{ij}\boldsymbol{\rho}_j$$

we will use the following formula, see Simo and Hughes [160]

$$\frac{d\mathbf{e}}{d\mathbf{T}} = \frac{1}{\|\mathbf{T}\|}[\mathbf{I} - \mathbf{e}\otimes\mathbf{e}] \tag{4.100}$$

and the chain rule

$$\frac{d\mathbf{e}}{dt} = \frac{d\mathbf{e}}{d\mathbf{T}}\frac{d\mathbf{T}}{dt} = \frac{1}{\|\mathbf{T}\|}[\mathbf{I} - \mathbf{e}\otimes\mathbf{e}]\frac{d\mathbf{T}}{dt}. \tag{4.101}$$

Here the full time derivative of the tangential traction $\dfrac{d\mathbf{T}}{dt}$ is given by the evolution equation (4.73). The tensor operations are considered on the tangent plane:

$$[\mathbf{I} - \mathbf{e}\otimes\mathbf{e}]\frac{d\mathbf{T}}{dt} =$$

$$= \left[a^{ij}\boldsymbol{\rho}_i\otimes\boldsymbol{\rho}_j - \frac{T_kT_l a^{ik}a^{jl}}{\|\mathbf{T}\|^2}\boldsymbol{\rho}_i\otimes\boldsymbol{\rho}_j\right]\left((-\epsilon_T a_{mn} + \Gamma_{mn}^r T_r)\dot{\xi}^n - h_m^r T_r\dot{\xi}^3\right)\boldsymbol{\rho}^m =$$

$$= \left(-\epsilon_T \dot{\xi}^i + a^{ik}T_l\Gamma^l_{kj}\dot{\xi}^j - h^k_j a^{ij}T_k\dot{\xi}^3\right)\boldsymbol{\rho}_i + \tag{4.102a}$$

$$+\frac{T_kT_l a^{ik}}{\|\mathbf{T}\|^2}\left(\epsilon_T\dot{\xi}^l - a^{jl}T_m\Gamma^m_{jn}\dot{\xi}^n + a^{jl}T_m h^m_j\dot{\xi}^3\right)\boldsymbol{\rho}_i. \tag{4.102b}$$

The time derivative of the normal force $N^{(n+1)}$ gives:

$$\frac{dN^{(n+1)}}{dt} = \frac{d}{dt}(\epsilon_N|\xi^3|) = -\epsilon_N\dot{\xi}^3, \tag{4.103}$$

where the minus sign is a result from the conditions that the contact integral is computed only if $\xi^3 < 0$. Summarizing, we get:

$$D_v(\delta W^T_c) =$$

$$\int_s \left(-\frac{\epsilon_N\mu\dot{\xi}^3 T_i\delta\xi^i}{\|\mathbf{T}\|} - \frac{\epsilon_T\mu|N|a_{ij}\dot{\xi}^i\delta\xi^j}{\|\mathbf{T}\|}\right. \tag{4.104a}$$

$$+\frac{\mu|N|T_k\Gamma^k_{ij}\dot{\xi}^j\delta\xi^i}{\|\mathbf{T}\|} - \frac{\mu|N|T_i h^i_j\dot{\xi}^3\delta\xi^j}{\|\mathbf{T}\|} \tag{4.104b}$$

$$+\frac{\mu|N|T_sT_l\delta\xi^s}{\|\mathbf{T}\|^3}\left(\epsilon_T\dot{\xi}^l - a^{jl}T_m\Gamma^m_{jn}\dot{\xi}^n + a^{jl}T_m h^m_j\dot{\xi}^3\right) \tag{4.104c}$$

$$-\frac{\mu|N|T_i}{\|\mathbf{T}\|}\left[((\delta\mathbf{r}_s - \delta\boldsymbol{\rho})\,a^{il}a^{jk}\,\boldsymbol{\rho}_k\otimes\boldsymbol{\rho}_l\,\mathbf{v}_j + \delta\boldsymbol{\rho}_{,j}\,a^{ik}a^{jl}\,\boldsymbol{\rho}_k\otimes\boldsymbol{\rho}_l\,(\mathbf{v}_s - \mathbf{v})\right) \tag{4.104d}$$

$$+h^{ij}(\delta\mathbf{r}_s - \delta\boldsymbol{\rho})\cdot\left(\boldsymbol{\rho}_j\otimes\mathbf{n} + \mathbf{n}\otimes\boldsymbol{\rho}_j\right)(\mathbf{v}_s - \mathbf{v})+ \tag{4.104e}$$

$$\left.+h^i_n\dot{\xi}^3\delta\xi^n - \Gamma^i_{kj}\dot{\xi}^j\delta\xi^k\right]\right)ds \tag{4.104f}$$

The sum of the parts (4.104b) and (4.104f) is zero. After some tensor algebra the other parts can be grouped into the following form:

$$D_v(\delta W^T_c) =$$

$$-\int_s\left((\delta\mathbf{r}_s - \delta\boldsymbol{\rho})\frac{\epsilon_N\mu T_i a^{ij}}{\|\mathbf{T}\|}\boldsymbol{\rho}_j\otimes\mathbf{n}(\mathbf{v}_s - \mathbf{v})\right)ds \tag{4.105a}$$

$$-\int_s \left((\delta\mathbf{r}_s - \delta\boldsymbol{\rho}) \frac{\epsilon_T \mu |N| a^{ij}}{\|\mathbf{T}\|} \boldsymbol{\rho}_i \otimes \boldsymbol{\rho}_j (\mathbf{v}_s - \mathbf{v}) \right) ds \qquad (4.105b)$$

$$+\int_s \left((\delta\mathbf{r}_s - \delta\boldsymbol{\rho}) \frac{\epsilon_T \mu |N| T_i T_j a^{ik} a^{jl}}{\|\mathbf{T}\|^3} \boldsymbol{\rho}_k \otimes \boldsymbol{\rho}_l (\mathbf{v}_s - \mathbf{v}) \right) ds \qquad (4.105c)$$

$$-\int_s \frac{\mu |N| T_i}{\|\mathbf{T}\|} \left((\delta\mathbf{r}_s - \delta\boldsymbol{\rho})\, a^{il} a^{jk}\, \boldsymbol{\rho}_k \otimes \boldsymbol{\rho}_l\, \mathbf{v}_j + \delta\boldsymbol{\rho}_{,j}\, a^{ik} a^{jl}\, \boldsymbol{\rho}_k \otimes \boldsymbol{\rho}_l\, (\mathbf{v}_s - \mathbf{v}) \right) ds \qquad (4.105d)$$

$$+\int_s \left(\frac{\mu |N| T_i}{\|\mathbf{T}\|} h^{ij} (\delta\mathbf{r}_s - \delta\boldsymbol{\rho}) \cdot \left(\boldsymbol{\rho}_j \otimes \mathbf{n} + \mathbf{n} \otimes \boldsymbol{\rho}_j \right) (\mathbf{v}_s - \mathbf{v}) \right) ds \qquad (4.105e)$$

$$+\int_s \frac{\mu |N| T_s T_l \delta\xi^s}{\|\mathbf{T}\|^3} \left(-a^{jl} T_m \Gamma^m_{jn} \dot{\xi}^n + a^{jl} T_m h^m_j \dot{\xi}^3 \right) ds. \qquad (4.105f)$$

The matrix consists then of a constitutive non-symmetric part (4.105a), a constitutive symmetric part (4.105b) and (4.105c), a symmetric rotational part (4.105d), a symmetric curvature part (4.105e) and a non-symmetric part curvature part (4.105f) which is preserved for curved surfaces. All geometrical parameters are computed for the master surface.

Remark 2. One can find from comparison with Peric and Owen [139], that they have considered the tangent matrix which is represented by the the main parts of the full tangent matrix.

4.5 Global solution scheme. Summary of the results

Summarizing the theoretical discussion about the covariant description, we present the global solution scheme for the numerical implementation in Table 4.1 and 4.2. All parts of tangent matrices contain either a term $(\delta\mathbf{r}_s - \delta\boldsymbol{\rho})$, or a term $\delta\boldsymbol{\rho}_{,j}$, resp. terms $(\mathbf{v}_s - \mathbf{v})$ and $\mathbf{v}_{,j}$, and, therefore, can be algorithmically computed. For discretization of any surface only two position matrices $\mathbf{A}$ and $\mathbf{A}_\xi$ are necessary. The proposed approach has been implemented in FEAP code see [172], "solid-shell" elements are used for modelling of elastic structures, see [60] and [59]. For the details of the finite element implementation we refer to Konyukhov and Schweizerhof [86].

Table 4.1: **Global solution scheme. Summary of the results for numerical implementation.**

---

1. Initialization of convective coordinates ξ^i.
   The projection procedure in eqns. (4.23, 4.24, 4.25)
   with no external loads gives $\xi^i_{(0)}$.

2. Loop over load increments and Newton iterations
   for the contact integral

   $\delta W_c = \int_s N\delta g\, ds + \int_s T_j \delta\xi^j ds$ where $\delta\xi^j = a^{ij}(\delta\mathbf{r}_s - \delta\boldsymbol{\rho})\cdot\boldsymbol{\rho}_i$

3. Loop over all contact elements and all contact points

   - compute projection points $\xi^i_{(n)}$ eqns. (4.23, 4.24, 4.25)
   - Check penetration $g = (\mathbf{r}_s - \boldsymbol{\rho})\cdot\mathbf{n}$. If $g > 0$ **then exit loop 3.**
   - Compute contact tractions and corresponding tangent matrices

   Normal traction: $N = \epsilon_N g$

   Tangent matrix $\mathbf{K}^{\mathbf{N}}$ for normal traction is defined via

   $\int_s \epsilon_N (\delta\mathbf{r}_s - \delta\boldsymbol{\rho})\cdot(\mathbf{n}\otimes\mathbf{n})(\mathbf{v}_s - \mathbf{v})ds - \int_s \epsilon_N g \left(\delta\boldsymbol{\rho}_{,j}\cdot a^{ij}(\mathbf{n}\otimes\boldsymbol{\rho}_i)(\mathbf{v}_s - \mathbf{v}) + \right.$

   $+ \left. (\delta\mathbf{r}_s - \delta\boldsymbol{\rho})\cdot a^{ij}(\boldsymbol{\rho}_j\otimes\mathbf{n})\mathbf{v}_{,i}\right)ds \boxed{- \int_s \epsilon_N g (\delta\mathbf{r}_s - \delta\boldsymbol{\rho})\cdot h^{ij}(\boldsymbol{\rho}_i\otimes\boldsymbol{\rho}_j)(\mathbf{v}_s - \mathbf{v})ds}$

   Tangent traction T_i is defined via the return-mapping algorithm.

   Trial step: $T_i^{(n+1)} = T_k^{(n)} a^{kj}_{(n)} (\boldsymbol{\rho}_j^{(n)}\cdot\boldsymbol{\rho}_i^{(n+1)}) -$

   $-\epsilon_T \cdot \begin{cases} (\Delta\boldsymbol{\rho}\cdot\boldsymbol{\rho}_i) & \text{for node-to-surface (NTS) and} \\ & \text{surface-to-surface (STS) approaches, where} \\ & \Delta\boldsymbol{\rho} = \boldsymbol{\rho}_{C^{(n+1)}}\big|_{\xi^1_{(n+1)},\ \xi^2_{(n+1)}} - (\boldsymbol{\rho}_{C^{(n)}} + \mathbf{u}_{C^{(n)}})\big|_{\xi^1_{(n)},\ \xi^2_{(n)}} \\ -(\mathbf{u}\cdot\boldsymbol{\rho}_i) & \text{for segment-to-analytical surface (STAS) approach} \end{cases}$

   Coulomb friction law:

   $\Phi_{(n+1)} = \sqrt{T_i^{(n+1)}T_j^{(n+1)}a^{ij}_{(n+1)}} - \mu N^{(n+1)}$

   Return-mapping step see Table 4.2.

   - Compute residual $\mathbf{R}$ from the contact integral in 2
   - Compute the full contact tangent matrix $\mathbf{K} = \mathbf{K}^{\mathbf{N}} + \mathbf{K}^{\mathbf{T}}$

Table 4.2: **Return-mapping scheme and tangent matrices for tangential traction.**

$$\text{if } \Phi^{tr}_{(n+1)} \leq 0$$

sticking condition

$$T^{(n+1)}_{i\ stick} = T^{(n+1)}_{i}$$

Tangent matrix $\mathbf{K^T}$

$$-\varepsilon_T \int_s (\delta\mathbf{r}_s - \delta\boldsymbol{\rho})a^{ij}\boldsymbol{\rho}_i \otimes \boldsymbol{\rho}_j(\mathbf{v}_s - \mathbf{v})ds$$

$$- \int_s T_i \left((\delta\mathbf{r}_s - \delta\boldsymbol{\rho})\, a^{il}a^{jk}\, \boldsymbol{\rho}_k \otimes \boldsymbol{\rho}_l\, \mathbf{v}_j + \delta\boldsymbol{\rho}_j\, a^{ik}a^{jl}\, \boldsymbol{\rho}_k \otimes \boldsymbol{\rho}_l\, (\mathbf{v}_s - \mathbf{v}) \right) ds$$

$$\boxed{+ \int_s T_i h^{ij}(\delta\mathbf{r}_s - \delta\boldsymbol{\rho}) \cdot \left(\boldsymbol{\rho}_j \otimes \mathbf{n} + \mathbf{n} \otimes \boldsymbol{\rho}_j \right) (\mathbf{v}_s - \mathbf{v})ds.}$$

$$\text{if } \Phi^{tr}_{(n+1)} > 0$$

sliding condition

$$T^{(n+1)}_{i\ slide} = \mu N^{(n+1)} \frac{(T^{tr})^{(n+1)}_i}{\|\mathbf{T}^{tr}_{(n+1)}\|}$$

Tangent matrix $\mathbf{K^T}$

$$-\int_s \left((\delta\mathbf{r}_s - \delta\boldsymbol{\rho})\frac{\epsilon_N \mu T_i a^{ij}}{\|\mathbf{T}\|}\boldsymbol{\rho}_j \otimes \mathbf{n}(\mathbf{v}_s - \mathbf{v}) \right) ds$$

$$-\int_s \left((\delta\mathbf{r}_s - \delta\boldsymbol{\rho})\frac{\epsilon_T \mu |N| a^{ij}}{\|\mathbf{T}\|}\boldsymbol{\rho}_i \otimes \boldsymbol{\rho}_j(\mathbf{v}_s - \mathbf{v}) \right) ds$$

$$+\int_s \left((\delta\mathbf{r}_s - \delta\boldsymbol{\rho})\frac{\epsilon_T \mu |N| T_i T_j a^{ik} a^{jl}}{\|\mathbf{T}\|^3}\boldsymbol{\rho}_k \otimes \boldsymbol{\rho}_l(\mathbf{v}_s - \mathbf{v}) \right) ds$$

$$-\int_s \frac{\mu |N| T_i}{\|\mathbf{T}\|} \left((\delta\mathbf{r}_s - \delta\boldsymbol{\rho})\, a^{il}a^{jk}\, \boldsymbol{\rho}_k \otimes \boldsymbol{\rho}_l\, \mathbf{v}_j + \delta\boldsymbol{\rho}_j\, a^{ik}a^{jl}\, \boldsymbol{\rho}_k \otimes \boldsymbol{\rho}_l\, (\mathbf{v}_s - \mathbf{v}) \right) ds$$

$$\boxed{+ \int_s \left(\frac{\mu |N| T_i}{\|\mathbf{T}\|}h^{ij}(\delta\mathbf{r}_s - \delta\boldsymbol{\rho}) \cdot \left(\boldsymbol{\rho}_j \otimes \mathbf{n} + \mathbf{n} \otimes \boldsymbol{\rho}_j \right) (\mathbf{v}_s - \mathbf{v}) \right) ds}$$

$$\boxed{+ \int_s \frac{\mu |N| T_s T_l \delta\xi^s}{\|\mathbf{T}\|^3} \left(-a^{jl}T_m \Gamma^m_{jn}\dot{\xi}^n + a^{jl}T_m h^m_j \dot{\xi}^3 \right) ds.}$$

Remark. Curvature parts in boxes (Table 4.1 and 4.2) can be omitted with very little loss of efficiency

4.6 Numerical examples

4.6.1 Sliding of a block. Linear approximation of the contact surfaces. Two types of the contact frictional problem

During the solution of the frictional problem, it is necessary to solve the evolution equation (4.73) with a return-mapping algorithm, as described in (4.76), (4.77) and (4.78). As was mentioned in **Remark 2**, it is important to know the value of coordinate increments $\Delta\xi^i$ and, therefore, displacement increments Δu^i in order to capture the "sticking-sliding" zone correctly. As a representative example for a-priori estimation of the value of incremental displacements, the stresses in an infinite layer have to be considered, see Fig. (4.3). Both a vertical displacement h and a horizontal displacement u are applied at the upper boundary. During the deformation the rectangle $ABCD$ is changing into a parallelogram AB_1C_1D. Under the assumption of linear elasticity and a plane strain deformation, the stresses in the layer are obtained via superposition of the normal compressive stress σ and the pure shear stress τ:

$$\sigma = \varepsilon\frac{E}{1-\nu^2} = \frac{h}{b}\frac{E}{1-\nu^2}; \quad \tau = \gamma G = \frac{u}{b}\frac{E}{2(1+\nu)}. \tag{4.106}$$

Now we assume Coloumb friction with μ as a friction coefficient at the lower boundary. Sliding starts if the condition $\tau = \mu\sigma$ is fulfilled. Thus, the condition of sticking of the thin layer can estimated by the following ratio:

$$\frac{\gamma(1-\nu)}{2\varepsilon} \leq \mu, \tag{4.107}$$

from which we obtain the threshold value of the horizontal displacement u:

$$u_{cr} = \frac{2\mu h}{1-\nu}, \tag{4.108}$$

One can see from the infinite layer, that sliding starts immediately at the complete lower boundary. However, though this is not a case for a finite dimensional block, or an arbitrarily thin layer, where a developing zone of sticking and sliding exists, eqns. (4.107) and (4.108) can be used as a rough estimation of the presence of the sticking condition, and, therefore, for the estimation of the displacement increments.

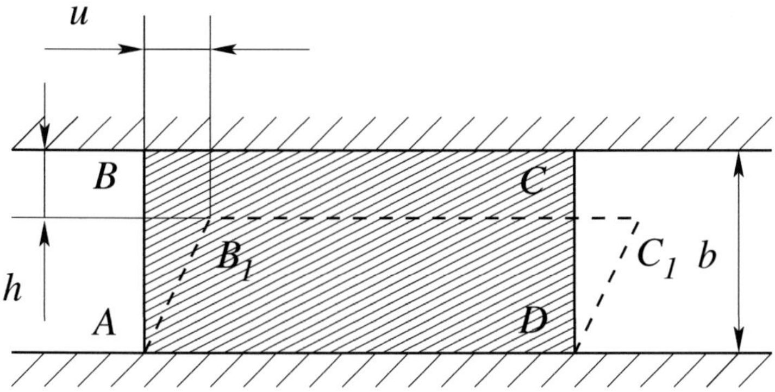

Figure 4.3: Plane deformation of a layer.

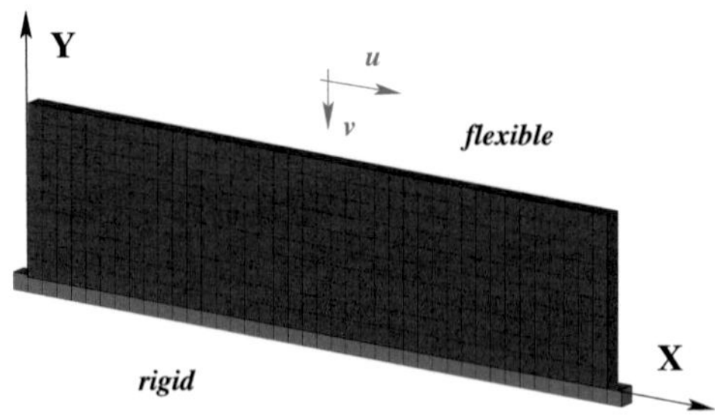

Figure 4.4: Sliding block on the base. Meshed surfaces. STS contact approach.

As an example for the computation, we consider a rectangular block (Fig. 4.4) with the following parameters: elasticity modulus $E = 2.1 \cdot 10^4$, Poisson ratio $\nu = 0.3$, length $a = 20$, height $b = 5$, thickness $c = 0.5$. The dimension system is assumed to be consistent. The lower supplementary block is added to model a rigid base. The Coloumb friction with a coefficient $\mu = 0.3$ is specified between two bodies. The contact surface of the upper block is assumed to be a "master", while the upper surface of the lower block is a "slave" surface within the "segment-to-segment" approach. The penalty parameters are chosen as $\varepsilon_N = \varepsilon_T = 2.1 \cdot 10^6$.

Since the problem is path-dependent, we will investigate a case when displacements at the upper edge are applied in two steps: at the first step, a vertical displacement $v = -7.0 \cdot 10^{-3}$ is applied, then, a horizontal

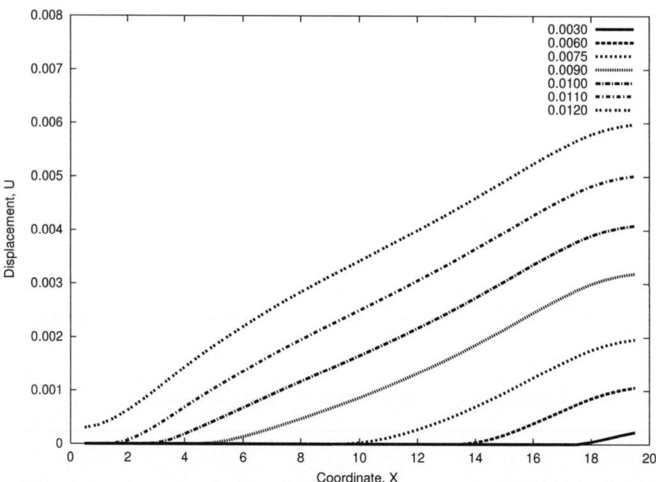

Figure 4.5: Horizontal displacements of the contact surface for various states of the displacement loading.

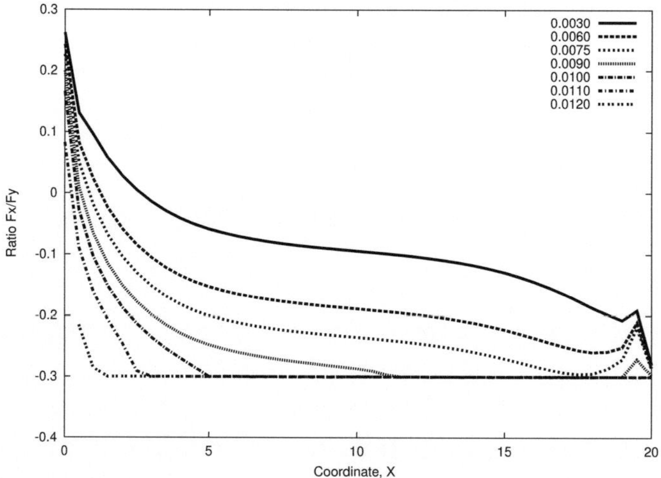

Figure 4.6: Reaction forces ratio F_x/F_y on the contact surface for various states of the displacement loading.

displacement is applied incrementally. Here, we should mention that initial conditions for the history variables are defined at the zero load step with zero external loads, i.e. the initial projection points are the sticking points, see step 1 in Table 4.1. An estimation of the critical horizontal displacement in eqn. (4.108) gives $u_{cr} = 6.0 \cdot 10^{-3}$, so in order to capture the sticking-sliding zone we choose a displacement increment $\Delta u = 2.5 \cdot 10^{-4}$ and apply it in 100 load steps. Our aim in the first computation is to show the development of the sticking-sliding zone. In order to verify this zone carefully we will consider a plot of the horizontal displacements and a plot of the reaction forces ratio on the boundary F_x/F_y. Of course, this zone is precisely specified by the return-mapping algorithm, but we are interested in various parameters. Fig. 4.5 contains the spatial distribution of the horizontal displacements at the lower boundary, if the following displacements $u = 3.0 \cdot 10^{-3}$; $6.0 \cdot 10^{-3}$; $7.5 \cdot 10^{-3}$; $9.0 \cdot 10^{-3}$; $10.0 \cdot 10^{-3}$; $11.0 \cdot 10^{-3}$; $12.0 \cdot 10^{-3}$ are applied at the upper boundary. As shown in the corresponding reaction forces ratio diagram in Fig. 4.6, sliding starts from $u = 7.5 \cdot 10^{-3}$, when the ratio $F_x/F_y = -0.3$ is reached. The block is considered to be sliding at the full lower boundary, when the applied displacement reaches the value $u = 11.0 \cdot 10^{-3}$. We can also conclude that the estimation of the threshold displacement given by eqn. (4.108) is a good approximation.

The spreading of the zone of sliding is found to be within a relatively short interval of loading. In some practical problems, as e.g. metal forming, the energy loss due to large sliding can be more important. Assume for the next discussion that stresses are approximated by eqn. (4.106) for the finite-dimensional block with size $AD = a$, e.g. for a very long block. Then the elastic energy accumulated at the critical state in the block has the following form:

$$E_{el} = \frac{\sigma \varepsilon ab}{2} + \frac{\tau \gamma ab}{2} = \frac{Eah^2}{2b(1 - \nu^2)}\left[1 + \frac{2\mu^2}{1 - \nu}\right]. \qquad (4.109)$$

If the sliding process is developing, when the block is dragged along the distance l, then the work of the critical sliding stresses τ_{sl}, is evaluated as:

$$E_{sl} = \frac{\tau_{sl} al}{2} = \frac{Eahl\mu}{b(1 - \nu^2)}. \qquad (4.110)$$

It is obvious, that during the large sliding a thin layer along a relatively large distance l, dissipation of energy due to sliding, eqn. (4.110) can be rather important then initial threshold value. Thus, frictional problems can be subdivided into two problems:

a) compute the global threshold value for sliding and the development of the distribution of the sticking-sliding zone;

b) compute forces which are necessary to drag the structure under the assumption of full sliding.

Obviously, for the first problem the evolution equation (4.73) must be computed with small steps within the return-mapping algorithm, but for the second problem the sticking zone is out of interest and for the analysis relatively large steps can be taken. Such problems are certainly present in forming processes with large plastic deformations. In order to show an example for the problem type **b)**, another analysis is performed with a displacement increment $\Delta u = 12.0 \cdot 10^{-3}$, which is even larger then the critical one and corresponds to the developed sliding zone, see Fig. 4.5. In order to compare the influence of the various parts of the tangent matrix we compute two cases
1) with the full tangent matrix;
2) only with the main part of the tangent matrix.
The penalty parameter is chosen as $\varepsilon_N = \varepsilon_T = 2.1 \cdot 10^5$. Table 4.3 shows the comparison of the numerical results between both cases by the number of iterations per load step. As we can see in the developed sliding region the full matrix in comparison with only the main matrix leads to a reduction of the number of equilibrium iterations per load step from 4 to 3. We should mention that for the previous example during the incremental horizontal loading there is no difference between the number of equilibrium iterations for both cases. Obviously this is due to the fairly small load steps. Thus, as expected, keeping all parts of the matrix appears to be only necessary in the case of large load increments.

Fig. 4.7 shows the spatial distribution of the relative horizontal displacements $u - u_{applied}$ at the lower boundary if the displacement at the upper boundary is taken exemplarily as $u = 0.012$, then $u = 0.048$ and finally $u = 0.120$. It is obvious, that the relative horizontal displacements hardly change during the fairly large sliding process.

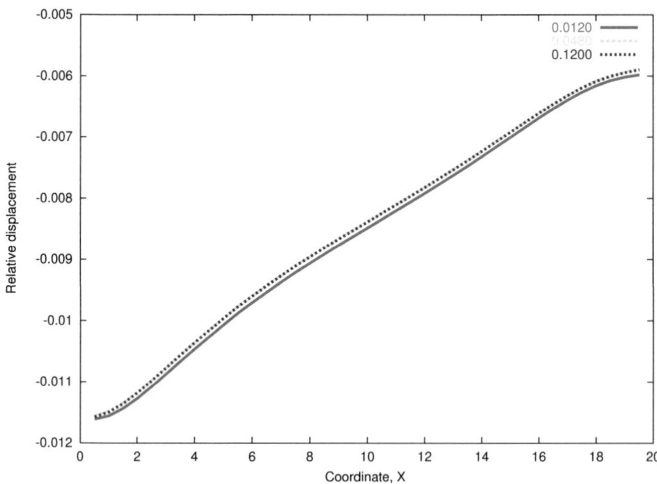

Figure 4.7: Relative horizontal displacements of the contact surface for various states of the displacement loading.

Case 1			Case 2		
No. l.s.	No. it./l.s.	Cum. No. it.	No. l.s.	No. it./l.s.	Cum. No. it.
1	4	4	1	5	5
2	6	10	2	6	11
3	5	15	3	5	16
4-20	3	66	4-20	4	84

Table 4.3: Sliding of a block. Bilinear elements. Segment-to-segment contact approach. Influence of various contact stiffness parts on convergence. Case 1: full matrix; case 2: only main matrix. Comparison of no. of iterations in all load steps (l.s.)

4.6.2 Sliding of a block. Quadratical approximation of the contact surfaces

Since general smoothing techniques for contact surfaces are out of the scope of this article, in this example we will use contact elements with quadratical approximation of the master surface together with a specially chosen geometry of both contact bodies in order to preserve C^1-continuity of the contact surfaces. Namely, we consider contact between a parabolical block sliding on a parabolical cylindrical base, see Fig. 4.8. The block is meshed with 18-node solid-shell elements with density $20 \times 1 \times 5$. Both master and slave contact surface geometries are

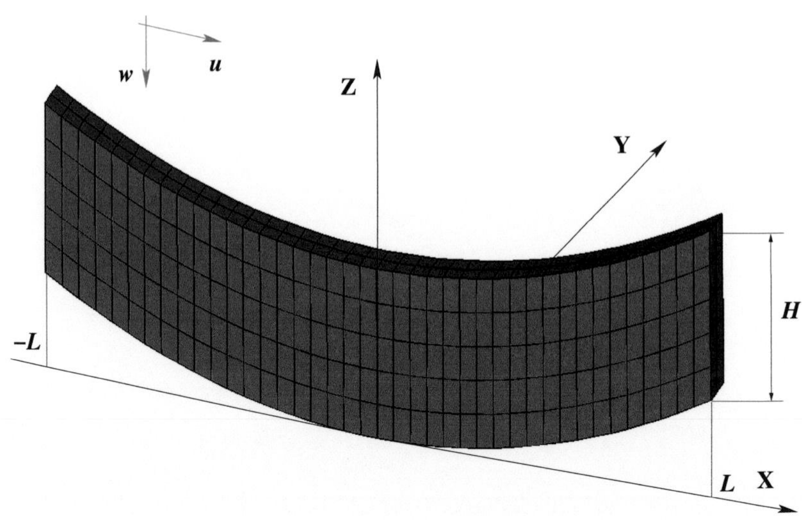

Figure 4.8: Sliding of a parabolical block on a parabolical base. Meshed block. NTS contact approach.

satisfying the equation

$$z = c \cdot x^2, \quad \text{with} \quad c = 0.03 \tag{4.111}$$

The contact is modeled by the node-to-surface approach with the master surface from the parabolical block. The parabolical slave surface of the fixed base is represented by slave nodes with the same mesh density as the master, which are not shown in Fig. 4.8. The geometrical parameters are $H = 5$, $L = 10$; the material is linear elastic with Young's modulus $E = 2.1 \cdot 10^4$, Poisson ratio $\nu = 0.3$, Coulomb friction coefficient $\mu = 0.3$.

In the case of contact with a curvilinear surface, even with homogeneous loading, zones with sticking and sliding can be present. One can expect from the rigid body mechanics that the sliding zone during vertical loading w in the current example is satisfying the following condition $|x| > 5$. From the friction cone for the parabolical cylinder follows that: $\tan \alpha|_{x=5} = z' = 2 \cdot 0.03x|_{x=5} = 0.3$. In order to inspect this ef-

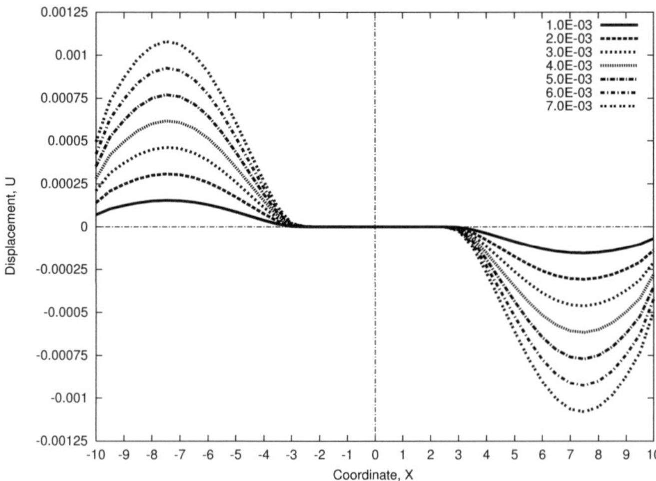

Figure 4.9: Distribution of the tangential displacements in OX direction. Parabolical cylinder. NTS contact approach. Incremental vertical loading.

fect in the deformable body, we apply at the upper edge the vertical displacement $w = 0.007$ in 7 load steps. In Fig.4.9 the distribution of the tangent displacement $u_\tau = u_x \cos \alpha + u_y \sin \alpha$ over x on the contact surface is depicted, where now and for the next example the angle α is computed in the reference configuration. For the further discussion we will distinguish based on the OZ axis, the left sliding zone with negative displacements and the right sliding zone with positive displacements. The tangential displacements from both zones are directed towards the OZ axis, therefore the distribution looks mirror-symmetric. One can see that the sticking zone is approximately satisfying the condition $|x| < 3$.

Equidistant motion on a cylinder. As continuation of the numerical example, we choose an equidistant motion of the upper edge of the parabolical block at the distance h from the generatrix of the parabolical base. The curve $\mathbf{r}$ of this motion satisfies the following equation:

$$\mathbf{r} = \boldsymbol{\rho} + h\mathbf{n}, \tag{4.112}$$

which for the parabola (4.111) can be written as:

$$
\mathbf{r} = \left\{ \begin{array}{c} x \\ 0 \\ cx^2 \end{array} \right\} + \frac{h}{\sqrt{1 + 4c^2x^2}} \left\{ \begin{array}{c} -2cx \\ 0 \\ 1 \end{array} \right\},
\tag{4.113}
$$

where $\mathbf{n}$ is a normal on the initial curve, and h is an initial vertical displacement. From eqn. (4.113) it is clear that the trajectory of the body is no longer a parabola. If the curvature of the cylinder is small, i.e. $c << 1$, then we consider a Taylor expansion with a linear term for the first coordinate and with a quadratic term for the second coordinate. Thus, we obtain as a first approximation of the trajectory in eqn. (4.113) a parabolical motion in the form:

$$
\mathbf{r} = \left\{ \begin{array}{c} x(1 - 2ch) \\ 0 \\ h + cx^2(1 - 2ch) \end{array} \right\}.
\tag{4.114}
$$

In this displacement driven problem the parabolical block is moving in the X-Z plane, providing an approximately constant compression.

Next, the loading is applied in two steps also: the first step is a vertical loading with $w = -h = 0.007$, then both a horizontal and a vertical loading are incrementally applied at the upper edge with $\Delta u = \Delta x = 2.5 \cdot 10^{-5}$ according to eqn. (4.114), providing the equidistant motion of the parabolical block. Now, two phases of the development of the sticking-sliding zone can be observed. The first phase corresponds to the situation when the right sliding zone disappears during horizontal loading, as presented in Fig. 4.10 for the following load steps: $u = 1.0 \cdot 10^{-3}$, $2.0 \cdot 10^{-3}$, $3.0 \cdot 10^{-3}$, $5.0 \cdot 10^{-3}$. Fig. 4.11 a) shows scaled deformed and undeformed states when only vertical displacements are applied and, therefore, the two sliding zones are symmetric. The configuration with the vanishing resp. vanished sliding zone on the right side is presented in Fig. 4.11 b). This moment can be detected also from the reaction forces ratio diagram F_τ/F_n in Fig. 4.13, where the right part of the sliding zone is also disappearing with $F_\tau/F_n = 0.3$. The second phase is the spreading of the left sliding zone through the contact surface shown in Fig. 4.12 exemplarily for $u = 5.0 \cdot 10^{-3}$, $10.0 \cdot 10^{-3}$, $15.0 \cdot 10^{-3}$, $20.0 \cdot 10^{-3}$, $25.0 \cdot 10^{-3}$, $30.0 \cdot 10^{-3}$. Here the zone without contact

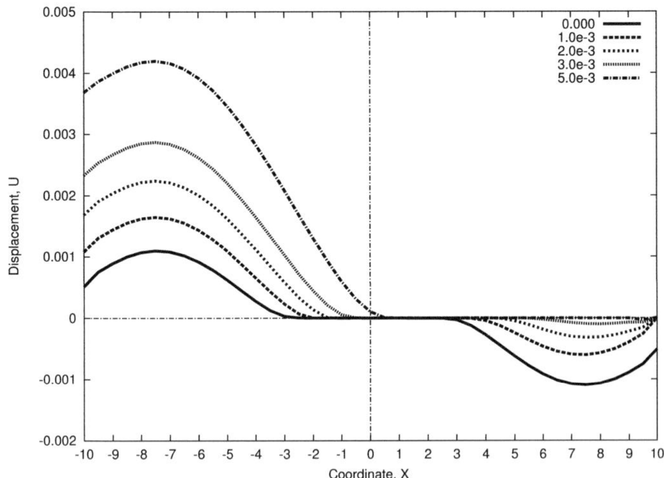

Figure 4.10: Parabolical cylinder. Horizontal loading. Distribution of the tangential displacements in X direction. Phase 1 — vanishing right sliding zone.

is detected as a zone with zero normal nodal forces $f_n = 0$ starting at a loading with $u \geq 1.52 \cdot 10^{-2}$. These sub-zones are marked with thicker lines in Fig. 4.12.

Again we now compare the influence of the various parts of the tangent matrix on the convergence rate when the applied displacements correspond to the developed sliding. Namely, the load is applied in 20 load steps with the displacement increment $\Delta u = 4.0 \cdot 10^{-2}$, providing a fully developed sliding motion from the first step on. The following cases are shown in table 4.4:

1) full tangent matrix;

2) without curvature parts;

3) only with main part of the tangent matrix.

We see that excluding the curvature matrix leads to a minor reduction of the convergence, while excluding the rotational part too causes a considerable increase of the number of equilibrium iterations per load step. We should also mention that during the analysis of the threshold value before full sliding the number of equilibrium iterations remains the same for each case due to a small load step. Thus, the computation with the rotational part is more important for the developed sliding problem of type **b**.

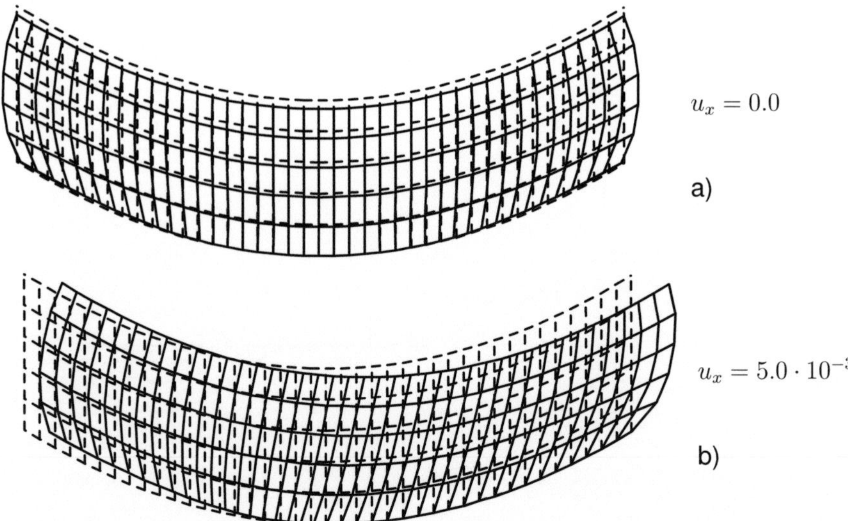

Figure 4.11: Parabolical cylinder. Initial vertical loading. Undeformed and deformed states with in addition applied horizontal displacement a) $u = 0.0$ – two sliding zones; b) $u = 5.0 \cdot 10^{-3}$ – only left sliding zone. (Displacements scaled: 250 times in x-direction, 40 times in z-direction.)

Case 1			Case 2			Case 3		
No. l.s.	No. it./l.s.	Cum. No. it.	No. l.s.	No. it./l.s.	Cum. No. it.	No. l.s.	No. it./l.s.	Cum. No. it.
1	6	6	1	6	6	1-9	6	54
2	5	11	2	5	11	10-12	7	75
3	5	16	3	5	16	13-15	8	99
4-18	4	76	4-16	4	68	15-17	9	126
19-20	5	86	17-20	3	88	18-20	10	156

Table 4.4: Full sliding of a parabolical block. Biquadratic elements. Node-to-segment contact approach. Influence of various contact stiffness parts on convergence. Case 1: full matrix; case 2: without curvature parts; case 3: only main matrix. Comparison of no. of iterations in all load steps (l.s.)

4.6.3 Large sliding on a rigid parabolical cylinder

As an example of a problem with a 3D spatial large sliding, we consider here a motion of a semi-circular cylinder on the surface of a paraboli-

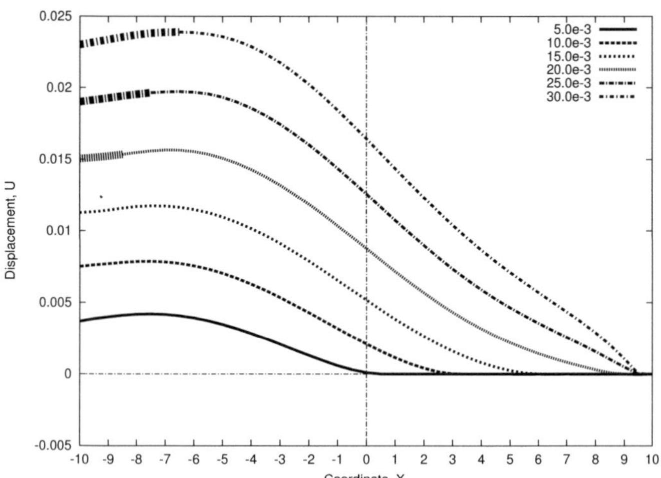

Figure 4.12: Parabolical cylinder. Horizontal loading. Distribution of the tangent displacement over X-coordinate. Phase 2 — spreading of the left sliding zone.

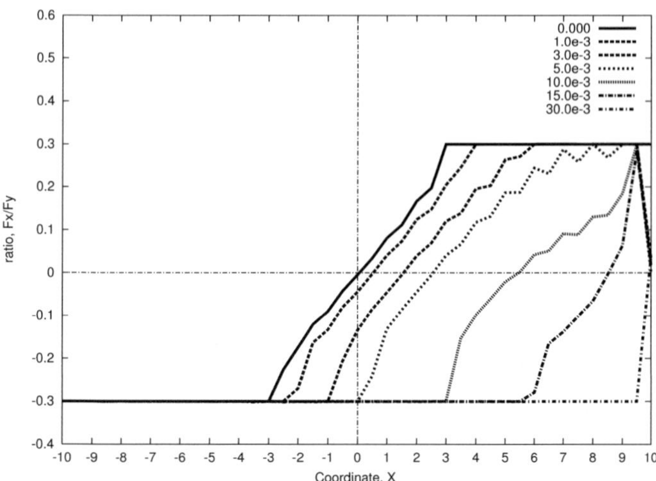

Figure 4.13: Parabolical cylinder. Horizontal loading. Reaction forces ratio F_τ/F_n on the contact surface for various states of loading.

cal cylinder in analogy to [100], see Fig. 4.14. The necessary details for the description of contact with rigid surfaces described by analytical functions is given in a short from [57].

4.6.3.1 Contact with a surface described by analytical functions

If a body contacts a rigid surface, the latter one is chosen as a "slave" surface in our description, but the integration is performed over the "master" surface. The rigid surface is then parameterized by internal coordinates α^1, α^2. Then a point $\mathbf{r}$ of this surface has to satisfy eqn. (4.26) as a point in the local coordinate system of the contact element too. This condition leads to the following equation

$$\mathbf{r}(\alpha^1, \alpha^2) = \boldsymbol{\rho}(\xi^1, \xi^2) + \xi^3 \mathbf{n}. \tag{4.115}$$

The 'slave' point projection procedure, which was necessary for the previous description with surface segments, now turns into the determination of the surface point defined by equation (4.115). Using a "segment-to-segment" type strategy for the computation of the contact integral, first integration points ξ_I^1, ξ_J^2 are defined on the "master" segment and then the corresponding internal coordinates α^1, α^2 of the rigid surface as well as the penetration ξ^3 are computed e. g. by the Newton method. For this algorithm we define a function $F(\alpha^1, \alpha^2, \xi^3)$ with the components given in eqn. (4.115)

$$\mathbf{F} = \begin{bmatrix} x_{s1} - x_1 - n_1 \xi^3 \\ x_{s2} - x_2 - n_2 \xi^3 \\ x_{s3} - x_3 - n_3 \xi^3 \end{bmatrix} \quad \text{with} \quad x_i = x_i(\xi^1, \xi^2). \tag{4.116}$$

Its derivative with respect to the coordinates $(\alpha^1, \alpha^2, \xi^3)$ is:

$$\mathbf{F}' = \begin{bmatrix} x_{s1,1} & x_{s1,2} & -n_1 \\ x_{s2,1} & x_{s2,2} & -n_2 \\ x_{s3,1} & x_{s3,2} & -n_3 \end{bmatrix}. \tag{4.117}$$

Then, the Newton iteration procedure reads as follows for iteration step n:

$$\Delta\boldsymbol{\alpha}_n = \begin{bmatrix} \Delta\alpha_n^1 \\ \Delta\alpha_n^2 \\ \Delta\xi_n^3 \end{bmatrix} = -(\mathbf{F}')_n^{-1}\,\mathbf{F}_n, \qquad (4.118)$$

$$\boldsymbol{\alpha}_{n+1} = \boldsymbol{\alpha}_n + \Delta\boldsymbol{\alpha}_n.$$

Parabolical cylinder. Consider a parabolical cylinder in the canonical form:

$$\begin{aligned} x_s &= \alpha \\ y_s &= c\alpha^2 \end{aligned} \qquad (4.119)$$

The Newton procedure in eqn. (4.118) in this case is reduced to the definition of α from the following iterative expression:

$$\alpha^{(n+1)} = \frac{c(\alpha^{(n)})^2 n_1 + n_1 x_2 - n_2 x_1}{2c\alpha^{(n)} n_1 - n_2} \qquad (4.120)$$

where an initial guess can be computed e.g. as

$$\alpha^{(0)} = x_1 \qquad (4.121)$$

The value of the penetration does not require an iterative procedure and can be computed after the definition of α as

$$\xi^3 = \frac{c\alpha^2 - 2c\alpha x_1 + x_2 - n_2 x_1}{2c\alpha^{(n)} n_1 - n_2}. \qquad (4.122)$$

Spiral equidistant motion to a cylinder. Now we consider a 3D motion on the surface of the parabolical cylinder. In order to generalize the equidistant motion in eqn. (4.114) into a spiral one we consider the parameterization in the form:

$$x = vt, \quad y = Ht/T, \qquad (4.123)$$

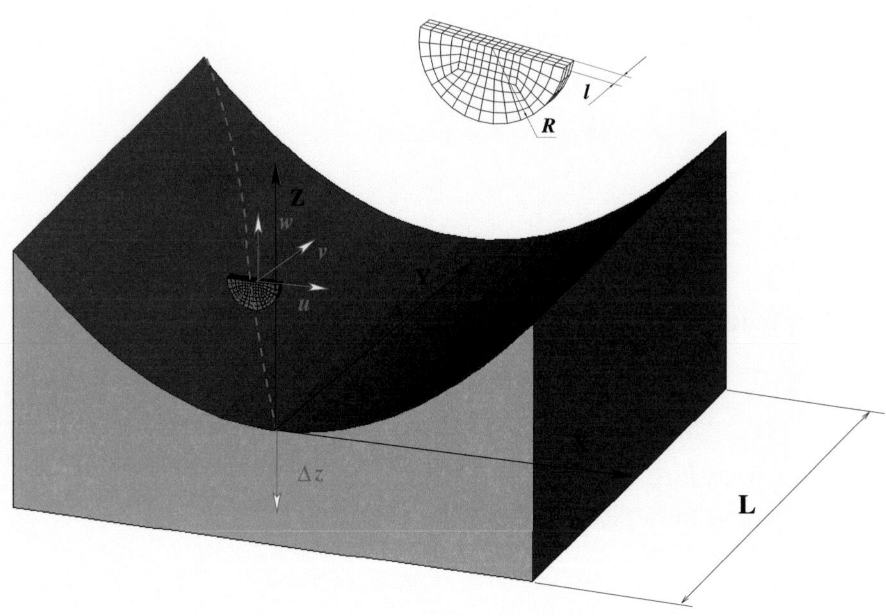

Figure 4.14: Spiral motion of a circular semi-cylinder on the parabolical cylinder. Segment-to-analytical surface approach.

where v is a loading rate, T is the final load step. Thus, the spiral motion with the trajectory shown in Fig. 4.14 is defined as

$$\mathbf{r} = \left\{ \begin{array}{c} vt(1 - 2ch) \\ Ht/T \\ h + c(vt)^2(1 - 2ch) \end{array} \right\}, \quad t = 0, 1, 2, ..., T. \tag{4.124}$$

For the numerical example, we chose the spiral motion of a short deformable circular semi-cylinder with radius $R = 1$ and $l = 0.4$ on the surface of the parabolical cylinder with parameters $c = 5 \cdot 10^{-2}$, $H = 20$, see Fig. 4.14. The semi-cylinder is made of the linear elastic material: $E = 2.10 \cdot 10^4$; $\nu = 0.3$ and meshed with linear "solid-shell" elements 16×3 as shown in Fig. 4.14. Coulomb friction with $\mu = 0.3$ is specified between the bodies.

The loading process consists of two stages:

1) the circular semi-cylinder positioned at the initial point is pressed into the parabolical cylinder with $\Delta z = h = -0.01$

2) then the circular semi-cylinder is moving on the parabolical surface, providing an equidistant motion of the central axis with the distance $R + \Delta z = 0.99$ according to eqn. (4.124) together with the upper surface parallel to the X-Z plane. Thus, the central axis of the semi-cylinder is moving according to the following equation:

$$\mathbf{r} = \left\{ \begin{array}{c} 1.001vt \\ 20t/T \\ h + 5.005 \cdot 10^{-2}(vt)^2 \end{array} \right\}. \tag{4.125}$$

The displacements are applied in 1000 load steps with $v = 10^{-2}$, providing an increment $\Delta y = 0.02$ which is larger than the critical threshold value in eqn. (4.108). The Gauss integration formula with 4×4 integration points is used to check the value of penetration in eqn. (4.115). Here we concentrate again on the investigation of the influence of the various parts of the matrix on the convergence. Since the contact geometry is linear only due to the discretization of the semi-cylinder, the cases with the main matrix and the tangent matrix without curvature part are compared, see Table 4.5 for the first 20 loads steps. Obviously, it is definitely advantageous to use the tangent matrix without curvature parts, but keeping the rotational parts in this problem.

4.7 Conclusions

In this contribution a fully convective description for frictional contact has been proposed. For this, a special local coordinate system according to the closest point procedure is used. The core of the description is to consider differential operations in the covariant form with expressing all values on the tangent plane. Thus, e.g. a penalty regularization of the Coloumb friction law leads to evolution equations expressed in the covariant derivatives. This approach has several advantages. First, the artificial non-symmetry of the sticking tangent matrix, which appeared

Case 1			Case 2		
No. l.s.	No. it./l.s.	Cum. No. it.	No. l.s.	No. it./l.s.	Cum. No. it.
1	9	9	1	10	10
2	8	17	2-18	13	231
3	5	22	19	12	243
4-20	4	90	20	11	254
...	...	...	...	...	...

Table 4.5: Sliding of a semi-cylinder on a parabolical block. Bilinear contact elements. Segment-to-analytical surface contact approach. Influence of various contact stiffness parts on convergence. Case 1: excluding only curvature matrix; case 2: only main matrix. Comparison of no. of iterations for the fist 20 load steps (l.s.)

in earlier publications, is removed. Second, the structure of each tangent matrix is more geometrical and algorithmic. It allows to distinguish between three parts of a tangent matrix, namely the "main" part, the "rotational" part and the "curvature" part. Further, the geometrical interpretation of the covariant derivatives leads to a continuous numerical integration algorithm which overcomes the discontinuities of the convective variables.

It was shown in the numerical examples that frictional contact problems can be subdivided into two types. The first type contains the development of a sticking-sliding zone. In this case, small loads steps, which can be estimated by considering an elastic layer under friction conditions, are necessary. In due course, it appears that then the differences in the convergence rate between computations with various tangent matrix are meaningless. For the second type a fully developed sliding is assumed and, therefore, fairly large steps beyond the threshold value can be applied. In this case it is important to compute with the matrix containing the rotational part. Keeping the curvature matrix leads only to a small improvement.

4.8   APPENDIX

A. Proof of the Weingarten formula.

Having taken the derivative of the unity equation $\mathbf{n}\cdot\mathbf{n} = 1$ with respect to surface coordinates ξ^i, we obtain $\mathbf{n}\cdot\mathbf{n}_i = 0$, from which follows that the vectors $\mathbf{n}_i$ are orthogonal to $\mathbf{n}$ and, therefore, lay on the tangent plane of the surface. Thus, $\mathbf{n}_i$ is expressed as a sum of the surface vectors $\boldsymbol{\rho}_i$

$$\mathbf{n}_i = c_{i.}^{.k}\boldsymbol{\rho}_k. \tag{4.126}$$

Computing a dot product with $\boldsymbol{\rho}_j$

$$(\mathbf{n}_i \cdot \boldsymbol{\rho}_j) = c_{i.}^{.k}(\boldsymbol{\rho}_k \cdot \boldsymbol{\rho}_j) \quad \rightarrow \quad (\mathbf{n}_i \cdot \boldsymbol{\rho}_j) = c_{i.}^{.k}a_{kj}, \tag{4.127}$$

a derivative of the orthogonality condition $\boldsymbol{\rho}_i\cdot\mathbf{n} = 0$, gives $\boldsymbol{\rho}_{ij}\cdot\mathbf{n} + \boldsymbol{\rho}_i\cdot\mathbf{n}_j = 0$. Thus

$$h_{ij} \equiv (\boldsymbol{\rho}_{ij} \cdot \mathbf{n}) = -(\boldsymbol{\rho}_i \cdot \mathbf{n}_j) \tag{4.128}$$

Therefore, the $c_{i.}^{.k}$ can be defined as

$$c_{i.}^{.k} = -h_{ij}a^{jk}, \tag{4.129}$$

from which **Weingarten's formula** (4.13) is obtained.

B. Proof of the Gauss-Codazzi formula.

In general, the derivatives of the coordinate surface vectors $\boldsymbol{\rho}_i$ are no longer on the surface, therefore, one should express them by the vectors $\boldsymbol{\rho}_1, \boldsymbol{\rho}_2, \mathbf{n}$

$$\boldsymbol{\rho}_{ij} = \Gamma_{ij}^k\boldsymbol{\rho}_k + h_{ij}\mathbf{n}. \tag{4.130}$$

Expressions for Γ_{ij}^k and h_{ij} follow from the computation of the dot product with the basis vectors $\boldsymbol{\rho}_j$ and the normal $\mathbf{n}$.

C. Covariant derivative of covariant components.

In the case of covariant components we need the derivative of a contravariant base vector $\dfrac{\partial \boldsymbol{\rho}^{i}}{\partial \xi^{j}}$ instead of ρ_{ij}, see eqn. (4.17). First, take the derivative of the mixed metric components:

$$\frac{\partial a_i^k}{\partial \xi^j} = \frac{\partial(\boldsymbol{\rho}_i \cdot \boldsymbol{\rho}^{\,k})}{\partial \xi^j} = (\boldsymbol{\rho}_{ij} \cdot \boldsymbol{\rho}^{\,k}) + (\boldsymbol{\rho}_i \cdot \frac{\partial \boldsymbol{\rho}^{\,k}}{\partial \xi^j}) = \Gamma_{ij}^k + (\boldsymbol{\rho}_i \cdot \boldsymbol{\rho}_{,j}^{\,k}) = 0, \quad (4.131)$$

therefore,

$$(\boldsymbol{\rho}_i \cdot \boldsymbol{\rho}_{,j}^{\,k}) = -\Gamma_{ij}^k \qquad (4.132)$$

and the covariant derivative for the covariant component gets the following form

$$\nabla_j T_i = \frac{\partial T_i}{\partial \xi^j} - T_k \Gamma_{ij}^k. \qquad (4.133)$$

D. Proof that the full time derivative is a Lie time derivative.

In order to prove eqn. (4.32) consider the vector $\mathbf{r}(\xi^i)$ in the reference Cartesian frame:

$$\mathbf{r} = X^k \mathbf{e}_k \qquad (4.134)$$

where $\mathbf{e}_k$ are unit vectors of the Cartesian reference frame. By definition of the reference frame the vectors $\mathbf{e}_k$ are time independent. Since here only the spatial case is considered, all indices are running from 1 to 3. The vector $\mathbf{r}$ is assumed to be time independent only for simplicity without loss of generality. The coordinate vectors $\mathbf{r}_i$ are defined as

$$\mathbf{r}_i = \frac{\partial X^k}{\partial \xi^i} \mathbf{e}_k = (F^{-1})_i^k \mathbf{e}_k \qquad (4.135)$$

where $(F^{-1})_i^k$ are components of the inverse gradient deformation tensor $\mathbf{F}$ with components

$$F_j^i = \frac{\partial \xi^i}{\partial X^j}, \qquad (4.136)$$

which are used for the vice versa transformation:

$$\mathbf{e}_k = F_k^i \mathbf{r}_i. \tag{4.137}$$

Eqns. (4.137) and (4.135) give the push-forward operator $\mathbf{F}$ and the pull-back operator $\mathbf{F}^{-1}$ respectively in a tensor form:

$$\mathbf{F} = F_j^i \mathbf{r}_i \otimes \mathbf{e}^{\,j}, \qquad \mathbf{F}^{-1} = (F^{-1})_j^i \mathbf{e}_i \otimes \mathbf{r}^{\,j} \tag{4.138}$$

The Lie time derivative of the spatial vector $\mathbf{T} = T^i \mathbf{r}_i$ is taken following the rule: pull-back to the reference configuration, take time derivative, push-forward to the current configuration:

$$
\begin{aligned}
\mathcal{L}_t \mathbf{T} \;&=\; \mathbf{F}\frac{d}{dt}(\mathbf{F}^{-1}\mathbf{T}) = \\[2mm]
&=\; F_n^k \mathbf{r}_k \otimes \mathbf{e}^{\,n} \cdot \frac{d}{dt}\left((F^{-1})_i^j \mathbf{e}_j \otimes \mathbf{r}^{\,i} \cdot T^m \mathbf{r}_m \right) = \\[2mm]
&=\; F_n^k \mathbf{r}_k \otimes \mathbf{e}^{\,n} \cdot \frac{d}{dt}\left((F^{-1})_i^j \mathbf{e}_j \delta_m^i T^m \right) = \\[2mm]
&=\; F_n^k \mathbf{r}_k \delta_j^n \frac{d}{dt}\left((F^{-1})_i^j T^i \right) = \frac{d((F^{-1})_i^j T^i)}{dt} F_j^k \mathbf{r}_k.
\end{aligned}
\tag{4.139}
$$

This is a full time derivative. It can be seen directly

$$\frac{d[(F^{-1})_i^j T^i]}{dt} F_j^k \mathbf{r}_k = \frac{d[(F^{-1})_i^j T^i]}{dt} \mathbf{e}_j = \frac{d[(F^{-1})_i^j T^i \mathbf{e}_j]}{dt} = \frac{d[T^i \mathbf{r}_i]}{dt} = \frac{d\mathbf{T}}{dt}. \tag{4.140}$$

For the proof only the time independence of the reference basis vectors $\mathbf{e}_j$ was used. Some algebraic manipulations of equation (4.139) are required to show this for the components.

$$
\begin{aligned}
\mathcal{L}_t \mathbf{T} \;&=\; \frac{d[(F^{-1})_i^j T^i]}{dt} F_j^k \mathbf{r}_k = \\[2mm]
&=\; \frac{\partial T^i}{\partial t}(F^{-1})_i^j F_j^k \mathbf{r}_k + \frac{\partial T^i}{\partial \xi^n}(F^{-1})_i^j F_j^k \dot{\xi}^n \mathbf{r}_k + \frac{\partial (F^{-1})_i^j}{\partial \xi^n} T^i F_j^k \dot{\xi}^n \mathbf{r}_k = \\[2mm]
&=\; \frac{\partial T^i}{\partial t}\delta_i^k \mathbf{r}_k + \frac{\partial T^i}{\partial \xi^n}\delta_i^k \dot{\xi}^n \mathbf{r}_k + \frac{\partial (F^{-1})_i^j}{\partial \xi^n} T^i F_j^k \dot{\xi}^n \mathbf{r}_k.
\end{aligned}
\tag{4.141}
$$

The last term contains Christoffel symbols. In order to elaborate this, their determination in the reference frame has to be considered. Equation (4.15) can be written as

$$\Gamma_{ij}^{k} a_{kl} = \mathbf{r}_{ij} \cdot \mathbf{r}_{l} \tag{4.142}$$

$$\Gamma_{ij}^{k} \frac{\partial X^{m}}{\partial \xi^{k}} \frac{\partial X^{m}}{\partial \xi^{l}} = \frac{\partial X^{n}}{\partial \xi^{i} \partial \xi^{j}} \frac{\partial X^{n}}{\partial \xi^{l}}$$

and, exploiting the chain rule,

$$\Gamma_{ij}^{k} \frac{\partial X^{m}}{\partial \xi^{k}} \frac{\partial X^{m}}{\partial \xi^{l}} \frac{\partial \xi^{k}}{\partial X^{p}} \frac{\partial \xi^{l}}{\partial X^{r}} = \frac{\partial X^{n}}{\partial \xi^{i} \partial \xi^{j}} \frac{\partial X^{n}}{\partial \xi^{l}} \frac{\partial \xi^{k}}{\partial X^{p}} \frac{\partial \xi^{l}}{\partial X^{r}} \longrightarrow \Gamma_{ij}^{k} \delta_{p}^{m} \delta_{r}^{m} = \frac{\partial X^{n}}{\partial \xi^{i} \partial \xi^{j}} \delta_{r}^{n} \frac{\partial \xi^{k}}{\partial X^{p}}$$

finally one obtains

$$\Gamma_{ij}^{k} = \frac{\partial X^{n}}{\partial \xi^{i} \partial \xi^{j}} \frac{\partial \xi^{k}}{\partial X^{n}} \tag{4.143}$$

Now the Lie derivative (eqn. (4.141)) can be written as:

$$\begin{aligned}
\mathcal{L}_{t} \mathbf{a} &= = \frac{\partial a^{i}}{\partial t} \mathbf{r}_{i} + \frac{\partial a^{i}}{\partial \xi^{n}} \dot{\xi}^{n} \mathbf{r}_{i} + \frac{\partial X^{j}}{\partial \xi^{i} \partial \xi^{n}} \frac{\partial \xi^{k}}{\partial X^{j}} a^{i} \dot{\xi}^{n} \mathbf{r}_{k} \\
&= \frac{\partial a^{i}}{\partial t} \mathbf{r}_{i} + \frac{\partial a^{i}}{\partial \xi^{n}} \dot{\xi}^{n} \mathbf{r}_{i} + \Gamma_{ij}^{k} \dot{\xi}^{j} a^{i} \mathbf{r}_{k}.
\end{aligned} \tag{4.144}$$

This is a full vector derivative (4.18) including the covariant derivative (4.19).

5

Computational aspects and implementation of the covariant approach for contact analysis

Abstract

This chapter is devoted to various numerical aspects arising in applications of the covariant approach with finite elements. Thus, the algorithms known in literature are reconsidered under the unified covariant description. These are "Node-To-Surface" (NTS), "Segment-To-Segment" (STS) and "Segment-To-Analytical (rigid) Surface" (STAS).

The closest point projection of the "slave" integration points onto the master segment forms the basis of the STS-algorithm. Thus, various numerical integration schemes based a) on simple increasing of the number of integration points of Lobatto or Gauss type; b) on subdivision of the integration area into subdomains with further application of the quadrature formula are developed to satisfy the patch test. A smoothing technique based on NURBS spline smoothing is considered in application with a covariant approach. The numerical example is chosen to show the behavior for the patch test. A set of closed form solutions for the penetration is obtained for the STAS approach – for a plane, for a cylinder, for a sphere, for a torus and for a cone. A reduced technique to compute penetration is considered for the surface of evolution. Special attention is given to discretization techniques in application with "Solid-Shell" elements with various orders of approximation. Algorithms are developed to improve various results such as quality of force-displacement or strain-displacement curves. Special attention is given to deep drawing problems.

Within the chapter the results partially published in articles and different conference proceedings as well as only presented in international conferences are summarized, somehow reconsidered and rewritten.

5.1 Computation of contact integrals – Mortar type contact*

Within the finite element method the contact integrals in eqn. (3.35), (4.60) leading to the residual, as well as the integrals leading to the consistent tangent matrix, see eqns. (3.40), (3.41), (3.42), (3.43) for non-frictional contact, or in Tables 4.5 and 4.5 from Chapter 4 for frictional contact, have to be computed using one or another quadrature formula. In the most common approach known as "node-to-surface" technique the value at the nodes from the finite element discretization of the "slave" part is taken directly. As is well known, this technique can be only directly applied in the case of linear approximations for both "slave" and "master" parts, see [194], and it does not satisfy the patch test, see [31]. This fact can be explained as under-integration of the contact integral, because the Lobatto quadrature formula with only two integration points gives exactly a nodal collocation formula in this case. As an improvement different quadrature formulae of higher order can be used. In this situation the question arrises: How many integration points have to be taken in order to achieve a certain error bound? The usual formula to estimate the integration error does not give a correct answer, because it requires differentiability of the integrand up to a certain order. This is not the case for the computation of the contact integral, which is discussed in the following: the function in the integral is defined on the master element, but the computation of the integral has to be done over the unknown slave surface. In practice, the penetration of "slave" points, e.g. integration points, from different "slave" segments into the master segment is checked, see Fig. 5.9. This can be considered as integration

---

*The section contains the selected sections from the article [57]: M. Harnau, A. Konyukhov, K. Schweizerhof. *Algorithmic aspects in large deformation contact analysis using "Solid-Shell" elements*, Computers and Structures, **83**:1804–1823, 2005.

of auxiliary functions over the known master surface which again define a function which is discontinuous on the master surface.

Remark

*The "Segment-To-Segment" approach based on the projection of integration points onto the master segment considered in the current section became known later as **the Mortar method** with a penalty regularization of contact tractions, see Fischer and Wriggers [41] (2005).*

5.1.1 Convergence test for the integration algorithm: computation of the energy associated with the penalty functional

As a representative example to show that the problem of integration of discontinuous functions arises during the "master-slave" approach we consider the classical Hertz problem. Assume that the contact problem of a cylinder and a semi-infinite elastic plane, see Fig. 5.1, is solved by the standard penalty approach for the finite element method. Let CD be a 2D contact element. Controlling the process with an applied vertical displacement h, the cylinder penetrates into the plane within the first iteration as shown in Fig. 5.1. A characteristic quantity for the satisfaction of contact is the value of the energy associated with the penalty form in e.g. the first iteration. It has the following form:

$$E_g = \frac{1}{2} \int_{AB} \varepsilon_N g_N^2 dx = \frac{\varepsilon_N}{2} J \ . \tag{5.1}$$

The value of the integral J in eqn. (5.1) can be evaluated in closed form, because both the penetration g_N and the contact zone AB are defined from the specific geometry as:

$$g_N = R - h - \sqrt{R^2 - x^2}, \quad AB = 2\sqrt{h(2R - h)} \ . \tag{5.2}$$

The integral J after evaluation and some transformations has the following form:

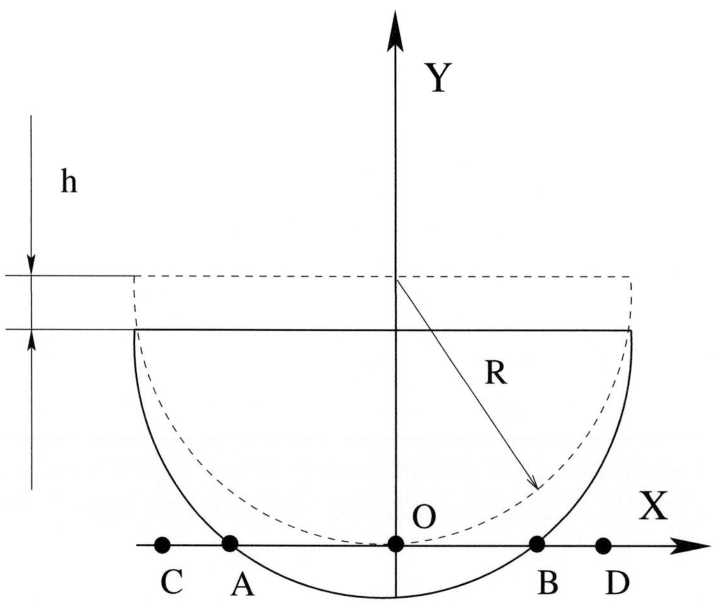

Figure 5.1: Cylinder during the first iteration

$$J = \int_{-\sqrt{h(2R-h)}}^{+\sqrt{h(2R-h)}} (R - h - \sqrt{R^2 - x^2})^2 dx = \tag{5.3}$$

$$= \frac{2}{3}(3R^2 - 2Rh + h^2)\sqrt{h(2R - h)} - 2R^2(R - h)\arcsin\frac{\sqrt{h(2R - h)}}{R}.$$

Within the finite element solution the value of the integral J becomes:

$$J = \int_{AC} \langle g_N \rangle^2 dx, \tag{5.4}$$

where $\langle\,\rangle$ denotes the Macauley brackets in the form

$$\langle g_N \rangle = \begin{cases} 0, & if \ \ g_N > 0 \\ g_N, & if \ \ g_N \leq 0 \end{cases} \tag{5.5}$$

The value of the integral in eqn. (5.4) is computed over the contact element CD, while the contact region AB is detected via the integration points, which then leads in general to a discontinuous function defined over the contact element CD. Before comparing the results, we describe in the following section one of the techniques to integrate discontinuous functions [33].

5.1.2 Integration schemes using a subdivision scheme into subdomains

The a-priori error estimation in the case of the application of Gauss quadrature rules for discontinuous functions is a rather complicated question, because it is necessary to know the behavior of the integrand, see e.g. [33]. However, this is in general not known in the considered cases of rather general contact surfaces. One can only expect, that increasing the number of integration points leads to a reduction of the integration error. As an improved and efficient technique to decrease the integration error a subdivision of the integration area into subdomains together with lower order integration in each subdomain can be used (see e.g. [33]). With the same number of integration points as for a standard Gauss integration this technique leads to a smaller integration error, as is shown in the following.

Let A be an area of element. In the case of a quadrilateral contact element with area A, this can be subdivided into non-overlapping subdomains A_{ij}:

$$A = \bigcup_{ij} A_{ij}. \tag{5.6}$$

Now we consider the subdivision of A with the local coordinate system ξ, η into rectangular subdomains, see Fig. 5.2, e.g. into m parts along the ξ axis and into n parts along η axis. In each subdomain a separate local coordinate system ξ_i, η_j, which has to satisfy the following conditions, is introduced:

$$\xi_i = -1 \quad if \quad \xi = \frac{2i}{m} - 1, \tag{5.7}$$

$$\xi_i = 1 \quad if \quad \xi = \frac{2(i+1)}{m} - 1, \quad i = 0, 1, 2, ...m - 1$$

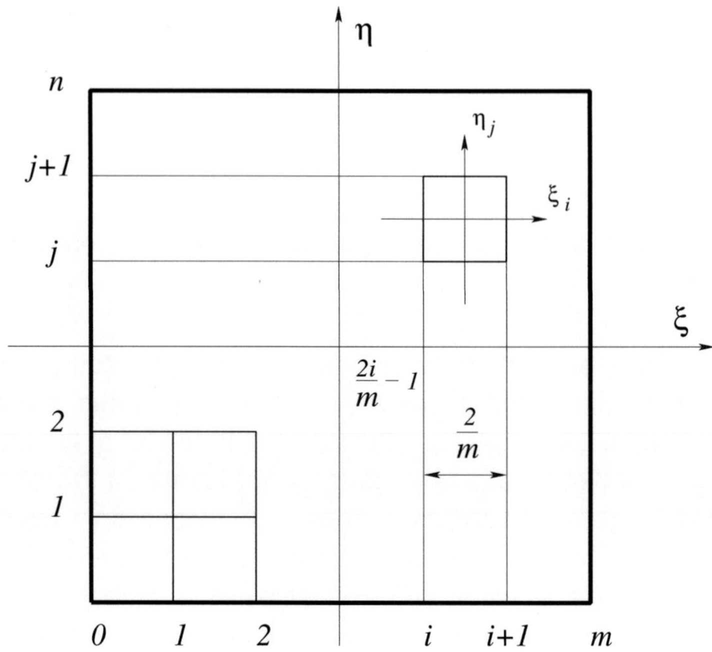

Figure 5.2: Subdivision of the contact segment into subdomains for integration

and

$$\eta_j = -1 \quad if \quad \eta = \frac{2j}{n} - 1, \tag{5.8}$$

$$\eta_j = 1 \quad if \quad \eta = \frac{2(j+1)}{n} - 1, \quad j = 0, 1, 2, \dots n - 1$$

Then, the transformation of the coordinates ξ_i, η_j into ξ, η can be written as:

$$\xi = \frac{\xi_i}{m} + \frac{2i+1}{m} - 1, \qquad \eta = \frac{\eta_j}{n} + \frac{2j+1}{n} - 1. \tag{5.9}$$

Finally, the integration of a function $f(\xi, \eta)$ over the area A in the local coordinate system ξ, η leads to a sum of integrals over each subdomain A_{ξ_i, η_j}:

$$\int_{-1}^{+1}\int_{-1}^{+1} f(\xi,\eta)d\xi d\eta = \sum_{i=0}^{m-1}\sum_{j=0}^{n-1}\int_{A_{\xi_i,\eta_j}} f(\xi,\eta)d\xi d\eta = \tag{5.10}$$

$$= \frac{1}{m\cdot n}\sum_{i=0}^{m-1}\sum_{j=0}^{n-1}\int_{-1}^{+1}\int_{-1}^{+1} f\left(\frac{\xi_i}{m}+\frac{2i+1}{m}-1, \frac{\eta_j}{n}+\frac{2j+1}{n}-1\right)d\xi_i d\eta_j.$$

Each integral in formula (5.10) is computed via standard quadrature formulae, e.g. Gauss integration.

In order to consider general features of the proposed approach, first the exact value of the integral J in eqn. (5.3) is compared with the computed value in (5.4) for a line contact. The following parameters are taken: radius of cylinder $R = 2.0$, vertical displacement $h = 0.1$, length of element $CD = 2.0$. In this case as exact value of the integral $J = 6.7099 \cdot 10^{-3}$ and as length of the contact zone $AB = 1.249$ are obtained. The following relative error e is used for comparison of the computed value J_{com} with the exact value J:

$$e = \frac{J_{exact} - J_{com}}{J_{exact}} \cdot 100\% . \tag{5.11}$$

Tab. 5.1 shows the relative error e in the case of various numbers of Gauss points and subdivisions.

As expected, the formula with subdivisions leads to a smaller error than the standard single domain Gauss formula. It is obvious that among the formulae with a fixed total number of integration points the smallest error is obtained by the formula that combines both the maximum number of subdivisions and the maximum number of Gauss points which can be independently chosen.

After the description of the contact elements in the following part we will show that the proposed approach allows first to diminish the error for the patch test and second to improve the quality of the results, e.g. the load-displacement curve.

No. of Gauss points	No. of subdivisions	$e\,\%$
2	1	93.4232
3	1	-32.4737
5	1	5.4618
6	1	-2.4670
3	2	4.7420
2	3	-0.1113
7	1	-1.0916
10	1	-0.6844
5	2	0.2399
2	5	0.2018
20	1	-0.0669
10	2	0.1147
5	4	0.0978
4	5	0.0153
2	10	0.0271
40	1	0.0114
20	2	0.0196
10	4	-0.0137
8	5	-0.0021
5	8	0.0001
4	10	0.0029
2	20	0.0142

Table 5.1: Relative error in energy of the contact integral for the penalty formulation; comparing standard Gauss integration with subdivisional Gauss integration in subdomains

5.2 Contact with rigid surfaces described by analytical functions – two strategies

Two strategies can be applied if one of the contacting bodies is rigid and can be analytically parameterized (directly by known analytical functions, or by suitable NURB splines in a CAD system). These strategies depend on the selection of the master or slave part in a surface coordinate system.

5.2.1 Rigid surface is a "slave" surface

If **the rigid surface is a "slave" surface** then we can write the following coordinate system:

$$\mathbf{r}_s(\alpha^1, \alpha^2) = \boldsymbol{\rho}(\xi^1, \xi^2) + \xi^3\,\mathbf{n}(\xi^1, \xi^2). \tag{5.12}$$

The rigid surface $\mathbf{r}_s$ is then parameterized by internal Gaussian coordinates α^1, α^2. $\boldsymbol{\rho}(\xi^1, \xi^2)$ is the parameterization of our "master" deformable segment from the finite element mesh and $\mathbf{n}$ is the normal to the master segment. In other words, a point $\mathbf{r}_s$ of this surface is observed in the local coordinate system of the contact master element. The standard Closest Point Projection procedure, which was necessary for the previous description with surface segments, now turns into the determination of the surface point defined by equation (5.12). Using a "segment-to-segment" type strategy for the computation of the contact integral, the integration points ξ_I^1, ξ_J^2 are defined on the "master" segment and then the corresponding internal coordinates α^1, α^2 of the rigid surface as well as the penetration ξ^3 are computed e. g. by the Newton method. For this algorithm we define a function $F(\alpha^1, \alpha^2, \xi^3)$ with the components given in eqn. (5.12)

$$\mathbf{F} = \begin{bmatrix} x_{s1} - x_1 - n_1\xi^3 \\ x_{s2} - x_2 - n_2\xi^3 \\ x_{s3} - x_3 - n_3\xi^3 \end{bmatrix} \quad with \ \ x_i = x_i(\xi^1, \xi^2). \tag{5.13}$$

Its derivative with respect to the coordinates $(\alpha^1, \alpha^2, \xi^3)$ is:

$$\mathbf{F}' = \begin{bmatrix} x_{s1,1} & x_{s1,2} & -n_1 \\ x_{s2,1} & x_{s2,2} & -n_2 \\ x_{s3,1} & x_{s3,2} & -n_3 \end{bmatrix}. \tag{5.14}$$

Then, the Newton iteration procedure reads as follows for iteration step n:

$$\Delta\boldsymbol{\alpha}_n = \begin{bmatrix} \Delta\alpha_n^1 \\ \Delta\alpha_n^2 \\ \Delta\xi_n^3 \end{bmatrix} = -(\mathbf{F}')_n^{-1}\,\mathbf{F}_n, \tag{5.15}$$

$$\boldsymbol{\alpha}_{n+1} = \boldsymbol{\alpha}_n + \Delta\boldsymbol{\alpha}_n.$$

Remark

A numerical example with such a strategy has been shown for the parabolic cylinder in Chapter 4, Section 4.6.3. Another example in which the iterative solution in eqn. (5.15) can be simplified is a surface of revolution.

5.2.1.1 Surface of revolution*

For a surface of revolution, given by an analytical function, or described by NURBS, see [38], a matrix form solution for the coordinate increments in eqn. (5.15) can be directly developed. In the simplest case $f(r)$ can be a plane curve uniquely projected onto the r axis, see Fig. 5.3. The revolution of the curve about the axis OZ gives a surface of revolution. In a Cartesian coordinate system it can be written as

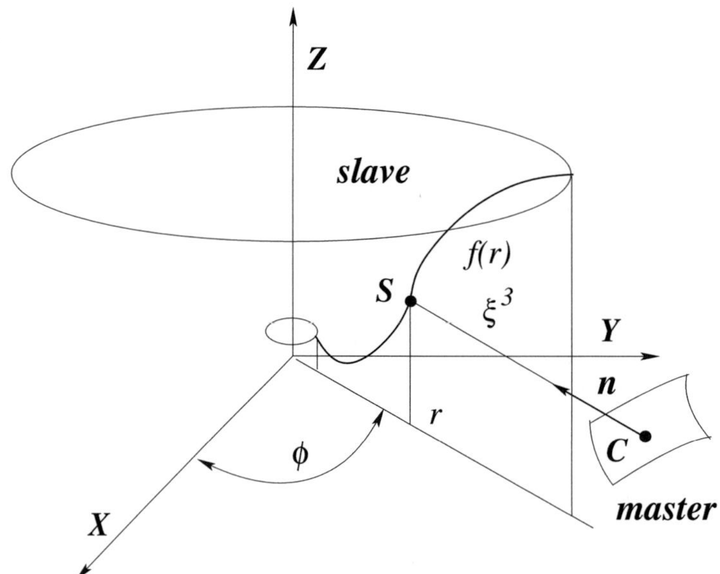

Figure 5.3: The surface of revolution

---

*The material has been partially included in [87]: A. Konyukhov, K. Schweizerhof. *Large Deformation Frictional Contact Formulation for Low Order "Solid Shell" Elements*, ECCOMAS-2004, Jyväskylä.

$$\mathbf{r}_s(r,\phi) = \begin{bmatrix} x_s \\ y_s \\ z_s \end{bmatrix} = \begin{bmatrix} r\cos\phi \\ r\sin\phi \\ f(r) \end{bmatrix}. \tag{5.16}$$

Then the iteration vector $\Delta\boldsymbol{\alpha}_n$ in eqn. (5.15) gets the following form:

$$\Delta\boldsymbol{\alpha}_n = \begin{bmatrix} \Delta r_n \\ \Delta\phi_n \\ \Delta\xi_n^3 \end{bmatrix} \tag{5.17a}$$

where

$$\Delta r_n = \frac{1}{D}\cdot((x_3 - f(r))(n_1\cos\phi + n_2\sin\phi) + n_3(r - x_1\cos\phi - x_2\sin\phi)) \tag{5.17b}$$

$$\Delta\phi_n = \frac{1}{Dr}\cdot\Big((f(r) - x_3 - rf'(r))(n_1\sin\phi - n_2\cos\phi)+ \tag{5.17c}$$

$$+ f'(r)(n_1 x_2 - n_2 x_1) + n_3(x_1\sin\phi - x_2\cos\phi)\Big)$$

$$\Delta\xi_n^3 = \frac{1}{Dr}\cdot\Big(f'(r)(x_1\cos\phi + x_2\sin\phi - r) + f(r) - x_3 \tag{5.17d}$$

$$+ \xi^3[f'(r)(n_1\cos\phi + n_2\sin\phi) - n_3]\Big)$$

and with the determinant

$$D = -n_3 + f'(r)(n_1\cos\phi + n_2\sin\phi). \tag{5.17e}$$

5.2.2 Rigid surface is a "master" surface

Preserving variables ξ^1, ξ^2 only for the finite element approximations for the case **the rigid surface is a "master" surface** we can write the following coordinate system:

$$\mathbf{r}_s(\xi^1,\xi^2) = \boldsymbol{\rho}(\alpha^1,\alpha^2) + p\,\mathbf{n}(\alpha^1,\alpha^2). \tag{5.18}$$

Now, an integration point $\mathbf{r}_s(\xi^1,\xi^2)$ from the deformable "slave" finite element segment is found in the direction of the normal $\mathbf{n}(\alpha^1,\alpha^2)$ to the rigid "master" surface $\boldsymbol{\rho}(\alpha^1,\alpha^2)$. This distance denoted as p plays role of the penetration. It is important to note that **the distance p between the master surface and slave point is not coinciding with the penetration** ξ^3 measured from the finite element segment, because, the normals from the master and the slave, in general, are not parallel. The situation is illustrated for the contact between the rigid sphere and a "slave" segment in Fig. 5.5. However, when two bodies are close to contact then the normals are almost parallel, in fact, they are enforced to be parallel by the contact algorithm. This leads to the possibility to use the current approach for the contact mechanics. *Another important remark is that the normal for any further computational algorithm (e.g. necessary for tangent matrices) must be computed from the deformable "slave" segment*, otherwise disconvergence will be obtained. The corresponding penetration is taken as $\xi^3 = p$.

The Newton method is exploited in this approach in order to solve eqn. (5.18) defining then a point with the coordinates α^1, α^2 on the rigid surface and the distance p between this surface and selected integration point on the "slave" segment $\mathbf{r}_s(\xi^1, \xi^2)$.

5.2.3 Surfaces allowing a closed form solution for the penetration

Here some simple analytical surfaces in both strategies are chosen for which it is not necessary to solve nonlinear equations (5.12) or (5.18) in order to compute the value of the penetration. We consider here plane, cylinder, sphere, torus and cone. For them it is possible to derive the closed form solution for the penetration.

5.2.3.1 Contact with a rigid plane

The simplest example with a closed form solution for *the rigid surface as a "slave"* strategy is a contact with a rigid plane. Consider a rigid "slave" plane given in the in analytical form as:

$$(\mathbf{r} - \mathbf{r}_0) \cdot \mathbf{N} = 0, \qquad (5.19)$$

where $\mathbf{r}_0$ is any point on the plane, $\mathbf{N}$ is a normal vector for the plane and $\mathbf{r}$ is a vector with Cartesian coordinates $\{x, y, z\}$. Assuming now that from one side the slave vector $\mathbf{r}_s$ belongs to the plane (eqn. (5.19)) and from another side it is observed from the master segment, i.e. it is satisfying eqn. (5.12) we can write the following system:

$$\begin{cases} (\mathbf{r}_s - \mathbf{r}_0) \cdot \mathbf{N} &= 0 \\ \mathbf{r}_s &= \boldsymbol{\rho}(\xi^1, \xi^2) + \xi^3 \mathbf{n}(\xi^1, \xi^2). \end{cases} \qquad (5.20)$$

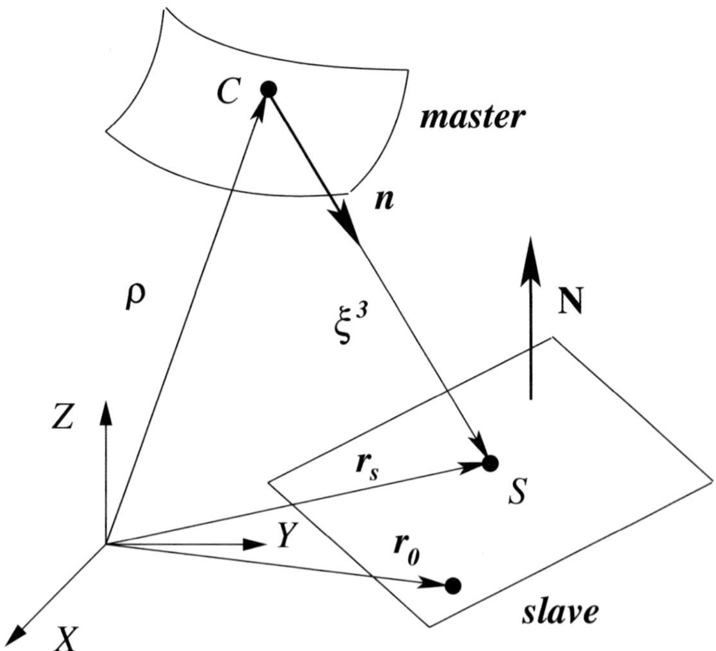

Figure 5.4: Contact with a rigid plane given analytically

Just inserting $\mathbf{r}_s$ from the second equation into the first one, we can obtain the value for the distance ξ^3, and therefore for the penetration as:

$$\xi^3 = -\frac{(\boldsymbol{\rho}(\xi^1_{i.p.}, \xi^2_{i.p.}) - \mathbf{r}_0) \cdot \mathbf{N}}{\mathbf{n}(\xi^1_{i.p.}, \xi^2_{i.p.}) \cdot \mathbf{N}} \tag{5.21}$$

It is easy to see, that the distance is always resolved if

$$\mathbf{n}(\xi^1_{i.p.}, \xi^2_{i.p.}) \cdot \mathbf{N} \neq 0, \tag{5.22}$$

which means the slave plane is not orthogonal to the tangent plane of the master segment.

For the further computations, as a vector $\boldsymbol{\rho}(\xi^1_{i.p.}, \xi^2_{i.p.})$ is given at integration points $\xi^1_{i.p.}, \xi^2_{i.p.}$ of the master segment, the normal $\mathbf{n}(\xi^1_{i.p.}, \xi^2_{i.p.})$ is also computed at the same points.

Remark

Since, the parameterization of the master segment is considered arbitrary (the segment in Fig. 5.4 is shown intentionally curved), the corresponding contact approach works with any kind of approximation for finite elements.

5.2.3.2   Contact with a rigid sphere

The computation of the penetration for contact with a rigid sphere is the most trivial case for *the rigid surface is a "master"* strategy, because the absolute value of a vector, and therefore, a distance in the Cartesian coordinate system is defined in a form of a sphere equation. However, we formally start with the definition of a coordinate system assigned to the spherical surface:

$$\mathbf{r}_s = \mathbf{R}_C + \boldsymbol{\rho}_{sph} + p\,\mathbf{n}_{sph}, \tag{5.23}$$

where $\mathbf{R}_C$ is a center of the sphere, $\boldsymbol{\rho}_{sph}$ is a vector of the sphere with radius R, satisfying $\|\boldsymbol{\rho}_{sph}\| = R$. Since $\boldsymbol{\rho}_{sph}$ is parallel to the unit normal $\mathbf{n}_{sph}$, see Fig. 5.5, we can rewrite eqn. (5.23) as

$$\mathbf{r}_s - \mathbf{R}_C = R\,\mathbf{n}_{sph} + p\,\mathbf{n}_{sph}, \tag{5.24}$$

Taking then the absolute value we obtain

$$\|\mathbf{r}_s - \mathbf{R}_C\| = |R + p|, \tag{5.25}$$

and the distance p as

$$p = \begin{cases} \|\mathbf{r}_s - \mathbf{R}_C\| - R & \text{– for an outward normal} \\ R - \|\mathbf{r}_s - \mathbf{R}_C\| & \text{– for an inward normal} \end{cases} \tag{5.26}$$

The first part in equation (5.26) is describing the simple geometrical fact $AS = CS - CA$ in Fig. 5.5 as a positive distance from the sphere in the outward direction of the sphere. A continuum body is the interior of the sphere in this case. The second equation describes a positive distance in the inward direction of the sphere. In this case the continuum is the exterior of the sphere.

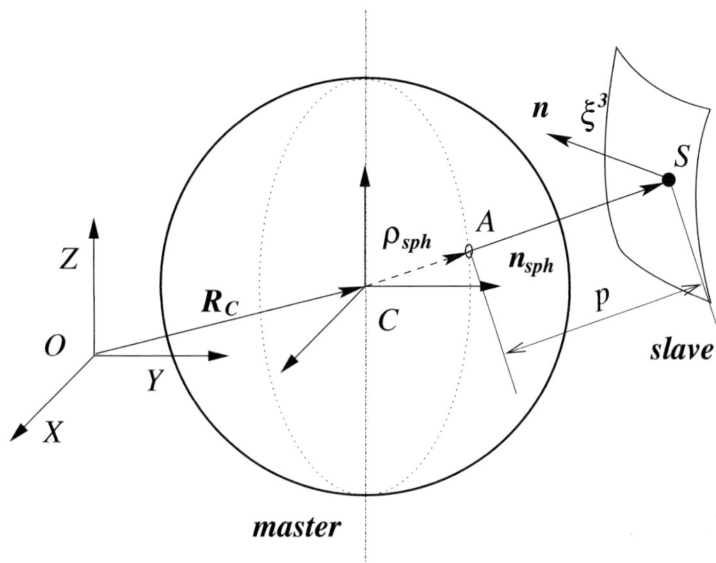

Figure 5.5: Contact with a rigid sphere given by an analytical equation

5.2.3.3 Contact with a rigid cylinder

Another example of the closed form solution within *the Rigid surface is a "master"* strategy is a distance between a cylinder and a point, see Fig. 5.6. The cylinder with a radius R can be given in the following form:

$$\boldsymbol{\rho}(\varphi, z) = \mathbf{R}_C + z\,\mathbf{e}_z + R\,\mathbf{e}_\varphi(\varphi) \qquad (5.27)$$

where $\mathbf{R}_C$ is an arbitrary point on the central axis of the cylinder CC_z, $\mathbf{e}_z$ is a unit vector of the central axis, $\mathbf{e}_\varphi(\varphi)$ is a unit vector in the radial direction of the polar coordinate system in the orthogonal plane, see Fig. 5.6. The unit coordinate vectors are orthogonal $(\mathbf{e}_\varphi(\varphi) \cdot \mathbf{e}_z) = 0$. Noting that the vector $\mathbf{e}_\varphi$ is normal to the cylinder, the slave point S is observed in the coordinate system assigned to the cylindrical surface as:

$$\mathbf{r}_s = \boldsymbol{\rho}(\varphi, z) + p\,\mathbf{e}_\varphi(\varphi), \qquad (5.28)$$

Substituting ρ in eqn. (5.28) from eqn. (5.27) we obtain:

$$\mathbf{R}_C - \mathbf{r}_s + z\,\mathbf{e}_z + (R + p)\,\mathbf{e}_\varphi(\varphi) = 0. \qquad (5.29)$$

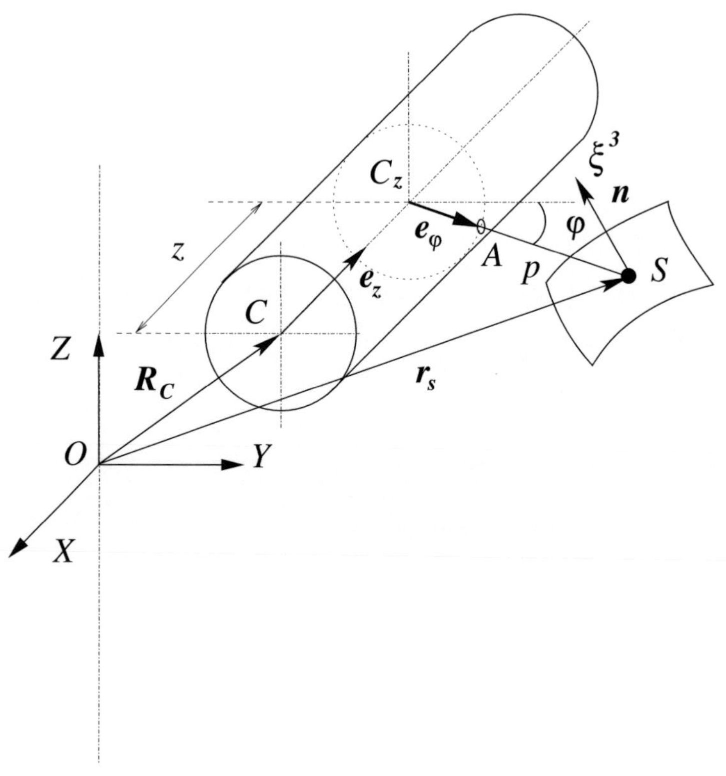

Figure 5.6: Contact with a rigid cylinder given by an analytical equation

First, a coordinate z is defined after taking the scalar product with $\mathbf{e}_z$:

$$z = -(\mathbf{R}_C - \mathbf{r}_s) \cdot \mathbf{e}_z; \tag{5.30}$$

afterwards the absolute value in eqn. (5.29) is taken as

$$\|\mathbf{R}_C - \mathbf{r}_s + z\,\mathbf{e}_z\| = |R + p|, \tag{5.31}$$

then the distance p is defined as

$$p = \begin{cases} \|\mathbf{R}_C - \mathbf{r}_s - \left((\mathbf{R}_C - \mathbf{r}_s) \cdot \mathbf{e}_z\right)\mathbf{e}_z\| - R & \text{– for an outward normal} \\ R - \|\mathbf{R}_C - \mathbf{r}_s - \left((\mathbf{R}_C - \mathbf{r}_s) \cdot \mathbf{e}_z\right)\mathbf{e}_z\| & \text{– for an inward normal} \end{cases} \tag{5.32}$$

The distance p is defined to be positive in both outward and inward direction, see the explanation in Section 5.2.3.2.

5.2.3.4   Contact with a rigid torus

Consider a torus as a result of rotating a circle given in the XOZ-plane along the OZ-axis by increasing the angular coordinate ψ in the XOY-plane, see Fig. 5.7. The torus can be given in the following form:

$$\boldsymbol{\rho}(\varphi, \psi) = R\,\mathbf{e}_R(\psi) + r\,\mathbf{e}_\varphi(\varphi), \tag{5.33}$$

where $\mathbf{e}_R$ and $\mathbf{e}_\varphi$ are radial unit vectors for the polar coordinate system given by two orthogonal planes. First, we define a scalar product $\mathbf{e}_R \cdot \boldsymbol{\rho}$. This product defines a projection of the vector $\overrightarrow{OA} = \boldsymbol{\rho} = (x, y, z)$ on the plane XOY, therefore,

$$\boldsymbol{\rho} \cdot \mathbf{e}_R = \sqrt{x^2 + y^2}. \tag{5.34}$$

This product allows to get rid of $\mathbf{e}_\varphi$ in eqn. (5.33), by taking an absolute value:

$$\|\boldsymbol{\rho} - R\,\mathbf{e}_R\|^2 = \|r\,\mathbf{e}_\varphi\|^2 \tag{5.35}$$

Transformation of the scalar products lead to a torus equation in the following form

$$\|\boldsymbol{\rho}\|^2 + R\,\|\mathbf{e}_R\|^2 - 2R\,(\boldsymbol{\rho} \cdot \mathbf{e}_R) = \|r\,\mathbf{e}_\varphi\|;$$

or written by coordinates:

$$x^2 + y^2 + z^2 + R^2 - 2R\sqrt{x^2 + y^2} = r^2. \tag{5.36}$$

Now we can proceed with the observation of the slave point $\mathbf{r}_s\ =$

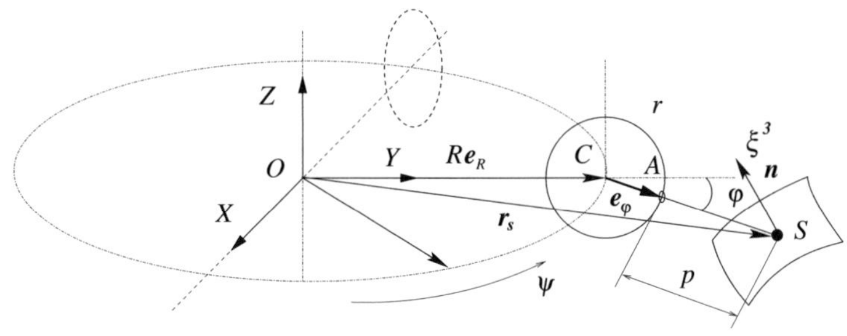

Figure 5.7: Contact with a rigid torus given by an analytical equation

(x_s, y_s, z_s) at the distance p in the toroidal coordinate system, remembering that e_φ is a normal vector to the torus surface

$$\mathbf{r}_s = R\,\mathbf{e}_R(\psi) + (r + p)\,\mathbf{e}_\varphi(\varphi), \qquad (5.37)$$

Similar transformations are leading to the following distance:

$$p = \begin{cases} \sqrt{x_s^2 + y_s^2 + z_s^2 + R^2 - 2R\sqrt{x_s^2 + y_s^2}} - r & \text{– for an outward normal} \\[2mm] r - \sqrt{x_s^2 + y_s^2 + z_s^2 + R^2 - 2R\sqrt{x_s^2 + y_s^2}} & \text{– for an inward normal} \end{cases} \qquad (5.38)$$

The distance p is defined to be positive in both outward and inward directions, see the explanation in Section 5.2.3.2.

5.2.3.5 Contact with a rigid cone

A cone can be considered also within *the rigid surface is a "master"* strategy. The cone, see Fig. 5.8, has the OZ-rotation axis and is defined as:

$$\boldsymbol{\rho}(r, \varphi) = r\,\mathbf{e}_r(\varphi) + r\tan\alpha\,\mathbf{e}_z, \qquad (5.39)$$

where $\mathbf{e}_r$ and $\mathbf{e}_z$ are unit vectors of the cylindrical coordinate system. The outward normal vector $\mathbf{n}$ is simply defined from the geometry of a triangle, see Fig. 5.8.

$$\mathbf{n} = \sin\alpha\,\mathbf{e}_r - \cos\alpha\,\mathbf{e}_z \qquad (5.40)$$

The slave point at the distance p from the cone surface is written first in the cone coordinate system and then in the cylindrical coordinate system as:

$$\begin{aligned} \mathbf{r}_s &= \boldsymbol{\rho}(r, \varphi) + p\mathbf{n}_{cone} \\ &= r\tan\alpha\,\mathbf{e}_z + r\,\mathbf{e}_r + p(\sin\alpha\,\mathbf{e}_r - \cos\alpha\,\mathbf{e}_z) \\ &= (r\tan\alpha - p\cos\alpha)\,\mathbf{e}_z + (r + p\sin\alpha)\,\mathbf{e}_r. \end{aligned} \qquad (5.41)$$

Definition of the distance p is tremendously simplified if we notice that the vectors $\boldsymbol{\rho}$ and $\mathbf{n}$ are orthogonal

$$\boldsymbol{\rho} \cdot \mathbf{n} = 0, \qquad (5.42)$$

then from eqn. (5.41) we have

$$p = (\mathbf{r}_s \cdot \mathbf{n}) = \sin\alpha \, (\mathbf{e}_r \cdot \mathbf{r}_s) - \cos\alpha \, (\mathbf{e}_z \cdot \mathbf{r}_s). \qquad (5.43)$$

It is more convenient to consider $\mathbf{r}_s$ in a cylindrical coordinate system in order to compute the scalar product in eqn. (5.43) as

$$\mathbf{r}_s = \sqrt{x_s^2 + y_s^2} \, \mathbf{e}_r(\varphi) + z_s \, \mathbf{e}_z, \qquad (5.44)$$

then the penetration in eqn. (5.43) becomes:

$$p = \sqrt{x_s^2 + y_s^2} \, \sin\alpha - z_s \, \cos\alpha. \qquad (5.45)$$

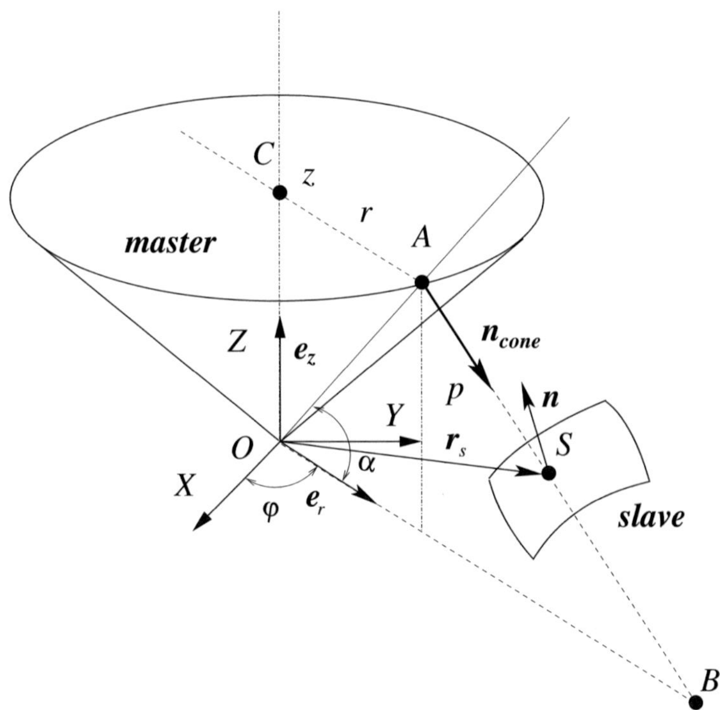

Figure 5.8: Contact with a rigid "master" cone surface

Remark

The rigid surface is a "master" strategy does not give the answer to the solvability of the contact algorithm. This criterion is appearing from the geometry analysis of the contact segment. Thus, considering the contact with a rigid cone described by *the rigid surface is a "slave"* strategy as a surface of revolution in Sect. 5.2.1.1 it is necessary for the solvability that the determinant in eqn. (5.17e) is not zero: $D \neq 0$. Taking $f'(r) = \tan \alpha$, the determinant is transformed as

$$D = \frac{\sin \alpha (n_1 \cos \phi + n_2 \sin \phi) - n_3 \cos \phi}{\cos \alpha}. \tag{5.46}$$

Using then the definition of the normal for the cone in eqn. (5.40), the determinant can be written as

$$D = \frac{\mathbf{n}_\xi \cdot \mathbf{n}_{cone}}{\cos \alpha} \neq 0. \tag{5.47}$$

Now, the criterion of the solvability is clear – a segment normal $\mathbf{n}_\xi$ should not be orthogonal to the cone normal $\mathbf{n}_{cone}$.

5.3 Finite element discretization for different contact approaches

The discretization techniques are considered in the current section. First, the well known "Node-To-Surface" (NTS) contact element is constructed using the covariant approach. Then the discretization for the "Segment-To-Segment" (STS) approach, (bi-linear and bi-quadratic STS elements), combined with various integration techniques (see Sect. 5.1) is considered. A smoothing technique based on a NURB spline approximation, see [38], is used to obtain C^1 continuous contact elements within the STS approach. Further a general discretization strategy is presented for the "Segment-To-Analytical-Surface" (STAS) approach.

5.3.1 Node-To-Surface (NTS) contact approach

The master "surface" is approximated and a "slave" node is taken directly from the finite element mesh in the case of the "Node-To-Surface" approach. A pair *approximation for the master segment and a separated node* leads to the Node-To-Surface contact element. If the approximation of the 'master' surface is defined by n nodes and the "slave" node is the $(n + 1)$'th node, then the nodal vector for the contact element can be written in the following form:

$$\mathbf{m}x^T = \{\mathbf{m}x^{(1)}, \mathbf{m}x^{(2)}, \ldots, \mathbf{m}x^{(n)}, \mathbf{m}x^{(n+1)}\}^T \tag{5.48}$$

$$= \{x_1^{(1)}, x_2^{(1)}, x_3^{(1)}, x_1^{(2)}, x_2^{(2)}, x_3^{(2)}, \ldots, x_1^{(n)}, x_2^{(n)}, x_3^{(n)}, x_1^{(n+1)}, x_2^{(n+1)}, x_3^{(n+1)}\}^T,$$

where $\mathbf{m}x^{(i)}$, $i = 1, 2, \ldots n + 1$ are vectors for nodal points. Let $N_{(k)}(\xi^1, \xi^2)$, $k = 1, 2, \ldots, n$ be the shape functions for the master surface parameterization. We then define the approximation matrix $\mathbf{A}(\xi^1, \xi^2)$ with $3 \times (n + 1)$ dimension as follows

$$\mathbf{A} = \begin{bmatrix} N_1 & 0 & 0 & N_2 & 0 & 0 & \ldots & N_{(n)} & 0 & 0 & -1 & 0 & 0 \\ 0 & N_1 & 0 & 0 & N_2 & 0 & \ldots & 0 & N_{(n)} & 0 & 0 & -1 & 0 \\ 0 & 0 & N_1 & 0 & 0 & N_2 & \ldots & 0 & 0 & N_{(n)} & 0 & 0 & -1 \end{bmatrix}. \tag{5.49}$$

On the basis of this notation, the relative displacement vector $\rho(\xi^1, \xi^2) - \mathbf{r}_s$ has the following form

$$\rho(\xi^1, \xi^2) - \mathbf{r}_s = \mathbf{A}\mathbf{m}x. \tag{5.50}$$

Both the relative variation, and the relative velocity vector have identical structure:

$$\delta\rho - \delta\mathbf{r}_s = \mathbf{A}\delta\mathbf{m}x, \quad \mathbf{v} - \mathbf{v}_s = \mathbf{A}\{\mathbf{v}\}. \tag{5.51}$$

For further developments only the matrix of the shape function derivatives $\mathbf{A}_{,j}$, $i = 1, 2$ has to be specified

$$\frac{\partial \mathbf{A}}{\partial \xi^j} = \mathbf{A}_{,j} = \begin{bmatrix} N_{1,j} & 0 & 0 & N_{2,j} & 0 & 0 & \ldots & N_{(n),j} & 0 & 0 & 0 & 0 & 0 \\ 0 & N_{1,j} & 0 & 0 & N_{2,j} & 0 & \ldots & 0 & N_{(n),j} & 0 & 0 & 0 & 0 \\ 0 & 0 & N_{1,j} & 0 & 0 & N_{2,j} & \ldots & 0 & 0 & N_{(n),j} & 0 & 0 & 0 \end{bmatrix}, \tag{5.52}$$

in order to describe the vectors $\delta\rho_{,j}$ and $\mathbf{v}_{,j}$

$$\delta\rho_{,j} = \mathbf{A}_{,j}\mathbf{m}x, \quad \mathbf{v}_{,j} = \mathbf{A}_{,j}\{\mathbf{v}\}. \tag{5.53}$$

5.3.2 Segment-To-Segment (STS) contact approach

Both "master" and "slave" segments should be approximated separately for a contact pair within the Segment-To-Segment (STS) approach, see Fig. 5.9. The STS contact element is constructed as follows. Since, the integration of all parameters is performed over the slave segment, we have to specify the location of the integration points on the slave element via the selected integration rule. The slave points are then projected onto the master segment. Thus, the contact area is approximated as a set of integration points which have penetrated into the master segment. All necessary geometrical contact parameters, such as the penetration ξ^3, the normal $\mathbf{n}$, the tangent vectors $\rho_1(\xi^1, \xi^2)$, etc. are computed from the master segment.

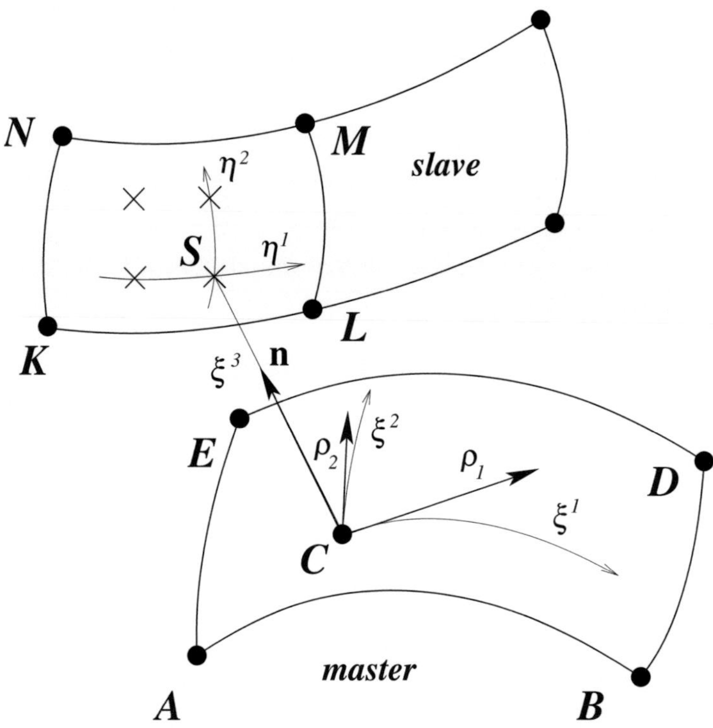

Figure 5.9: Segment-To-Segment (STS) contact approach

Let **ABDE** be a master segment (see Fig. 5.9) defined by n nodes

$$\mathbf{m}x_{master}^T = \{\mathbf{m}x^{(1)}, \mathbf{m}x^{(2)}, \ldots, \mathbf{m}x^{(n)}\}^T \tag{5.54}$$

$$= \{x_1^{(1)}, x_2^{(1)}, x_3^{(1)}, x_1^{(2)}, x_2^{(2)}, x_3^{(2)}, \ldots, x_1^{(n)}, x_2^{(n)}, x_3^{(n)}\}^T,$$

and **KLMN** be a slave segment defined by m nodes

$$\mathbf{m}y_{slave}^T = \{\mathbf{m}y^{(1)}, \mathbf{m}y^{(2)}, \ldots, \mathbf{m}y^{(m)}\}^T \tag{5.55}$$

$$= \{y_1^{(1)}, y_2^{(1)}, y_3^{(1)}, y_1^{(2)}, y_2^{(2)}, y_3^{(2)}, \ldots, y_1^{(m)}, y_2^{(m)}, y_3^{(m)}\}^T.$$

The shape functions $N_k(\xi^1, \xi^2)$, $k = 1, 2, ...n$ and $M_k(\eta^1, \eta^2)$, $k = 1, 2, ...m$ are defined for the "master" and the "slave" segment respectively. A displacement vector for the contact element is defined by $n + m$ nodes as

$$\mathbf{u}^T = \{u_1^{(1)}, u_2^{(1)}, u_3^{(1)}, u_1^{(2)}, u_2^{(2)}, u_3^{(2)}, ..., u_1^{(n)}, u_2^{(n)}, u_3^{(n)}, \tag{5.56}$$
$$u_1^{(n+1)}, u_2^{(n+1)}, u_3^{(n+1)}, u_1^{(n+2)}, u_2^{(n+2)}, u_3^{(n+2)}, ..., u_1^{(n+m)}, u_2^{(n+m)}, u_3^{(n+m)}\}^T.$$

The approximation matrix $\mathbf{A}$ now contains both $N_k(\xi^1, \xi^2)$, and $M_k(\eta^1, \eta^2)$ shape functions and has the dimension $3 \times (n + m)$:

$$\mathbf{A}(\xi^1, \xi^2, \eta^1, \eta^2) = \begin{bmatrix} N_1 & 0 & 0 & N_2 & 0 & 0 & ... & N_{(n)} & 0 & 0 \\ 0 & N_1 & 0 & 0 & N_2 & 0 & ... & 0 & N_{(n)} & 0 \\ 0 & 0 & N_1 & 0 & 0 & N_2 & ... & 0 & 0 & N_{(n)} \end{bmatrix} \tag{5.57}$$

$$\begin{bmatrix} -M_1 & 0 & 0 & -M_2 & 0 & 0 & ... & -M_{(m)} & 0 & 0 \\ 0 & -M_1 & 0 & 0 & -M_2 & 0 & ... & 0 & -M_{(m)} & 0 \\ 0 & 0 & -M_1 & 0 & 0 & -M_2 & ... & 0 & 0 & -M_{(m)} \end{bmatrix}.$$

The slave points are found on the slave segment after the selection of the integration rule, and therefore, are computed as interpolation over the slave nodal points $\mathbf{m}y^{(k)}$, see eqn. (5.55):

$$\mathbf{r}_s(\eta_I^1, \eta_J^2) \,|_{(\eta^1 = \eta_I^1, \, \eta^2 = \eta_J^2)} = \sum_{k=1}^{m} M_k(\eta_I^1, \eta_J^2)\mathbf{m}y^{(k)}, \tag{5.58}$$

where η_I^1, η_J^2 are integration points. Their projections onto the master segment are computed then as

$$\boldsymbol{\rho}(\xi_c^1, \xi_c^2) = \sum_{k=1}^{n} N_k(\xi_c^1, \xi_c^2)\mathbf{m}x^{(k)}, \tag{5.59}$$

where the coordinates ξ_c^1, ξ_c^2 are found as the solution of the closest point projection procedure.

The relative displacement vector is approximated via the matrix $\mathbf{A}$ for the full coordinate vector $\mathbf{m}x = \{\mathbf{m}x_{master}^T, \mathbf{m}y_{slave}^T\}^T$

$$\boldsymbol{\rho}(\xi_c^1, \xi_c^2) - \mathbf{r}_s(\eta_I^1, \eta_J^2) = \mathbf{A}(\xi_c^1, \xi_c^2, \eta_I^1, \eta_J^2)\mathbf{m}x, \tag{5.60}$$

Since the differential operations for all parts of the tangent matrix, see e.g. for non-frictional contact eqns. (3.40), (3.41), (3.42) and (3.43) are defined only on the master surface, the matrix of the shape function derivatives is computed only for the ξ^i derivatives as

$$\frac{\partial \mathbf{A}}{\partial \xi^j} = \mathbf{A}_{,j} = \begin{bmatrix} N_{1,j} & 0 & 0 & N_{2,j} & 0 & 0 & \ldots & N_{(n),j} & 0 & 0 \\ 0 & N_{1,j} & 0 & 0 & N_{2,j} & 0 & \ldots & 0 & N_{(n),j} & 0 \\ 0 & 0 & N_{1,j} & 0 & 0 & N_{2,j} & \ldots & 0 & 0 & N_{(n),j} \end{bmatrix}$$

$$\begin{bmatrix} 0 & 0 & 0 & \ldots & 0 & 0 & 0 \\ 0 & 0 & 0 & \ldots & 0 & 0 & 0 \\ 0 & 0 & 0 & \ldots & 0 & 0 & 0 \end{bmatrix} . \tag{5.61}$$

Based on the matrices defined in eqn. (5.57) and (5.61), the relative vectors for the variation and for the velocity will have the same form as in eqns. (5.51) and (5.53).

Remark:

An integration rule should be specified by the user for the "Segment-To-Segment" contact approach. It should be noted that the Lobatto quadrature formula (see e.g. [33] for a further discussion about numerical integration) with only 2×2 integration points leads exactly to the result of the "Node-To-Surface" element.

5.3.3 Segment-To-Analytical Surface (STAS) contact approach

The STAS approach is used in combination with one of the rules to compute the penetration for the analytically given surfaces, described in Sect. 5.2. Thus, an approximation is given only for the contact segment covering the surface of the meshed deformed body. Consider then the nodal displacement vector with n nodes

$$\mathbf{m}u^T = \{u_1^{(1)}, u_2^{(1)}, u_3^{(1)}, u_1^{(2)}, u_2^{(2)}, u_3^{(2)}, \ldots, u_1^{(n)}, u_2^{(n)}, u_3^{(n)}\}^T, \tag{5.62}$$

which is taken from the finite element discretization of the deformable body.

The approximation matrix $\mathbf{A}$ contains then only shape functions

$N_{(k)}$, $k = 1, 2, \ldots n$:

$$\mathbf{A} = \begin{bmatrix} N_1 & 0 & 0 & N_2 & 0 & 0 & \ldots & \ldots & \ldots & N_{(n)} & 0 & 0 \\ 0 & N_1 & 0 & 0 & N_2 & 0 & \ldots & \ldots & \ldots & 0 & N_{(n)} & 0 \\ 0 & 0 & N_1 & 0 & 0 & N_2 & \ldots & \ldots & \ldots & 0 & 0 & N_{(n)} \end{bmatrix}; \quad (5.63)$$

the matrix of the shape function derivatives $\mathbf{A}_{,j}$, $i = 1, 2$ has the following form

$$\frac{\partial \mathbf{A}}{\partial \xi^j} = \mathbf{A}_{,j} = \begin{bmatrix} N_{1,j} & 0 & 0 & N_{2,j} & 0 & 0 & \ldots & N_{(n),j} & 0 & 0 \\ 0 & N_{1,j} & 0 & 0 & N_{2,j} & 0 & \ldots & 0 & N_{(n),j} & 0 \\ 0 & 0 & N_{1,j} & 0 & 0 & N_{2,j} & \ldots & 0 & 0 & N_{(n),j} \end{bmatrix}. \quad (5.64)$$

All geometrical parameters are computed, of course, based on the segment geometry, see also the discussion in Sect. 5.2.2.

5.4 Various approximations of contact surfaces defined by finite elements

As shown in Sect. 5.3, the approximation matrix $\mathbf{A}$ and its derivatives should be supplied for all contact approaches. A huge variety of approximations is known in finite element practice, starting from the classical results in Zienkiewicz and Taylor [209] extended then to high-order finite element techniques as in Solin et.al. [164] or the isogeometrical finite element approximations via NURBS in Hughes et.al. [72]. We consider here only quadrilateral elements with linear and quadratic approximation and smoothing techniques based on simplified NURB splines for a quadrilateral area.

5.4.1 Quadrilateral segment with linear approximation

The geometry for the quadrilateral segment with linear approximation, see Fig. 5.10, is given by 4 nodes, and therefore, 4 shape functions are necessary to define the segment. Shape functions and their first derivatives for $-1 \leq \xi^1, \xi^2 < +1$ are given in Table 5.2.

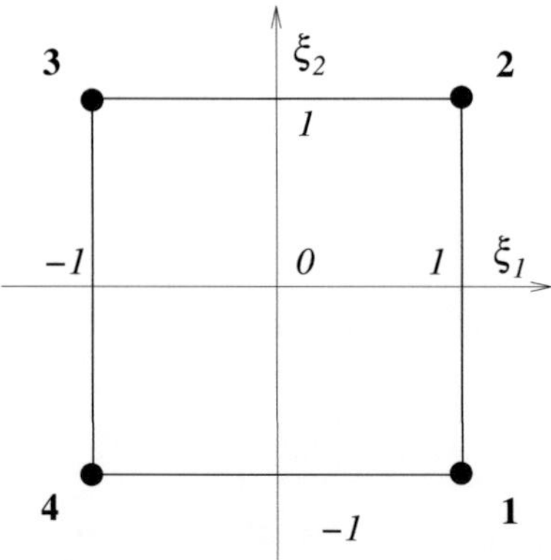

Figure 5.10: Quadrilateral segment with linear approximation

	$N^k(\xi^1,\xi^2)$	$\dfrac{\partial N^k}{\partial \xi^1}$	$\dfrac{\partial N^k}{\partial \xi^2}$
N^1	$\frac{1}{4}(1+\xi^1)(1-\xi^2)$	$\frac{1}{4}(1-\xi^2)$	$-\frac{1}{4}(1+\xi^1)$
N^2	$\frac{1}{4}(1+\xi^1)(1+\xi^2)$	$\frac{1}{4}(1+\xi^2)$	$-\frac{1}{4}(1+\xi^1)$
N^3	$\frac{1}{4}(1-\xi^1)(1+\xi^2)$	$-\frac{1}{4}(1+\xi^2)$	$\frac{1}{4}(1+\xi^1)$
N^4	$\frac{1}{4}(1-\xi^1)(1-\xi^2)$	$-\frac{1}{4}(1-\xi^2)$	$-\frac{1}{4}(1-\xi^1)$

Table 5.2: Shape functions for a quadrilateral segment with linear approximation

5.4.2 Quadrilateral segment with quadratic Lagrangian approximation

The geometry for the quadrilateral segment with quadratic Lagrangian approximation, see Fig. 5.11, is given by 9 nodes. The corresponding 9 shape functions and their first derivatives for $-1 \leq \xi^1, \xi^2 < +1$ are given in Table 5.3

5.4.3 Surface smoothing techniques in a covariant approach

Since linear finite elements have been mostly used for FE models, it was soon recognized that faceted geometries are causing problems

	$N^k(\xi^1, \xi^2)$	$\dfrac{\partial N^k}{\partial \xi^1}$	$\dfrac{\partial N^k}{\partial \xi^2}$
N^1	$\xi^1(\xi^1 - 1)\xi^2(\xi^2 - 1)/4$	$(2\xi^1 - 1)\xi^2(\xi^2 - 1)/4$	$\xi^1(\xi^1 - 1)(2\xi^2 - 1)/4$
N^2	$\xi^1(\xi^1 + 1)\xi^2(\xi^2 - 1)/4$	$(2\xi^1 + 1)\xi^2(\xi^2 - 1)/4$	$\xi^1(\xi^1 + 1)(2\xi^2 - 1)/4$
N^3	$\xi^1(\xi^1 + 1)\xi^2(\xi^2 + 1)/4$	$(2\xi^1 + 1)\xi^2(\xi^2 + 1)/4$	$\xi^1(\xi^1 + 1)(2\xi^2 + 1)/4$
N^4	$\xi^1(\xi^1 - 1)\xi^2(\xi^2 + 1)/4$	$(2\xi^1 - 1)\xi^2(\xi^2 + 1)/4$	$\xi^1(\xi^1 - 1)(2\xi^2 + 1)/4$
N^5	$(1 - \xi^1\xi^1)\xi^2(\xi^2 - 1)/2$	$-\xi^1\xi^2(\xi^2 - 1)$	$(1 - \xi^1\xi^1)(2\xi^2 - 1)/2$
N^6	$\xi^1(\xi^1 + 1)(1 - \xi^2\xi^2)/2$	$(2\xi^1 + 1)(1 - \xi^2\xi^2)/2$	$-\xi^1(\xi^1 + 1)\xi^2$
N^7	$(1 - \xi^1\xi^1)\xi^2(\xi^2 + 1)/2$	$-\xi^1\xi^2(\xi^2 + 1)$	$(1 - \xi^1\xi^1)(2\xi^2 + 1)/2$
N^8	$\xi^1(\xi^1 - 1)(1 - \xi^2\xi^2)/2$	$(2\xi^1 - 1)(1 - \xi^2\xi^2)/2$	$-\xi^1(\xi^1 - 1)\xi^2$
N^9	$(1 - \xi^1\xi^1)(1 - \xi^2\xi^2)$	$-2\xi^1(1 - \xi^2\xi^2)$	$-2(1 - \xi^1\xi^1)\xi^2$

Table 5.3: Shape functions for a quadrilateral segment with quadratic Lagrangian approximation

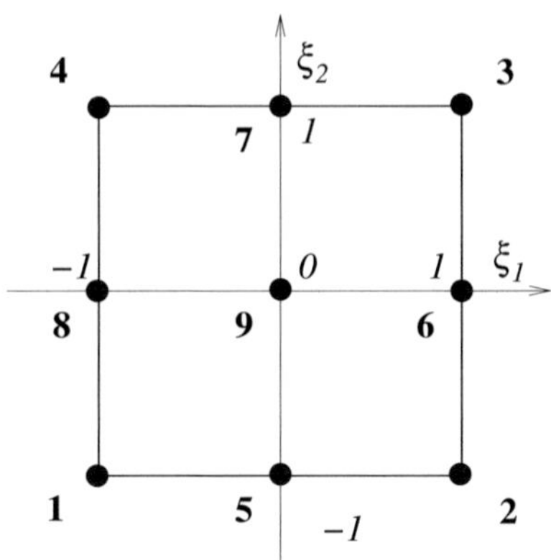

Figure 5.11: Quadrilateral segment with quadratic Lagrangian approximation

within the contact algorithm due to "jumping" normals on contact boundaries. Thus, the idea of smoothing for linear contact surfaces appeared. The first publications have been focusing on smoothing for rigid obstacles with splines, see Schweizerhof and Hallquist [155] (1992), Shimizu and Sano [159] (1995)), Heege and Alart [65] (1996) for forming problems. Pietrzak and Curnier [144] (1998) discussed a problem resulting from the non-smoothness of the contact surfaces with the example of two coaxial cylinders and used different splines for covering the linear meshes. The major difficulties using a smooth approximation are the algorithm for the computation of sliding forces and the linearization procedure, especially for the 3D case. Thus, a special approach based on a so-called moving friction cone description for the computation of sliding forces in combination with various spline smoothing techniques have been proposed for 2D cases in Wriggers at.el. [190] (2001) and in Krstulovic-Opara and Wriggers [101] (2002), then generalized for 3D cases in Krstulovic-Opara et.al. [102] (2002). The difficulties in the linearization have then been resolved for numerical examples by using an automatic differentiation software.

The effect of C^2 and C^1 smoothing for frictionless 2D problems was investigated in Stadler et.al. [169] (2003) and a subdivision scheme for smooth contact surfaces, describing a procedure to sew different meshes (various quadrilateral patches, triangular + quadrilateral) has been considered in Holzapfel and Korelc [168] (2004). Again an automatic differentiation software has been used in both publications.

An approach with convective coordinates together with different approximations including splines and Hermite interpolations for the contact surface were used to enforce C^1 continuity in the 2D case in Padmanabhan and Laursen [134] (2001). A smoothing procedure for the 3D case, based on the usage of Gregory patches was suggested in Puso and Laursen [148] (2002). The problem with the linearization for curvilinear surfaces has been overcome by taking the Mortar approach for the linear geometry. The application of a control spline polygon for the searching procedure has been discussed in El-Abbasi et.al. [37] (2001) and in Al-Dojayli and Meguid [2] (2002)

Since, in this thesis the linearization problems are resolved by the covariant derivation, we are free now to choose any kind of approximations. Here, Hermitian approximation as a simple subclass of the NURB spline approximation is taken.

However, a very short information about spline and NURBS surfaces is necessary to describe even the simple structure of the smooth element. Advanced approximation techniques are described in special monographs for CAD surfaces, see e.g. Farin [38], Piegl and Tiller [143]; they can be applied within the covariant description without any limitation.

5.4.3.1   Spline interpolation of curves

We start with the construction of splines for curves and consider the interpolaton problem with different splines.

Bernstein Polynomials   are forming a basis function space to construct polynomial splines and can be generalized into **N**on-**U**niform **R**ational **B**-**S**pline (**NURBS**). The Bernstein Polynomials are defined as follows:

$$B_i^n(t) = C_i^n t^i (1-t)^{n-i}, \quad i = 0, 1, ..., n, \tag{5.65}$$

where $C_i^n = \dfrac{n(n-1)...(n-i+1)}{i!}$ are binomial coefficients. The first three polynomial sets up to the third order contain the following polynomials:

$$
\begin{array}{lll}
B_0^0(t) = 1, & B_0^1(t) = 1-t, & B_1^1(t) = t; \\[4pt]
B_0^2(t) = (1-t)^2, & B_1^2(t) = 2t(1-t), & B_2^2(t) = t^2; \\[4pt]
B_0^3(t) = (1-t)^3, & B_1^3(t) = 3t(1-t)^2, & B_2^3(t) = 3t^2(1-t), \quad B_3^3(t) = t^3.
\end{array}
\tag{5.66}
$$

The first problem of spline interpolation   is specified as a construction of a spline which is passing through selected points. Let $\mathbf{x}_0, \mathbf{x}_1, ..., \mathbf{x}_n \in R^3$ be given points with parameters $u_0, u_1, ..., u_n$.
Construct a curve $\mathbf{r}(u)$ which satisfies the following conditions:

$$\mathbf{r}(u_k) = \mathbf{x}_k, \quad k \in \{0, 1, ...n\}. \tag{5.67}$$

The second problem of spline interpolation   is specified as a construction of a spline line which is passing through selected points and is possessing certain tangent vectors. Let $\mathbf{x}_0, \mathbf{x}_1, ..., \mathbf{x}_n \in R^3$ be given points with parameters $u_0, u_1, ..., u_n$ and $\mathbf{m}_0, \mathbf{m}_1, ..., \mathbf{m}_n \in R^3$ be given tangent vector

with the same parameters.
Construct a curve $\mathbf{r}(u)$ which satisfies the following conditions:

$$\mathbf{r}(u_k) = \mathbf{x}_k, \quad \left.\frac{d\mathbf{r}(u)}{du}\right|_{u=u_k} = \mathbf{m}_k, \quad k \in \{0, 1, ...n\}. \tag{5.68}$$

The solution of the second problem allows to obtain then the C^1-continuous spline for the first problem.

Parameterization of the spline curve. A global spline should be parameterized based on the coordinates of the interpolation points. One of the simplest parameterization is to use the distance between points as a parameter (a chord length parameterization):

$$u_0 = 0, \ u_1 = ||\mathbf{x}_1 - \mathbf{x}_0||, \ ... \ , \ u_i = u_{i-1} + ||\mathbf{x}_i - \mathbf{x}_{i-1}||, \ i = 2, ..., n. \tag{5.69}$$

The second interpolation problem – solution in the form of piecewise cubic C^1 interpolation. We consider only piecewise interpolations, because this case is more suitable for further finite element construction. In the first problem, interpolation points together with their tangent vectors are known. Both the continuity of tangent vectors (C^1 continuity), and the local support are satisfied. The solution can be constructed in the form of a Bezier curve, see Farin [38]. The global Bezier curve consists of local Bezier curves of third order; it passes through the points $\mathbf{x}_0, \mathbf{x}_1, ..., \mathbf{x}_n$ $\in R^3$ and satisfies C^1 continuity being a Bezier curve of third order on each interval $(\mathbf{x}_i, \mathbf{x}_{i+1}), \ i \in \{0, 1, ..., n-1\}$. It should be mentioned that this problem does not have a unique solution. As an additional constraint equation, the normalization of the tangent vector can be taken

$$||\mathbf{m}_i|| = 1, \quad i \in \{0, 1, ..., n\}. \tag{5.70}$$

Boundary Bezier points for each interval $(\mathbf{x}_i, \mathbf{x}_{i+1})$ are defined from the interpolation condition:

$$\mathbf{b}_{3i} = \mathbf{x}_i, \quad i = 0, 1, ..., n; \tag{5.71}$$

the inner Bezier points are defined as

$$\mathbf{b}_{3i+1} = \mathbf{b}_{3i} + \alpha_i \mathbf{m}_i, \quad \mathbf{b}_{3i-1} = \mathbf{b}_{3i} - \beta_i \mathbf{m}_i. \tag{5.72}$$

Several choices for α_i and β_i values are discussed in Farin [38], recommended ones are

$$\alpha_i = \beta_i = 0.4||\Delta\mathbf{x}_i||. \tag{5.73}$$

The global Bezier curve is then defined on the interval between points $\mathbf{x}_i$ and $\mathbf{x}_{i+1}$ by local Bezier curves:

$$\begin{aligned}
\mathbf{x}(u) &= \mathbf{b}_{3i}B_0^3(t) + \mathbf{b}_{3i+1}B_1^3(t) + \mathbf{b}_{3i+2}B_2^3(t) + \mathbf{b}_{3i+3}B_3^3(t) = \\
&= \sum_{j=0}^{3} \mathbf{b}_{3i+j}B_j^3(t);
\end{aligned} \tag{5.74}$$

where the local variable $t \in [0,1]$ is given via the global parameter u, see in eqn. (5.69):

$$t = (u - u_i)/\Delta_i, \quad \Delta_i = u_{i+1} - u_i, \quad i = 0, 1, ..., n-1. \tag{5.75}$$

The first interpolation problem – solution in the form of piecewise cubic C^1 interpolation. The solution of the first problem as piecewise cubic C^1 interpolation in Bezier form is, in general, not unique – all tangent vectors $\mathbf{m}_i$ have to be defined to achieve uniqueness. One of the possible approximations of the tangent vectors is a vector describing a chord through the points $\mathbf{x}_{i-1}$ and $\mathbf{x}_{i+1}$, i.e.

$$\mathbf{m}_i = \mathbf{x}_{i-1} - \mathbf{x}_{i+1}, \tag{5.76}$$

thus leading to the inner Bezier points in the form:

$$\begin{aligned}
\mathbf{b}_{3i-1} &= \mathbf{b}_{3i} - \frac{\Delta_{i-1}}{3(\Delta_{i-1} + \Delta_i)}\mathbf{m}_i \\
\mathbf{b}_{3i+1} &= \mathbf{b}_{3i} + \frac{\Delta_i}{3(\Delta_{i-1} + \Delta_i)}\mathbf{m}_i
\end{aligned} \tag{5.77}$$

This method is known as the Catmull-Rom spline.

This construction of the spline fails at boundary points $\mathbf{x}_0$ and $\mathbf{x}_n$. Several methods are known to overcome this problem. The idea behind the Bessel method is to find an interpolating parabola $\mathbf{q}(u)$ through the starting points $\mathbf{x}_0, \mathbf{x}_1, \mathbf{x}_2$ (or through the ending points $\mathbf{x}_0, \mathbf{x}_1, \mathbf{x}_2$) and to take the spline tangent direction as the tangent direction for this parabola:

$$\mathbf{m}_0 = \frac{d\mathbf{q}}{du} \qquad (5.78)$$

Another method (Overhauser spline) is to construct a special cubic interpolant in the form:

$$\mathbf{x}(u) = \frac{u_{i+1} - u}{\Delta_i}\mathbf{q}_i(u) + \frac{u - u_i}{\Delta_i}\mathbf{q}_{i+1}(u), \quad i = 1, ..., n - 2 \qquad (5.79)$$

where $\mathbf{q}_i(u)$ is the parabola passing through three points, then at the beginning points one sets $\mathbf{x}(u_0) = \mathbf{q}_0(0)$.

Piecewise cubic C^1 interpolation in Hermite form. The Hermite form of the spline is one of the most straightforward implementations into the finite element method. One can show that the Hermite interpolation is simply another form of the Bezier interpolation. Consider, as example, the first spline problem. The boundary Bezier points are the interpolation points. In order to obtain the inner Bezier points, the derivative formula for Bezier curves has to be used:

$$\frac{d}{du}\mathbf{x}(u_i) = \frac{3}{\Delta_{i-1}}(\mathbf{b}_{3i} - \mathbf{b}_{3i-1}) = \frac{3}{\Delta_i}(\mathbf{b}_{3i+1} - \mathbf{b}_{3i}) = \mathbf{m}_i \qquad (5.80)$$

Thus, the inner Bezier points are given as

$$\mathbf{b}_{3i+1} = \mathbf{b}_{3i} + \frac{\Delta_i}{3}\mathbf{m}_i, \quad i = 0, 1, ..., n - 1$$

$$\mathbf{b}_{3i-1} = \mathbf{b}_{3i} - \frac{\Delta_{i-1}}{3}\mathbf{m}_i, \quad i = 0, 1, ..., n \qquad (5.81)$$

The spline in Bezier form can be written after some algebra in the form:

$$\mathbf{x}(u) = \mathbf{x}_i H_0^3(u) + \mathbf{m}_i H_1^3(u) + \mathbf{m}_{i+1} H_2^3(u) + \mathbf{x}_{i+1} H_3^3(u), \qquad (5.82)$$

where $H_i^3(u)$ are Hermite polynomials. They can be expressed in terms of Bernstein polynomials as follows:

$$
\begin{aligned}
H_0^3(u) &= B_0^3(t) + B_1^3(t) \\
H_1^3(u) &= \frac{\Delta_i}{3} B_1^3(t) \\
H_2^3(u) &= -\frac{\Delta_i}{3} B_2^3(t) \\
H_3^3(u) &= B_2^3(t) + B_3^3(t)
\end{aligned}
\tag{5.83}
$$

where $t = (u - u_i)/\Delta_i \in [0,1]$ is the local parameter of the interval $[u_i, u_{i+1}]$.

Equations (5.82) defines the relationship between the Hermite and the Bezier form of the spline. If the Hermite form is known, then the inner Bezier points can be calculated and vice versa.

Piecewise cubic C^2 interpolation in Bezier form. The requirement of C^2-continuity leads to a linear system of equations with a unique solution, however, the local support property is lost then which makes this scheme inconvenient in FEM. Another strategy of keeping the C^2-continuity is to use full NURBS definitions.

Rational B-spline – NURBS. Rational B-splines are one of the standard curve and surface descriptions in CAD and computer graphics systems. A generalization of the polynomial curve is a rational curve with order n of Bernstein's polynomials. The Bezier form of it can be written as

$$
\mathbf{x}(u) = \frac{w_0 \mathbf{b}_0 B_0^n(t) + w_1 \mathbf{b}_1 B_1^n(t) + ... w_n \mathbf{b}_n B_n^n(t)}{w_0 B_0^n(t) + w_1 B_1^n(t) + ... w_n B_n^n(t)}.
\tag{5.84}
$$

This description of a spline has many advantages due to the additional degrees of freedom because of additional weights w_i. It leads to multiple possibilities to control the geometry. Rational B-splines constructed in this way are called **N**on-**U**niform **R**ational **B**-**S**pline or **NURBS**. Special literature, however, is available containing all NURBS properties in detail, see e.g. Farin [38], Piegl and Tiller [143].

5.4.3.2 Spline surfaces interpolation

The theory outlined for the spline curve interpolation is directly applied for the surface interpolation problem. However, the situation with surfaces is far more complicated: e.g. it is necessary to distinguish a regular or an irregular net of interpolation points, rules for mixed derivatives should be additionally developed etc. All these problems are forming a huge area of research, and we will take only the simplest structures out of it.

Our main assumption is

- Interpolating points are defining a regular quadrilateral surface mesh in a Cartesian coordinate system.

Thus, we directly can use the tensor product of Bezier splines, considered in Sect. 5.4.3.1.

Composite surfaces consist of patches. A patch is a part of the global surface and has its own parameterization. This property is called a `local support property`. **The first interpolation problem** is formulated then for surfaces as follows:

Let the matrix

$$\begin{bmatrix} \mathbf{x}_{00} & \mathbf{x}_{01} & \cdots & \mathbf{x}_{0m} \\ \mathbf{x}_{10} & \mathbf{x}_{11} & \cdots & \mathbf{x}_{1m} \\ \cdots & \cdots & \cdots & \cdots \\ \mathbf{x}_{n0} & \mathbf{x}_{n1} & \cdots & \mathbf{x}_{nm} \end{bmatrix} \tag{5.85}$$

be a given as a regular net of points $\mathbf{x}_{ij} \in R^3$ with parameters $u_0, u_1, ..., u_n$ and $v_0, v_1, ..., v_m$. These points are vertices of the quadrilateral mesh. Construct the smooth C^1 surface parameterization $\mathbf{x}(u, v)$ that satisfies the following condition:

$$\mathbf{x}(u_i, v_j) = \mathbf{x}_{ij}, \quad i = 0, 1, ..., n; \; j = 0, 1, ..., m \tag{5.86}$$

The interpolation problem can then be solved with patches constructed as tensor products of $1D$ bicubic Bezier curves. However, additional equations for the twist estimation $\mathbf{x}_{uv}$ on the vertices are necessary. This leads to an additional degree of freedom, and therefore to different models of the composite surface.

A bicubic Hermite patch   is constructed as a tensor product of C^1-splines given in Hermite form. The surface interpolation vector $\mathbf{x}(u, v)$ gets then the following form:

$$\mathbf{x}(u, v) = \sum_{i=0}^{3} \sum_{j=0}^{3} \mathbf{h}_{ij} H_i^3(u) H_j^3(v); \quad 0 \leq u, v \leq 1, \tag{5.87}$$

where $H_i^3(u)$ are the cubic Hermite functions from eqn. (5.83) and the terms $\mathbf{h}_{i,j}$ are defining the following matrix

$$[\mathbf{h}_{ij}] = \begin{bmatrix} \mathbf{x}(0,0) & \mathbf{x}_v(0,0) & \mathbf{x}_v(0,1) & \mathbf{x}(0,1) \\ \mathbf{x}_u(0,0) & \mathbf{x}_{uv}(0,0) & \mathbf{x}_{uv}(0,1) & \mathbf{x}_u(0,1) \\ \mathbf{x}_u(1,0) & \mathbf{x}_{uv}(1,0) & \mathbf{x}_{uv}(1,1) & \mathbf{x}_u(1,1) \\ \mathbf{x}(1,0) & \mathbf{x}_v(1,0) & \mathbf{x}_v(1,1) & \mathbf{x}(1,1) \end{bmatrix} \tag{5.88}$$

Additional information for the mixed derivative $\mathbf{x}_{uv}$ at vertices is required leading to `the so-called twist estimation problem`. Various methods include then the Coons patch, the Adini twist, the Bessel twist, the Gregory patch and others, see Farin [38]. In our numerical examples we further use the simplest **Zero twist** just setting all mixed derivatives to zero at the vertices

$$\mathbf{x}_{uv}(u_i, v_j) = 0. \tag{5.89}$$

Finally, after a lengthy introduction into spline interpolation theory we are ready to construct a simple smooth 3D contact element.

5.4.3.3   A quadrilateral smooth contact surface

The idea of surface smoothing discussed in computational contact mechanics is to cover the given linear surface mesh describing a possible contact surface with a smooth spline surface. This is a mesh dependent task for a certain type of spline. Thus, for simplicity we consider here only a quadrilateral area with a regular quadrilateral mesh, see Fig. 5.12. This area can be covered by a composite spline surface constructed via regular patches as discussed in Sect. 5.4.3.2. Three different types of patches are necessary for the quadrilateral surface: **a corner patch** for four corners – No. 1-4 in Fig. 5.12, **a boundary patch** for boundaries – No. 5-8, and **a central patch** for the inner region – No. 9.

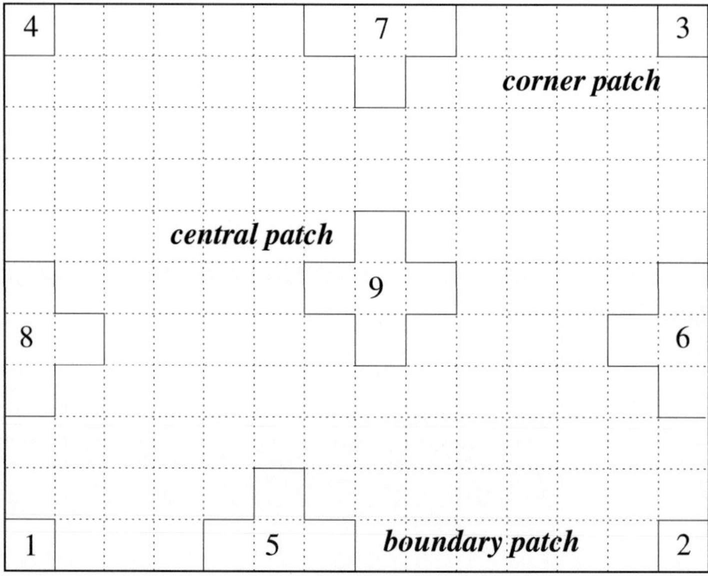

Figure 5.12: A quadrilateral regularly meshed surface. 3 types of patches are necessary for a smooth contact surface: central, boundary and corner patches.

Following the technique introduced in Sect. 5.4.3.2, the structure of the matrix $[\mathbf{h}_{ij}]$ in eqn. (5.88) should be defined depending on the position of the nodal vector. Assuming that nodal points from the regular net in Fig. 5.12 are given in the form $\mathbf{x}_{ij} \in R^3$ $i = 0, 1, ..., n$; $j = 0, 1, ..., m$ we have to introduce parameters $u_0, u_1, ..., u_n$ and $v_0, v_1, ..., v_m$ in order to obtain the parameterization in the form of eqn. (5.86). This parameterization with u_i, v_j is constructed as a tensor product using the chord length for lines in eqn. (5.69). Start with the line $\mathbf{x}_{i0}, i = 0, 1, ..., n$ and choose parameters $u_0, u_1, ..., u_n$

$$u_0 = 0, \quad u_1 = ||\mathbf{x}_{10} - \mathbf{x}_{00}||, \quad ... \quad , u_i = u_{i-1} + ||\mathbf{x}_{i0} - \mathbf{x}_{i-10}||, \quad i = 2, ..., n; \tag{5.90}$$

then, parameters $v_0, v_1, ..., v_n$ are chosen with the line $\mathbf{x}_{0j}, j = 0, 1, ..., m$ in the form

$$v_0 = 0, \quad v_1 = ||\mathbf{x}_{01} - \mathbf{x}_{02}||, \quad ... \quad , v_j = v_{j-1} + ||\mathbf{x}_{0j} - \mathbf{x}_{0j-1}||, \quad i = 2, ..., m. \tag{5.91}$$

If the cells of the mesh have approximately the same size, it is even better to choose the simplest parameterization:

$$\begin{cases} u_i = i, & i = 0, \ldots, n; \\ v_j = j, & j = 1, \ldots, m. \end{cases} \qquad (5.92)$$

The central patch is defined by 12 nodal points and has the pattern of nodal points presented in Fig. 5.13. For simplicity, zero twist $\mathbf{x}_{uv} = 0$ is taken further for the finite elements. Taking into account the simplest

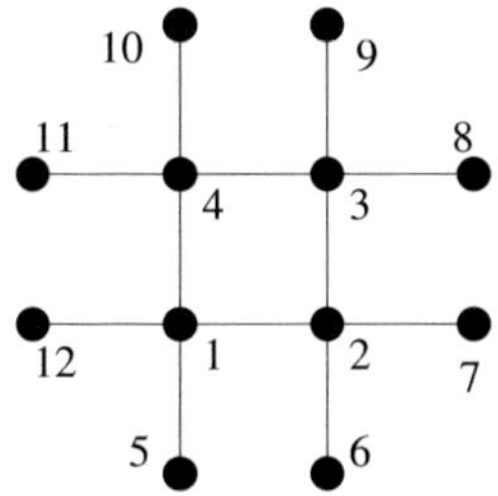

Figure 5.13: Central patch – a pattern of nodal points.

parameterization in eqn. (5.92) *(in this case the computation rule for the first partial derivatives is coinciding with the finite difference scheme)*, the matrix $[\mathbf{h}_{ij}]$ has the following form:

$$[\mathbf{h}_{ij}] = \begin{bmatrix} \mathbf{x}_1 & \frac{1}{2}(\mathbf{x}_4 - \mathbf{x}_5) & \frac{1}{2}(\mathbf{x}_{10} - \mathbf{x}_1) & \mathbf{x}_4 \\ \frac{1}{2}(\mathbf{x}_2 - \mathbf{x}_{12}) & 0 & 0 & \frac{1}{2}(\mathbf{x}_3 - \mathbf{x}_{11}) \\ \frac{1}{2}(\mathbf{x}_7 - \mathbf{x}_1) & 0 & 0 & \frac{1}{2}(\mathbf{x}_8 - \mathbf{x}_4) \\ \mathbf{x}_2 & \frac{1}{2}(\mathbf{x}_3 - \mathbf{x}_6) & \frac{1}{2}(\mathbf{x}_9 - \mathbf{x}_2) & \mathbf{x}_3 \end{bmatrix}. \qquad (5.93)$$

Using the formulae for the Hermite patch, the tensor form can be rewritten in the vector form as follows:

$$\mathbf{x}(u, v) = \sum_{i=0}^{3} \sum_{j=0}^{3} \mathbf{h}_{i,j} H_i^3(u) H_j^3(v) = \sum_{i=1}^{12} N^i(u, v) \mathbf{x}_i, \quad 0 \le u, v \le 1. \qquad (5.94)$$

This vector form is used with the shape functions $N^i(u, v)$ and corresponds to a nodal vector defined by nodes $\mathbf{x}_i$. This form is standard for the finite element method. Thus, the approximation of the vector is given as

$$
\mathbf{x}(u,v) \;=\; \sum_{i=1}^{12} N^i(u,v)\mathbf{x}_i =
\tag{5.95}
$$

$$
\begin{aligned}
=\; & \mathbf{x}_1\Big[H_1(u)H_1(v) - \frac{1}{2}H_1(u)H_3(v) - \frac{1}{2}H_3(u)H_1(v)\Big] \\
& + \mathbf{x}_2\Big[\frac{1}{2}H_2(u)H_1(v) + H_4(u)H_1(v) - \frac{1}{2}H_4(u)H_3(v)\Big] \\
& + \mathbf{x}_3\Big[H_4(u)H_4(v) + \frac{1}{2}H_4(u)H_2(v) + \frac{1}{2}H_2(u)H_4(v)\Big] \\
& + \mathbf{x}_4\Big[\frac{1}{2}H_1(u)H_2(v) + H_1(u)H_4(v) - \frac{1}{2}H_3(u)H_4(v)\Big] \\
& + \mathbf{x}_5\frac{1}{2}\big(-H_1(u)H_2(v)\big) + \mathbf{x}_6\frac{1}{2}\big(-H_4(u)H_2(v)\big) \\
& + \mathbf{x}_7\frac{1}{2}H_3(u)H_1(v) + \mathbf{x}_8\frac{1}{2}H_3(u)H_4(v) \\
& + \mathbf{x}_9\frac{1}{2}H_4(u)H_3(v)) + \mathbf{x}_{10}\frac{1}{2}H_1(u)H_3(v)) \\
& + \mathbf{x}_{11}\frac{1}{2}\big(-H_2(u)H_4(v)\big) + \mathbf{x}_{12}\frac{1}{2}\big(-H_2(u)H_1(v)\big).
\end{aligned}
$$

The corner patch is defined by 8 nodes and has the pattern of nodal points presented in Fig. 5.14. Now, the rules discussed in Sect. 5.4.3.1

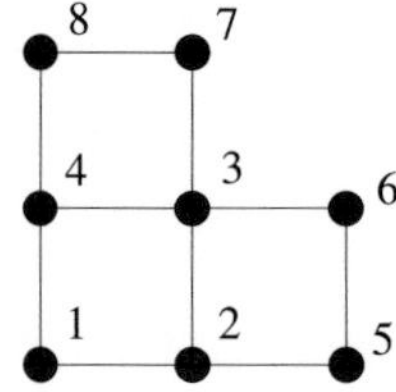

Figure 5.14: Corner patch – a pattern of nodal points.

eqn. (5.78) for boundary points for curves are used to compute the derivatives for the boundary points for a surface spline. Derivatives for the inner points are computed similar to the central patch. The following approximations coinciding with the finite difference scheme of order

$\mathcal{O}(h^2)$ is obtained:

$$
\begin{aligned}
\mathbf{x}_u(0,0) &= -\frac{3}{2}\mathbf{x}_1 + 2\mathbf{x}_2 - \frac{1}{2}\mathbf{x}_5 & \mathbf{x}_v(0,0) &= -\frac{3}{2}\mathbf{x}_1 + 2\mathbf{x}_4 - \frac{1}{2}\mathbf{x}_8 \\
\mathbf{x}_u(0,1) &= -\frac{3}{2}\mathbf{x}_4 + 2\mathbf{x}_3 - \frac{1}{2}\mathbf{x}_6 & \mathbf{x}_v(0,1) &= -\frac{1}{2}\mathbf{x}_1 + \frac{1}{2}\mathbf{x}_8 \\
\mathbf{x}_u(1,0) &= \frac{1}{2}\mathbf{x}_5 - \frac{1}{2}\mathbf{x}_1 & \mathbf{x}_v(1,0) &= -\frac{3}{2}\mathbf{x}_2 + 2\mathbf{x}_3 - \frac{1}{2}\mathbf{x}_7 \\
\mathbf{x}_u(1,1) &= \frac{1}{2}\mathbf{x}_6 - \frac{1}{2}\mathbf{x}_4 & \mathbf{x}_v(1,1) &= -\frac{1}{2}\mathbf{x}_2 - \frac{1}{2}\mathbf{x}_7.
\end{aligned}
\tag{5.96}
$$

Then the matrix $[\mathbf{h}_{ij}]$ gets the following form:

$$
[\mathbf{h}_{ij}] =
\begin{bmatrix}
\mathbf{x}_1 & -\frac{3}{2}\mathbf{x}_1 + 2\mathbf{x}_4 - \frac{1}{2}\mathbf{x}_8 & -\frac{1}{2}\mathbf{x}_1 + \frac{1}{2}\mathbf{x}_8 & \mathbf{x}_4 \\
-\frac{3}{2}\mathbf{x}_1 + 2\mathbf{x}_2 - \frac{1}{2}\mathbf{x}_5 & 0 & 0 & -\frac{3}{2}\mathbf{x}_4 + 2\mathbf{x}_3 - \frac{1}{2}\mathbf{x}_6 \\
\frac{1}{2}\mathbf{x}_5 - \frac{1}{2}\mathbf{x}_1 & 0 & 0 & \frac{1}{2}\mathbf{x}_6 - \frac{1}{2}\mathbf{x}_4 \\
\mathbf{x}_2 & -\frac{3}{2}\mathbf{x}_2 + 2\mathbf{x}_3 - \frac{1}{2}\mathbf{x}_7 & -\frac{1}{2}\mathbf{x}_2 - \frac{1}{2}\mathbf{x}_7 & \mathbf{x}_3
\end{bmatrix}.
\tag{5.97}
$$

The vector form of the approximation of the surface vector $\mathbf{x}(u,v)$ is given as

$$
\mathbf{x}(u,v) = \tag{5.98}
$$

$$
\begin{aligned}
& \mathbf{x}_1 \left[H_1(u)H_1(v) - \frac{3}{2}H_1(u)H_2(v) - \frac{1}{2}H_1(u)H_3(v) \right. \\
& \qquad \left. - \frac{3}{2}H_2(u)H_1(v) - \frac{1}{2}H_3(u)H_1(v) \right] \\
+ \; & \mathbf{x}_2 \left[2H_2(u)H_1(v) + H_4(u)H_1(v) - \frac{3}{2}H_4(u)H_2(v) - \frac{1}{2}H_4(u)H_3(v) \right] \\
+ \; & \mathbf{x}_3 \left[2H_2(u)H_4(v) + 2H_4(u)H_2(v) + H_4(u)H_4(v) \right] + \\
+ \; & \mathbf{x}_4 \left[2H_1(u)H_2(v) + H_1(u)H_4(v) - \frac{3}{2}H_2(u)H_4(v) - \frac{1}{2}H_3(u)H_4(v) \right] \\
+ \; & \mathbf{x}_5 \frac{1}{2} \left[-H_2(u)H_1(v) + H_3(u)H_1(v) \right] \\
+ \; & \mathbf{x}_6 \frac{1}{2} \left[-H_2(u)H_4(v) + H_3(u)H_4(v) \right]
\end{aligned}
$$

$$+ \quad \mathbf{x}_7 \frac{1}{2} \left[-H_4(u)H_2(v) + H_4(u)H_3(v) \right]$$

$$+ \quad \mathbf{x}_8 \frac{1}{2} \left[-H_1(u)H_2(v) + H_1(u)H_3(v)) \right]$$

The boundary patch is defined by 10 nodes and has the pattern of nodal points presented in Fig. 5.15. The partial derivatives for boundary points

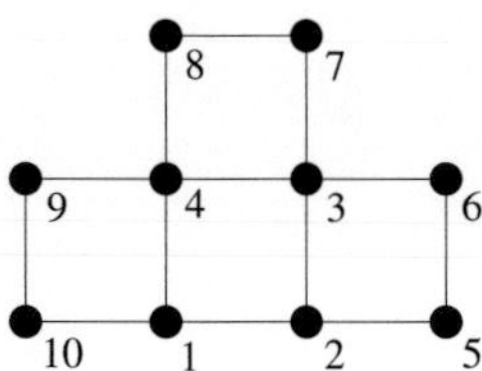

Figure 5.15: Boundary patch – a pattern of nodal points.

are computed as

$$
\begin{aligned}
\mathbf{x}_u(0,0) &= \frac{1}{2}\mathbf{x}_2 - \frac{1}{2}\mathbf{x}_{10} & \mathbf{x}_v(0,0) &= -\frac{3}{2}\mathbf{x}_1 + 2\mathbf{x}_4 - \frac{1}{2}\mathbf{x}_8 \\
\mathbf{x}_u(0,1) &= \frac{1}{2}\mathbf{x}_3 - \frac{1}{2}\mathbf{x}_9 & \mathbf{x}_v(0,1) &= \frac{1}{2}\mathbf{x}_8 - \frac{1}{2}\mathbf{x}_1 \\
\mathbf{x}_u(1,0) &= \frac{1}{2}\mathbf{x}_5 - \frac{1}{2}\mathbf{x}_1 & \mathbf{x}_v(1,0) &= -\frac{3}{2}\mathbf{x}_2 + 2\mathbf{x}_3 - \frac{1}{2}\mathbf{x}_7 \\
\mathbf{x}_u(1,1) &= \frac{1}{2}\mathbf{x}_6 - \frac{1}{2}\mathbf{x}_4 & \mathbf{x}_v(1,1) &= \frac{1}{2}\mathbf{x}_7 - \frac{1}{2}\mathbf{x}_2.
\end{aligned}
\tag{5.99}
$$

The matrix $[\mathbf{h}_{ij}]$ has the following form:

$$
[\mathbf{h}_{ij}] =
\begin{bmatrix}
\mathbf{x}_1 & -\frac{3}{2}\mathbf{x}_1 + 2\mathbf{x}_4 - \frac{1}{2}\mathbf{x}_8 & \frac{1}{2}\mathbf{x}_8 - \frac{1}{2}\mathbf{x}_1 & \mathbf{x}_4 \\
\frac{1}{2}\mathbf{x}_2 - \frac{1}{2}\mathbf{x}_{10} & 0 & 0 & \frac{1}{2}\mathbf{x}_3 - \frac{1}{2}\mathbf{x}_9 \\
\frac{1}{2}\mathbf{x}_5 - \frac{1}{2}\mathbf{x}_1 & 0 & 0 & \frac{1}{2}\mathbf{x}_6 - \frac{1}{2}\mathbf{x}_4 \\
\mathbf{x}_2 & -\frac{3}{2}\mathbf{x}_2 + 2\mathbf{x}_3 - \frac{1}{2}\mathbf{x}_7 & \frac{1}{2}\mathbf{x}_7 - \frac{1}{2}\mathbf{x}_2 & \mathbf{x}_3
\end{bmatrix}
\tag{5.100}
$$

The vector form approximation of the boundary patch vector $\mathbf{x}(u,v)$ is given as

$$
\begin{aligned}
\mathbf{x}(u,v) \;=\; & \mathbf{x}_1 \left[H_1(u)H_1(v) - \frac{3}{2}H_1(u)H_2(v) - \frac{1}{2}H_1(u)H_3(v) - \frac{1}{2}H_3(u)H_1(v) \right] \\[2mm]
+ \;& \mathbf{x}_2 \left[\frac{1}{2}H_2(u)H_1(v) + H_4(u)H_1(v) - \frac{3}{2}H_4(u)H_2(v) - \frac{1}{2}H_4(u)H_3(v) \right] \\[2mm]
+ \;& \mathbf{x}_3 \left[\frac{1}{2}H_2(u)H_4(v) + 2H_4(u)H_2(v) + H_4(u)H_4(v) \right] \\[2mm]
+ \;& \mathbf{x}_4 \left[2H_1(u)H_2(v) - \frac{1}{2}H_3(u)H_4(v) + H_1(u)H_4(v) \right] \\[2mm]
+ \;& \mathbf{x}_5 \frac{1}{2}H_3(u)H_1(v) + \mathbf{x}_6 \frac{1}{2}H_3(u)H_4(v) \\[2mm]
+ \;& \mathbf{x}_7 \frac{1}{2}(-H_4(u)H_2(v) + H_4(u)H_3(v)) \\[2mm]
+ \;& \mathbf{x}_8 \frac{1}{2}(-H_1(u)H_2(v) + H_1(u)H_3(v)) \\[2mm]
- \;& \mathbf{x}_9 \frac{1}{2}H_2(u)H_4(v) - \mathbf{x}_{10} \frac{1}{2}H_2(u)H_1(v).
\end{aligned}
\tag{5.101}
$$

Writing all necessary shape functions in vector form together with their derivatives is an absolutely tedious task. During implementation it is organized via subroutines containing the Hermite functions $H_1(t), H_2(t), H_3(t), H_4(t)$ used for all shape functions. These functions are just consequently renumbered after their definition in eqn. (5.83). The functions and their derivatives are presented in Table 13.1.

5.5 Numerical examples

In the current section various computational aspects and arising problems are shown discussing some numerical examples. Integration schemes using subdivision into subdomains developed in Sect. 5.1.2 are applied together with STS and STAS contact approaches with various order of approximations for finite elements. Thus, STS approach with various numerical schemes is applied for the standard patch test with linear approximations, then the STS approach with smooth contact surfaces is studied. The influence of the integration scheme on the

	$H_k(t)$	$\dfrac{\partial H_k}{\partial t}$	$\dfrac{\partial^2 H_k}{\partial t^2}$
$H_1(t)$	$1 - 3t^2 + 2t^3$	$-6t + 6t^2$	$-6 + 12t$
$H_2(t)$	$t - 2t^2 + t^3$	$1 - 4t + 3t^2$	$-4 + 6t$
$H_3(t)$	$t^3 - t^2$	$3t^2 - 2t$	$6t - 2$
$H_4(t)$	$3t^2 - 2t^3$	$6t - 6t^2$	$6 - 12t$

Table 5.4: Hermite functions $H_i(t)$, $t \in [0, 1]$ and their derivatives are forming a shape function space for smooth contact elements.

force-displacement curve is presented for a deep drawing example – various approximations of solid shell elements are discussed. An example with deep drawing of a plate into a pot is chosen to illustrate the STAS algorithm with various combinations of rigid surfaces. A special example with a deep drawing of a strip is devoted to the selection of the approximation order for shell elements and contact approach in order to obtain the correct *thickness strain vs. loading displacement* curve.

5.5.1 Classical contact patch test - linear approximations

One of the important problems for contact algorithm is the ability to transfer the stresses correctly through the contact surface. This problem leads to the patch test. The contact patch test serves to check the ability to transfer an uniform stress state through the contact surface. Different techniques were proposed to pass the patch test. Taylor and Papadopoulos [173] proposed the two-pass algorithm based on interchanging the master and slave parts to pass the patch test in the case of a linear approximation for 2D problems. Zavarise and Wriggers [199] proposed the integration over overlapping regions in the 2D case in order to more accurately treat the contact conditions. Crisfield [31] considered contact elements with higher order approximation to satisfy a patch test in the 2D case. Also in the 2D case the integration of the contact integral over the overlapping zone, which is constructed by projection was investigated in El-Abbasi and Bathe [36] to satisfy the patch test. Jones and Papadopoulos [80] considered a special pressure interpolation in the overlapping region of the two contact elements to pass the patch test in 3D. Heinstein and Laursen [66] developed an algorithm based on the construction of a special 3D element in the overlapping region for mesh-matching problems as well as to pass the patch test.

Here we consider the application of the Segment-To-Segment contact approach with various integration schemes to the modified patch test problem, originally proposed in Crisfield [31] for the 2D patch test. The upper block with the dimensions $1 \times 1 \times 0.5$ is meshed with a regular rectangular $3 \times 3 \times 2$ mesh. The lower block has the same geometry as the upper block, and a finer, but distorted $6 \times 6 \times 2$ mesh is used, see Fig. 5.16. Both blocks are made of elastic material with the following parameters: Young's modulus $E = 1.0 \cdot 10^5$, Poisson ratio $\nu = 0.3$. The value of the penalty is chosen as $\varepsilon = 1.0 \cdot 10^7$. During contact the upper block is considered as a slave. An uniform vertical displacement of $\Delta = 0.05$ is applied on the top surface.

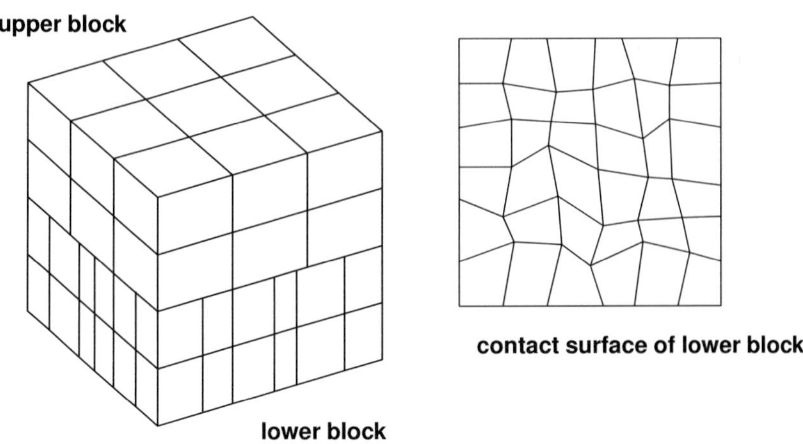

Figure 5.16: Blocks for the patch test; Upper block – regular mesh; Lower block – distorted mesh

The integration algorithm based on integration of subdomains is an approximate approach to integrate discontinuous functions. Here we show, that with this technique it is possible to construct a sequence of

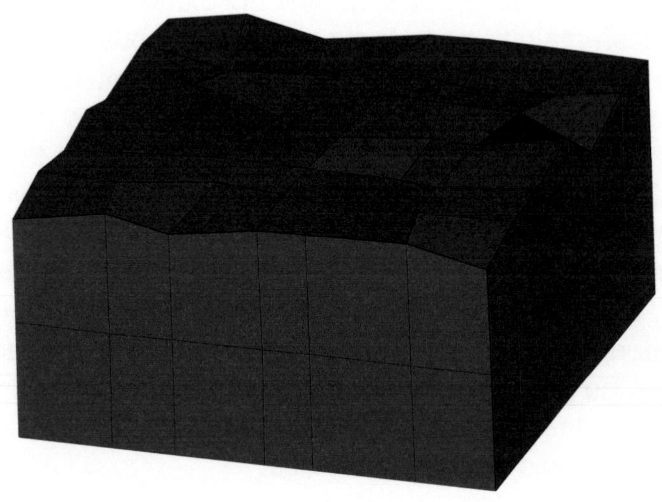

Figure 5.17: Node-To-Surface (NTS) approach fails to pass the patch test; "roof-like" contact surface

results with diminishing error to finally satisfy the patch test. In order to investigate in the case of uniform stresses the normal contact traction N, the normal stress σ_z and the vertical nodal displacement u_z of the contact surface of the lower block are checked. Their values are controlled by parameters used in statistics: by the mean value $\bar{x}$ and by the standard deviation σ and the coefficient of variation $C_v = 100\% \cdot \sigma/\bar{x}$ in order to estimate the variation. In Table 5.5 the results concerning the mean value and the coefficient of variation for the following quantities are given: sum of contact tractions $N = \varepsilon_p g_N$ over the surface, computed at Gauss points of the contact surface; normal stresses σ_z, computed for the upper and lower surface of each element of the lower block; nodal vertical displacements u_z for the contact surface. For comparison, the first computation was made for the node-to-surface approach with a of penalty value of 10^5 which was reduced due to the convergence problems. As is known, this approach fails the patch test. Fig. 5.17 shows the "roof-like" contact surface of the lower block for this case. As expected this approach leads to the maximum of the coefficient of variation.

From Table 5.5 it becomes clear that also the integration with subdomains leads to – though only slightly – smaller variations than an algorithm with standard Gauss integration. It appears rather remarkable that the variations of the tractions remain constant while the variation of the stresses and displacements falls below one percent. Visually, the application of the STS approach together with higher order of integration rules, and integration with subdomains leads then **to a flat surface**, and therefore, disappearing of **the "roof-like" structure**.

No. Gpt	No. sbd.	tractions N		stress σ_z		displ. u_z	
		$\bar{x} \cdot 10^4$	$v\ \%$	$\bar{x} \cdot 10^3$	$v\ \%$	$\bar{x} \cdot 10^{-2}$	$v\ \%$
NTS^*	1	-1.3084	-68.087	-1.5248	-120.039	-0.9150	-157.93
2	1	-1.7767	-21.474	-4.6787	-2.5772	-2.4286	-7.7430
6	1	-1.6738	-19.851	-4.7077	-1.1663	-2.4143	-1.5571
3	2	-1.6210	-16.760	-4.7137	-0.8570	-2.4180	-1.2770
2	3	-1.6354	-17.331	-4.7115	-0.8789	-2.4170	-1.4131
10	1	-1.6614	-19.696	-4.7109	-0.8595	-2.4150	-1.3121
2	5	-1.6477	-15.793	-4.7113	-0.7798	-2.4166	-1.2097
5	2	-1.6408	-17.572	-4.7139	-0.7942	-2.4160	-1.2383
20	1	-1.6537	-19.226	-4.7124	-0.7791	-2.4164	-1.1597
10	2	-1.6408	-16.667	-4.7128	-0.7585	-2.4159	-1.1304
4	5	-1.6299	-16.790	-4.7141	-0.7136	-2.4158	-1.0822
5	4	-1.6447	-16.395	-4.7125	-0.7239	-2.4154	-1.1190
2	10	-1.6337	-16.341	-4.7126	-0.7578	-2.4170	-1.1335

Table 5.5: Influence of different integration schemes; patch test; mean value and coefficient of variation for the following quantities: contact tractions N on the contact surface, normal stresses σ_z in the lower block and vertical nodal displacements u_z on the contact surface; NTS^* - node-to-surface approach

5.5.2 Contact patch test with smooth surfaces

As we have seen the STS approach either with increasing the number of integration points, or with combining integration with a subdomain technique allows to improve the patch test result. This visually leads to **the flattening of the roof-like surfaces** in the case of linear approximations. This example is illustrating the situation for the smooth surface.

For the patch test two elastic blocks are taken, see Fig. 5.18. The lower block consists of nine regular elements with linear approximation. The upper surface is covered with a smooth contact surface – four corner patches, four boundary patches and one central patch, as discussed in Sect. 5.4.3.3. The upper block is only one element with linear approximation. The size of the upper element is chosen such that the nodes are projected in the middle of corner elements for the lower block, see Fig. 5.18. The upper surface of the lower block is a master surface, and the lower surface of the upper block is a slave one. As in the previous example, the vertical displacements are applied to the four upper nodes of the upper block.

From the meshed geometry and loading it is obvious that the NTS contact approach with linear approximation will completely fail the patch test leading to *"the roof-like"* surface, see Fig. 5.19 a). For the second test, the surface has been covered with smooth patch, but the NTS algorithm has been applied – the result is shown in Fig. 5.19 b) – the contact surface is smooth, but it is erroneously pointed upward. Only a combination of STS approach (here with 3×3 Gauss integration points) with a smooth contact surface leads to the correct deformed surface positioned downward, see Fig. 5.19 c).

5.5.3 Free bending of a metal sheet on two cylinders

A second example is the free bending problem of a metal sheet (thickness t). Geometry and deformed configuration are shown in Fig. 5.20 for the bilinear and biquadratic elements. The material is taken to be elasto-plastic of Henky's type with isotropic plastic hardening with the following data: Bulk modulus $\kappa = 1.75 \cdot 10^4 \frac{kN}{cm^2}$; Shear modulus $\mu = 8.077 \cdot 10^3 \frac{kN}{cm^2}$; $\sigma_i 16.0 \frac{kN}{cm^2}$; $\sigma_\infty = 40.0 \frac{kN}{cm^2}$; $\delta = 20.0$; $H = 20.0$.

At the beginning the metal sheet with thickness $t = 0.25\ cm$ is positioned on two cylindrical rigid bodies. As loading a displacement u is prescribed in the center of the sheet. Due to symmetry only one half of the system has to be modeled and discretized using 12 bilinear resp. 6 biquadratic elements and a rather fine mesh with 100 bilinear resp. 50 biquadratic elements. Concerning the application of different contact approaches the following variations were investigated:

Figure 5.18: Geometry for the patch test. The upper surface of the lower block is covered with a smooth spline surface.

1) The rigid cylinder is modeled by linear finite elements with 48 elements in the circumferential direction. The metal sheet is treated as "master" part, the cylinder as "slave" part. Contact is modeled by the "Node-To-Surface" (NTS) approach.

2) The model is the same as in 1, but contact is modeled by the "Segment-To-Segment" (STS) approach.

3) The surface of the rigid cylinder is described analytically (STAS approach), the metal sheet is then the "master" part. The contact integral is computed by a nodal collocation formula with additional gap interpolation over the element surface.

4) The model is the same as in 3. The quadrature formula is chosen either with different numbers of integration Gauss points, or with subdomains on the "master" part.

5.5.3.1 Case with bilinear solid-shell elements

In Fig. 5.21 the results for the global central reaction force for case 1, case 2 with 2×2 Gauss points and case 3 each with a mesh of 12 bilinear

Failure of the patch test for the NTS contact approach with linear approximation

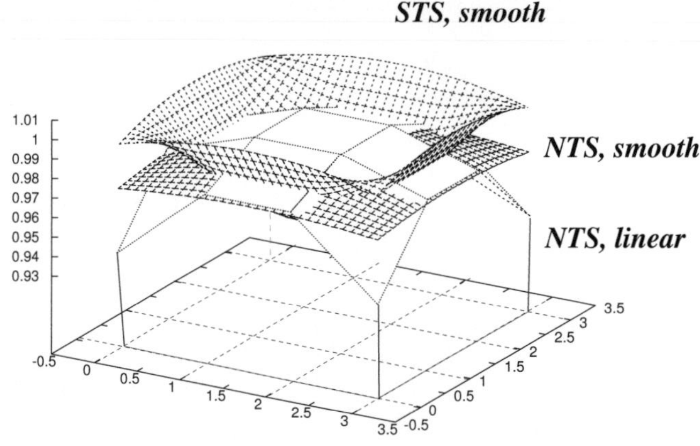

Figure 5.19: Upper surface for three computations: a) NTS with linear approximation (see above the whole block); b) NTS with smooth spline approximation; c) STS with smooth spline approximation of the master contact surface. Vertical displacements are scaled.

solid-shell elements are plotted, see the structure of solid-shell elements in [57]. The result with the fine mesh of 100 bilinear solid-shell elements of elements with one integration point for contact evaluation is taken for comparison as "exact" solution. The nodal collocation formula of case 3 shows the largest oscillations, because the value of the penetration is checked only at the nodes of the sheet and the mesh is relatively coarse in comparison with the geometrical size of the cylinder despite the an-

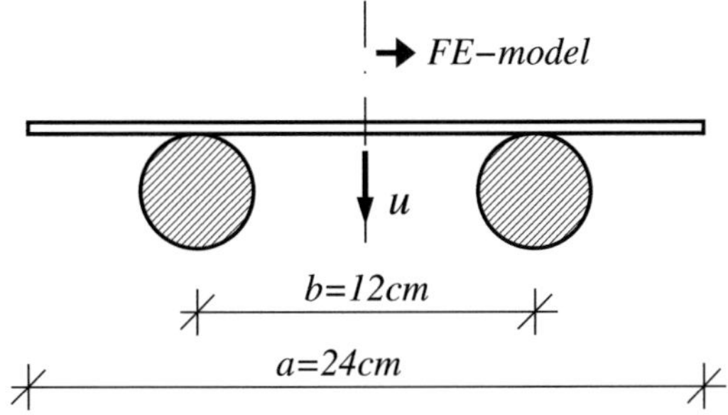

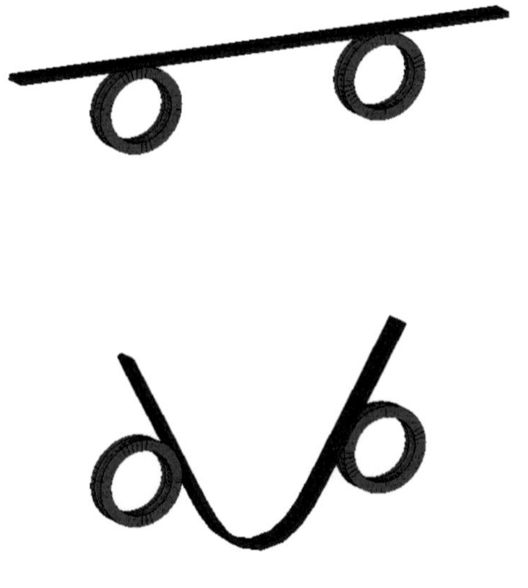

197

Figure 5.20: Geometry and loading process for a metall sheet on two rigid cylinders. Case with biquadratic elements is shown.

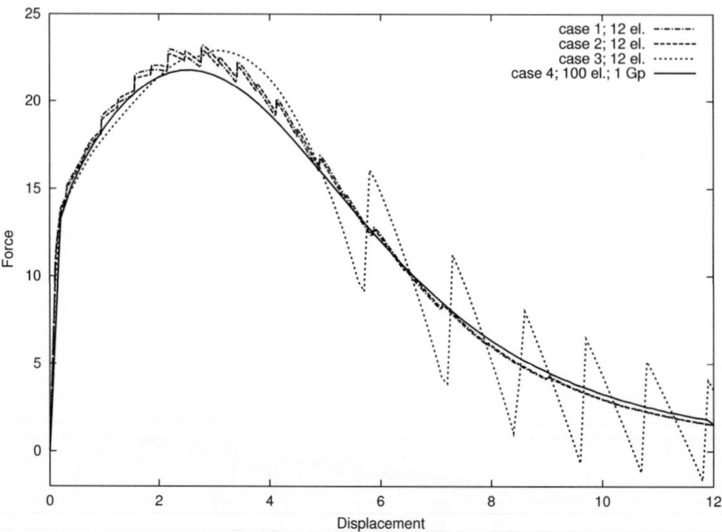

Figure 5.21: Force-deflection curves for free bending problem; bilinear solid-shell elements; comparing quadrature formulae of low order and mesh refinement

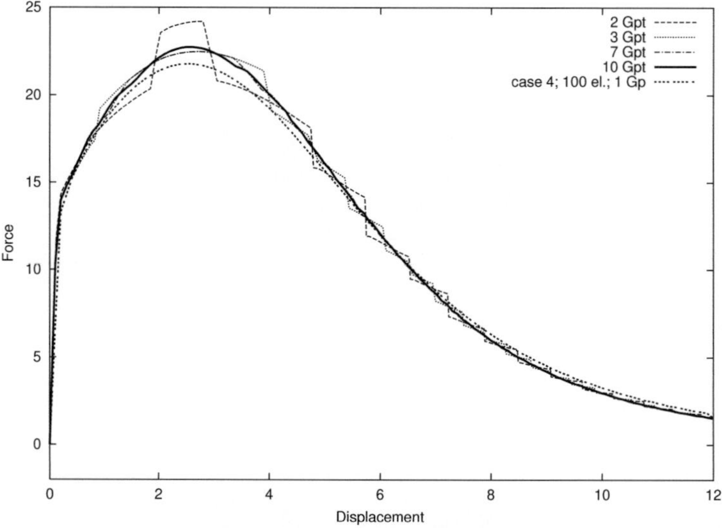

Figure 5.22: Force-deflection curves for free bending problem; bilinear solid-shell elements; contact against analytically defined contact surface; influence of the number of Gauss points

alytical description of the cylinder. If contact is checked at the nodes of the mesh of the rigid cylinder, in case 1 of the "Node-To-Surface" approach, jumps appear with smaller amplitudes. This is an obvious consequence of the finer mesh for the cylinder in comparison with the sheet mesh. As is well known, this would improve with a finer mesh on both sides. Using a Gauss quadrature for the "Segment-To-Segment" strategy leads to only slightly reduced jumps, because in both "Node-To-Surface" and "Segment-To-Segment" approaches the rigid cylinder is modeled still with a rather coarse finite element mesh. In order to investigate the influence of the order of the Gauss integration for the STAS approach when the rigid cylinder is given exactly by an analytical function, case 4 was extended with 2×2, 3×3, 7×7 and 10×10 integration points and, finally, compared to the results with a refined mesh of 100 elements, but with 1 Gauss point only, see Fig. 5.22. Obviously, the quadrature formulae with 2×2 integration points leads to rather large oscillations. This is due to the fact that rather non-smooth contact check-

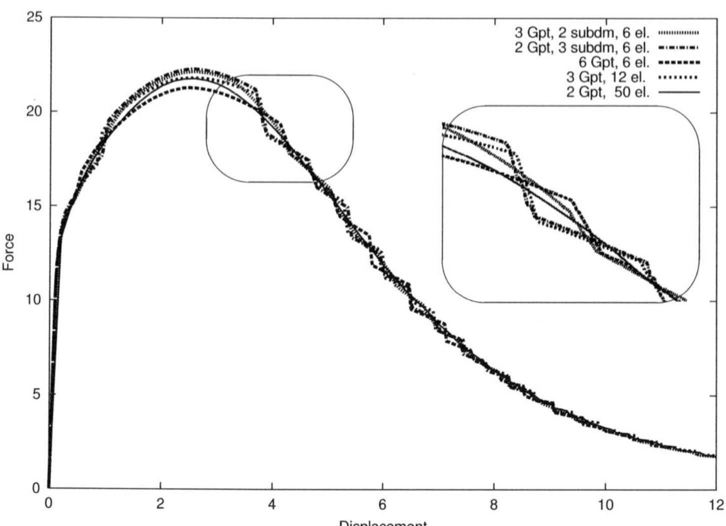

Figure 5.23: Force-deflection curves for free bending problem; biquadratic solid-shell elements; Influence of the number of Gauss points and refined mesh

ing is performed, which can be improved using more contact points. As mentioned earlier, checking contact at Gauss points can be interpreted as integration of a discontinuous function for which no a-priori error estimation is avaliable. One can see that a convergent sequence of curves is achieved if the number of integration points is increased even in the case of a coarse mesh for the sheet. The influence of the mesh refinement with a softer response of the sheet is also obvious.

5.5.3.2 Case with biquadratic solid-shell elements*

The next step is to consider the influence of the number of Gauss points for the sheet meshed with biquadratic elements. Fig. 5.23 shows the

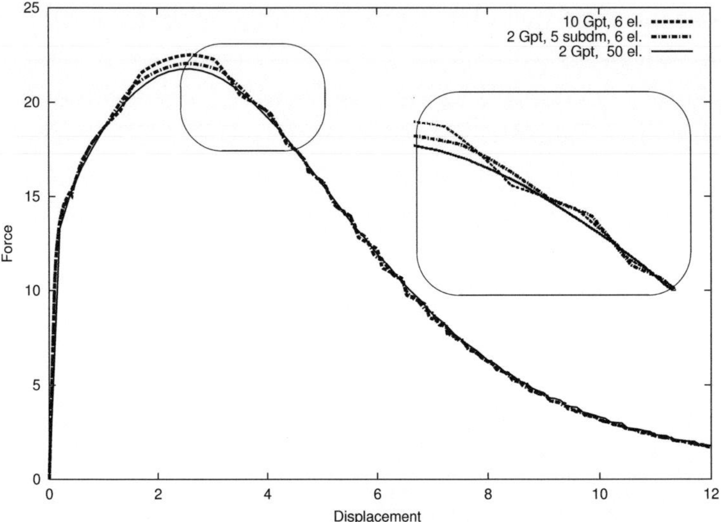

Figure 5.24: Force-deflection curves for free bending problem; biquadratic solid-shell elements; subdivision of the integration domain into subdomains results in smaller oscillations

result of the computation for the beam meshed with 6 elements, if the integration formula with 6×6 integration points is used and if as an alternative the integration formula with 2 subdomains and 3×3 integration points in each and with 3 subdomains and 2×2 integration points is used. The result is compared with a refined mesh of 12 elements for the beam meshed with 12 elements with 3×3 integration points. For comparison a 50 element mesh and 2×2 integration points per element

---

*The material has been reported at [88]: A. Konyukhov, K. Schweizerhof. *Application of a covariant description to the contact of shells with different approximation*, 5th IASS-IACM,Conference Proceedings, Salzburg, Austria, 2005.

is chosen. The density of the Gauss points to check the penetration is identical in the first three cases, but as a consequence of the smallest a-priori integration error for the algorithm with subdomains, the latter leads to a reduction of the oscillations. In Fig. 5.24 it is shown that even a relatively large number of 10×10 Gauss points per element still preserves oscillations. The integration with 5 subdomains and 2 Gauss points for the 6 element mesh leads again to a solution with a smaller deviation.

5.5.3.3 Frictional case with bilinear solid-shell elements

A comparison with the frictional case for the following friction coefficients: $\mu = 0.1$, $\mu = 0.2$, $\mu = 0.3$ is presented in Fig. 5.25, where the computation was performed with the "best" 10×10 integration formula. All contact pairs with rigid surfaces are modeled with the STAS approach. As an obvious result, the reaction force is increasing following the modification of the friction coefficient.

   Remark

Here the STAS approach has been applied successfully for the C^1-continuous rigid surfaces. If the rigid boundary is not smooth, or another approach is favorable (e.g. boundaries are not rigid surfaces and STS is necessary), then the full history transfer algorithm is absolutely necessary to resolve the frictional contact problem, see the study of the example *"drawing of an elastic strip into a channel with sharp corners"* in Sect. 6.7.2.

5.5.4 Deep drawing of a cylindrical pot – combination of STAS contact elements[*]

In this numerical example the deep drawing process of a circular plate into a cylindrical pot with counter die is simulated, see the geometry in Fig. 5.26. The material of a circular plate is elasto-plastic with the following parameters: $\kappa = 1.75 \cdot 10^4 \ kN/cm^2$; $\mu = 8077 \ kN/cm^2$; $\sigma_i = 16 \ kN/cm^2$. The circular sheet has a uniform thickness $t = 1 \ mm$ and a diameter $D = 16.0 \ cm$. The geometry of the tools is shown in Fig. 5.26. Due to symmetry only a quarter of the structure is discretized using

---

[*]The material has been reported at [87]: A. Konyukhov, K. Schweizerhof. *Large Deformation Frictional Contact Formulation for Low Order "Solid Shell" Elements*, ECCOMAS-2004, Jyväskylä.

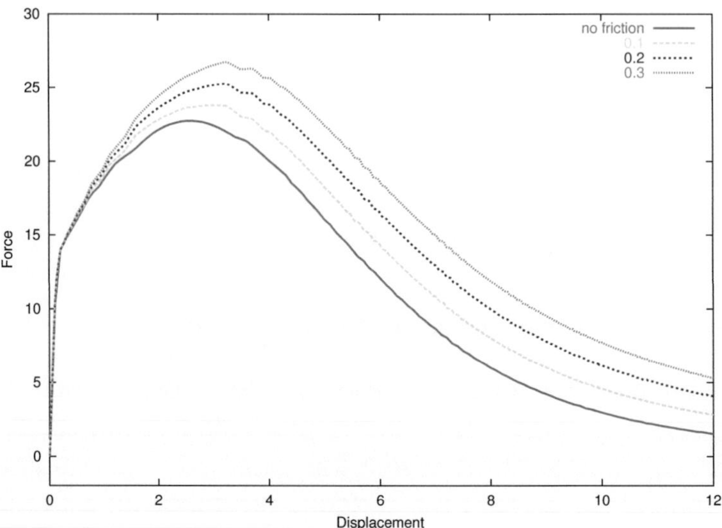

Figure 5.25: Force-deflection curves for free bending problem; influence of the choice for the friction coefficient; 10×10 Gauss points

a mesh with 175 bilinear elements. As loading the displacement u is applied incrementally with $\Delta u = 0.0025$ for the punch, as well as for the counter die, however, for the latter after the punch is contacting the counter die.

The contact is defined with rigid surfaces using the STAS approach. The example is illustrating the ability of the STAS approach to model a rigid tool via a combination of planes, cylinders and toruses. Only the non-frictional case is simulated here.

It can be seen from the deformation in Fig. 5.27 that the blank is dominantly drawn along the upper part of the die and the punch.

5.5.5 Deep drawing – test for the quality of shell elements as well as for the quality of contact algorithm*

One of the advantages of the "solid-shell" being 3D continuum formulation is the ability to obtain the thickness strain. However, **obtaining**

---

*The material has been reported at [156]: K. Schweizerhof, A. Konyukhov, *Contact with shells*, 7th World Congress on Computational Mechanics. Los-Angeles, 2007.

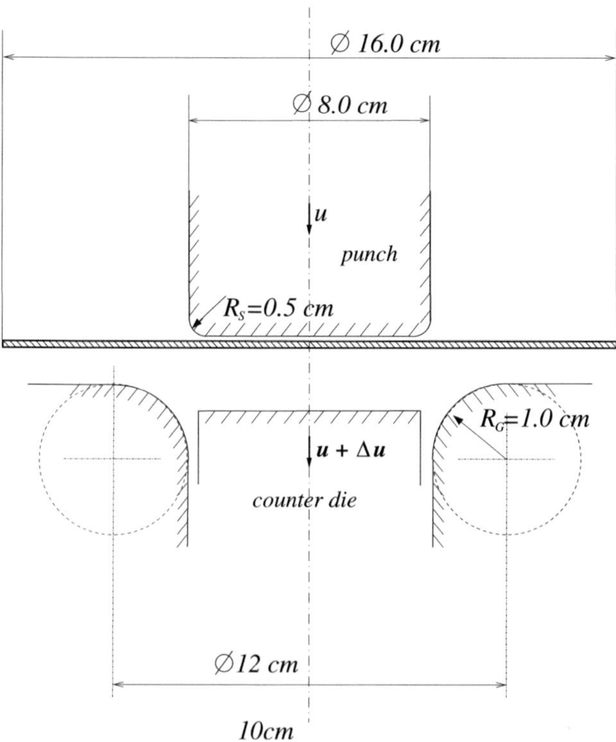

Figure 5.26: Geometry for deep drawing process.

such characteristic curves as thickness strain vs. loading displacement can be a crucial test for the quality of both finite elements for shell problems and of contact elements. Thus, the necessity of a correct algorithm to transport the history variables has been reported for special geometrical cases such as *drawing of an elastic strip into a channel with sharp corners*, see [93] and also Sect. 6.7.2.

In order to study this case in a purified situation a deep drawing with a counter die is modeled for one strip with thickness $h = 0.125$, see the geometry in Fig. 5.28. Only half of the object is modeled due to symmetry: $a = 2.750 \times 2$, $b = 2.875 \times 2$, $c = 3.000 \times 2$, $a = 2.750 \times 2$, $l = 10.000 \times 2$. The punch and the counter die are modelled via the STAS approach as a combination of planes and cylinders with $r = 0.500$, $R = 1.000$, see the geometry in Fig. 5.28. The geometry is chosen such that during drawing the distance between the punch and the counter die remains equal to

a) $u = 0.00\ cm$

b) $u = 1.25\ cm$

c) $u = 2.50\ cm$

d) $u = 3.75\ cm$

d) $u = 5.00\ cm$

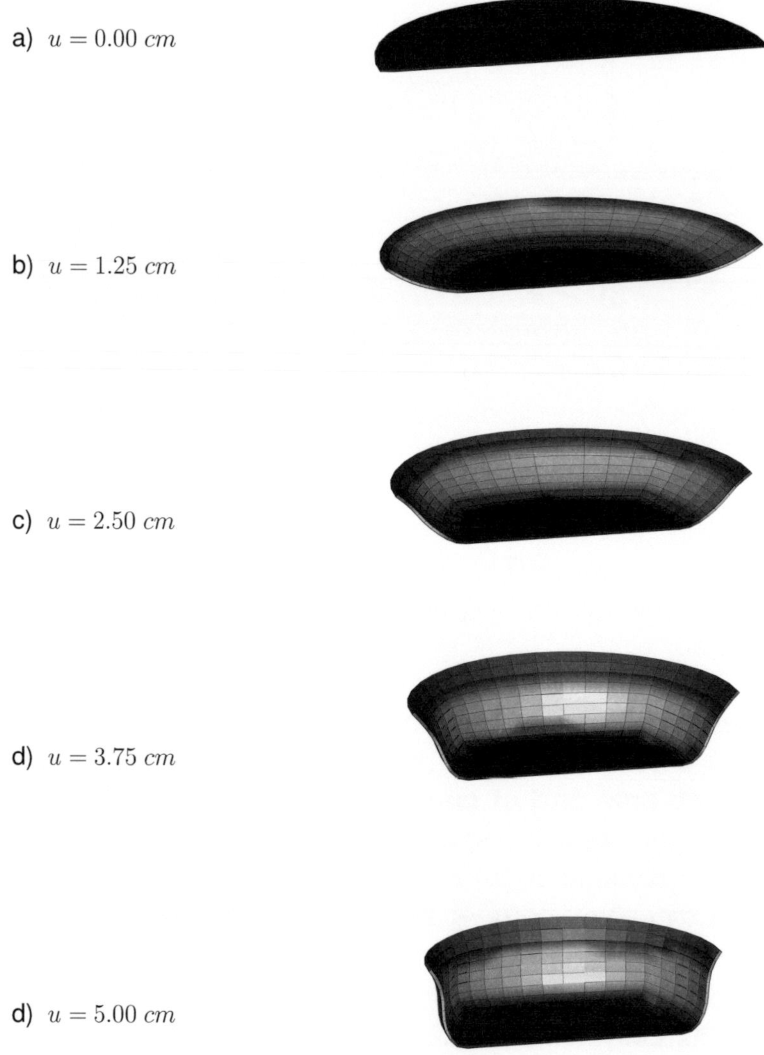

Figure 5.27: Deep drawing process. Blank at various deformation states.

the thickness of the blank $h = c - b$. As in the example from Sect. 5.5.4, the blank (the strip) is drawn by an applied vertical displacement $\mathbf{u}$ to both the punch, and the counter die – though, small distances $\delta_1 = 0.02$ and $\delta_2 = 0.01$ are supplied at the beginning, see Fig. 5.28. Both, bi-linear and biquadratic solid-shell elements are used for comparison of the thickness strain – loading displacement curve. Thus, first 40 bilinear elements are applied, then 20 biquadratic elements are applied, see mesh and deformed configuration in Fig. 5.29.

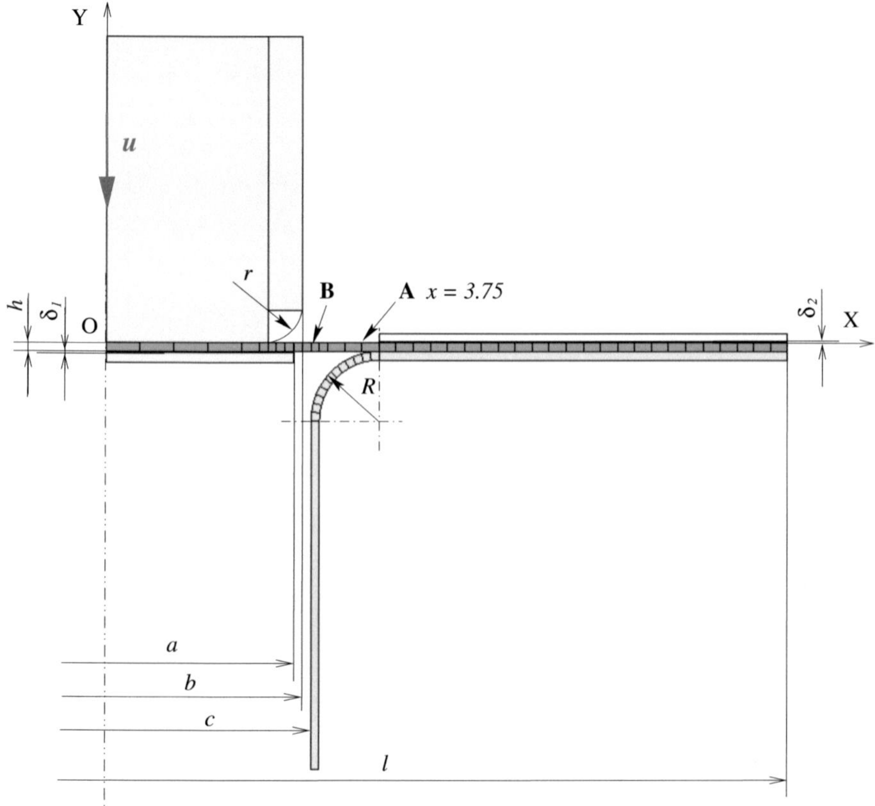

Figure 5.28: Geometry for the deep drawing process.

The problem finally has been resolved using only the STAS approach with 10×10 integration points, however, only a perfect smooth geometry of rigid body allows to overcome the problem with the correct transfer of history variables, see example in Sect. 6.7.2. Thus, the combination of the STS approach to model rigid bodies leads to difficulties with convergence. The example with bilinear solid shell elements requires smaller load steps, and therefore, more incremental load steps than the example with biquadratic solid shell elements: 1000 incremental load steps for bilinear and only 100 incremental load steps until the strip is reaching the configuration at $u = 5.000$ presented in Figure 5.29. The advantage of the biquadratic solid shell elements becomes more pronounced if we consider the thickness strain computed at points during the deformation process. In Fig. 5.30 the strain vs. loading displacement curve is given for the non-frictional case at point **A** with $x = 3.750$ (see geometry in Fig. 5.28). The highly oscillatory behavior of the curve for the bilinear mesh is due to enforcing a circular geometry by linear meshes. Each peak in the figure is reflecting the rolling of a linear segment from the strip on a circular part with radius R, see Fig. 5.28. The frictional case with bilinear meshes is causing even more severe convergence problem, therefore, only a case with biquadratic meshes is shown in Fig. 5.31 for different points **A** $(x = 3.750)$ and **B** $(x = 3.000)$ for a small coefficient of friction $\mu = 0.01$.

5.6 Conclusions

The current section deals with various computational aspects arising during the solution of contact problems in application together with a covariant approach. First, the standard known approaches as Node-To-Segment (NTS) and Segment-To-Segment (STS) are reconsidered in a covariant fashion. Thus, finite element approximations of contact parameters including tangent matrices are considered for linear, quadratic and spline surface approximations. A special integration rules based on integration of discontinuous functions such as increasing the number of integration points and integration over subdomains with selected either Gauss or Lobatto rules are developed. It is shown that the application of these rules within the STS approach allows to improve the results for the contact patch test. A combination of smooth surfaces based on NURBS interpolation together with the STS approach is studied for the patch test. The Segment-To-Analytical-Surface approach is developed for cases of contact with rigid surfaces. Within this approach there are two cases: *a rigid surface is a slave* and *a rigid surface is a master* – and the closed form solutions for penetration are derived for contact with a rigid plane, sphere, cylinder, torus and cone. The simplified procedure is derived for surfaces of revolution.

In the present chapter continuum finite "Solid-Shell" elements have been investigated for applications in large deformation contact analysis including specific deep drawing situations. Thus, a large number of examples is focusing on the comparison of the contact algorithms used especially for problems in sheet metal forming. The discussed contact elements with different degree of geometrical approximation of the contact surfaces were tested in the example of a free bending of metal sheet with large sliding in contact. A force-deflection curve was chosen to represent the main characteristics of the results. Large oscillations appeared if relatively coarse meshes together with a low order integration are used. It was found that increasing the number of integration points leads to an improved reduction of these oscillations, but an integration procedure with an additional subdivision into subdomains leads to a sufficiently further reduction of oscillations with the same total number of Gauss points over the contact area. It was shown that it is possible within the STS approach together with an integration over subdomains to efficiently improve the quality of different characteristics such as force-displacement and strain displacement curve. It was shown that the higher than linear order finite elements (here biquadratic ones) are superior concerning the results and using the same number of degrees of freedom, especially for modeling of deep drawing processes.

Summarizing the discussion over various techniques one may conclude that the higherst quality can be achieved only within the smooth contact finite element with higher then linear approximation of finite elements together with the full transfer of history variables algorithm, however, the skilled user of contact algorithms can achieve the good result depending on the type of modeling with lower efforts.

Model with bilinear solid shell elements.

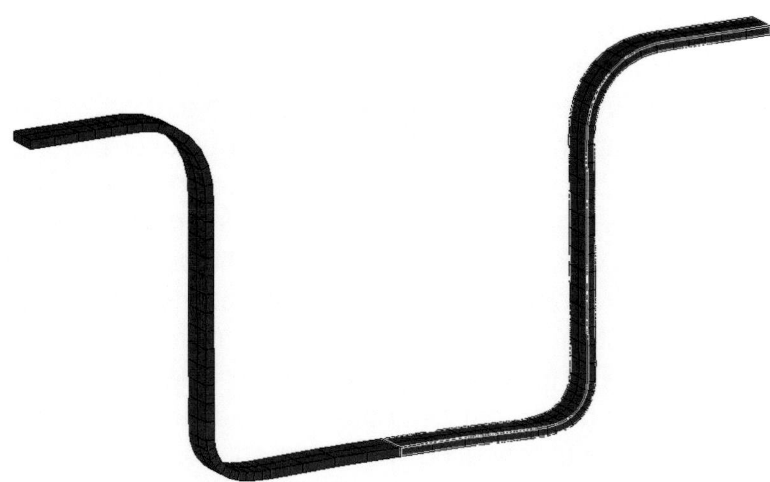

Model with biquadratic solid shell elements.

Figure 5.29: Deformed configuration for both meshes at applied vertical displacements $\mathbf{u} = 5.000$

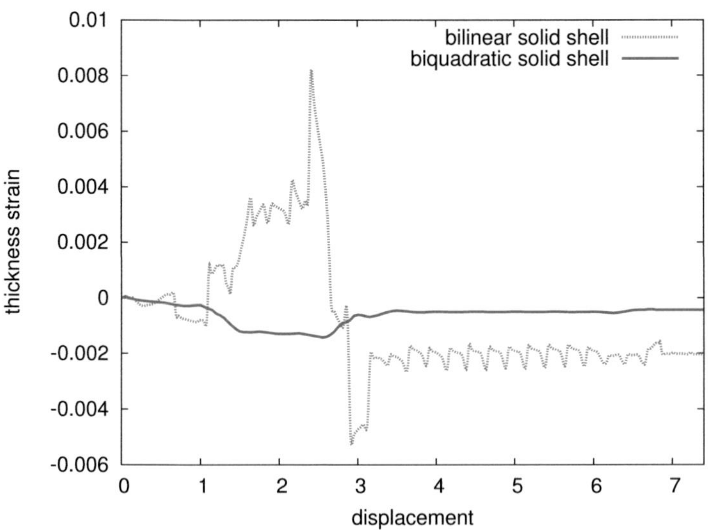

Figure 5.30: Thickness strain vs. loading displacement. Comparison of bilinear and biquadratic solid shell elements.

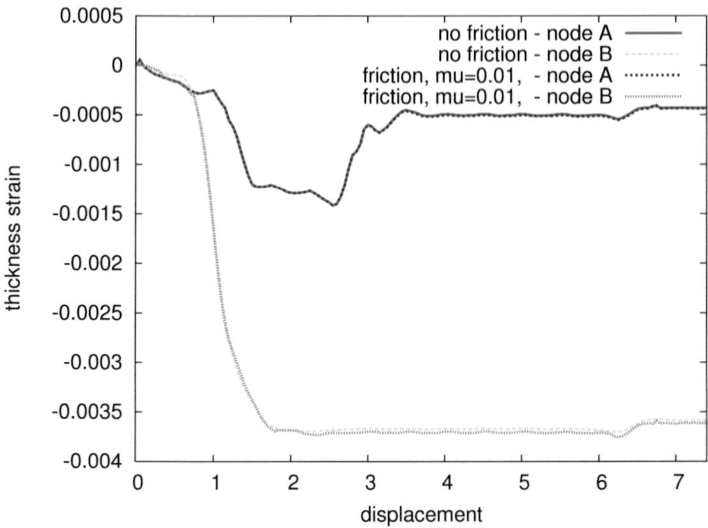

Figure 5.31: Thickness strain vs. loading displacement at different points. Frictional case. Biquadratic solid shell elements.

6

A special focus on 2D formulations for contact problems using a covariant description[*]

Abstract
A fully covariant description, based on the consideration of contact conditions especially for the 2D case is proposed. The description is based on a reconsideration of contact kinematics and all necessary operations such as derivatives in a specially chosen curvilinear coordinate system based on a curved geometry in plane. In addition, details of the finite element implementation are presented for the simple linear contact element. Special cases, requiring the update of history variables as well as their careful transfer over the element boundaries are illustrated by numerical examples. With these procedures artificial jumps in the contact forces can be avoided.

Keywords
covariant description   contact problem   friction

6.1   INTRODUCTION

In the literature various contact descriptions for an effective finite element implementation are available, which can be basically characterized by the following: from 2D to 3D formulations, from non-frictional

---

[*]The chapter is published in [92]: A. Konyukhov, K. Schweizerhof. *A special focus on 2D formulations for contact problems using a covariant description*, International Journal for Numerical Methods in Engineering, **66**:1432–1465, 2006.

to frictional contact. Some major references are cited in the following: Wriggers et. al. [194] used an elasto-plastic analogy and the penalty regularization for 2D frictional problems restricted to piecewise linear contact elements. Parisch [137] considered non-frictional 3D contact and Parisch and Lübbing (1997) [138] revised the procedure for frictional contact within the penalty method for piecewise bilinear surface elements. Peric and Owen [139] used the penalty method for 3D frictional contact problems with small deformations. The main characteristics of the cited investigations are that the penalty functional as well as its linearization were considered in a global coordinate system and restricted to linear resp. bilinear surface elements. Laursen and Simo [109], however, formulated the penalty based contact conditions and the return mapping algorithm via convective surface coordinates, but the following linearization performed in the global coordinate system led to an artificial non-symmetry of the tangent matrix in the case of sticking. Wriggers [188] could overcome this artefact using the idea of mesh tying functionals. General overviews over contact conditions and contact algorithms which are nowadays used in practice, are covered by the books of Wriggers [188] and Laursen [106], while the theoretical aspects of the regularization methods in contact mechanics can be found in Kikuchi and Oden [84]. Beyond that the covariant description proposed in Konyukhov and Schweizerhof [86], [89] allows a unified description of contact problems within the penalty method independently of the surface discretization. The method contains rather complicated mathematical transformations in the local 3D coordinate system, however, finally leading to the consistent formulation of frictional contact. Some advantages of this approach were shown, e.g. the sticking matrix preserves necessarily its symmetry.

In the current contribution, we aim to present the development in a more simple comparative manner for both 2D and 3D formulations. We will show the unity of 2D and 3D formulations, where the 2D case can be derived, from one hand, as a simplified case of the particular 3D geometry of contact surfaces and, from the other hand, can be constructed separately based on the differential geometry of 2D plane curves. This consideration has additional advantages, e.g. the subdivision of the contact tangent matrices into the "main", the "rotational" part and the "curvature" part has a pure geometrical meaning. It is also possible to distinguish a-priori various cases, where some of the parts are necessary or can be omitted.

The article is organized as follows. We start with 2D kinematics based on a curved geometry in a plane. The main results concerning the 3D covariant description can then be presented without extensive involvement into mathematics. For further details of the 3D description we refer to [89]. Two-dimensional contact will also be considered separately in 2D as well as a reduction of the 3D developments. In addition, we will compare to known formulations and present some numerical examples. A particular focus is on problems concerning contact points traversing edges of contact segments and on problems with reversible loading.

6.2 Geometry and Kinematics of Contact

Considering a special contact case – contact between two cylindrical infinite bodies with plane strain deformations, see Fig. 6.1, leads to a definition of a 2D contact. In this case a generatrix **GH** of the first cylindrical body is a contact line and corresponds to a contact line **G'H'** which is also a generatrix but of the second cylindrical body. Thus, 3D contact which can be seen as an interaction between two surfaces is reduced to an interaction between two boundary curves in the 2D case, see Fig. 6.2. One of boundary curves is chosen as the master curve. A coordinate system is considered on the boundary, either for a surface in 3D or for a curve in 2D. Thus contact occurs or two bodies are coming into contact, if a slave point belonging to the second body S penetrates into the master body, where penetration is defined as the shortest distance between the surfaces of the two bodies. For simplicity we assume now that the parameterization of the boundaries is sufficiently smooth.

6.2.1 Nomenclature of the used symbols

Throughout the article a tensor notation with regard to both, surface and curve geometry, are used, therefore, a short notation used in the contribution is provided:

ξ – arbitrary parameterization of a curve, convective coordinate.

ζ – the normal coordinate for 2D bodies, if the description is based on a cylindrical geometry. The value ζ describes the penetration.

s – length parameterization of a curve.

$\mathbf{r}_s$ – position vector of the master point.

$\rho(\xi)$ – position vector of the projection point.

ρ_ξ, $\rho_{\xi\xi}$ – the first resp. second derivative of the position vector in the case of an arbitrary parameterization.

τ – tangent normal vector in the case of the length parameterization of the curve.

ν – normal vector in the case of both, arbitrary and length parameterizations of the curve.

a_{ij}, h_{ij} – components of the metrics resp. of the curvature tensor in the case of an arbitrary parameterization of the curve.

For geometrical applications of the covariant derivation we refer to [47], and for mechanical applications to [121].

6.2.2 Definition of penetration.
Closest point projection procedure

Let the boundary of the master body be a smooth curve, parameterized by the parameter ξ: $\rho = \rho(\xi)$. The vector $\mathbf{r}_s$ describes the location of a slave point **S**, see Fig. 6.2. Then the problem to find the shortest distance between the curve $\rho(\xi)$ and the slave point **S** is defined via the minimum of the function:

$$F := \|\mathbf{r}_s - \rho(\xi)\| \longrightarrow \min. \tag{6.1}$$

The necessary condition for the minimum is the requirement of the first derivative to be zero:

$$F' = (\mathbf{r}_s - \rho(\xi)) \cdot \frac{d\rho}{d\xi} = 0. \tag{6.2}$$

Eqn. (6.2) is identical to the orthogonality condition between the vector $\mathbf{r}_s - \rho(\xi)$ and the tangent vector $\dfrac{d\rho}{d\xi} \equiv \rho_\xi$, and serves to define a projection point **C**, see Fig. 6.2. The solution can be obtained e.g. by an iterative Newton scheme. For the latter the second derivative is necessary:

$$F'' = (\mathbf{r}_s - \boldsymbol{\rho}(\xi)) \cdot \frac{d^2\boldsymbol{\rho}}{d\xi^2} - \frac{d\boldsymbol{\rho}}{d\xi} \cdot \frac{d\boldsymbol{\rho}}{d\xi}, \tag{6.3}$$

which is finally shown to be positive to specify the minimum distance. The iterative scheme is then defined as:

$$\begin{cases} \Delta\xi = -F'/F'' = -\dfrac{(\mathbf{r}_s - \boldsymbol{\rho}) \cdot \boldsymbol{\rho}_\xi}{(\mathbf{r}_s - \boldsymbol{\rho}) \cdot \boldsymbol{\rho}_{\xi\xi} - (\boldsymbol{\rho}_\xi \cdot \boldsymbol{\rho}_\xi)} \\[2ex] \xi^{(n+1)} = \xi^{(n)} + \Delta\xi \end{cases} \tag{6.4}$$

We will show in the finite implementation section, that for a 2D contact element with a linear approximation the general iterative scheme is reduced to an exact definition of the projection point.

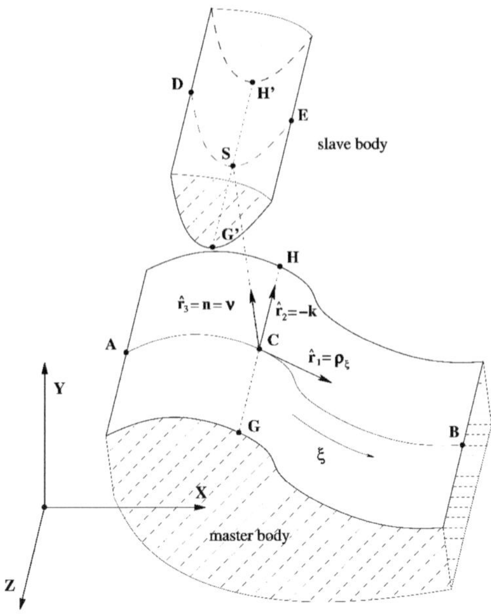

Figure 6.1: Two dimensional contact as a special case of three dimensional contact – contact between cylindrical surfaces with parallel axes Z. Local surface coordinate system on smooth master contact surface.

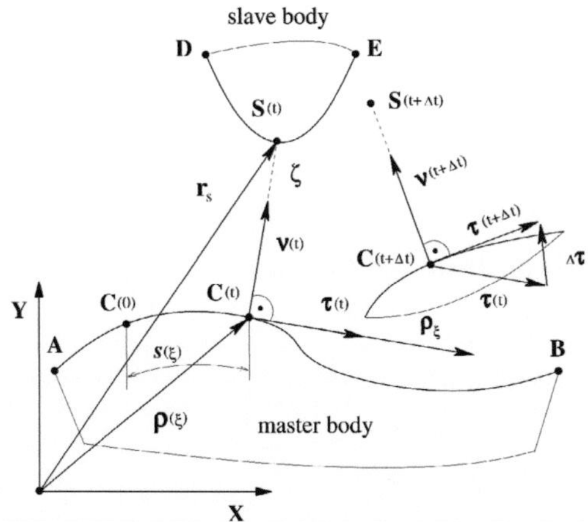

Figure 6.2: Two dimensional contact. Definitions. Contact boundaries are smooth curves in 2D.

6.2.3   2D contact kinematics

In the 2D case contact bodies are bounded by plane curves, therefore, one can take advantage of their geometry. The geometrical properties of the contact quantities can be defined in a very straightforward manner, if we use the natural parameter length, i.e. $\rho = \rho(s)$ with $s = s(\xi)$. On the plane we define a curvilinear coordinate system associated with the curve by introducing two principal vectors as a basis: the tangent vector $\rho_\xi = \dfrac{\partial \rho}{\partial \xi}$ and the unit normal vector ν

$$\mathbf{r}_s(\xi, \zeta) = \rho(\xi) + \zeta \nu(\xi). \tag{6.5}$$

Looking at the following implementation of the algorithm in a FE program, the introduction of a natural parameter s would lead to additional numerical effort, because the length of a boundary is changing during deformation. Thus, we will only show the geometrical properties using the parameter s, whereas for a finite element implementation we then turn to the Lagrangian coordinate ξ. As the local coordinate system is associated with the slave point **S**, then the closest point procedure eqn. (6.2) is already fulfilled by this definition. The second coordinate ζ is an exact (not scaled!) value of penetration often also known as gap and used for the formulation of the non-penetrability condition.

The normal unit vector $\boldsymbol{\nu}$ in the case of arbitrary Lagrangian parameterizations with ξ can be defined via a cross product in a Cartesian coordinate system as:

$$\boldsymbol{\rho}_\xi = \frac{\partial \boldsymbol{\rho}}{\partial \xi}; \quad \Longrightarrow \quad \boldsymbol{\nu} = \frac{[\mathbf{k} \times \boldsymbol{\rho}_\xi]}{\sqrt{\boldsymbol{\rho}_\xi \cdot \boldsymbol{\rho}_\xi}}, \tag{6.6}$$

where $\mathbf{k}$ is the third unit vector in this Cartesian coordinate system. The definition of the normal vector in eqn. (6.6) comes from the standard definition of the surface unit normal for cylindrical surfaces, see the contact between cylinders in Fig. 6.1, as

$$\boldsymbol{\nu} = \frac{[\hat{\mathbf{r}}_1 \times \hat{\mathbf{r}}_2]}{||[\hat{\mathbf{r}}_1 \times \hat{\mathbf{r}}_2]||}, \quad \hat{\mathbf{r}}_1 = \boldsymbol{\rho}_\xi, \quad \hat{\mathbf{r}}_2 = \mathbf{k}, \tag{6.7}$$

where $\hat{\mathbf{r}}_2$ is a unit vector of the cylinder generatrix. The definition in eqn. (6.6) gives a set of covariant basis vectors $\boldsymbol{\rho}_\xi$, $\mathbf{k}$ for a cylindrical surface. The surface metric tensor can then be defined by the following matrix a_{ij}:

$$[a_{ij}] = \begin{bmatrix} (\boldsymbol{\rho}_\xi \cdot \boldsymbol{\rho}_\xi) & 0 \\ 0 & 1 \end{bmatrix} \tag{6.8}$$

This matrix allows to define a contravariant basis for the cylindrical surface $\boldsymbol{\rho}^1, \boldsymbol{\rho}^2$, where only the first vector is changing its length:

$$\begin{pmatrix} \boldsymbol{\rho}^1 \\ \boldsymbol{\rho}^2 \end{pmatrix} = \begin{bmatrix} (\boldsymbol{\rho}_\xi \cdot \boldsymbol{\rho}_\xi) & 0 \\ 0 & 1 \end{bmatrix}^{-1} \begin{pmatrix} \boldsymbol{\rho}_\xi \\ \boldsymbol{\nu} \end{pmatrix} \quad \Longrightarrow \quad \boldsymbol{\rho}^1 = \frac{\boldsymbol{\rho}_\xi}{(\boldsymbol{\rho}_\xi \cdot \boldsymbol{\rho}_\xi)}, \quad \boldsymbol{\rho}^2 = \boldsymbol{\nu}. \tag{6.9}$$

We note here, that the dot product of the covariant basis with the contravariant basis $\boldsymbol{\rho}^1, \boldsymbol{\rho}^2$ leads to a unit matrix (a mixed metrics tensor):

$$[a_i^j] = [(\boldsymbol{\rho}_i \cdot \boldsymbol{\rho}^j)] = \begin{bmatrix} 1 & 0 \\ 0 & 1 \end{bmatrix}. \tag{6.10}$$

6.2.3.1 Derivatives of the basis vectors

Derivatives of the covariant basis vectors $\boldsymbol{\rho}_\xi$ and $\boldsymbol{\nu}$ are necessary for the further formulation and for the linearization. The derivative of the tangent vector ρ_ξ can be expressed via the covariant basis vectors

$$\rho_{\xi\xi} := \frac{\partial \rho_\xi}{\partial \xi} = \Gamma \rho_\xi + h_{11}\nu, \tag{6.11}$$

where Γ and h_{11} are defined after taking a scalar product of eqn. (6.11) with ρ_ξ and ν:

$$\Gamma = \frac{\rho_{\xi\xi} \cdot \rho_\xi}{\rho_\xi \cdot \rho_\xi}, \quad h_{11} = \rho_{\xi\xi} \cdot \nu. \tag{6.12}$$

In order to compute the derivative of the unit normal ν, first, we have to take the derivative of the following identity:

$$\frac{\partial}{\partial \xi}(\nu \cdot \nu = 1) \implies \frac{\partial \nu}{\partial \xi} \cdot \nu = 0, \tag{6.13}$$

which leads to the orthogonality condition of the vectors ν and $\dfrac{\partial \nu}{\partial \xi}$. Thus, this derivative can be expressed via the tangent vector ρ_ξ as:

$$\nu_\xi = a\rho_\xi. \tag{6.14}$$

Then the scalar a is obtained after taking the dot product of eqn. (6.14) and ρ_ξ:

$$a = (\nu_\xi \cdot \rho_\xi)/(\rho_\xi \cdot \rho_\xi) = -(\rho_{\xi\xi} \cdot \nu)/(\rho_\xi \cdot \rho_\xi). \tag{6.15}$$

The last equation is obtained via the derivative of the equation $\rho_\xi \cdot \nu = 0$. Finally, we get

$$\nu_\xi = -\frac{(\rho_{\xi\xi} \cdot \nu)}{(\rho_\xi \cdot \rho_\xi)}\rho_\xi = -\frac{h_{11}}{a_{11}}\rho_\xi. \tag{6.16}$$

In eqns. (6.12) and (6.16) the scalar value $h_{11} = (\rho_{\xi\xi} \cdot \nu)$ is a curvature coefficient and $a_{11} = (\rho_\xi \cdot \rho_\xi)$ a metric coefficient for a cylindrical surface. Eqn. (6.16) represents the Weingarten formula and eqn. (6.11) resp. the Gauss-Codazzi formula for a cylindrical surface. Now, we can obtain the basis vectors for the coordinate system in eqn. (6.5):

$$\mathbf{r}_1 = \frac{\partial \mathbf{r}}{\partial \xi} = (1 - \frac{h_{11}}{a_{11}}\zeta)\rho_\xi, \tag{6.17}$$
$$\mathbf{r}_2 = \nu.$$

The geometrical properties in 2D contact are easily found, if they are reconsidered from the plane curve geometry with a natural parameterization. This parameterization is based on a length parameter s, i.e. $\rho = \rho(s)$. The length s represents then the full path length passed by the projection point C on the master curve during contact interaction, see Fig. 6.2. The tangent vector τ in this case has a unit length

$$\tau = \frac{d\rho}{ds}.$$
(6.18)

The relation between the parameter ξ and the length parameter s is defined as:

$$ds = \sqrt{(\rho_\xi \cdot \rho_\xi)}d\xi.$$
(6.19)

For the plane curve in natural parameterization **the Serret-Frenet formulae** are used to define derivatives of the basis vectors in analogy to eqns. (6.11) and (6.16):

$$\frac{d\tau}{ds} = \kappa\nu; \quad \frac{d\nu}{ds} = -\kappa\tau,$$
(6.20)

where κ is a curvature of the curve. In this case the normal unit vector ν is defined from eqn. (6.20) and is pointing into the convex part of a body. Then the curvature κ can be computed from eqn. (6.20) by premultiplying the first equation with ν and taking the chain rule into account:

$$\kappa = \frac{d\tau}{ds} \cdot \nu = \frac{(\rho_{\xi\xi} \cdot \nu)}{(\rho_\xi \cdot \rho_\xi)} = \frac{h_{11}}{a_{11}}.$$
(6.21)

6.2.3.2 Covariant derivative of a tangent vector T.

We define a tangent vector field $\mathbf{T}$ – later taken as a friction force – as a covariant vector field in the spatial coordinate system:

$$\mathbf{T}(\xi) = T(\xi)\mathbf{r}^1|_{\zeta=0} = T(\xi)\rho^1 = T(\xi)\frac{\rho_\xi}{(\rho_\xi \cdot \rho_\xi)}.$$
(6.22)

The definition in the form of eqn. (6.22) has the advantage that the weak form – used later – becomes rather simple. The absolute value of the covariant vector eqn. (6.22) is computed as:

$$\|\mathbf{T}(\xi)\| = \frac{|T(\xi)|}{\|\boldsymbol{\rho}_\xi\|} = \frac{|T(\xi)|}{\sqrt{a_{11}}}. \tag{6.23}$$

The full time derivative of this vector field is determined taking the changing metric into account and then considering its value at $\zeta = 0$:

$$\left.\frac{d\mathbf{T}}{dt}\right|_{\zeta=0} = \left[\underbrace{\left(\frac{\partial T}{\partial t} + \frac{\partial T}{\partial \xi}\dot{\xi}\right)\mathbf{r}^1 + T\left(\dot{\xi}\frac{\partial \mathbf{r}^1}{\partial \xi} + \dot{\zeta}\frac{\partial \mathbf{r}^1}{\partial \zeta}\right)}_{\dfrac{dT}{dt}} \right]_{\zeta=0}. \tag{6.24}$$

The derivative is expressed then via the contravariant basis vectors as $\boldsymbol{\rho}^1$, $\boldsymbol{\rho}^2$

$$\left.\frac{d\mathbf{T}}{dt}\right|_{\zeta=0} = \frac{D_1 T}{dt}\boldsymbol{\rho}^1 + \frac{D_2 T}{dt}\boldsymbol{\rho}^2.$$

Here a full time derivative in the covariant form is introduced. Its value on the tangent line is computed using the scalar product with $\boldsymbol{\rho}_\xi$ as (see Appendix):

$$\frac{D_1 T}{dt} := \left(\frac{d\mathbf{T}}{dt}\cdot\boldsymbol{\rho}_\xi\right)_{\zeta=0} = \frac{dT}{dt} - \frac{\boldsymbol{\rho}_{\xi\xi}\cdot\boldsymbol{\rho}_\xi}{(\boldsymbol{\rho}_\xi\cdot\boldsymbol{\rho}_\xi)}\dot{\xi} + \frac{h_{11}}{a_{11}}\dot{\zeta}. \tag{6.25}$$

The second term on the right hand side contains the Christoffel symbol and the last term contains the curvature for a cylindrical surface.

6.2.3.3 Convective velocities

An important part for the formulation as well as for the linearization of the weak form is a time derivative of the vector of a slave point $\mathbf{S}$ in eqn. (6.5)

$$\frac{d\mathbf{r_s}}{dt} = \frac{\partial \boldsymbol{\rho}}{\partial t} + \dot{\xi}\boldsymbol{\rho}_\xi + \zeta\left(\frac{\partial \boldsymbol{\nu}}{\partial t} + \dot{\xi}\frac{\partial \boldsymbol{\nu}}{\partial \xi}\right) + \dot{\zeta}\boldsymbol{\nu}. \tag{6.26}$$

With $\mathbf{v}_s = \dfrac{d\mathbf{r}_s}{dt}$ as the absolute velocity of the slave point **S** resp. $\mathbf{v} = \dfrac{\partial \boldsymbol{\rho}}{\partial t}$ as the velocity of its projection on the master surface. The dot product with the normal vector $\boldsymbol{\nu}$ leads to the rate of the penetration

$$\dot{\zeta} = (\mathbf{v}_s - \mathbf{v}) \cdot \boldsymbol{\nu}. \qquad (6.27)$$

Considering a value of the convective tangent velocity $\dot{\xi}$ on the tangent line, i.e. at $\zeta = 0$, we need the dot product of eqn. (6.26) with $\boldsymbol{\rho}_\xi$:

$$\dot{\xi} = \frac{(\mathbf{v}_s - \mathbf{v}) \cdot \boldsymbol{\rho}_\xi}{(\boldsymbol{\rho}_\xi \cdot \boldsymbol{\rho}_\xi)}. \qquad (6.28)$$

In the case of a length parameterization with $s = \xi$, eqn. (6.28) leads to the following convective velocity $\dot{s}$:

$$\dot{s} = (\mathbf{v}_s - \mathbf{v}) \cdot \boldsymbol{\tau}. \qquad (6.29)$$

From the kinematical equation (6.26) we can obtain an equation for the variations by changing the time derivative operator into the variation operator δ. This equation is also considered on the tangent line, i.e. at $\zeta = 0$:

$$\delta\mathbf{r}_s - \delta\boldsymbol{\rho} = \delta\xi\boldsymbol{\rho}_\xi + \delta\zeta\boldsymbol{\nu}. \qquad (6.30)$$

Eqn. (6.30) gives a variation of the displacement field for the expression of the virtual work of contact tractions on the contact surface.

6.2.3.4 Evolution equations for contact tractions

The evolution equations can be regarded as rate equations for the contact tractions. The contact traction vector $\mathbf{R}_s$ is defined for the slave point s in the local coordinate system on the master curve in the covariant form as:

$$\mathbf{R}_s = T\boldsymbol{\rho}^1 + N\boldsymbol{\rho}^2 = \mathbf{T} + N\boldsymbol{\rho}^2 = T\frac{\boldsymbol{\rho}_\xi}{(\boldsymbol{\rho}_\xi \cdot \boldsymbol{\rho}_\xi)} + N\boldsymbol{\nu}. \qquad (6.31)$$

For the normal traction N, the following regularized equation is applied

$$N = \epsilon_N \langle \zeta \rangle, \qquad (6.32)$$

where ϵ_N is a penalty parameter for the normal interaction and $\langle\rangle$ are Macauley brackets in the form

$$\langle\zeta\rangle = \begin{cases} 0, & if \ \zeta > 0 \\ \zeta, & if \ \zeta \leq 0 \end{cases}.$$

According to eqn. (6.27), the rate of a normal traction can be computed as

$$\dot{N} = \epsilon_N\dot{\zeta}H(-\zeta) = \epsilon_N H(-\zeta)(\mathbf{v}_s - \mathbf{v})\cdot\boldsymbol{\nu}, \tag{6.33}$$

where $H(-\zeta)$ is the Heaviside function.

As a reasonable equation for the regularization of the tangent traction vector $\mathbf{T}$ we choose a proportional relation between the full time derivative $\dfrac{d\mathbf{T}}{dt}$ and the relative velocity vector expressed on the tangent line $\zeta = 0$:

$$\frac{D_1 T}{dt}\boldsymbol{\rho}^1 = -\epsilon_T\dot{\xi}\boldsymbol{\rho}_\xi, \tag{6.34}$$

in component form written as

$$\frac{D_1 T}{dt} = -\epsilon_T\dot{\xi}(\boldsymbol{\rho}_\xi \cdot \boldsymbol{\rho}_\xi), \tag{6.35}$$

where ϵ_T is a penalty parameter for the tangential interaction. Applying the results from eqn. (6.25) leads to the evolution equations in the form of covariant derivatives:

$$\frac{dT}{dt} = -\epsilon_T(\boldsymbol{\rho}_\xi \cdot \boldsymbol{\rho}_\xi)\dot{\xi} + \frac{\boldsymbol{\rho}_{\xi\xi} \cdot \boldsymbol{\rho}_\xi}{(\boldsymbol{\rho}_\xi \cdot \boldsymbol{\rho}_\xi)}\dot{\xi} - \frac{h_{11}}{a_{11}}\dot{\zeta}. \tag{6.36}$$

which is used to compute a trial tangent traction.

6.3 Weak formulation in the spatial coordinate system

Next we consider the contact tractions $\mathbf{R}_s$ and $\mathbf{R}_m$ on both the slave and the master contact curves with corresponding lengths l_s and l_m in the current configuration. Let $\delta\mathbf{u}_s$ resp. $\delta\mathbf{u}_m$ be variations of the displacement field on the curves l_s resp. l_m, then the work of the contact forces is determined in the following integral

$$\delta W_c = \int_{l_s} \mathbf{R}_s \cdot \delta \mathbf{u}_s dl_s + \int_{l_m} \mathbf{R}_m \cdot \delta \mathbf{u}_m dl_m, \qquad (6.37)$$

which must be added to the global work of the internal and external forces. Due to equilibrium at the contact boundary $\mathbf{R}_s dl_s = -\mathbf{R}_m dl_m$, equation (6.37) can be also written as

$$\delta W_c = \int_{l_s} \mathbf{R}_s \cdot (\delta \mathbf{u}_s - \delta \mathbf{u}_m) dl_s. \qquad (6.38)$$

The integral in (6.38) is considered in the local coordinate system. We redefine now the variations $\delta \mathbf{u}_s = \delta \mathbf{r}_s$ for a slave point and $\delta \mathbf{u}_m = \delta \boldsymbol{\rho}$ for a projection of the slave point onto the master curve.

Substituting the variation $\delta \mathbf{u}_s - \delta \mathbf{u}_m = \delta \mathbf{r}_s - \delta \boldsymbol{\rho}$ from eqn. (6.30) and also the full contact traction vector eqn. (6.31) into the integral (6.38) we obtain

$$\delta W_c = \int_l (N \delta \zeta + T \delta \xi) dl. \qquad (6.39)$$

A closer look reveals that the contact integral (6.39) contains the work of the contact tractions T and N defined on the master contact curve and is computed along the slave curve $l \equiv l_s$.

6.4 Linearization process

Here, we show the derivation of the normal contact matrices using the geometry of plane curves. For all other results, we give a sketch of the linearization procedure with a comparative discussion of results available in the literature, in order to avoid the repetition of complicated mathematics.

6.4.1 Necessary operations. Linearization of convective variations

Since in the contact integral the linearization of the contact tractions is directly given by the evolution equations, it is only necessary to find derivatives of the convective variations $\delta\zeta$ and $\delta\xi$ to fulfill all steps in the preparation for further forms.

6.4.1.1 Linearization of $\delta\zeta$

The result will be obtained assuming a natural parameterization of the corresponding boundary curve.

$$\frac{d}{dt}\delta\zeta = \frac{d}{dt}\left[(\delta\mathbf{r}_s - \delta\boldsymbol{\rho}) \cdot \boldsymbol{\nu}\right] =$$

$$= \frac{\partial(\delta\mathbf{r}_s - \delta\boldsymbol{\rho})}{\partial s} \cdot \boldsymbol{\nu}\dot{s} + (\delta\mathbf{r}_s - \delta\boldsymbol{\rho}) \cdot \frac{\partial\boldsymbol{\nu}}{\partial t} + (\delta\mathbf{r}_s - \delta\boldsymbol{\rho}) \cdot \frac{\partial\boldsymbol{\nu}}{\partial s}\dot{s}. \tag{6.40}$$

The first term can be rewritten, taking into account eqn. (6.29) for a convective velocity in the case of a natural parameterization, as follows:

$$\delta\underbrace{\frac{\partial(\mathbf{r}_s - \boldsymbol{\rho})}{\partial s}}_{-\boldsymbol{\tau}} \cdot \boldsymbol{\nu}\dot{s} = -(\delta\boldsymbol{\tau} \cdot \boldsymbol{\nu})\underbrace{\left((\mathbf{v}_s - \mathbf{v}) \cdot \boldsymbol{\tau}\right)}_{\dot{s}} = -\delta\boldsymbol{\tau} \cdot (\boldsymbol{\nu}\otimes\boldsymbol{\tau})(\mathbf{v}_s - \mathbf{v}). \tag{6.41}$$

In order to rewrite the second term, we have to take first a partial time derivative of the orthogonality condition:

$$\boldsymbol{\tau} \cdot \boldsymbol{\nu} = 0 \implies \frac{\partial(\boldsymbol{\tau} \cdot \boldsymbol{\nu})}{\partial t} = \frac{\partial^2\boldsymbol{\rho}}{\partial s\partial t} \cdot \boldsymbol{\nu} + \frac{\partial\boldsymbol{\nu}}{\partial t} \cdot \boldsymbol{\tau} = 0, \tag{6.42}$$

leading to the expression

$$\frac{\partial\boldsymbol{\nu}}{\partial t} \cdot \boldsymbol{\tau} = -\frac{\partial\mathbf{v}}{\partial s} \cdot \boldsymbol{\nu}. \tag{6.43}$$

From the other side, using the unity condition of the vector $\boldsymbol{\nu}$, we can express the time derivative in terms of the tangent vector $\boldsymbol{\tau}$ by analogy to eqns. (6.14), (6.15), as

$$\frac{\partial\boldsymbol{\nu}}{\partial t} = a\boldsymbol{\tau} = \left(\boldsymbol{\tau} \cdot \frac{\partial\boldsymbol{\nu}}{\partial t}\right)\boldsymbol{\tau} \tag{6.44}$$

and substituting eqn. (6.43) we obtain

$$\frac{\partial \boldsymbol{\nu}}{\partial t} = -\left(\frac{\partial \mathbf{v}}{\partial s} \cdot \boldsymbol{\nu}\right)\boldsymbol{\tau}. \qquad (6.45)$$

Eqn. (6.45) allows to transform the second term in eqn. (6.40) as follows

$$(\delta \mathbf{r}_s - \delta\boldsymbol{\rho}) \cdot \frac{\partial \boldsymbol{\nu}}{\partial t} = -(\delta \mathbf{r}_s - \delta\boldsymbol{\rho}) \cdot \left(\frac{\partial \mathbf{v}}{\partial s} \cdot \boldsymbol{\nu}\right)\boldsymbol{\tau} = \qquad (6.46)$$

introducing a tensor product $\boldsymbol{\tau} \otimes \boldsymbol{\nu}$ in order to transform a dot product

$$= -(\delta \mathbf{r}_s - \delta\boldsymbol{\rho}) \cdot (\boldsymbol{\tau} \otimes \boldsymbol{\nu})\frac{\partial \mathbf{v}}{\partial s} = -(\delta \mathbf{r}_s - \delta\boldsymbol{\rho}) \cdot (\boldsymbol{\tau} \otimes \boldsymbol{\nu})\frac{\partial \boldsymbol{\tau}}{\partial t}. \qquad (6.47)$$

The last term in eqn. (6.47) is obtained reversing the order of differentiation as

$$\frac{\partial \mathbf{v}}{\partial s} = \frac{\partial}{\partial s}\frac{\partial \boldsymbol{\rho}}{\partial t} = \frac{\partial}{\partial t}\frac{\partial \boldsymbol{\rho}}{\partial s} = \frac{\partial \boldsymbol{\tau}}{\partial t}. \qquad (6.48)$$

The third term in (6.40) is reorganized into a tensor form with a second Serret-Frenet formula and with equation (6.29) for the convective velocity $\dot{s}$:

$$(\delta \mathbf{r}_s - \delta\boldsymbol{\rho}) \cdot \frac{\partial \boldsymbol{\nu}}{\partial s}\dot{s} = -(\delta \mathbf{r}_s - \delta\boldsymbol{\rho}) \cdot \kappa\boldsymbol{\tau} \otimes \boldsymbol{\tau}(\mathbf{v}_s - \mathbf{v}). \qquad (6.49)$$

Therefore, combining eqn. (6.41), (6.46) and (6.49), we obtain a final formula for the linearization of $\delta\zeta$:

$$\frac{d}{dt}\delta\zeta = -\left(\delta\boldsymbol{\tau} \cdot \boldsymbol{\nu} \otimes \boldsymbol{\tau}(\mathbf{v}_s - \mathbf{v}) + (\delta \mathbf{r}_s - \delta\boldsymbol{\rho}) \cdot \boldsymbol{\tau} \otimes \boldsymbol{\nu}\frac{\partial \boldsymbol{\tau}}{\partial t}\right) \qquad (6.50)$$
$$- (\delta \mathbf{r}_s - \delta\boldsymbol{\rho}) \cdot \kappa\boldsymbol{\tau} \otimes \boldsymbol{\tau}(\mathbf{v}_s - \mathbf{v}).$$

6.4.1.2 Linearization of the convective variation $\delta\xi$

The linearization of the convective variations $\delta\xi^i$ in a 3D formulation is the most complicated part of the process. First, see Parisch [137], Laursen and Simo [109], Wriggers [187] the convective variations $\delta\xi^i$ were introduced via an iterative Newton scheme, see eqn. (6.4). A kinematical definition of $\delta\xi^i$ can be found in the books of Wriggers [188] and Laursen [106]. The full linearization of these terms combined with the contact integral defined only on the surface led to an artificial non-symmetry of the tangent matrix for the sticking case, which was mentioned in Laursen and Simo [109]. Wriggers [188] could avoid this by looking at it as a mesh tying procedure. In [89] the variations of $\delta\xi^i$ were defined kinematically and expressed on the tangent plane of the contact surface. In addition, the linearization process was performed in the covariant form on the tangent plane. For the sticking case this leads directly to a symmetric matrix and allows to avoid the artificial non-symmetry.

Here only the main points of the linearization process are depicted, for the full derivation we refer to [89].

1. The convective variations are defined on the tangent plane of the spatial coordinate system via consideration of the slave point velocity as $\dot{\xi}^j = a^{ij}(\mathbf{v}_s - \mathbf{v}) \cdot \boldsymbol{\rho}_i$.

2. During the linearization of $\delta\xi^i$ the derivative of the metric tensor is obtained as derivative of the spatial metric tensor considering its value on the tangent plane.

The final result for the 3D case is then:

$$\frac{d}{dt}(\delta\xi^i) =$$

$$= -(\delta\mathbf{r}_s - \delta\boldsymbol{\rho})\, a^{il}a^{jk}\, \boldsymbol{\rho}_k \otimes \boldsymbol{\rho}_l\, \mathbf{v}_j - \delta\boldsymbol{\rho}_{,j}\, a^{ik}a^{jl}\, \boldsymbol{\rho}_k \otimes \boldsymbol{\rho}_l\, (\mathbf{v}_s - \mathbf{v}) \quad (6.51\text{a})$$

$$+ h^{ij}(\delta\mathbf{r}_s - \delta\boldsymbol{\rho}) \cdot \left(\boldsymbol{\rho}_j \otimes \mathbf{n} + \mathbf{n} \otimes \boldsymbol{\rho}_j\right)(\mathbf{v}_s - \mathbf{v}) + \quad (6.51\text{b})$$

$$+ h^i_n \dot{\xi}^3 \delta\xi^n - \Gamma^i_{kj}\dot{\xi}^j \delta\xi^k. \quad (6.51\text{c})$$

The reduction into the specific plane geometry in the current contribution leads to:

$$\frac{d}{dt}(\delta\xi) =$$

$$= -\frac{(\delta \mathbf{r}_s - \delta\boldsymbol{\rho}) \cdot \boldsymbol{\rho}_\xi \otimes \boldsymbol{\rho}_\xi \, \mathbf{v}_j + \delta\boldsymbol{\rho}_\xi \cdot \boldsymbol{\rho}_\xi \otimes \boldsymbol{\rho}_\xi \, (\mathbf{v}_s - \mathbf{v})}{(\boldsymbol{\rho}_\xi \cdot \boldsymbol{\rho}_\xi)^2} \qquad (6.52a)$$

$$+ \frac{(\boldsymbol{\rho}_{\xi\xi} \cdot \boldsymbol{\nu})}{(\boldsymbol{\rho}_\xi \cdot \boldsymbol{\rho}_\xi)^2}(\delta \mathbf{r}_s - \delta\boldsymbol{\rho}) \cdot \left(\boldsymbol{\rho}_\xi \otimes \boldsymbol{\nu} + \boldsymbol{\nu} \otimes \boldsymbol{\rho}_\xi\right)(\mathbf{v}_s - \mathbf{v})+ \qquad (6.52b)$$

$$+ \frac{h_{11}}{a_{11}}\dot{\zeta}\delta\xi - \frac{\boldsymbol{\rho}_{\xi\xi} \cdot \boldsymbol{\rho}_\xi}{(\boldsymbol{\rho}_\xi \cdot \boldsymbol{\rho}_\xi)}\dot{\xi}\delta\xi. \qquad (6.52c)$$

The non-symmetric part in eqn. (6.52c) is intentionally kept in untransformed form, because it will give a zero in sum with similar terms in the evolution equation (6.36) during the forthcoming linearization.

6.4.2 Tangent matrices

We derive the tangent matrix for the normal part in the case of a natural parameterization. In order to avoid the complexity for the sticking – sliding cases for the tangential part, the derivation is given as a reduction of the 3D case.

6.4.2.1 Tangent matrix for the normal part

The normal part is defined by the following integral:

$$\delta W_c^N = \int_l N\delta\zeta \, dl. \qquad (6.53)$$

The integral is computed over the slave surface l, while all functions are defined on the master surface. Thus, a linearization of dl is not necessary within the process:

$$D(\delta W_c^N) =$$

$$\int_l \left(\frac{dN}{dt}\delta\zeta + N\frac{d\delta\zeta}{dt}\right) dl =$$

then the application of eqn. (6.33) and eqn. (6.50) leads to

$$= \int_l \epsilon_N(\delta \mathbf{r}_s - \delta\boldsymbol{\rho}) \cdot (\boldsymbol{\nu} \otimes \boldsymbol{\nu})(\mathbf{v}_s - \mathbf{v}) \, dl - \qquad (6.54a)$$

$$- \int_l \epsilon_N \zeta \left(\delta \boldsymbol{\tau} \cdot (\boldsymbol{\nu} \otimes \boldsymbol{\tau})(\mathbf{v}_s - \mathbf{v}) + (\delta \mathbf{r}_s - \delta \boldsymbol{\rho}) \cdot (\boldsymbol{\tau} \otimes \boldsymbol{\nu}) \frac{\partial \boldsymbol{\tau}}{\partial t} \right) dl - \quad (6.54b)$$

$$- \int_l \epsilon_N \zeta \kappa (\delta \mathbf{r}_s - \delta \boldsymbol{\rho}) \cdot (\boldsymbol{\tau} \otimes \boldsymbol{\tau})(\mathbf{v}_s - \mathbf{v}) dl. \quad (6.54c)$$

Remark.

The contact matrix obtained via eqn. (6.54) is computed only for the case $\zeta < 0$ – this simplification allows us to exclude the usage of the Heaviside function.

The form in natural coordinates allows a simple geometrical interpretation of each part in eqn. (6.54) and even allows to determine situations where some of them are zero. The first part eqn. (6.54a) is called main part and defines the constitutive relation for normal contact conditions. The second part eqn. (6.54b) is called rotational part and defines the geometrical stiffness due to the rotation of the tangent vector of the master curve. It disappears when a master segment is moving in parallel, because only in this case the derivative of a unit vector $\boldsymbol{\tau}$ becomes zero, see Fig. 6.2. The third part eqn. (6.54c) is called curvature part. This part disappears when the curvature κ of a master segment is zero, i.e. in the case of linear approximations of the master segment.

6.4.2.2 Tangent matrix for tangential traction

The part of the contact integral which includes the effect of the tangential interaction is given as:

$$\delta W_c^T = \int_l T \delta \xi \, dl \quad (6.55)$$

The linearized equation has to be subdivided into a part for sticking and another part for sliding, which differ concerning the return-mapping scheme.

Sticking. In this case, the tangential force T has to be computed from the solution of the evolution equation eqn. (6.36), e.g. via the backward Euler scheme. The simplest case with linear approximations will be presented in the following section concerning the finite element implementation.

Sticking is fulfilled according to Coulomb's friction law, i.e. the inequality $\|\mathbf{T}\| \leq \mu|N|$ has to be valid in each load step. The linearized contact integral has then the following form:

$$D_v(\delta W_c^T) = \int_l \left(\frac{dT}{dt} \delta\xi + T \frac{d\delta\xi}{dt} \right) dl =$$

$$- \int_l \frac{\varepsilon_T}{(\boldsymbol{\rho}_\xi \cdot \boldsymbol{\rho}_\xi)} (\delta\mathbf{r}_s - \delta\boldsymbol{\rho}) \cdot \boldsymbol{\rho}_\xi \otimes \boldsymbol{\rho}_\xi (\mathbf{v}_s - \mathbf{v}) dl \tag{6.56a}$$

$$- \int_l \frac{T}{(\boldsymbol{\rho}_\xi \cdot \boldsymbol{\rho}_\xi)^2} \left[(\delta\mathbf{r}_s - \delta\boldsymbol{\rho}) \cdot \boldsymbol{\rho}_\xi \otimes \boldsymbol{\rho}_\xi \, \mathbf{v}_\xi + \delta\boldsymbol{\rho}_\xi \cdot \boldsymbol{\rho}_\xi \otimes \boldsymbol{\rho}_\xi \, (\mathbf{v}_s - \mathbf{v}) \right] dl \tag{6.56b}$$

$$+ \int_l \frac{T h_{11}}{(\boldsymbol{\rho}_\xi \cdot \boldsymbol{\rho}_\xi)^2} (\delta\mathbf{r}_s - \delta\boldsymbol{\rho}) \cdot \left(\boldsymbol{\rho}_\xi \otimes \boldsymbol{\nu} + \boldsymbol{\nu} \otimes \boldsymbol{\rho}_\xi \right) (\mathbf{v}_s - \mathbf{v}) dl. \tag{6.56c}$$

Sliding. If sliding is detected, i.e. if $\|\mathbf{T}\| > \mu|N|$, then the sliding force is computed according to Coulomb's friction law. We also keep a covariant form:

$$T^{sl} = \mu|N| \frac{T_{tr}}{\|\mathbf{T}_{tr}\|} = \mu|N| \, \|\boldsymbol{\rho}_\xi\| sgn(T_{tr}), \tag{6.57}$$

with

$$\|\boldsymbol{\rho}_\xi\| = (\boldsymbol{\rho}_\xi \cdot \boldsymbol{\rho}_\xi)^{1/2} = \sqrt{a_{11}}.$$

The linearized contact integral gets the following form:

$$D_v(\delta W_c^T) =$$

$$- \int_l \frac{\epsilon_N \mu \, sgn(T_{tr})}{(\boldsymbol{\rho}_\xi \cdot \boldsymbol{\rho}_\xi)^{1/2}} (\delta\mathbf{r}_s - \delta\boldsymbol{\rho}) \cdot (\boldsymbol{\rho}_\xi \otimes \boldsymbol{\nu})(\mathbf{v}_s - \mathbf{v}) dl \tag{6.58a}$$

$$- \int_l \frac{\mu|N| \, sgn(T_{tr})}{(\boldsymbol{\rho}_\xi \cdot \boldsymbol{\rho}_\xi)^{3/2}} \left((\delta\mathbf{r}_s - \delta\boldsymbol{\rho}) \cdot \boldsymbol{\rho}_\xi \otimes \boldsymbol{\rho}_\xi \, \mathbf{v}_\xi + \delta\boldsymbol{\rho}_\xi \cdot \boldsymbol{\rho}_\xi \otimes \boldsymbol{\rho}_\xi \, (\mathbf{v}_s - \mathbf{v}) \right) dl \tag{6.58b}$$

$$+ \int_l \frac{\mu h_{11} |N| \, sgn(T_{tr})}{(\boldsymbol{\rho}_\xi \cdot \boldsymbol{\rho}_\xi)^{3/2}} (\delta \mathbf{r}_s - \delta \boldsymbol{\rho}) \cdot \left(2\boldsymbol{\rho}_\xi \otimes \boldsymbol{\nu} + \boldsymbol{\nu} \otimes \boldsymbol{\rho}_\xi\right) (\mathbf{v}_s - \mathbf{v}) dl. \quad (6.58c)$$

The non-symmetric part eqn. (6.58c) now is resulting from the last term of the evolution equation (6.36).

Remark:

The derivations for the two dimensional case allow to describe all parts of the tangent matrix and to find all cases, when some of them become zero. The main parts, eqns. (6.56a) and (6.58a), the so-called constitutive parts, contain a penalty parameter and describe the stiffness of the contact interaction due to the chosen interface model. This is based on an allowable elastic deformation due to the regularization in the case of sticking resp. due to the applied sliding force $\mu|N|$ in the tangential direction in the case of sliding. The rotational parts eqns. (6.56b) and (6.58b), contain a metric coefficient $a_{11} = (\boldsymbol{\rho}_\xi \cdot \boldsymbol{\rho}_\xi)$ and a vector $\boldsymbol{\rho}_\xi$. A metric coefficient is a measure of the tensile deformation of the contact master line, e.g. a component of the Cauchy-Green tensor for the contact line can be written as $\varepsilon_{11} = (a_{11} - 1)/2$. The vector $\boldsymbol{\rho}_\xi$ is a measure of the rotation of the master segment, which becomes obvious if the length is chosen as a coordinate $s = \xi$. In this case we find $\delta \boldsymbol{\rho}_\xi = \delta \boldsymbol{\tau}$ and $\mathbf{v}_\xi = \frac{\partial \boldsymbol{\tau}}{\partial t}$. The vector $\boldsymbol{\tau}$ is a unit vector, therefore the vectors $\delta \boldsymbol{\tau}$ and $\frac{\partial \boldsymbol{\tau}}{\partial t}$ are describing the rotation of the unit vector $\boldsymbol{\tau}$ (see Fig. 6.2). It is identical to zero only in the case of parallel motions. Thus, the rotational part is obviously negligible in the case of small deformations and small rotations of the master line. The curvature parts eqns. (6.56c) and (6.58c) describe the stiffness of the contact interaction due to the curvature of the contact surface. If the surface has zero curvature or is approximated by linear elements, then the curvature part becomes zero.

6.5   Finite element implementation

The structure of all parts of the tangent matrix is algorithmic. It is sufficient for the discretization to define only approximations of a relative displacement vector $(\mathbf{r}_s - \boldsymbol{\rho})$ and its derivative with respect to ξ. Following the standard iso-parametric technique as for finite elements, we consider a contact surface element with the same order of approximation for the geometry as for the displacement field. The boundary curve can be given with any curve description (spline, NURB, etc.). For simplicity, we consider only the node-to-segment contact approach. Let a boundary curve or a master segment of it be defined by n nodes with $\mathbf{x}^{(1)}, \mathbf{x}^{(2)}, ..., \mathbf{x}^{(n)}$; and $\mathbf{x}^{(n+1)}$ for a slave node. A standard grouping of a displacement vector can then be written as

$$\mathbf{u}^T = \{u_1^{(1)}, u_2^{(1)}, u_1^{(2)}, u_2^{(2)}, ..., u_1^{(n)}, u_2^{(n)}, u_1^{(n+1)}, u_2^{(n+1)}\}^T, \tag{6.59}$$

where the first n nodes resp. $2n$ displacements belong to the master surface, while the $(n+1)$'th term is belonging to the "slave" node resp. is describing the "slave" displacements.

We introduce a matrix of shape functions $\mathbf{A}$

$$\mathbf{A} = \begin{bmatrix} -N_1 & 0 & 0 & -N_2 & 0 & 0 & ... & -N_n & 0 & 1 & 0 \\ 0 & -N_1 & 0 & 0 & -N_2 & 0 & ... & 0 & -N_n & 0 & 1 \end{bmatrix}, \tag{6.60}$$

where N_i, $i = 1, 2, ..., n$ are shape functions. The matrix of the derivatives of the shape functions $\mathbf{A}'$ is defined then as

$$\mathbf{A}' = - \begin{bmatrix} N_1' & 0 & 0 & N_2' & 0 & 0 & ... & N_n' & 0 & 0 & 0 \\ 0 & N_1' & 0 & 0 & N_2' & 0 & ... & 0 & N_n' & 0 & 0 \end{bmatrix}, \tag{6.61}$$

The relative vector of variations $(\delta\mathbf{r}_s - \delta\boldsymbol{\rho})$ and the relative velocity vector $(\mathbf{v}_s - \mathbf{v})$ are then written as

$$\delta\mathbf{r}_s - \delta\boldsymbol{\rho} = \mathbf{A}\delta\mathbf{u}, \quad \mathbf{v}_s - \mathbf{v} = \mathbf{A}\dot{\mathbf{u}}, \tag{6.62}$$

where $\dot{\mathbf{u}}$ is the nodal velocity vector. With the matrix of the derivatives $\mathbf{A}'$ e.g. a derivative $\delta\boldsymbol{\rho}_\xi$ of a vector $\delta\boldsymbol{\rho}$ can be defined as

$$\delta\boldsymbol{\rho}_\xi = -\mathbf{A}'\delta\mathbf{u}. \tag{6.63}$$

We note that the matrices $\mathbf{A}$ and $\mathbf{A}'$ are sufficient to build the tangent matrices as well as the residual vector for any arbitrary contact surface.

6.5.1 Linear contact element

The simplest approximation is a linear contact element within the node-to-segment approach. This linear contact element has 3 nodes: the first two nodes $\mathbf{x}^1, \mathbf{x}^2$ are approximating a contact boundary, while the third node $\mathbf{x}^S$ is the slave node. The approximation on the master element is defined as

$$\boldsymbol{\rho}(\xi) := \frac{1-\xi}{2}\mathbf{x}^1 + \frac{1+\xi}{2}\mathbf{x}^2 = \frac{1-\xi}{2}\begin{pmatrix} x_1 \\ y_1 \end{pmatrix} + \frac{1+\xi}{2}\begin{pmatrix} x_2 \\ y_2 \end{pmatrix}. \tag{6.64}$$

The tangent vector $\boldsymbol{\rho}_\xi$ is then given as:

$$\boldsymbol{\rho}_\xi = \frac{\mathbf{x}^2 - \mathbf{x}^1}{2} = \frac{1}{2}\begin{pmatrix} x_2 - x_1 \\ y_2 - y_1 \end{pmatrix}. \tag{6.65}$$

A single metric coefficient becomes:

$$a_{11} = \boldsymbol{\rho}_\xi \cdot \boldsymbol{\rho}_\xi = 0.25 \cdot \left((x_1 - x_2)^2 + (y_1 - y_2)^2\right), \tag{6.66}$$

which is the square of the length of the vector $\boldsymbol{\rho}_\xi$. The unit normal vector $\boldsymbol{\nu}$ to the contact segment is defined in a Cartesian coordinate system via a cross product of the tangent vector $\boldsymbol{\rho}_\xi$ and the third unit vector $\mathbf{k}$ which is normal to the plane:

$$\boldsymbol{\nu} := \frac{[\mathbf{k} \times \boldsymbol{\rho}_\xi]}{(\boldsymbol{\rho}_\xi \cdot \boldsymbol{\rho}_\xi)} = \frac{1}{2a_{11}}\begin{pmatrix} y_2 - y_1 \\ x_1 - x_2 \end{pmatrix}. \tag{6.67}$$

In this definition according to eqn. (6.7) and to Fig. 6.1, we assume that the solid body occupies the lower part relative to the contact element in Fig. 6.3. The matrix of the shape functions $\mathbf{A}$

$$\mathbf{A} = \begin{bmatrix} -\dfrac{1-\xi}{2} & 0 & -\dfrac{1+\xi}{2} & 0 & 1 & 0 \\ 0 & -\dfrac{1-\xi}{2} & 0 & -\dfrac{1+\xi}{2} & 0 & 1 \end{bmatrix}, \tag{6.68}$$

and the matrix of the derivatives $\mathbf{A}'$

$$\mathbf{A}' = \frac{1}{2}\begin{bmatrix} 1 & 0 & -1 & 0 & 0 & 0 \\ 0 & 1 & 0 & -1 & 0 & 0 \end{bmatrix}, \tag{6.69}$$

are used to approximate the displacement field as well as the derivatives. The displacement vector $\mathbf{u}$ is defined as

$$\mathbf{u}^T = \{u_x^{(1)}, u_y^{(1)}, u_x^{(2)}, u_y^{(2)}, u_x^{(S)}, u_y^{(S)}\}^T. \tag{6.70}$$

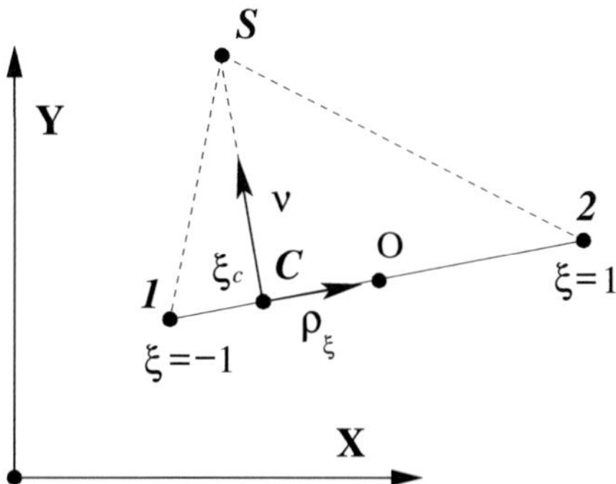

Figure 6.3: Linear contact element. Node-to-segment approach.

6.5.1.1 Closest point projection procedure

Consider now the increment $\Delta\xi$ in load step (m) for the closest point procedure in eqn. (6.4) in order to define the internal coordinate ξ_C of the projection point C. Due to the linear approximation in the contact element we get $\rho_{\xi\xi} = 0$.

$$\begin{aligned}
\Delta\xi &= \frac{(\mathbf{r}_s - \boldsymbol{\rho}) \cdot \boldsymbol{\rho}_\xi}{(\boldsymbol{\rho}_\xi \cdot \boldsymbol{\rho}_\xi)} = \\
&= \frac{4}{\|\mathbf{x}^2 - \mathbf{x}^1\|^2} \left(\mathbf{x}^S - \frac{\mathbf{x}^1 + \mathbf{x}^2}{2} - \frac{\mathbf{x}^2 - \mathbf{x}^1}{2}\xi^{(m)} \right) \cdot \frac{\mathbf{x}^2 - \mathbf{x}^1}{2} = \\
&= \frac{2\mathbf{x}^S \cdot (\mathbf{x}^2 - \mathbf{x}^1)}{\|\mathbf{x}^2 - \mathbf{x}^1\|^2} - \frac{\|\mathbf{x}^2\|^2 - \|\mathbf{x}^1\|^2}{\|\mathbf{x}^2 - \mathbf{x}^1\|^2} - \xi^{(m)}.
\end{aligned} \tag{6.71}$$

It is obvious that the Newton update scheme $\xi^{(m+1)} = \xi^{(m)} + \Delta\xi$ in eqn. (6.4) is independent of the initial guess $\xi^{(m)}$, therefore, the last expression in eqn. (6.71) together with this scheme leads to the exact value for the internal coordinate ξ_C:

$$\xi_C = \frac{2\mathbf{x}^S \cdot (\mathbf{x}^2 - \mathbf{x}^1)}{\|\mathbf{x}^2 - \mathbf{x}^1\|^2} - \frac{\|\mathbf{x}^2\|^2 - \|\mathbf{x}^1\|^2}{\|\mathbf{x}^2 - \mathbf{x}^1\|^2}. \qquad (6.72)$$

The result can be also interpreted as convergence in one iteration. Since the geometry of this element is quite simple, the same result can be also obtained directly via the triangle in Fig. 6.3.

A simple searching algorithm leads to computations of all further components for a contact element at the projection point ξ_C only if the slave point S is projected onto this element, i.e. if $|\xi_C| \leq 1$.

6.5.1.2 Return-mapping scheme

In order to define the tangent traction vector $\mathbf{T}$, we apply the standard return-mapping scheme based on the elasto-plastic analogy, as is e.g. presented in the books of Wriggers [188] and Laursen [106]. The trial tangential traction is computed via the evolution equation (6.36), which in the case of a linear approximation is reduced to:

$$\dot{T} = -\epsilon_T(\boldsymbol{\rho}_\xi \cdot \boldsymbol{\rho}_\xi)\dot{\xi}. \qquad (6.73)$$

We consider here the simplest case for frictional problems: a quasi-statical motion leading to the development from sticking to sliding. In addition, we assume that during motion a slave point is not crossing an element boundary. Thus, for our case we need only one additional variable ξ^0 per contact element in order to define the initial position of the slave point on the master segment. The trial tangent traction is obtained via the application of the backward Euler scheme within the evolution equation (6.73). Since we have a linear approximation for the geometry, we have

$$T^{(m+1)} = T^{(m)} - \epsilon_T a_{11}(\xi^{(m+1)} - \xi^{(m)}) = \qquad (6.74)$$

and continuing recursively, we obtain

$$= T^{(m-1)} - \epsilon_T a_{11}(\xi^{(m+1)} - \xi^{(m-1)}) = \ldots = T^{(0)} - \epsilon_T a_{11}(\xi^{(m+1)} - \xi^0).$$

Assuming in addition, that at the initial position the tangential traction

$T^{(0)}$ was zero, we obtain

$$T_{tr}^{(m+1)} = -\epsilon_T a_{11}(\xi^{(m+1)} - \xi^0). \qquad (6.75)$$

The last eqn. (6.75) serves now to compute the trial tangential reaction.

The return-mapping following the Coulomb friction condition leads then with:

$$N^{(m+1)} = \epsilon_N \zeta^{(m+1)},$$

to

$$T^{(m+1)} = \begin{cases} T_{tr}^{(m+1)} & if \;\; |T_{tr}^{(m+1)}| < \mu|N^{(m+1)}|\sqrt{a_{11}} \;\; (sticking) \\[2mm] \mu|N^{(m+1)}|\sqrt{a_{11}}\,sgn(T_{tr}^{(m+1)}) & if \;\; |T_{tr}^{(m+1)}| \geq \mu|N^{(m+1)}|\sqrt{a_{11}} \;\; (sliding) \end{cases} \qquad (6.76)$$

The inequality condition in eqn. (6.76) is obtained from the following:

$$\|\mathbf{T}_{tr}^{(m+1)}\| < \mu|N| \implies \frac{T_{tr}^{(m+1)}}{\sqrt{a_{11}}} < \mu|N|, \qquad (6.77)$$

where eqn. (6.23) for the absolute value of the covariant vector is taken into account.

The global solution scheme for the simplest case discussed here is presented in Table 6.1.

6.6 Treatment of special cases

In this section, we consider, how to treat some particularities, which were mentioned and excluded in section 6.5.1.2. The first problem is arising when the applied load is not simply modified proportionally. In this situation a trial load can not be computed only via eqn. (6.75), because the attraction point ξ^0 must be updated. Thus we have to extend the algorithm as is shown in the following. The second problem is arising when the projection point is crossing an element boundary during the incremental loading. In this case, the computation according to eqn. (6.75) will produce a jump, because the convective coordinate ξ belongs to different elements, see Wriggers [188] and Laursen [106].

6.6.1 Update of the sliding displacements in the case of reversible loading

We consider a geometrical interpretation of the return-mapping scheme in eqn. (6.76) together with the evolution equation (6.75), see Fig. 6.4.

$$|T_{tr}^{(m)}| < \mu|N^{(m)}|\sqrt{a_{11}} \quad \Longrightarrow \quad \epsilon_T|\xi^{(m)} - \xi^0| < \mu|N^{(m)}| \tag{6.78a}$$

$$|\xi^{(m)} - \xi^0| < R_\xi^{(m)}, \quad R_\xi^{(m)} = \frac{\mu|N^{(m)}|}{\epsilon_T} \tag{6.78b}$$

Eqn. (6.78b) describes an allowable elastic region $\mathbf{A}^{(m)}\mathbf{B}^{(m)}$ with a center of attraction $\mathbf{O}^{(m)}$. All points inside this domain are in "sticking condition". If now a point $\xi^{(m+1)}$ appears to be outside of the domain at load step $(m + 1)$, then its only admissible position is on the boundary of the domain, i.e. must coincide with $\mathbf{B}^{(m+1)}$. A sliding force is applied then at the contact point, see eqn. (6.76). As long as we have a motion of the contact point only in one direction the sign function for the sliding force $sgn(T_{tr}^{(m+1)}) = sgn(\Delta\xi^{(m+1)})$ does not change and the computation will be correct. However, when a reversible load is applied which forces the contact point to move forward or backward, the attraction point $\mathbf{O}^{(m)}$ must be updated, in order to define the sign function for the sliding force correctly. This update can be defined geometrically from Fig. 6.4:

$$|\Delta\xi^{(m+1)}| = |\Delta\xi_{sl}| + R_\xi^{(m+1)} = |\Delta\xi^{(m+1)}| - \frac{\mu|N^{(m)}|}{\epsilon_T}. \tag{6.79}$$

The absolute value of the sliding displacement is then computed at load step $(m + 1)$ as:

$$|\Delta\xi_{sl}| = |\xi^{(m+1)} - \xi^{(0)}| - \frac{\mu|N^{(m+1)}|}{\epsilon_T}, \tag{6.80}$$

and the updated center of the elastic domain becomes:

$$\xi_c^{(up)} = \xi^{(0)} + sgn(\xi^{(m+1)} - \xi^{(0)})|\Delta\xi_{sl}|. \tag{6.81}$$

For the next step, the evolution equation (6.75) is corrected as

$$T^{(m+2)} = -\epsilon_T a_{11}(\xi^{(m+2)} - \xi_c^{up}).$$

(6.82)

Remark.

As an alternative procedure the back-substitution of the evolution equation (6.75) into eqn. (6.80) gives the updated scheme via the trial force:

$$|\Delta\xi_{sl}| = \frac{1}{\epsilon_T}\left(|T_{tr}^{(m+1)}| - \mu|N^{(m+1)}|\right),$$

(6.83)

which can be found e.g. in Wriggers [188].

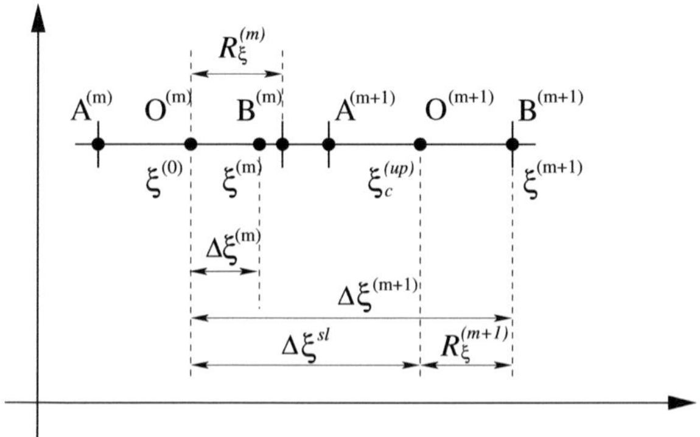

Figure 6.4: Coulomb friction. Updating of sliding displacements in convective coordinates. Motion of friction cone and center of attraction.

6.6.2 Crossing an element boundary – continuous integration scheme

Consider two adjacent elements $\mathbf{I}^{(m)}\mathbf{J}^{(m)}$ and $\mathbf{J}^{(m)}\mathbf{L}^{(m)}$ at load step (m), see Fig. 6.5. Let the contact point $\mathbf{S}^{(m)}$ be projected onto the element $\mathbf{I}^{(m)}\mathbf{J}^{(m)}$ and its projection is a point $\mathbf{K}^{(m)}$ with the convective coordinate $\xi^{(m)}$. At load step $(m+1)$ these two elements move into the position $\mathbf{I}^{(m+1)}\mathbf{J}^{(m+1)}$ and $\mathbf{J}^{(m+1)}\mathbf{L}^{(m+1)}$, but the contact point is moved into position $\mathbf{S}^{(m+1)}$ and is projected now onto element $\mathbf{J}^{(m+1)}\mathbf{L}^{(m+1)}$. We assume that the angular deformation of these elements is small in comparison with a rigid body motion, thus the elements are drawn as straight lines. It is obvious, that the direct computation according to the evolution equation (6.75) results in a jump. Let e.g. the point $\mathbf{K}^{(m)}$ have the convective coordinate $\xi^{(m)} = 0.9$ close to the right element boundary, (see parameterization in Fig. 6.3), and then point $\mathbf{K}^{(m+1)}$ has $\xi^{(m+1)} = -0.9$ close to the left element boundary. The distance between them is only $\Delta\xi = 0.2$, but the evolution equation gives:

$$T_{tr}^{(m+1)} = -\epsilon_T a_{11}(\xi^{(m+1)} - \xi^{(m)}) = 1.8\epsilon_T. \tag{6.84}$$

The maximum possible jump following this straightforward action is easily determined from the limit values of the convective coordinates:

$$T_{jump} = -\epsilon_T a_{11}\left(\lim_{\xi\to-1+0}(\xi)_{\xi\in\mathbf{IJ}} - \lim_{\xi\to+1-0}(\xi)_{\xi\in\mathbf{JL}} \right) = 2\epsilon_T a_{11}. \tag{6.85}$$

This jump appears only due to the different approximation of the adjacent elements. In order to overcome this, we can compute the force in geometrical form. The incremental tangential displacements $\Delta\rho$ can be expressed in the metrics of the second element $\mathbf{J}^{(m+1)}\mathbf{L}^{(m+1)}$ at time step $(m+1)$:

$$\Delta\boldsymbol{\rho} = \Delta\xi\boldsymbol{\rho}_\xi^{(m+1)}, \tag{6.86}$$

and alternatively, it can be geometrically defined via the incremental displacement vector $\Delta\mathbf{u}$:

$$\Delta\boldsymbol{\rho} = \boldsymbol{\rho}(\xi^{(m+1)})_{\xi\in\mathbf{JL}} - \left(\boldsymbol{\rho}(\xi^{(m)}) + \Delta\mathbf{u}(\xi^{(m)}) \right)_{\xi\in\mathbf{IJ}}. \tag{6.87}$$

Then $\Delta\xi$ is defined as

$$\Delta\xi = \frac{\left(\boldsymbol{\rho}(\xi^{(m+1)})_{\xi\in\mathbf{JL}} - \left(\boldsymbol{\rho}(\xi^{(m)}) + \Delta\mathbf{u}(\xi^{(m)}) \right)_{\xi\in\mathbf{IJ}} \right) \cdot \boldsymbol{\rho}_\xi^{(m+1)}}{a_{11}^{(m+1)}} \tag{6.88}$$

and the evolution equation becomes

$$T_{tr}^{(m+1)} = T^{(m)} - \epsilon_T a_{11}^{(m+1)}\Delta\xi. \tag{6.89}$$

Modifications of the global solution scheme given in Table 6.1 are represented in Table 6.2 according to the special cases.

6.6.3 Remarks on additional developments

One can see, that the continuous integration scheme as presented in Section 6.6.2 leads to an increasing number of history variables, in fact, in addition to $\xi^{(m)}$ the vector $\rho(\xi^{(m)})$ must be stored. Moreover, other history variables such as the updated sliding displacements must be transfered in a similar fashion. The continuous integration scheme, of course, is particularly important for contact problems with singularities, e.g. sliding of an edge along a curve as also shown in Fig. 6.5. For other cases with nonsingular geometry it is more efficient to exclude contact points, once they appear outside the master element, but then, as a compensation, introduce additional contact points in the slave segment, e.g. integration points, within the so-called segment-to-segment approach, for details see Zavarise and Wriggers [199] especially for the 2D case and Harnau, Konyukhov and Schweizerhof [57] for the 3D case. Another approach to increase the number of contact points is the mortar method, see e.g. in McDevitt and Laursen [123] and recent developments in Puso and Laursen [150].

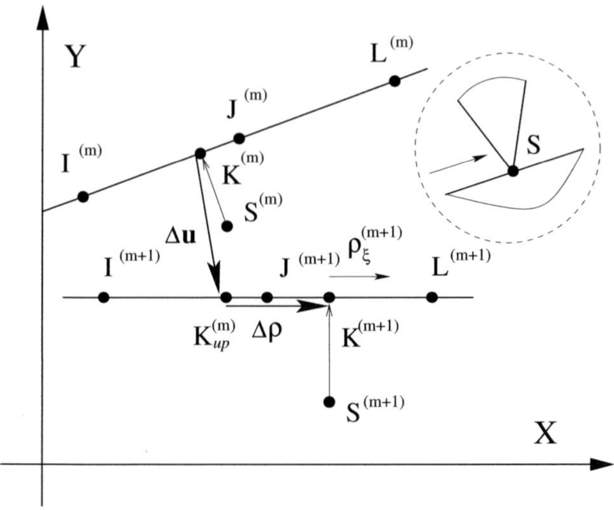

Figure 6.5: Crossing an element boundary within a load increment. Typical case for the continuous integration scheme.

Another important problem arises due to the non-smoothness of contact boundaries, if for the boundary a low order finite element mesh is used. This leads to jumps in both the normal and the tangential characteristics when crossing the element boundary see Fig. 6.6, because neither a normal vector $\mathbf{n}$, nor a tangent vector ρ_ξ are defined at the edge point B. If the real boundary is an edge then e.g. adaptive methods can improve the result for a straight geometry, see Wriggers and Scherf [191] and Wriggers [188]. If the real boundary is smooth, then various smoothing techniques based on the approximation of the boundary with e.g. splines can be used. There are numerous publications on this subject, see Wriggers et. al. [190], Padmanabhan and Laursen [134] and Stadler et.al. [169] especially for 2D problems, and then in Puso and Laursen [148], Krstulovic-Opara et. al. [102], Stadler and Holzapfel [168] for 3D problems. In this case, the geometrical singularity is removed, i.e. the normal $\mathbf{n}$ and the tangent vector ρ_ξ are uniquely defined at point B. However, the continuous integration scheme in eqn. (6.88-6.89) is still necessary, as the smooth patches have in all above publications local support, i.e. their convective coordinate is defined separately.

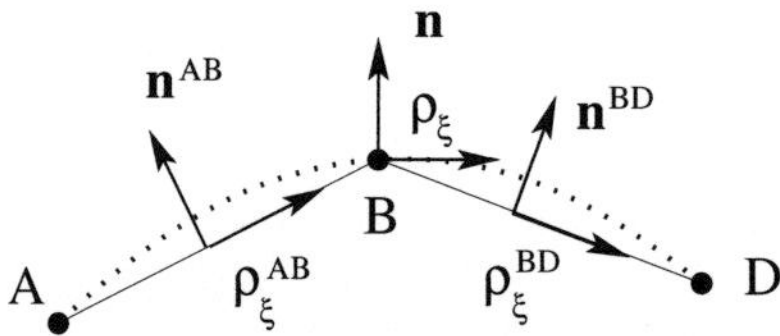

Figure 6.6: Low-order finite element approximation. Approximation of a real boundary with a smooth curve.

6.7 Numerical examples

6.7.1 Sliding of a block. Linear approximation of the contact surfaces. Reversible loading process

We consider the sliding of an elastic block similar as shown in [89], however, here the block will be loaded with horizontally prescribed reversible displacements. The main point is to show the update procedure for sliding displacements and investigate the development of the

Table 6.1: Global solution scheme for a linear contact element.

1. Initialization of the convective coordinate ξ^0.
   Initial condition for the evolution equation for all contact elements:
   Compute projection point ξ_C in eqn. (6.72), with no external loads $\longrightarrow \xi^0 = \xi_C$.

2. Loop over load increments m, $m = 1, ..., m_{end}$.

3. Loop over global Newton iterations i, $i = 1, ..., i_{end}$.

4. Loop over all contact elements

   - compute projection points $\xi_C^{(i)}$ eqn. (6.72). **If** $|\xi_C^{(i)}| \geq 1$ **then exit loop 4.**
   - check penetration $\zeta^{(i)} = (\mathbf{r}_s - \rho) \cdot \boldsymbol{\nu}$. **If** $\zeta^{(i)} > 0$ **then exit loop 4.**
   - compute contact tractions and corresponding tangent matrices at contact point ξ_C.

   Normal traction: $N^{(i)} = \epsilon_N \zeta^{(i)}$

   Tangent matrix $\mathbf{K^N}$ for normal traction

   $$\mathbf{K^N} = \epsilon_N \, \mathbf{A}^T (\boldsymbol{\nu} \otimes \boldsymbol{\nu}) \mathbf{A} + \frac{\epsilon_N \, \zeta^{(i)}}{(\rho_\xi \cdot \rho_\xi)} \left(\mathbf{A}'^T (\boldsymbol{\nu} \otimes \rho_\xi) \mathbf{A} + \mathbf{A}^T (\rho_\xi \otimes \boldsymbol{\nu}) \mathbf{A}' \right)$$

   Trial tangent traction: $T_{tr}^{(i)} = -\epsilon_T (\rho_\xi \cdot \rho_\xi)(\xi_C^{(i)} - \xi^0)$
   Real tangent traction T and corresponding matrices
   are defined via the return-mapping algorithm

<table>
<tr><td>

if $|T_{tr}^{(i)}| \leq \mu |N^{(i)}| \|\rho_\xi\|$

sticking

$T^{(i)} = T_{tr}$

Tangent matrix $\mathbf{K^T}$

$\mathbf{K^T} = -\dfrac{\varepsilon_T}{(\rho_\xi \cdot \rho_\xi)} \mathbf{A}^T (\rho_\xi \otimes \rho_\xi) \mathbf{A} +$

$+ \dfrac{\varepsilon_T}{(\rho_\xi \cdot \rho_\xi)^2} \left[\mathbf{A}'^T (\rho_\xi \otimes \rho_\xi) \mathbf{A} + \mathbf{A}^T (\rho_\xi \otimes \rho_\xi) \mathbf{A}'^T \right]$

</td><td>

if $|T_{tr}^{(i)}| > \mu |N^{(i)}| \|\rho_\xi\|$

sliding

$T^{(i)} = \mu |N^{(i)}| \|\rho_\xi\|$

Tangent matrix $\mathbf{K^T}$

$\mathbf{K^T} = -\dfrac{\varepsilon_N \, \mu \, sgn(T_{tr}^{(i)})}{(\rho_\xi \cdot \rho_\xi)^{1/2}} \mathbf{A}^T (\rho_\xi \otimes \boldsymbol{\nu}) \mathbf{A} +$

$+ \dfrac{\mu |N| sgn(T_{tr}^{(i)})}{(\rho_\xi \cdot \rho_\xi)^{3/2}} \left[\mathbf{A}'^T (\rho_\xi \otimes \rho_\xi) \mathbf{A} + \right.$

$\left. + \mathbf{A}^T (\rho_\xi \otimes \rho_\xi) \mathbf{A}'^T \right]$

</td></tr>
</table>

   - Compute the full contact tangent matrix $\mathbf{K} = \mathbf{K^N} + \mathbf{K^T}$
   - Compute residual $\mathbf{R}$

   $$\mathbf{R}_N = N^{(i)} \mathbf{A}^T \boldsymbol{\nu}; \qquad \mathbf{R}_T = \frac{T^{(i)}}{(\rho_\xi \cdot \rho_\xi)} \mathbf{A}'^T \rho_\xi,$$

   $$\mathbf{R} = \mathbf{R}_N + \mathbf{R}_T.$$

end loop over contact elements
end loop over global Newton iterations
end loop over load increments

Table 6.2: Modifications of the global solution scheme according to special cases.

1. Initialization of the convective coordinate ξ^0.
   Initial condition for the evolution equation for all contact elements:
   Compute projection point ξ_C in eqn. (6.72), with no external loads $\longrightarrow \xi^0 = \xi_C$.

2. Loop over load increments m, $m = 1, ..., m_{end}$.

3. Loop over global Newton iterations i, $i = 1, ..., i_{end}$.

4. Loop over all contact elements

   - compute projection points $\xi_C^{(i)}$ eqn. (6.72). **If** $|\xi_C^{(i)}| \geq 1$ **then exit loop 4.**

   - check penetration $\zeta^{(i)} = (\mathbf{r}_s - \rho) \cdot \boldsymbol{\nu}$. **If** $\zeta^{(i)} > 0$ **then exit loop 4.**

   - compute contact tractions and corresponding tangent matrices at contact point ξ_C.

   Compute **normal traction**: $N^{(i)}$ and corr. matrix, see Table 6.1

   Trial tangent traction $T_{tr}^{(i)}$ according to the specific algorithm:

a) **reversible loading**	b) **continuous integration**
$T^{(i)} = -\epsilon_T a_{11}(\xi^{(i)} - \xi_c^{up})$.	$\Delta\xi$ see eqn. (6.88) $T_{tr}^{(i)} = T^{(m-1)} - \epsilon_T a_{11}^{(i)} \Delta\xi$.

   - Compute **real tangent traction** $T^{(i)}$ and corr. tangent matrices $\mathbf{K^T}$ according to the return mapping scheme, see Table 6.1

   - Compute the full contact tangent matrix $\mathbf{K} = \mathbf{K^N} + \mathbf{K^T}$, see Table 6.1

   - Compute residual $\mathbf{R}$, see Table 6.1

end loop over contact elements
end loop over global Newton iterations

    Update and store necessary history variables

a) **reversible loading**	b) **continuous integration**
Compute and store the update center $\xi_c^{(up)}$ according to eqn. (6.81)	Store history variables for $\Delta\xi$ according to eqn. (6.88) and $T^{(m)}$.

end loop over load increments

sticking-sliding zone. All numerical investigations are performed with FEAP-MeKa [172] including the implementation of the presented algorithms.

As an example for the computation, we consider a rectangular block (Fig. 6.7) resting on a surface with the following parameters: elasticity modulus $E = 2.1 \cdot 10^4$, Poisson ratio $\nu = 0.3$, length $a = 20$, height $b = 5$. The dimensions are assumed to be consistent. The block is uniformly meshed by linear finite elements: 40 elements in horizontal direction and 10 elements in vertical direction. The lower surface represents a rigid base. Coulomb friction with a coefficient $\mu = 0.3$ is specified between the surfaces. The contact surface of the deformable block is assumed to be a "master", while the surface of the rigid base is a "slave" surface within the "node-to-segment" approach. The penalty parameters are chosen as $\varepsilon_N = \varepsilon_T = 2.1 \cdot 10^6$. The loading is applied as prescribed displacements at the top side of the deformable block. This example is chosen to show the robustness of the contact algorithm and the update scheme within a covariant description, though the current results can be achieved certainly with other known techniques.

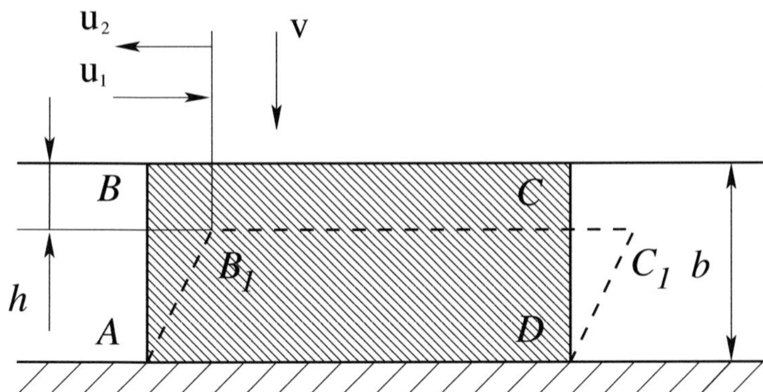

Figure 6.7: Plane deformation of a block. Applied displacement loading at top of the block.

The main question arising before the computation is, which size of the load step is allowed, as it is important to capture the spreading of the sticking-sliding zone correctly. We use the concept of the critical displacement u_{cr}, introduced in [89], namely, a value of the applied displacements after which the block fully slides. As shown for the infinite layer this value is given as

$$u_{cr} = \frac{2\mu h}{1 - \nu}.$$

(6.90)

The estimation of the critical horizontal displacement in eqn. (6.90) gives $u_{cr} = 6.0 \cdot 10^{-3}$, thus in order to investigate the sticking-sliding zone properly we choose a displacement increment $\Delta u = 1.0 \cdot 10^{-4}$. The displacements are applied according to the loading process given in Table 6.7.1.

No. l.s.	Δu	$u \cdot 10^{-3}$	$\Delta v \cdot 10^{-3}$	v	Loading
0	0.0	0.0	0.0	0	initialization of conv. coord.
1	0.0	0.0	$7.0 \cdot 10^{-3}$	$0.0 - 7.0$	vertical displ. v
2–80	10^{-4}	$0.0 - 8.0$	0.0	7.0	forward horizontal displ. u_1
81–84	$-2.5 \cdot 10^{-5}$	$8.0 - 7.9$	0.0	7.0	backward horizontal displ. u_2
85–163	$-1.0 \cdot 10^{-4}$	$7.9 - 0.0$	0.0	7.0	backward horizontal displ. u_2
164–280	$-1.0 \cdot 10^{-4}$	$0.0 - (-1.17)$	0.0	7.0	backward horizontal displ. u_2

Table 6.3: Plane deformation of a block. Loading procedure with prescribed displacements on the top side of the deformable block.

As a consequence of the reversible loading a hysteresis curve as shown in Fig. 6.8 is developed. The applied displacement at point **C**, see Fig. 6.7, is depicted along the x-axis, and the computed horizontal displacement at point **D** is depicted along the y-axis. We obtain a spreading of the sliding zone during the forward loading process (curve **OF**) as well as during the backward loading process (curve **FG**). The horizontal displacements u along the contact line **AD**, see Fig. 6.7, together with the distribution of the stress ratio T/N allow us to define the sticking-sliding zone during the loading process.

Forward loading. Path OF on the hysteresis curve. The horizontal displacement distribution as well as the stress ratio T/N distribution on the contact boundary are shown in Fig. 6.9 and Fig. 6.10 for the following loading points:

a) Load step 1, see Table 6.7.1, i.e. only vertical displacements are applied $v = 0.007$. This is the starting point **O** on the hysteresis curve Fig. 6.8;

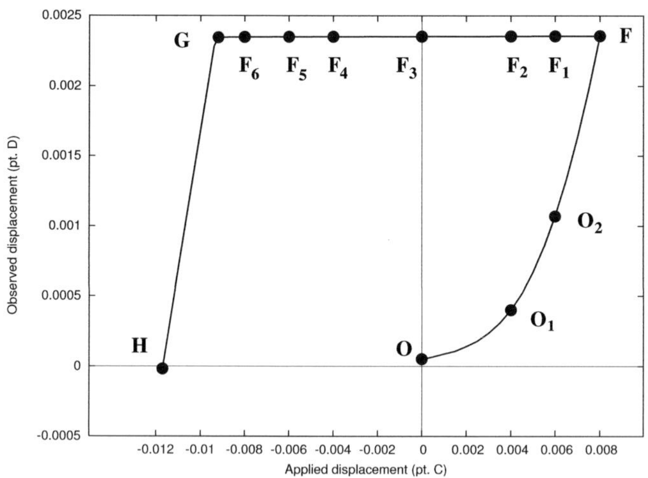

Figure 6.8: Plane deformation of a block. Hysteresis curve. Observed horizontal displacement at point D vs. applied horizontal displacement at point C.

b) Intermediate points with applied horizontal displacements $u = 0.0040$ and $u = 0.0060$, resp. points O_1 and O_2 on the hysteresis curve Fig. 6.8;

c) The load step No. 80 with $u = 0.0080$, see Table 6.7.1, is chosen as a final point of the forward loading, see also the point $\mathbf{F}$ on the hysteresis curve Fig. 6.8.

The development of the sliding zone with increasing displacement loading u is given in Fig. 6.9 and the development of the stress ratio T/N in Fig. 6.10. At the end of the forward loading the sliding zone is increased to about $8 \leq x \leq 20$.

Backward loading. Part FGH on the hysteresis curve. The horizontal displacement distribution and the contact stress ratio T/N distribution on the contact boundary are shown in Fig. 6.11 and in Fig. 6.12 for the following loading points:

a) Last load step of the forward loading with $u = 0.0080$, resp. point $\mathbf{F}$ on the hysteresis curve Fig. 6.8 as a starting point;

b) Intermediate points on the unloading part $\mathbf{FG}$ with applied horizontal displacements $u = 0.0040$, $u = 0.000$, $u = -0.0040$, $u = -0.0060$,

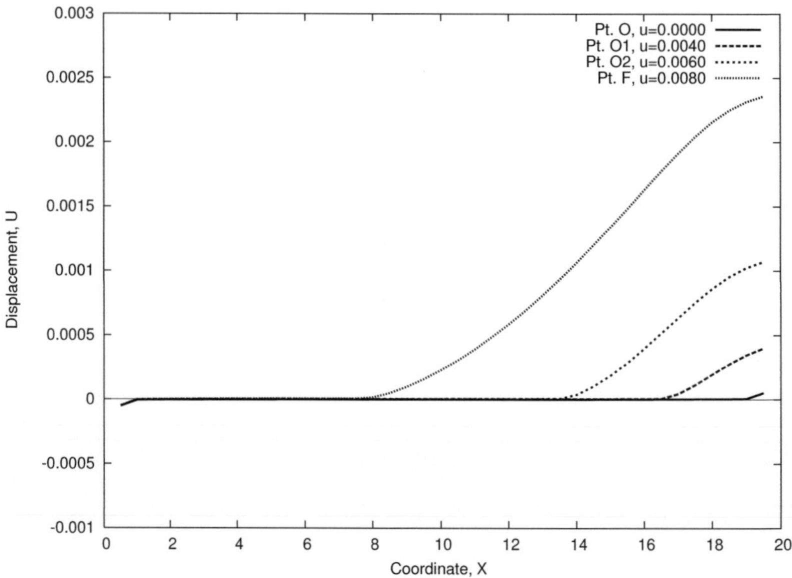

Figure 6.9:   Plane deformation of a block. Horizontal displacement of the contact boundary. Forward loading.

$u = -0.0080$, resp. points $\mathbf{F}_2$, $\mathbf{F}_3$, $\mathbf{F}_4$, $\mathbf{F}_5$, $\mathbf{F}_6$ on the hysteresis curve;

c) Selected point with $u = 0.010$ on the full backward sliding part $\mathbf{GH}$ of the hysteresis curve.

When the load is reversed, starting from load step No. 81, see Table 6.7.1, all points on the contact boundary are sticking. On the contact boundary we have so-called residual horizontal displacements (in analogy to plasticity). During the following unloading back to $u = 0.000$ the whole boundary is still sticking: points $\mathbf{F}_1$, $\mathbf{F}_2$, $\mathbf{F}_3$. The latter we can only detect from the stress ratio diagram in Fig. 6.12, where the curves vary inside the layer $-0.3 \leq T/N \leq 0.3$. The residual displacements are not changing, see diagram 6.11, until sliding is beginning. Starting from the applied displacements $u = -0.0040$ (point $\mathbf{F}_4$) we can detect the beginning of sliding at the left corner of the block, as the stress ratio curve is approaching its limit ratio 0.3. The final part of the hysteresis diagram, from point $\mathbf{F}_4$ to point $\mathbf{G}$, is responsible for the development of the sliding zone in the backward direction, which can be observed either

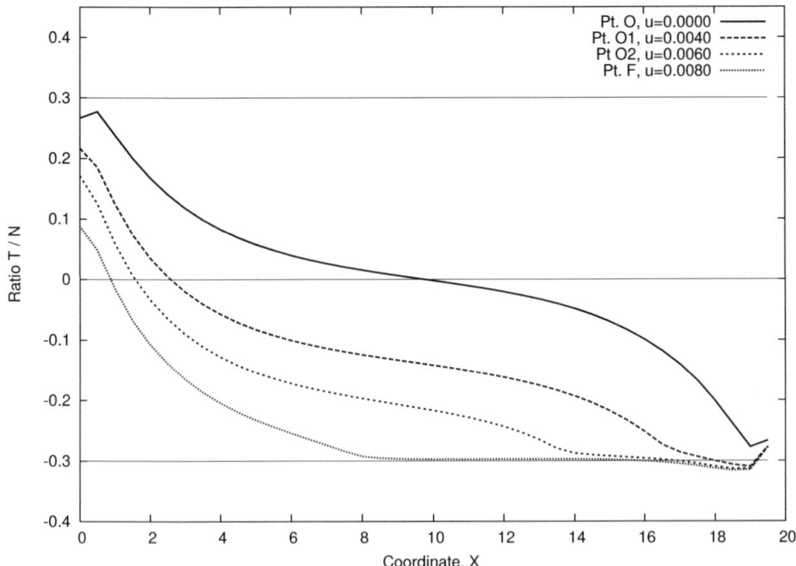

Figure 6.10: Plane deformation of a block. Stress ratio T/N on the master contact surface. Forward loading.

by the development of the horizontal displacements in the left part of the displacement diagram 6.11, or by the development of the zone with a stress ratio $T/N = 0.3$ in diagram 6.12. Full sliding of the block starts at point **G** with $u \approx -0.0092$. Beyond that, horizontal displacements on the contact are changing proportionally to the applied displacement loading, i.e. linearly, as we see from the linear part **GH** in the hysteresis curve.

Remark.

In the presented example the displacements are small and slaves nodes never cross the element boundaries, therefore the continuous integration scheme is not necessary. In the next example we show a particular case with large sliding in which the application of the continuous integration scheme is necessary.

6.7.2 Drawing of an elastic strip into a channel with sharp corners

In this section, we consider a special contact case, for which the application of the continuous integration scheme described in sect. 6.6.2 is absolutely necessary. An elastic strip **AD**, see Fig. 6.13 ($E = 2.1 \cdot 10^4$, $\nu = 0.3$, thickness $h = 0.5$, length $L = 24$) is positioned at the beginning of a channel with width $a = 13$. The corners of the channel are rather sharp. The channel itself is modeled by two rigid blocks B_1 and B_2. The strip is loaded incrementally by a prescribed displacement v at the center until it is inserted into the channel, see Fig. 6.14. The strip here is modeled with 24 linear solid-shell elements, see Hauptmann and Schweizerhof [60] and Hauptmann et.al. [59], and due to symmetry only one half of the system is modeled. The crucial point during the analysis is the sliding of a sharp corner **C** over the element boundaries **1**, **2**, **3**, see Fig. 6.14. A load-displacement curve computed for the loading point is chosen as the representative parameter to compare various contact approaches. The following variations were investigated:

1. Non-frictional case with the "node-to-surface" approach without the continuous integration scheme proposed in eqn. (6.88) and (6.89).

2. Non-frictional case with the "segment-to-segment" approach. Here the number of integration points in the contact segment is varied.

3. Frictional case with the "node-to-surface" approach without the continuous integration scheme.

4. Frictional case with the "node-to-surface" approach with the continuous integration scheme.

5. Frictional case with the "segment-to-segment" approach. Here the number of integration points in the contact segment are varied.

We start the investigation with the non-frictional problem (case 1, 2) applying the load increment $v = 0.005$ with the penalty parameter $\varepsilon_N = 2.1 \cdot 10^5$. The elements from the strip are chosen to be a master, while the sharp corner is a singular slave node. The load-displacement curve for the "node-to-surface" approach contains a jump when the sharp corner is crossing the boundary nodes **1** resp. **2**, see Fig. 6.14.

The solution process is no longer converging after the sharp corner is crossing the boundary node **3**. As an alternative for improvements, we chose the "segment-to-segment" approach, described in Harnau, Konyukhov and Schweizerhof [57]. The sharp corner **C** is modeled with two slave segments which are orthogonal to each other and take contact points as the Lobatto integration points. Taking 2 integration points with 2 sub-domains, or e.g. 5 integration points allows to compute the full load-displacement curve without the jumps obtained with NTS scheme, see Fig. 6.15. The smoothing effect in the last cases happens because the contact is checked not only against the single edge node, but also against the set of contact points which covers fairly densely the sharp edge.

The next study is devoted to the frictional problem (case 3, 4) with the load increment $v = 0.0025$, the penalty parameters $\varepsilon_N = 2.1 \cdot 10^5, \varepsilon_T = 2.1 \cdot 10^5$ and a friction coefficient $\mu = 0.2$. The straightforward analysis without the continuous integration scheme (case 3) leads to a jump in the force-displacement curve when the sharp corner is crossing the first boundary node **1**. The solution is no longer converging when the sharp corner is crossing the second boundary node **2**, see Fig. 6.16. The application of the continuous integration scheme, however, allows to obtain the full force-displacement curve, even in the part when the strip is fully inserted into the channel, see the straight part of the curves in Fig. 6.16. For this case a side part of the channel is modeled as a rigid surface described by an analytical function. For comparison, the analysis is carried out with various friction coefficients $\mu = 0.1$ and $\mu = 0.3$, see Fig. 6.16.

It would be favorable also for the frictional case to perform the analysis without the continuous integration scheme avoiding to store a lot of information about slave nodes and apply the "segment-to-segment" approach just increasing the number of integration points (case 2). The results of such an analysis using the same loading parameters as for the analysis in Fig. 6.16 are shown in Fig. 6.17. 5 and 10 Lobatto integration points are taken. Despite the fact that the solution is converging, the load-displacement curve shows large oscillations in the sliding regions after passing node **2**. This result is due to the fact, that from one load step to the other the history variables are not transported correctly. Only the upper envelope of the oscillatory curve could be used as rep-

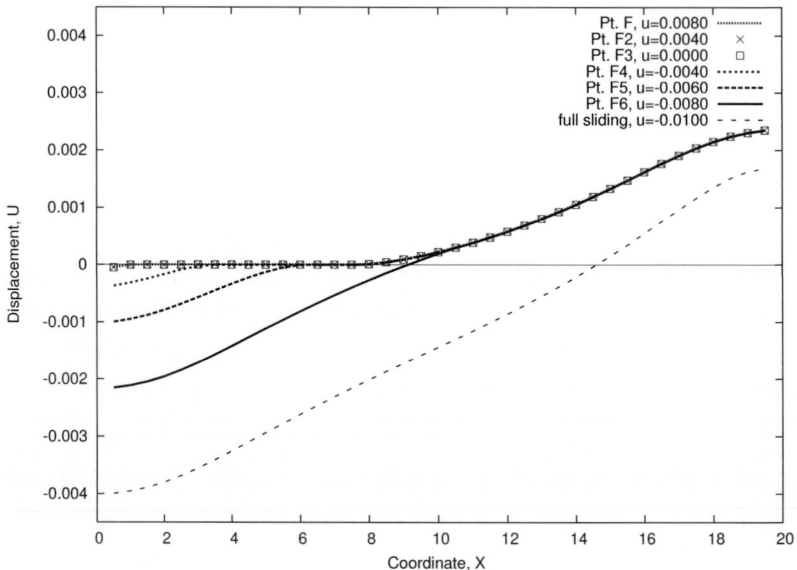

Figure 6.11: Plane deformation of a block. Backward loading. Horizontal displacement of the contact boundary.

resentation of the correct load-displacement curve. This confirms the necessity of the continuous integration scheme in particular for frictional contact.

Remark. We have to note that reversing the master surface, or the so-called symmetric treatment of the contact, in the current example would also not resolve the problem in a sufficient manner.

6.8 Conclusions

In this contribution a convective description was reconsidered for the 2D quasi-statical frictional contact problem. Special attention is paid to the derivation of the necessary equations either as a reduction of the known 3D covariant formulation, or directly from the special 2D cylindrical geometry of the contact surfaces. The algorithmic linearization in the covariant form allows to obtain the tangent matrices before the linearization process. Thus, an implementation can be easily carried out without providing any special attention to the approximation of the

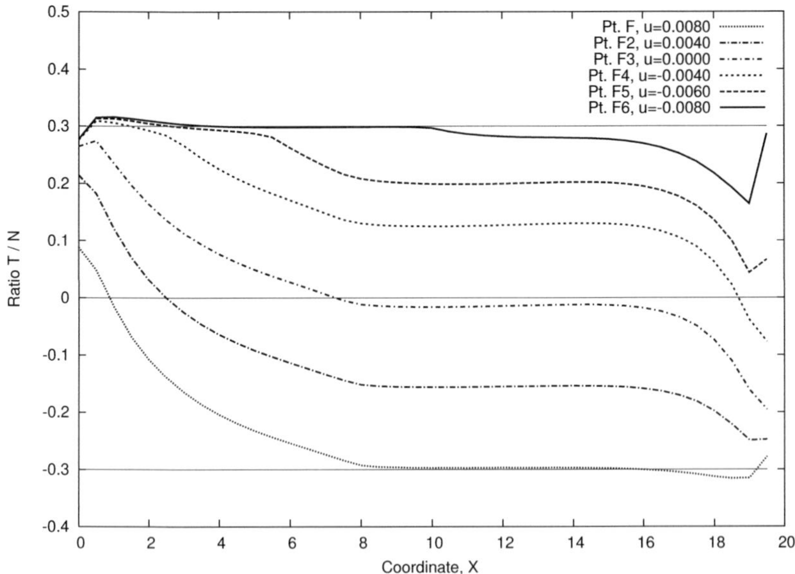

Figure 6.12: Plane deformation of a block. Backward loading. Stress ratio T/N on the master contact surface.

contact surfaces (e.g. finite element mesh, CAD surface etc.). A simple linear contact element is chosen to illustrate the algorithmic implementation into a FE code.

Different situations requiring the application of more advanced techniques, such as an update of sliding displacements and a continuous integration scheme for the frictional case are discussed and illustrated with numerical examples. Thus, the update technique is absolutely necessary for the simulation of reversible loading processes, as the residual deformations have to be described correctly. The continuous integration technique allows to transport all history variables correctly over the contact segment boundaries, however, additional storage is required. In the particular example of a sliding edge on a surface, it is shown that for the non-frictional contact problem the "segment-to-segment" approach with different integration schemes can improve the result, but for the frictional contact problem the continuous integration scheme is absolutely necessary independent of the approaches NTS and STS.

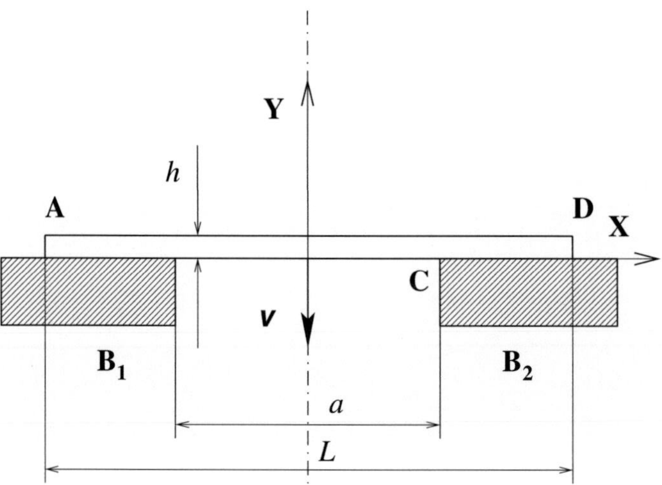

Figure 6.13: Drawing of an elastic strip into a channel with sharp corners. Geometrical parameters.

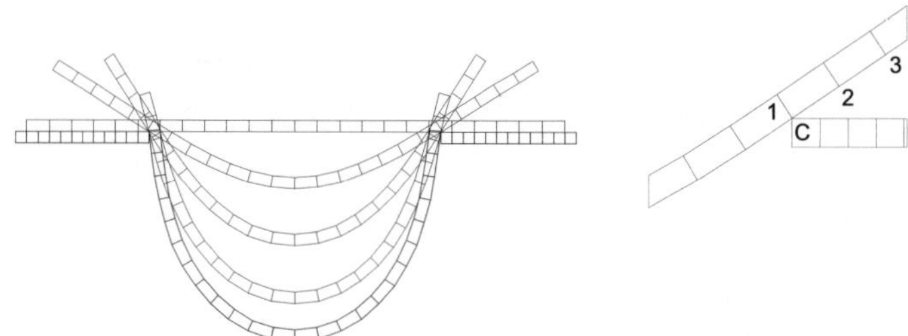

Figure 6.14: Sequence of deformations for the elastic strip. Nodes are sliding over the sharp corner **C**.

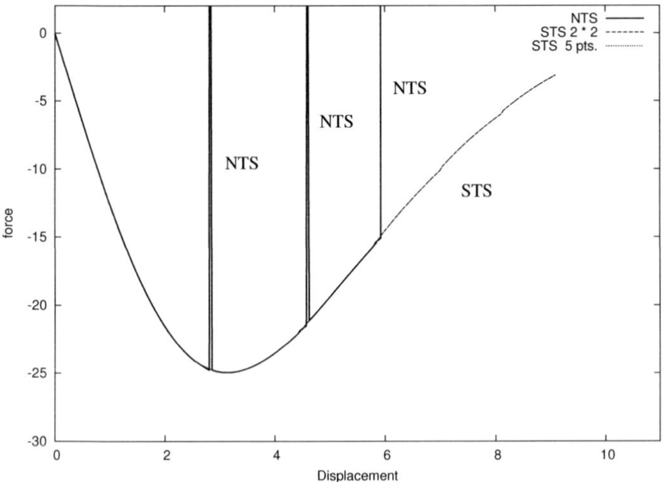

Figure 6.15: Drawing of an elastic strip over a sharp corner. Load-displacement curve. Non-frictional case. NTS and STS approaches with different Lobatto integration – a) 2×2 int. points with 2 sub-domains and b) 5×5 int. points.

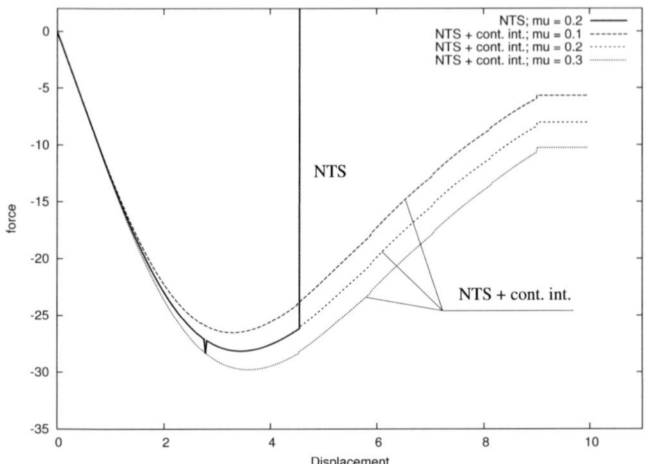

Figure 6.16: Drawing of an elastic strip over a sharp corner. Load-displacement curve. Frictional case – various friction coefficients. Straightforward NTS approach and continuous integration schemes.

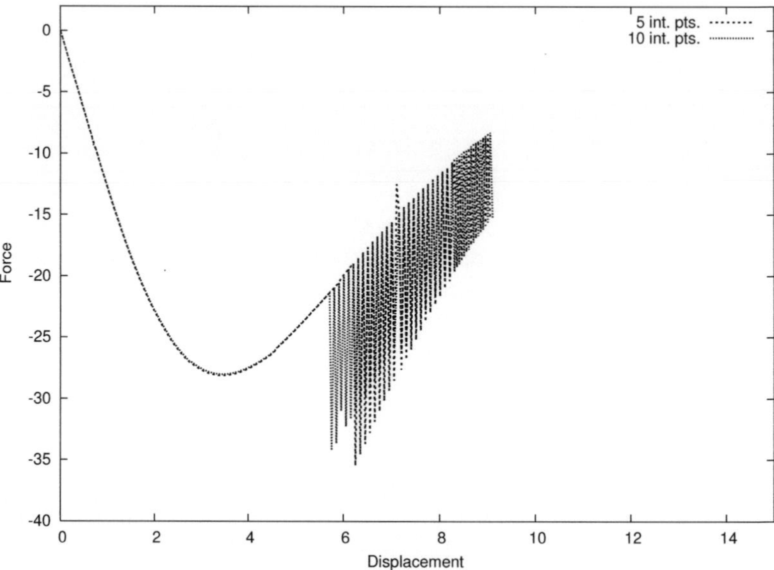

Figure 6.17: Drawing of an elastic strip over a sharp corner. Load-displacement curve. Frictional case. Straightforward application of the "segment-to-segment" scheme without the continuous integration scheme leads to high oscillations also for a larger number of integration points.

6.9 APPENDIX.
Covariant derivative of tangent vector $\mathbf{T}$

The full time derivative of the tangent vector $\mathbf{T}$ is considered in the contravariant basis $\mathbf{r}^1, \mathbf{r}^2$

$$\frac{d\mathbf{T}}{dt} = \frac{d}{dt}\left(T\mathbf{r}^1\right) = \frac{dT}{dt}\mathbf{r}^1 + T\frac{d\mathbf{r}^1}{dt}. \tag{6.91}$$

For the computation we assume, that the component T is a scalar function of t, ξ, i.e. $T = T(t, \xi)$, and the basis vector $\mathbf{r}^1$ depends implicitly on time via the convective coordinates ξ and ζ, see the definition in eqn. (6.17). Then we obtain:

$$\frac{d\mathbf{T}}{dt} = \left(\frac{\partial T}{\partial t} + \frac{\partial T}{\partial \xi}\dot{\xi}\right)\mathbf{r}^1 + T\left(\dot{\xi}\frac{\partial \mathbf{r}^1}{\partial \xi} + \dot{\zeta}\frac{\partial \mathbf{r}^1}{\partial \zeta}\right). \tag{6.92}$$

The partial derivatives of the contravariant basis vectors are expressed via the Christoffel symbols Γ_{ij}^k, see e.g. Marsden and Hughes [121],

$$\begin{aligned}
\frac{\partial \mathbf{r}^1}{\partial \xi} &= \Gamma_{11}^1 \mathbf{r}^1 + \Gamma_{12}^1 \mathbf{r}^2, \\
\frac{\partial \mathbf{r}^1}{\partial \zeta} &= \Gamma_{21}^1 \mathbf{r}^1 + \Gamma_{22}^1 \mathbf{r}^2.
\end{aligned} \tag{6.93}$$

Using the orthogonality of the covariant and contravariant basis vectors $\mathbf{r}_i$ and $\mathbf{r}^i$, we can write the following:

$$\frac{d\mathbf{T}}{dt} = \left(\frac{d\mathbf{T}}{dt}\cdot\mathbf{r}_1\right)\mathbf{r}^1 + \left(\frac{d\mathbf{T}}{dt}\cdot\mathbf{r}_2\right)\mathbf{r}^2. \tag{6.94}$$

For the formulation of the evolution equations, it is necessary to obtain only a covariant component of the full time derivative on the tangent line, therefore, from the expansion (6.94) we need only the first term computed at $\zeta = 0$. We introduce this derivative as follows

$$\frac{D_1 T}{dt} \equiv \left(\frac{d\mathbf{T}}{dt}\cdot\mathbf{r}_1\right)_{\zeta=0} = \left(\frac{d\mathbf{T}}{dt}\cdot\boldsymbol{\rho}_\xi\right) = \frac{\partial T}{\partial t} + \frac{\partial T}{\partial \xi}\dot{\xi} + T(\Gamma_{11}^1\dot{\xi} + \Gamma_{21}^1\dot{\zeta})_{\zeta=0}. \tag{6.95}$$

One can recognize in eqn. (6.95) the full time derivative via the covariant derivatives. We compute the value of the necessary Christoffel symbols in brackets, according to their definition in eqn. (6.93) and to the definition of $\mathbf{r}_1$ in (6.17).

$$
\begin{aligned}
(\Gamma^1_{11})_{\zeta=0} &= \left(\frac{\partial \mathbf{r}^1}{\partial \xi} \cdot \mathbf{r}_1\right)_{\zeta=0}, \\
(\Gamma^1_{21})_{\zeta=0} &= \left(\frac{\partial \mathbf{r}^1}{\partial \zeta} \cdot \mathbf{r}_1\right)_{\zeta=0}.
\end{aligned}
\tag{6.96}
$$

In order to avoid the derivative of the contravariant vector in (6.96), we take a derivative of the following identity

$$
\mathbf{r}^1 \cdot \mathbf{r}_1 = 1 \implies \frac{\partial}{\partial(\ldots)}(\mathbf{r}^1 \cdot \mathbf{r}_1) = 0 \implies \frac{\partial \mathbf{r}^1}{\partial(\ldots)} \cdot \mathbf{r}_1 + \mathbf{r}^1 \cdot \frac{\partial \mathbf{r}_1}{\partial(\ldots)} = 0 \tag{6.97}
$$

$$
\implies \frac{\partial \mathbf{r}^1}{\partial(\ldots)} \cdot \mathbf{r}_1 = -\mathbf{r}^1 \cdot \frac{\partial \mathbf{r}_1}{\partial(\ldots)}.
$$

Now, we can compute the Christoffel symbols directly

$$
(\Gamma^1_{11})_{\zeta=0} = -\left(\frac{\partial \mathbf{r}_1}{\partial \xi} \cdot \mathbf{r}^1\right)_{\zeta=0} = -\frac{\partial}{\partial \xi}\left((1 - \frac{h_{11}}{a_{11}}\zeta)\boldsymbol{\rho}_\xi\right)_{\zeta=0} \cdot \boldsymbol{\rho}^1 = \tag{6.98}
$$

the contravariant basis vector $\boldsymbol{\rho}^1$ is expressed via the covariant one in eqn. (6.9)

$$
= -\left(\frac{\partial}{\partial \xi}(1 - \frac{h_{11}}{a_{11}}\zeta)\boldsymbol{\rho}_\xi + (1 - \frac{h_{11}}{a_{11}}\zeta)\boldsymbol{\rho}_{\xi\xi}\right)_{\zeta=0} \cdot \frac{\boldsymbol{\rho}_\xi}{(\boldsymbol{\rho}_\xi \cdot \boldsymbol{\rho}_\xi)} = -\frac{\boldsymbol{\rho}_{\xi\xi} \cdot \boldsymbol{\rho}_\xi}{\boldsymbol{\rho}_\xi \cdot \boldsymbol{\rho}_\xi} = -\Gamma.
$$

From eqns. (6.11) and (6.12) the identity mit $-\Gamma$ is found.
The last necessary Christoffel symbol is computed analogously

$$
(\Gamma^1_{21})_{\zeta=0} = -\left(\frac{\partial \mathbf{r}_1}{\partial \zeta} \cdot \mathbf{r}^1\right)_{\zeta=0} = -\frac{\partial}{\partial \zeta}\left((1 - \frac{h_{11}}{a_{11}}\zeta)\boldsymbol{\rho}_\xi\right)_{\zeta=0} \cdot \boldsymbol{\rho}^1 = \tag{6.99}
$$

$$
= \frac{h_{11}}{a_{11}}\boldsymbol{\rho}_\xi \cdot \frac{\boldsymbol{\rho}_\xi}{(\boldsymbol{\rho}_\xi \cdot \boldsymbol{\rho}_\xi)} = \frac{h_{11}}{a_{11}}.
$$

Finally, the tangent component of the full derivative in eqn. (6.95) gets the following form:

$$
\frac{D_1 T}{dt} := \frac{\partial T}{\partial t} + \frac{\partial T}{\partial \xi}\dot{\xi} + T(\Gamma^1_{11}\dot{\xi} + \Gamma^1_{21}\dot{\zeta})_{\zeta=0} = \frac{dT}{dt} - T\left(\frac{\boldsymbol{\rho}_{\xi\xi} \cdot \boldsymbol{\rho}_\xi}{\boldsymbol{\rho}_\xi \cdot \boldsymbol{\rho}_\xi}\dot{\xi} - \frac{h_{11}}{a_{11}}\dot{\zeta}\right). \tag{6.100}
$$

7

Incorporation of contact for high order finite elements in covariant form[*]

Abstract

The covariant contact description is applied to incorporate the treatment of contact problems into a high-order finite element technique. A hierarchical enrichment of the shape functions space allows to construct a contact layer finite element combining both exact geometry representation for contact surfaces, and a mesh with linear shape functions for the interior of the contacting bodies. The developed contact approach can be viewed as a smoothing technique for linear meshes as well as a general application of high order FEM. The good approximation property for the developed contact layer elements is shown for the classical Hertz problem even within a few contact elements discretizing the contact zone.

Keywords

smooth contact   covariant description   consistent linearization   high order FE   anisotropic refinement   contact layer FE

7.1   Introduction

One of the big advantages of high order finite elements along with the high accuracy is the possibility to describe the given geometry of surfaces exactly. The high quality for shell structure analysis has been

---

[*]The chapter is published in [97]: A. Konyukhov, K. Schweizerhof. *Incorporation of contact for high-order finite elements in covariant form*, Computer Methods in Applied Mechanics and Engineering, **198**(13-14):1213 - 1223, 2009.

shown e.g. in Rank et.al [152] (2005). Also high order FE has became a basis for a, so-called, iso-geometrical analysis for arbitrary 3D structures, see Hughes et.al. [72] (2005). The exact geometry naturally leads also to improved results for contact problems, however, some difficulties occur due to the nonlinear nature of contact problems.

For small displacement problems an hp-version of the finite element method has been discussed in Paczelt et.al. [133] (1999). A gap function defined as a difference between approximated normal displacements from two contacting bodies and corresponding to a penetration in the direction of an initial normal vector $\mathbf{n}_c$ is used as a measure of contact interaction. The penalty method and the augmented Lagrangian method have been used to enforce contact conditions within the simple iteration method possessing only a linear rate of convergence. For large displacement problems simulated with linear finite elements the gap has been introduced via the closest point projection procedure taking into account an updated normal vector earlier in Wriggers and Simo [192] (1985). The problem was solved via an iterative solver of Newton's type possessing quadratic rate of convergence for which a consistent linearization was necessary. A linearization procedure via convective coordinates has been used for 3D problems in Laursen and Simo [109] (1993), though the linearization has been carried out in the global coordinate system. A fully covariant description in the local surface coordinate system corresponding to the closest point procedure has been developed in Konyukhov and Schweizerhof [86] (2004), [89] (2005). Since all operations such as formulation of a weak form in accordance with the applied method (penalty, augmented Lagrangian), the return-mapping algorithm and the linearization are carried out in the local surface coordinate system via convective coordinates i.e. in the fully covariant form, the description is applicable to any parametric approximations of the surfaces. A straightforward geometrical interpretation of various contact stiffness parts, see [92] and the possibility of generalization into coupled anisotropy including tangential adhesion and friction, see [90], [91] are among other advantages of the fully covariant description.

Non-smoothness of linear approximations for contact surfaces (namely facetted surfaces) is causing artificial oscillations in contact tractions and also can lead to dis-convergence of the iterative solution. This has been recognized in numerical contact mechanics and several approaches have been proposed: earlier proposals are dealing with a smoothing of a rigid surface, see Schweizerhof and Hallquist [155] (1992) and in Heege and Alart [65] (1996). Then various aspects of smoothing techniques for master contact surfaces for both non-frictional and frictional cases have been extensively studied in Pietrzak and Curnier [144] (1999), Padmanabhan and Laursen [134] (2001), Puso and Laursen [148] (2002), Krstulovic-Opara and Wriggers [102] (2002), Stadler, Holzapfel and Korelc [169] (2003) and others. Despite the number of publications all developments have been concerned predominantly with the geometrical smoothing of the given linear mesh. The discussion about the unknown influence of the void between the smoothed surfaces and the linear mesh surface remains open and a high-order finite element technique seems to be a key to fill this void. Thus, the current contribution is aimed at considerations of the covariant contact description together with high-order finite elements with exact representation of the contact boundaries as a smoothing technique for piecewise linear meshes. A special anisotropically refined p-finite element adopting refinement only in a single layer is constructed to take advantage both from the covariant contact description, and from the p-finite element technique. Since the contact description itself is considered independently from the p-finite element techniques it can be easily applied for other general cases with high-order FE.

The article is organized as follows. After the introduction, the kinematics of contact interaction, the weak formulation in accordance with the full Lagrangian and the penalty method is considered in Section 2. In addition to the developed earlier consistent tangent matrices, the linearization procedure with regard to the exact geometry should be fulfilled. Implementation details for contact with a rigid surface leading to a "Contact Layer-to-Rigid Surface" contact element as well as for contact between two deformable bodies leading to a "Contact Layer-to-Contact Layer" contact element are considered in Section 3. Extensive testing and some numerical problems for the Hertz problem are discussed in Section 4.

7.2 Contact interaction in covariant form

A contact interaction in the covariant form, [86], [89] is observed as a pointwise interaction between contacting surfaces. One of the contacting surfaces is selected as a parameterized reference surface and is called historically a "master" surface, the other contacting surface (it is called then a "slave surface") is represented by points (resp. "slave points"). The interaction is observed then in the local surface coordinate attached to the master surface.

7.2.1 Kinematics of contact. Measures of contact

A closest distance between the slave point and the master surface is a natural measure of the contact interaction and is defined via the closest point projection (CPP) procedure, see Fig. 7.1:

$$||\mathbf{r}_S - \boldsymbol{\rho}(\xi^1, \xi^2)|| \rightarrow \min, \quad \longrightarrow (\mathbf{r}_S - \boldsymbol{\rho}) \cdot (\mathbf{r}_S - \boldsymbol{\rho}) \rightarrow \min, \tag{7.1}$$

where $\mathbf{r}_S$ is a vector of the slave point $\mathbf{S}$, $\boldsymbol{\rho}(\xi^1, \xi^2)$ is a parameterization of the master surface. The fundamental questions for the CPP procedure such as existence and uniqueness of the solution of eqn. (7.1) for the surfaces of arbitrary geometry are studied in Konyukhov and Schweizerhof [95]. The closest point projection procedure gives rise to a local curvilinear 3D coordinate system defined via surface tangent vectors $\boldsymbol{\rho}_i = \dfrac{\partial \boldsymbol{\rho}}{\partial \xi^i}$, $i = 1, 2$ and a normal vector $\mathbf{n}$. All measures of the contact interaction are defined in this coordinate system: the coordinate increments $\Delta \xi^i$, $i = 1, 2$ are measures of tangential interaction and the third coordinate ξ^3 is a measure of normal interaction. The vector of the slave point $\mathbf{r}_S$ is defined as:

$$\mathbf{r}_S = \boldsymbol{\rho}(\xi^1, \xi^2) + \xi^3 \mathbf{n}. \tag{7.2}$$

Measures of the rate of deformation are separately defined after consideration of the relative velocity $\mathbf{v}_S - \mathbf{v}$ of a slave point $\mathbf{S}$ with respect to the local coordinate system. On the tangent plane and, therefore, at the

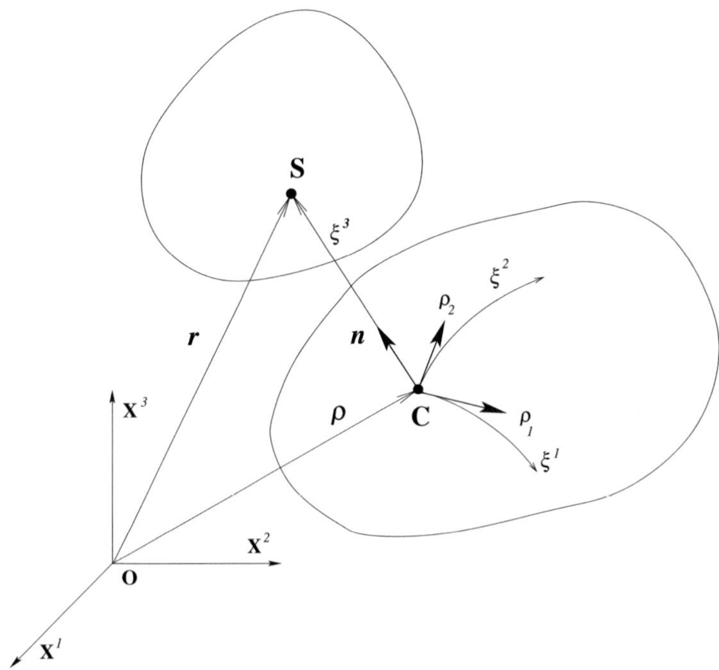

Figure 7.1: Closest point procedure and definition of the spatial coordinate system.

projection point **C** it has the following values for the normal interaction

$$\dot{\xi}^3 = (\mathbf{v}_S - \mathbf{v}) \cdot \mathbf{n} \tag{7.3}$$

and for the tangential interaction

$$\dot{\xi}^i = a^{ij}(\mathbf{v}_S - \mathbf{v}) \cdot \boldsymbol{\rho}_j, \tag{7.4}$$

where a^{ij} are contravariant components of the metric tensor. They are defined as inverse of a matrix with covariant components $a_{ij} = (\boldsymbol{\rho}_i \cdot \boldsymbol{\rho}_j)$. Using kinematical analogy, all rate terms are further directly used in the following variational formulation, i.e. $\dot{\xi}^3$ leads to $\delta\xi^3$, $(\mathbf{v}_S - \mathbf{v})$ leads to $(\delta\mathbf{r}_S - \delta\boldsymbol{\rho})$.

7.2.2 Weak formulation

Due to the additivity of the energy it is sufficient to consider only a part δW_c of the full virtual work arising from contact interaction between surfaces. Let $\mathbf{R}$ be a contact traction vector acting on an infinitesimal part of the master surface ds and $\mathbf{R}_S$ be a contact traction vector acting respectively on an infinitesimal part of the slave surface ds_S. Then the work of the contact tractions δW_c is computed as

$$\delta W_c = \int_S \mathbf{R} \cdot \delta\boldsymbol{\rho}\, ds + \int_{S_S} \mathbf{R}_S \cdot \delta \mathbf{r}_S\, ds_S, \tag{7.5}$$

where the integrals should be taken over the mutual contacting area. Taking into account the pointwise equilibrium on this area $\mathbf{R}\, ds + \mathbf{R}_S\, ds_S = 0$, the work in eqn. (7.5) can be represented as an integral taken over the slave surface

$$\delta W_c = \int_S \mathbf{R}_S \cdot (\delta \mathbf{r}_S - \delta\boldsymbol{\rho})ds_S. \tag{7.6}$$

The next step is to represent the traction acting on the slave surface in the local coordinate system i.e. via master surface coordinate vectors

$$\mathbf{R}_S = T^i \boldsymbol{\rho}_i + N\mathbf{n}. \tag{7.7}$$

Inserting then eqn. (7.7) into eqn. (7.6) and taking into account eqns. (7.3), (7.4) together with the kinematical analogy, we can obtain the following weak form:

$$\delta W_c = \underbrace{\int_s N\delta\xi^3 ds_S}_{\delta W_c^N} + \underbrace{\int_s T_j\delta\xi^j ds_S}_{\delta W_c^T}, \tag{7.8}$$

where the normal part δW_c^N represents the normal non-penetration contact condition and δW_c^T represents a frictional interaction. N is the normal contact traction and T_j, $j = 1,2$ are covariant components of the tangential contact traction.

Remark.
In the full two-body contact algorithm the contact integral (7.8) is computed via the quadrature formula where the integration points from the

slave surface s_S are projected first to the master surface. Afterwards, an iterative method of Newton's type is applied to reach the equilibrium condition. Due to this sequence (global solution after the projection) the full derivative of the functional (7.8) resp. the linearization procedure is taken fully in the metrics of the master surface and, therefore, the linearization of the slave surface term is zero $\mathcal{L}[ds_S] = 0$.

7.2.3 Constitutive equations for the contact traction

Constitutive equations should be generally supplied in the local coordinate system separately for normal and tangent tractions leading to a variety of models between contact interfaces (e.g. elastic, viscoelastic, plastic, adhesive etc. laws in both normal and tangential directions). A pure geometrical enforcement of the non-penetration condition does not require additional constitutive relations for normal traction and is usually specified via the Kuhn-Tucker conditions:

$$\xi^3 > 0; \qquad N_{master} = 0; \qquad \text{no contact} \qquad (7.9)$$
$$\xi^3 = 0; \quad N_{master} < 0 \to N > 0; \quad \text{non penetration} \qquad (7.10)$$
$$\xi^3 N = 0; \qquad\qquad \text{complementary condition} \quad (7.11)$$

The standard Kuhn-Tucker condition in eqn. (7.10) formulated on the master surface is modified here according to the representation (7.7) of the contact traction on the slave surface.

If the full Lagrange multiplier method is further involved to satisfy the conditions (7.9-7.11) then the normal traction N is independently approximated leading to the exact satisfaction of the non-penetration condition (7.10) at the integration points (the method is also known as a Mortar method, see Fischer and Wriggers [41]). The penalty method from the mathematical point of view leads to an approximately satisfied non-penetration condition and from a mechanical point of view it is *a constitutive equation for normal elastic compliance*. This condition is formulated as

$$N = \begin{cases} 0 & \text{if } \xi^3 > 0 \quad \text{no contact} \\ \epsilon_N \xi^3 & \text{if } \xi^3 \leq 0 \quad \text{penetration.} \end{cases} \qquad (7.12)$$

For further linearization the rate equation is obtained considering the time derivative of eqn. (7.12):

$$\dot{N} = \epsilon_N \dot{\xi}^3 \text{ for penetration } \xi^3 \leq 0. \tag{7.13}$$

In the more general case of tangential interactions the contact traction depends on the relative velocity of contacting bodies. This leads to the necessity to formulate constitute relations for contacting bodies. In the current contribution the mostly applied relation in engineering practice, the Coulomb friction law, is considered. According to this law a "sticking" case and a "sliding" case are identified via the following yield function Φ

$$\Phi := \sqrt{a^{ij} T_i T_j} - \mu|N|, \tag{7.14}$$

formulated here via the master surface metrics with μ as a friction coefficient. The sticking-sliding conditions can be defined in the Kuhn-Tucker form with regard to the aforementioned measures $\Delta\xi^i$ in the case of contact $\xi^3 = 0$, see eqn. (7.11)

$$\Delta\xi^i = 0,\ i = 1, 2; \quad \text{if } \Phi < 0 \quad \text{sticking;} \tag{7.15}$$

$$\exists \lambda > 0 \text{ that } T_i \boldsymbol{\rho}^i = -\lambda \dot{\xi}^i \boldsymbol{\rho}_i; \quad \text{if } \Phi \geq 0 \quad \text{sliding.} \tag{7.16}$$

The last eqn. (7.16) shows that the tangential traction is acting in the direction opposite to the relative velocity. During the recent years it has been recognized in the numerical community that the full Lagrange multiplier scheme with exact enforcement of the sticking-sliding conditions is only robust for the rather small displacement case, see Jones and Papadopoulos [79], Solberg and Papadopoulos [163]. In the case of large displacements an augmented Lagrangian method is more applicable, see Pietrzak and Curnier [144], Laursen and Simo [109] and recently Hüeber et.al. [69] for the Mortar method and Konyukhov and Schweizerhof [94] for anisotropic friction models. However, in both cases the return-mapping algorithm is employed to check whether the real tangent traction belongs to the sticking domain or to the sliding domain. For both, penalty and augmented Lagrangian methods the sticking traction is computed assuming elastic compliance in the tangential direction. *The full description in a covariant form is then identical to a formulation of 2D elasto-plasticity for the surface interface.* Namely, the full incremental displacements are additively decomposed into elastic and plastic, or sliding, parts as $\Delta\xi^i = \Delta\xi^i_{el} + \Delta\xi^i_{sl}$. The principle of maximum dissipation is used to derive then the sliding force as well as the sliding displacements. For derivation details even in a more general case of anisotropic contact interfaces including coupling of adhesion and friction we refer an interested reader to Konyukhov and Schweizerhof [90], [91]. In the current contribution only the necessary formulae are presented. The elastic tangential traction is given in the rate form as proportionality of the covariant derivative of a tangential traction vector $\mathbf{T}^{el}$ to a relative velocity vector $\mathbf{v}_S - \mathbf{v}$ is assumed. Components in the local coordinate system are given via the following evolution equation:

$$\frac{dT_i^{el}}{dt} = (-\epsilon_T a_{ij} + \Gamma_{ij}^k T_k^{el})\dot{\xi}^j - h_{ij}a^{jk}T_k^{el}\dot{\xi}^3, \quad i, j, k = 1, 2. \tag{7.17}$$

where Γ_{ij}^k are the Christoffel symbols; a_{ij} resp. hij are covariant components of the metric resp. curvature tensor of the master surface; ϵ_T is a penalty parameter representing a tangent stiffness of the contact interfaces. For implementation eqn. (7.17) is simplified and then solved numerically via the backward Euler finite difference scheme. The full contact algorithm and computational issues will be further discussed in Section 7.3. The real tangent traction is computed via the return-mapping scheme with regard to the yield function Φ^{el} in eqn. (7.14) computed by the elastic trial tractions T_i^{el}

$$T_i = \begin{cases} T_i^{el} & \text{if } \Phi^{el} \leq 0 \quad \text{sticking} \\ T_i^{sl} = \dfrac{T_i^{el}}{\sqrt{a^{ij}T_i^{el}T_j^{el}}} & \text{if } \Phi^{el} > 0 \quad \text{sliding} \end{cases} \tag{7.18}$$

The return-mapping scheme for the isotropic case is systematically studied in the monographs of Wriggers [188] and Laursen [106].

7.2.4 Consistent linearization of the weak form

The full Newton iterative method will be applied to solve the global equilibrium equations. This requires the full linearization of the functional in eqn. (7.8) representing the equilibrium conditions on the contact boundaries. Linearization for the penalty approach is obtained using the covariant derivation in the local surface coordinate system, where derivatives of contact tractions are given in eqns. (7.13), (7.17) and derivatives of convective coordinates are given in eqns. (7.3) and (7.4). The complete derivation procedure requires the linearization of convective coordinate variations as well as extensive tensor transformations and are omitted here. The interested reader is referred to the articles of Konyukhov and Schweizerhof [86], [89], from which further results are represented in Appendix 7.6.

7.3 Finite element implementation

In order to implement the possibility of contact for high order finite elements all elements on the potential contact boundary are modified into the Contact Layer contact element shown in Fig. 7.2 according to the following scheme:

1. Each element on the boundary is constructed as an anisotropically refined element, see the terminology in Solin et.al. [164]. Thus, the shape function space is hierarchically constructed via the Lobatto shape functions with the enrichment possibility for the edge degree of freedom (DOF) p_{edge}, the side DOF p_{side} and the bubble DOF p_{bubble} possessing necessary conformity requirements to the internal elements (for further numerical examples the layer element is satisfying conformity conditions with the linear mesh i.e. $p_{int.mehs} = 1$).

2. The shape function space is modified according to the blending function method in order to represent the boundary surface exactly (e.g. linear edge shape functions which have support at vertices V_5, V_6, V_7, V_8 are modified).

3. The truncated shape function space organized via all shape functions with a support on a curvilinear boundary $V_5 V_6 V_7 V_8$ becomes a shape function space for a contact element. Thus, all derived tangent matrices and residuals are assembled with regard to the DOF's of the corresponding high order element.

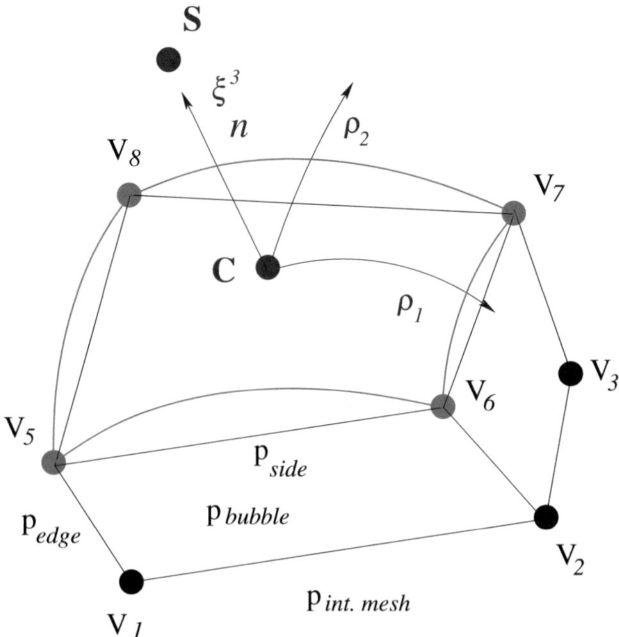

Figure 7.2: Structure of contact layer finite element

According to the described scheme it is possible then to subtract the approximation of the master surface given by the vector $\rho(\xi^1, \xi^2)$ together with the approximation of the slave surface $\mathbf{r}_s(\eta^1, \eta^2)$. Since all possible contacting pairs "master-slave" should be tested it leads to the necessity of a global searching routine which is out of scope of the current contribution. One of the most robust scheme for linear "master-slave" segments allowing self-contact is described in Benson and Hallquist [11]. The approximation operator $\mathbf{A}$ for a relative displacement vector $\rho(\xi^1, \xi^2) - \mathbf{r}_s(\eta^1, \eta^2)$ in the standard finite element method also for the p-version of finite elements is defined as a linear operator with regard to a DOF vector $\mathbf{x}$

$$\mathbf{x} = \{x_{ms}^{(1)}, y_{ms}^{(1)}, z_{ms}^{(1)}, ..., x_{ms}^{(m)}, y_{ms}^{(m)}, z_{ms}^{(m)}, x_{sl}^{(1)}, y_{sl}^{(1)}, z_{sl}^{(1)}, ..., x_{sl}^{(n)}, y_{sl}^{(n)}, z_{sl}^{(n)}\}^T, \qquad (7.19)$$

where the index ms indicates correspondence to the master surface and the index sl indicates correspondence to the slave surface. The number of the degree of freedom depending on the involved method can be related either to the nodal DOF for the classical nodal version of finite element, or to the generalized DOF including a hierarchical structure of the vertex, edge, side and bubble degree of freedom for a p-element.

The linear operator is described then via the matrix

$$
\mathbf{A} =
\begin{bmatrix}
M^1 & 0 & 0 & M^2 & 0 & 0 & \dots & \dots & \dots & M^m & 0 & 0 \\
0 & M^1 & 0 & 0 & M^2 & 0 & \dots & \dots & \dots & 0 & M^m & 0 \\
0 & 0 & M^1 & 0 & 0 & M^2 & \dots & \dots & \dots & 0 & 0 & M^m
\end{bmatrix}
\tag{7.20}
$$

$$
\begin{bmatrix}
-N^1 & 0 & 0 & -N^2 & 0 & 0 & \dots & \dots & \dots & -N^n & 0 & 0 \\
0 & -N^1 & 0 & 0 & -N^2 & 0 & \dots & \dots & \dots & 0 & -N^n & 0 \\
0 & 0 & -N^1 & 0 & 0 & -N^2 & \dots & \dots & \dots & 0 & 0 & -N^n
\end{bmatrix}.
$$

with $M^k = M^k(\xi^1, \xi^2)$, $k = 1, m$ as shape functions for a master surface; $N^k = N^k(\eta^1, \eta^2)$, $k = 1, n$ are shape functions for a slave surface. A relative virtual displacement or velocity vector is given then as

$$
\begin{aligned}
\delta\boldsymbol{\rho}(\xi^1, \xi^2) - \delta\mathbf{r}_s(\eta^1, \eta^2) &= \mathbf{A}\delta\mathbf{x} \quad \text{for a virtual displacement vector;} \\
\mathbf{v}(\xi^1, \xi^2) - \mathbf{v}_s(\eta^1, \eta^2) &= \mathbf{A}\mathbf{v} \quad \text{for a velocity vector.}
\end{aligned}
\tag{7.21}
$$

In addition only derivatives with respect to the master convective coordinates ξ^i should be computed for further representation of the tangent matrices:

$$
\delta\boldsymbol{\rho}_{,i} = \frac{\partial\mathbf{A}}{\partial\xi^i}\delta\mathbf{x} = \mathbf{A}_{,i}\delta\mathbf{x}.
\tag{7.22}
$$

7.3.1 Representation of tangent matrices for FE implementation

Since all parts of the tangent matrices contain either a vector of virtual relative displacements $\delta\boldsymbol{\rho} - \delta\mathbf{r}_s$ or a relative velocity $\mathbf{v} - \mathbf{v}_s$, and/or their derivatives with respect to the master convective variable ξ^i only the main part (7.36) and the rotational part (7.37) for the linearized normal part are considered. The other parts of the tangent matrices are algorithmically constructed.

All parts except the rotational parts are transformed using the approximation operator $\mathbf{A}$ in eqn. (7.21) directly. Taking the main part (7.36) as an example one can obtain

$$
\begin{aligned}
\mathcal{L}_N^m &= \int_{s_S} \epsilon_N(\delta\mathbf{r}_S - \delta\boldsymbol{\rho}) \cdot (\mathbf{n} \otimes \mathbf{n})(\mathbf{v}_S - \mathbf{v})ds_S \\
&= \delta\mathbf{x} \cdot \int_{s_S} \epsilon_N \mathbf{A}^T(\mathbf{n} \otimes \mathbf{n})\mathbf{A}ds_S\mathbf{v} = \delta\mathbf{x} \cdot [\mathbf{K}_N^m]\mathbf{v}
\end{aligned}
\tag{7.23}
$$

where $[\mathbf{K}_N^m]$ is the corresponding main part of the tangent matrix.

The rotational part is transformed using the operator $\mathbf{A}_{,i}$ in eqn. (7.22). Here the part $\mathcal{L}_N^r$ from eqn. (7.37) is taken

$$
\begin{aligned}
\mathcal{L}_N^r &= -\int_{s_S} N \left[\delta\boldsymbol{\rho}_{,j} \cdot a^{ij}(\mathbf{n} \otimes \boldsymbol{\rho}_i)(\mathbf{v}_S - \mathbf{v}) + (\delta\mathbf{r}_S - \delta\boldsymbol{\rho}) \cdot a^{ij}(\boldsymbol{\rho}_j \otimes \mathbf{n})\mathbf{v}_{,i} \right] ds_S \\
&= \delta\mathbf{x} \cdot \int_{s_S} N \left[\mathbf{A}_{,j}^T a^{ij}(\mathbf{n} \otimes \boldsymbol{\rho}_i)\mathbf{A} + \mathbf{A}^T a^{ij}(\boldsymbol{\rho}_j \otimes \mathbf{n})\mathbf{A}_{,i} \right] ds_S \mathbf{v} \\
&= \delta\mathbf{x} \cdot [\mathbf{K}_N^r]\mathbf{v}
\end{aligned}
\tag{7.24}
$$

7.3.2 Tangent matrices for the non-linear approximation operator

For complex situations with an exact representation of boundaries via **the blended function method** in general nonlinear approximations are involved (e.g. NURB representation of the contact surfaces). Thus the approximation operator $\mathbf{A}$ is nonlinear with respect to variables (DOF's) for the finite element approximation (or knot variables for the NURB, iso-geometrical FE etc.)

$$
\boldsymbol{\rho}(\xi^1, \xi^2) - \mathbf{r}_s(\eta^1, \eta^2) = \mathbf{A}(\mathbf{x}, \xi^i, \eta^i).
\tag{7.25}
$$

In such a case, the approximation operator itself should be linearized with respect to the global variable vector $\mathbf{x}$ in order to represent the globally consistent tangent matrices. We introduce $D\mathbf{A}$ in the sense of the Frechet derivative with respect to a global variable vector $\mathbf{x}$. Considering the Taylor expansion for the relative virtual displacement vector in eqn. (7.21) and taking into account the approximation property we obtain $\mathbf{A}(\mathbf{0}, \xi^i, \eta^i) = \mathbf{0}$

$$
\begin{aligned}
\delta\boldsymbol{\rho}(\xi^1, \xi^2) - \delta\mathbf{r}_s(\eta^1, \eta^2) &= \mathbf{A}(\delta\mathbf{x}, \xi^i, \eta^i) \\
&= \mathbf{A}(\mathbf{0}, \xi^i, \eta^i) + D\mathbf{A}(\mathbf{0}, \xi^i, \eta^i)\delta\mathbf{x} + \dots \\
&= D\mathbf{A}(\mathbf{0}, \xi^i, \eta^i)\delta\mathbf{x} + \dots
\end{aligned}
\tag{7.26}
$$

Taking a partial derivative with respect to the convective variables ξ^i the operator $D\mathbf{A}_{,i}$ is introduced as:

$$\delta\boldsymbol{\rho}_{,i} = \frac{\partial D\mathbf{A}(\mathbf{0},\xi^i,\eta^i)}{\partial \xi^i}\delta\mathbf{x} + ... = D\mathbf{A}_{,i}\delta\mathbf{x} + ... \qquad (7.27)$$

Thus, in the case of nonlinear approximations for contact boundaries all parts of tangent matrices are computed following the pattern presented in eqns. (7.23), (7.24) taking into account the linearized operator $D\mathbf{A}$ in eqn. (7.26) and its derivative $D\mathbf{A}_{,i}$ in eqn. (7.27). It leads formally to the substitution of $\mathbf{A}$ with $D\mathbf{A}$ and respectively of $\mathbf{A}_{,i}$ with $D\mathbf{A}_{,i}$.

7.3.3 Computation of contact integrals

Two situations should be distinguished for computations of all integrals for equilibrium conditions as well as for corresponding tangent matrices: contact between deformable and rigid bodies (so-called Signorini problems); and contact between two deformable bodies. The first problem leads to *the Contact Layer - Rigid Surface* (CLRS) finite element and the second problems leads to *Contact Layer - Contact Layer* (CLCL) finite. element. The structure of these elements is considered here for the 2D case as layer-wise enrichment of the initially linear mesh. However, the general contact algorithm is not limited to this case.

7.3.4 Contact Layer - Rigid Surface (CLRS) finite element

An initially linear element, see Fig. 7.3, is enriched hierarchically with edge $p^{e_l} - 1$ shape functions for the left side, $p^{e_r} - 1$ edge shape functions for the right side, $p^{e_b} - 1$ edge shape functions for the bottom side and $(p^b - 1) \times (p^b - 1)$ bubble shape functions, where p^* is the highest polynomial degree of the corresponding shape function. An exemplary element with quadratic degree $p = 2$ of freedom in 2D leads to $2 \times (4\ vertex\ + 3\ edge\ + 1\ bubble) = 16$ DOF. The linear boundary is represented then exactly by the blending function method (e.g. an exact circular arch in the following numerical examples). The contact with a rigid surface can be resolved assuming that the parameterization of the rigid surface (becoming then a "slave" surface) is given as $\mathbf{r}_S(\eta^1,\eta^2)$.

All contact integrals are computed then over the master surface in the following sequence:

1. A master segment is covered by a set of integration points C (e.g. by Gauss points) ξ_i^1, ξ_j^2 $i, j = 1, N_p$ according to the chosen order N_p.

2. The CPP procedure eqn. (7.1) is interpreted now vice versa: both the penetration ξ^3 and the convective coordinate η^1, η^2 of the point S on the slave surface where a normal $\mathbf{n}$ (computed at the integration point C of the master surface) is "penetrating" the slave surface. The numerical solution of the system is then derived from eqn. (7.2)

$$\mathbf{F}(\eta^1, \eta^2, \xi^3) \equiv \mathbf{r}_S(\eta^1, \eta^2) - \boldsymbol{\rho}(\xi_C^1, \xi_C^2) - \xi^3 \mathbf{n}(\xi_C^1, \xi_C^2) = 0. \qquad (7.28)$$

Remark.
For some surfaces such as plane, cylinder, sphere and torus the solution of eqn. (7.28) can be obtained in a closed form, see further information in Harnau et.al. [57].

3. All contact integrals (for residuals as well as for tangent matrices) are formally computed via the quadrature formula, where a component with an integration point ξ_i^1, ξ_j^2 is included only if the corresponding penetration is less then zero $\xi^3|_{\xi_i^1, \xi_j^2} \leq 0$.

$$\int f(\xi^1, \xi^2) ds = \sum_{i,j} H(-\xi^3) f(\xi_i^1, \xi_j^2) det J(\xi_i^1, \xi_j^2) A_i A_j, \qquad (7.29)$$

where A_i, A_j are weights of the quadrature formula, J is the Jacobian of the transformation $ds \rightarrow d\xi^1 d\xi^2$ for a master segment, $H(-\xi^3)$ is a Heaviside function.

Remark.
All matrices reduced for the 2D case for the residual as well as for the tangent matrices are presented in Konyukhov and Schweizerhof [92]

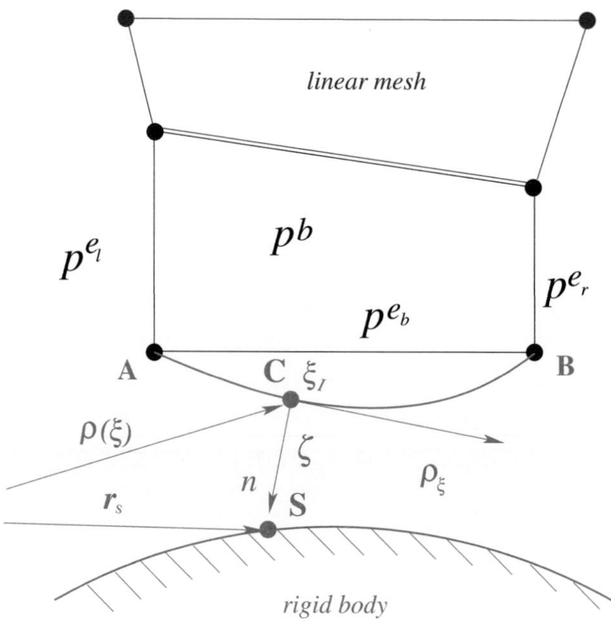

Figure 7.3: *Contact Layer - Rigid Surface* contact element. 2D-representation.

7.3.5 Contact Layer - Contact Layer (CLCL) finite element

Both master and slave elements are enriched hierarchically in a similar fashion as described in the previous Section 7.3.4. An exemplary element with quadratic degree $p = 2$ of freedom in 2D leads to $2 \times 2 \times (4\,vertex + 3\,edge + 1\,bubble) = 32$ DOF. All contact integrals are computed over the slave surface in the following sequence:

1. A slave segment is covered by a set of integration points C (e.g. by Gauss points) $\eta_i^1, \eta_j^2\ i, j = 1, N_p$ according to the chosen order N_p and the corresponding vectors $\mathbf{r}_s(\eta_i^1, \eta_j^2)$ are computed.

2. The CPP procedure eqn. (7.1) is executed to find out the convective coordinate ξ_i^1, ξ_j^2 on the master segment corresponding to projections of $\mathbf{r}_s(\eta_i^1, \eta_j^2)$. Points with coordinates ξ_i^1, ξ_j^2 not laying inside the master segment are excluded *(a, so-called, local searching procedure)*.

3. All contact integrals are formally computed via the quadrature for-

mula, where a component with an integration point ξ_i^1, ξ_j^2 is included only if the corresponding penetration is less then zero $\xi^3|_{\xi_i^1, \xi_j^2} \leq 0$,

$$\int f(\xi^1, \xi^2)ds = \sum_{i,j} H(-\xi^3)f(\xi_i^1, \xi_j^2)det J(\eta_i^1, \eta_j^2)A_i A_j, \qquad (7.30)$$

where A_i, A_j are weights of the quadrature formula, J is the Jacobian of the transformation $ds_S \rightarrow d\eta^1 d\eta^2$ for a slave element, $H(-\xi^3)$ is a Heaviside function.

The presented approach to compute integrals is also known as a Mortar method discussed in Fischer and Wriggers [41], where a mortar part is related to the master part, and a non-mortar part is related to the slave part, see Fig. 7.4. This approach allows also to satisfy the contact patch-test, see Harnau et.al. [57].

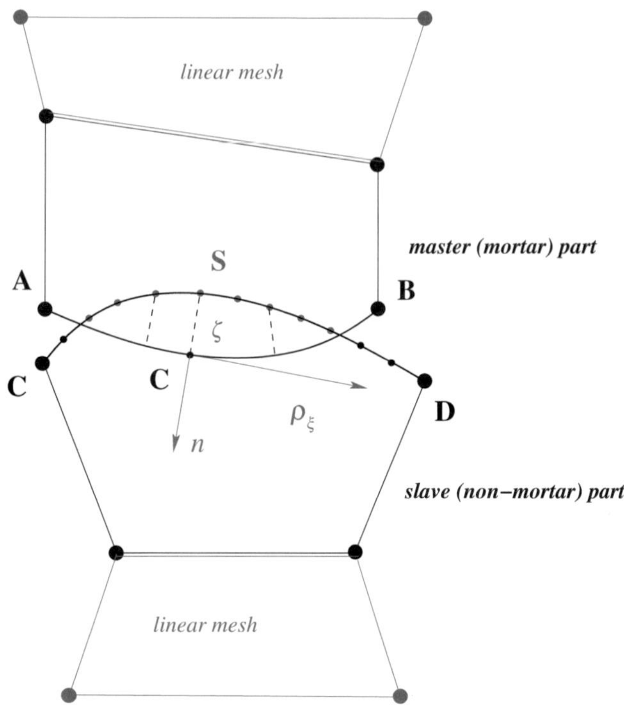

Figure 7.4: *Contact Layer - Contact Layer* contact element. 2D-representation.

7.3.6 Lagrange multiplier method for normal traction

As discussed before a robust scheme based on the full Lagrange multiplier method for arbitrary large sliding problems would include the enforcement of only normal non-penetration conditions combined with the augmented Lagrangian method for tangential condition. In a case of contact with a rigid surface (CLRS) the finite element is modified according to the following scheme:

1. Both, displacement vector and normal traction are approximated (mixed formulation). Let $\mathbf{A}$ be an approximation matrix for a relative displacement vector corresponding to the DOF vector $\mathbf{x}$, see eqn. (7.21), and $\mathbf{B}$ $[1 \times N_\lambda]$ be an approximation matrix for the normal traction N corresponding to vector with N_λ DOF's for a normal traction λ. An element vector for the mixed formulation $\mathcal{X}$ contains then both, displacement DOF's $\mathbf{x}$ and variables for normal traction λ:

$$\left.\begin{array}{ccc} \delta\rho - \delta\mathbf{r}_s & = & \mathbf{A}\delta\mathbf{x} \\ N & = & \mathbf{B}\lambda \end{array}\right\} \rightarrow \{\mathcal{X}\} = \left\{\begin{array}{c} \mathbf{x} \\ \lambda \end{array}\right\} \tag{7.31}$$

2. The consistent linearization shown in eqn. (7.35) is derived according to both independent variables $\mathbf{x}$ and λ. Thus, eqn. (7.35) leads to the following matrix

$$[\mathbf{K}] = \left[\begin{array}{cc} \mathbf{K}_{\mathbf{xx}} & \mathbf{K}_{\mathbf{x}\lambda} \\ \mathbf{K}_{\lambda\mathbf{x}} & \mathbf{O} \end{array}\right], \tag{7.32}$$

where the matrix $\mathbf{K}_{\mathbf{xx}}$ is derived from the rotational part $\mathcal{L}_N^r$ and the curvature $\mathcal{L}_N^c$ part in eqns. (7.37)-(7.38), see also the FE representation in Section 7.3.1.

$$\mathbf{K}_{\mathbf{xx}} = [\mathbf{K}_N^r] + [\mathbf{K}_N^c]. \tag{7.33}$$

Since the constitutive equation for normal traction is not involved into the linearization the main part $\mathcal{L}_N^r$ in eqn. (7.36) is representing then a coupled part $\mathbf{K}_{\mathrm{x}\lambda}$

$$\mathbf{K}_{\mathrm{x}\lambda} = \int_{s_S} \mathbf{A}^T \mathbf{n} \mathbf{B} ds_S. \tag{7.34}$$

7.3.6.1 Discussion about BB-stability of the proposed Lagrange multiplier approach

As is known, a mixed formulation within the finite element method should satisfy the discrete Babuska-Brezzi (BB) condition in order to ensure convergence and stability with respect to the mesh size and discretized constraints. The Lagrange method described in this section in combination with both the approximation of the stress field and the integration rule of corresponding contact integrals via a set of projected slave integration points over the master segment (not via the set of slave nodes!) is falling into the case **"one-pass pressure interpolation method"** which has been intensively studied in Solberg and Papadopoulos [163] and proved to be BB-stable *upon sufficiently accurate integration of contact boundary terms P.2755 [163]*. The BB-stability in due course leads to the satisfaction of the contact patch test. Various integration techniques including combinations of Gauss or Lobatto quadrature formulae of higher order together with subdivisions of the integrand into subdomains were studied e.g. in [57] in order to improve the patch-test.

7.3.7 Global solution scheme

The global solution scheme in the case of a *Contact Layer -Contact Layer* finite element is presented in Table 7.1. Some remarks are necessary for better understanding.

- Since tangential tractions are defined in the rate form, see eqn. (7.17), initial conditions should be supplied for further numerical solutions. This is done in Step 0 of the global scheme, in which initial coordinates ξ_0^i of the sticking point are computed and initial tangential tractions are assigned $T_{i,0} = 0$.

- For a *Contact Layer-Rigid Surface* finite element the CPP procedure in a) is changed as discussed in Section 7.3.4.

- For the mixed formulation discussed in Section 7.3.6 condition b) of *No contact* in loop 4 is changed into $N = \mathbf{B}\boldsymbol{\lambda} > 0$, see definition in eqn. (7.31).

- The simplest backward Euler scheme is employed in step d) to solve the evolution equation (7.17) numerically. The improved numerical scheme based on a geometrical interpretation of the covariant derivative allows to overcome jumps occurring in the contact segments due to different approximations, for a detailed discussion also concerning contact with rigid surfaces see in Konyukhov and Schweizerhof [92].

- In many cases the curvature of the master surface is often changing only slightly during deformation and then curvature parts can be omitted without large influence on convergence, see also the discussion in [86].

7.4   Numerical examples

Contact between a deformable cylinder and a rigid plane, the classical Hertz problem possessing an analytical solution, see e.g. in Johnson [77], is chosen to test the ability of the anisotropically p-refined layered element to approximate the exact solution. The problem is well known also as a benchmark-test for h-adaptive schemes in contact mechanics, see Wriggers and Scherf [191]. For the computation only a quarter of the cylinder in Fig. 7.5 is meshed with 1120 bilinear finite elements. All parameters are chosen in order to satisfy the assumptions for the Hertz problem. The material of the cylinder is linear elastic with Young's modulus $E = 1.0 \cdot 10^5 N/mm^2$ and Poisson ratio $\nu = 0.3$; radius of the cylinder $R = 100mm$. The quarter of cylinder is loaded by a distributed pressure at the upper side equivalent to the global load $P/2$. The level of the load P is intentionally chosen to keep the contact radius a small in order to stay in the frame of the Hertz hypothesis, namely $a \ll R$. Thus, only six bounding elements are modified into *Contact Layer-Rigid Surface* (CLRS) finite elements (the sixth element satisfies additionally the conformity condition with a linear mesh on the

Table 7.1: Global solution scheme for Contact Layer – Contact Layer (CLCL) finite element

0. Initialization of convective coordinates ξ_0^i and tangential tractions $T_{i,0} = 0$. The CPP procedure in eqn. (7.1) is computed without any external applied load.
1. Loop over load increments I
2. Newton iteration loop k:
   Iterative solution of global equilibrium equations.

> 3. Loop over contact pairs "master-slave"
> 4. Loop over integration points η^1, η^2 of the slave segment
>
> > a) compute projection ξ^1, ξ^2 onto the master segment, eqn. (7.1).
> >
> >   **if** ξ^1, ξ^2 do not belong to the master segment *(Local searching procedure)* **then** exit loop 4.
> >
> > b) compute penetration ξ^3 at the projection point
> >
> >   **if** $\xi^3 > 0$ **then** exit loop 4. *(No contact)*.
> >
> > c) compute the normal traction $N = \epsilon_N \xi^3$ and the corresponding tangent matrix $[\mathbf{K}_N]$ in eqns. (7.36–7.38).
> >
> > d) compute the trial tangential traction T_i^{el}
> >
> > $$T_i^{el} = T_i^{I-1} - \epsilon_T(\xi^j - \xi_0^j)a^{ij},$$
> >
> > where T_i^{I-1} are real tangential tractions from the previous converged load step $I - 1$.
> >
> > e) *return-mapping algorithm*
> >   compute the trial yield function
> >
> > $$\Phi^{el} = \sqrt{a^{mn}T_m^{el}T_n^{el}} - \mu|N|$$
> >
> > | if $\Phi^{el} \leq 0$ | if $\Phi^{el} > 0$ |
> > |---|---|
> > | **sticking condition** | **sliding condition** |
> > | $T_{i\ stick} = T_i^{el}$ | $T_{i\ slide} = \mu|N|\dfrac{T_i^{el}}{\sqrt{T_m^{el}T_n^{el}a^{mn}}}$ |
> > | Compute the corresponding tangent matrix $[\mathbf{K}_{T,st}]$ in eqns. (7.39–7.41). | Compute the corresponding tangent matrix $[\mathbf{K}_{T,sl}]$ in eqns. (7.42–7.48). |
> >
> > e) Compute residual arising from eqn. (7.8)
> >
> > $$\mathbf{R} = N\mathbf{A}^T\mathbf{n} + T^i\mathbf{A}^T\boldsymbol{\rho}_i$$
> >
> > e) Assemble residual and tangent matrices
>
> Check global convergence

right side), see Fig. 7.5. The contact radius $a = \sqrt{\dfrac{4PR(1-\nu^2)}{\pi E}}$ and the contact pressure $p = \dfrac{2P}{\pi a^2}\sqrt{a^2 - x^2}$ (see derivation e.g. in [77]) are main parameters for the verification. Several incremental loading states corresponding to the extension of the contact zone through several elements are studied.

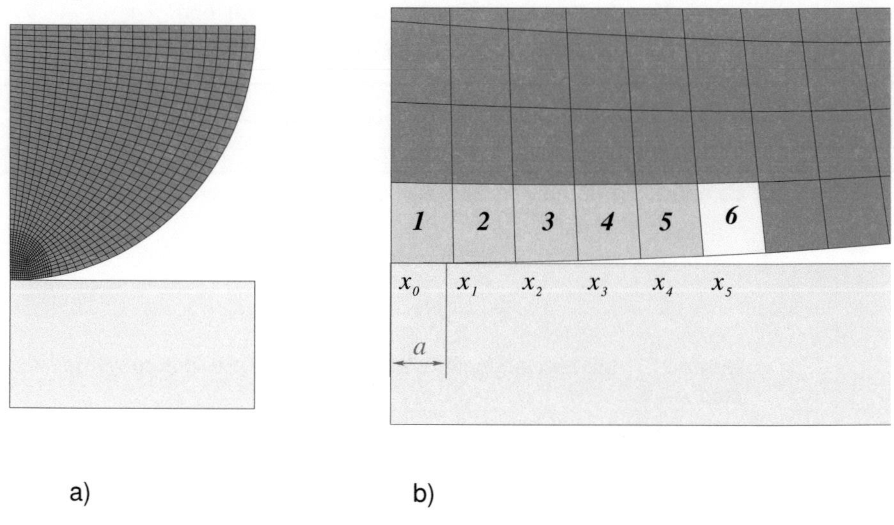

Figure 7.5: a) Initial linear mesh. b) Anisotropic p-refinement in contact layer. Only six elements are modified into *Contact Layer-Rigid Surface (CLRS)* elements.

7.4.1 Loading case 1. Contact zone within one element

The first load level is taken as $P = 0.657\,kN$ corresponding to the contact radius $a = 0.857\,mm$ laying completely inside the first contacting element, see Fig. (7.5) b). The computation is performed first with an initially linear mesh employing a node-to-segment contact approach and then with CLRS elements with increasing polynomial degree $p = 2, 3, 4$. The penalty method has been used to enforce the contact conditions. As shown in Harnau et.al. [57] a problem of integration of non-smooth functions appears during the integration of contact integrals. This leads to the impossibility to set up the order of the quadrature formula, a-priori, based only on knowledge of the polynomials involved into the shape function space. Various integration techniques such as combination of Gauss, or Lobatto quadrature formulae of higher order together with subdivision of the integrand into sub-domains were successfully applied in [57] for contact problems e.g. in order to improve the patch-test. The current computation is performed only with 10 Gauss points. A comparison with the analytical solution is given in Fig. 7.6.

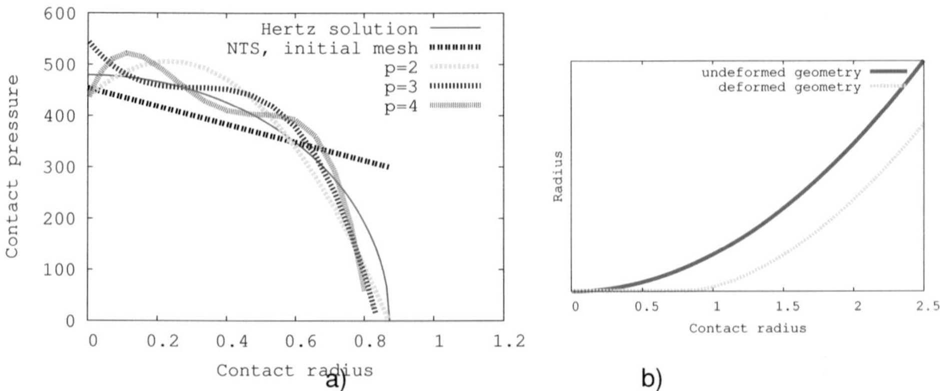

Figure 7.6: Contact zone is within a single element. a) Contact pressure vs. radius. Variation of the polynomial degree for the CLRS element. b) Deformed shape.

Obviously a good approximation is achieved even with the second order polynomial $p = 2$; also the deformed shape (Fig. 7.6 b) is represented exactly. It is obvious that this result is impossible to achieve with a linear mesh and a node-to-segment approach.

7.4.2 Loading case 2. Contact zone within several elements

If the contact zone is spreading through several elements a specific problem is arising: *if the the contact zone is ending inside a contact element then a highly oscillatory solution occurs for the contact stresses.* This phenomenon is shown in Fig. 7.7. In this case the load $P = 7.5\,kN$ is applied resulting in the contact zone ending inside the fourth element. The amplitude of the oscillations is even higher than the maximum of stresses which makes the computed results inadmissible for applications. Moreover, oscillations occur not only inside the bounding element (fourth element in the example), but also in neighboring elements. Neither increasing the polynomial degree, nor application of the mentioned above composed integration rules of higher order could reduce the oscillations. Oscillations are present for both penalty and Lagrange multiplier method. Only if the contact zone is ending in the vicinity of the element boundary (e.g. represented by a node for 2D) then the contact stresses are well approximated. The problem has been recognized for high-order FE already in Paczelt et.al. [133] and as remedy the re-meshing with the goal to move the node to the end of the contact zone was performed.

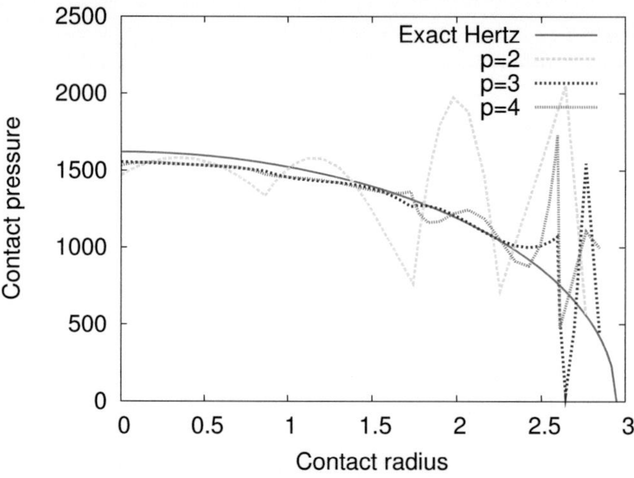

Figure 7.7: Oscillations for contact pressure occur at $P = 7.5kN$ when the contact zone is ending inside the element. Neither increasing the polynomial degree p, nor application of the composed integration rules of higher order could reduce the oscillations.

In the case of arbitrary loading contact detection is performed via the CPP procedure, therefore, it would be advantageous to find out techniques without moving a node. It was found in computations that the under-integration of the CLRS elements together with the lowering of the order of polynomial degree only for bounding elements, e.g. where the contact zone is ending, leads to a reduction of the oscillations. This effect is shown in Fig. 7.8, where the polynomial degree is taken to be

$p = 2$ and contact integrals only for the bounding element have been computed with only 2 Gauss points (neither the exact function representing the circular arch, nor the quadratic shape functions are fully integrated). Algorithmically this leads to the contact detection in only two points, see Table 7.1, thus the size of the contact zone is underestimated.

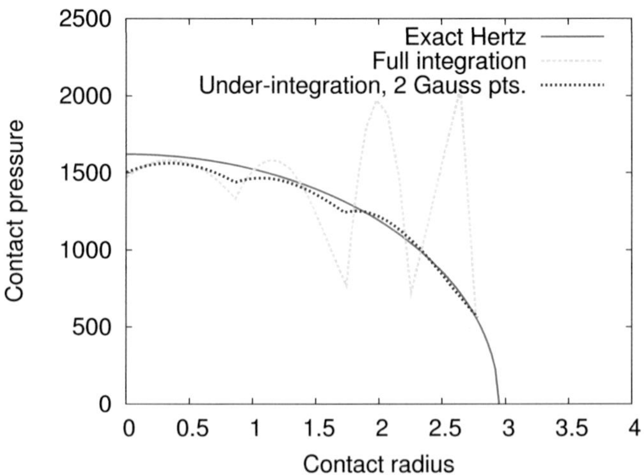

Figure 7.8: Contact pressure at $P = 7.5kN$. Under-integration of the bounding element with $p = 2$ leads to a reduction of oscillations.

This technique is tested then for the development of the contact zone through the element for $p = 2$. The applied force P is then rising in the range $5.90\,kN \leq P \leq 10.5\,kN$ while the contact zone is spreading through the fourth element, see Fig. 7.9. Obviously, the oscillations are not eliminated completely, but their amplitudes are sufficiently reduced.

Remark.

A fairly good correlation with the analytical Hertz solution even for the cases where only the contact layer is hierarchically enriched with polynomial order p allows to consider the proposed contact layer elements as a generalization of smoothing techniques known in contact mechanics, where only contact surfaces are smoothed. However, the full p-refinement combined with contact can be applied without loss of generality including the refinement the internal zone as well, which can lead even to a better correlation.

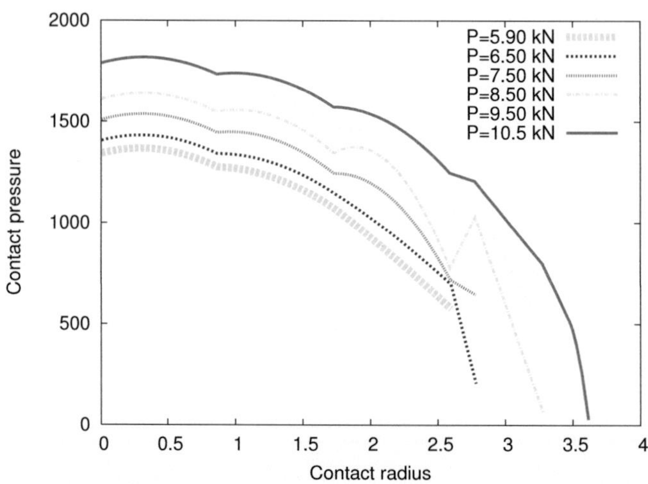

Figure 7.9: Contact pressure. Under-integration of the bounding element with $p = 2$. The contact zone is spreading through the bounding element while the load P is increasing.

7.5 Conclusions and outlook

In this contribution a covariant contact description is applied to derive all parameters which are necessary for numerical iterative solutions of Newton's type in combination with high order finite element techniques allowing an exact representation of the contact boundary (e.g. involved by the blended function method). In the latter case, additional linearization of the approximation operator is only necessary to fulfill the consistent linearization procedure. The structure of the derived tangent matrices remains algorithmic. If the penalty method is involved to enforce the normal contact condition then the matrices can be subdivided into a main, or constitutive part, a rotational part and a curvature part. The mixed method with Lagrange multiplier enforcement for the normal traction leads to a block structure of the corresponding tangent matrix, where a diagonal block is represented by the sum of the rotational and the curvature parts. The tangential conditions are enforced by the penalty method leading to similarly structured matrices as well.

Special contact layer finite elements are constructed as anisotropically refined p-finite elements allowing refinement only in contact layers.

Contact with rigid bodies leads then to a *Contact Layer - Rigid Surface* finite element, and contact between deformable bodies leads to a *Contact Layer - Contact Layer* finite element. The elements are verified with the classical Hertz contact problem. It is shown, that good correlations can be achieved even within a few elements. However, oscillations can occur if the contact zone appears inside the master contact segment. In such cases, the reduction of the polynomial order together with under-integration of only these bounding elements leads to improved acceptable results.

From a general point of view the proposed remedy for the oscillations has limitations. However, from the mathematical point of view the source of oscillations lays *in the inability to approximate non-smooth functions* (contact pressure) via the set of functions given *on the fixed domain* represented by the master segment, known as the Gibbs phenomenon. This explains the effect of under-integration as a simple elimination of the singular point (end of the contact zone as a point of non-smoothness) from the computation. A more advanced technique would fall into the local enrichment of the space function space, i.e. into the partition of unity method which is the focus of our further developments.

7.6 APPENDIX

7.6.1   Linearized normal part δW_c^N

If the Lagrange multiplier method is applied to enforce non-penetration conditions, then a normal traction is an independent variable and is included into the linearization as an independent rate variable $\dot{N}$. The rate equation (7.13) is taken into account only for the penalty method. In this case the linearized equation is written as

$$\mathcal{L}[\delta W_c^N] = \int_{s_S} \dot{N}\delta\xi^3 ds_S + \int_{s_S} N\mathcal{L}[\delta\xi^3] ds_S = \tag{7.35}$$

$$\underbrace{\int_{s_S} \epsilon_N(\delta\mathbf{r}_S - \delta\boldsymbol{\rho}) \cdot (\mathbf{n}\otimes\mathbf{n})(\mathbf{v}_S - \mathbf{v}) ds_S}_{\mathcal{L}_N^m} \tag{7.36}$$

$$\underbrace{- \int_{s_S} N\left[\delta\boldsymbol{\rho}_{,j} \cdot a^{ij}(\mathbf{n}\otimes\boldsymbol{\rho}_i)(\mathbf{v}_S - \mathbf{v}) + (\delta\mathbf{r}_S - \delta\boldsymbol{\rho}) \cdot a^{ij}(\boldsymbol{\rho}_j \otimes \mathbf{n})\mathbf{v}_{,i}\right] ds_S}_{\mathcal{L}_N^r} \tag{7.37}$$

$$\underbrace{- \int_{s_S} N(\delta\mathbf{r}_S - \delta\boldsymbol{\rho}) \cdot h^{ij}(\boldsymbol{\rho}_i \otimes \boldsymbol{\rho}_j)(\mathbf{v}_S - \mathbf{v}) ds_S}_{\mathcal{L}_N^c} . \tag{7.38}$$

Here, a comma stands for a partial derivative operator, namely $(\ldots)_{,i} \equiv \dfrac{\partial(\ldots)}{\partial\xi^i}$. One of the advantages of the covariant description is the representation of the linearized functional by parts: $\mathcal{L}^m$ is a main part depending on the constitutive relation (this part represents an elastic compliance with stiffness ϵ_N); $\mathcal{L}^r$ is a rotational part depending on the rotation of the master segment and $\mathcal{L}^c$ is a curvature part depending on the curvature of the master segment. Other geometrical properties of contact operators are discussed in Konyukhov and Schweizerhof [92]. Another advantage is that for further finite element approximation *only a single approximation operator* for the relative displacement vector $\mathbf{r}_s - \mathbf{r}$ should be introduced which makes the implementation procedure algorithmic for any parameterization of the surface.

7.6.2 Linearization of the tangential part δW_c^T

The expression for the tangential part δW_c^T in eqn. (3.35) is varying with regard to the return-mapping procedure in eqn. (7.18).

For **the sticking case** the evolution equation (4.73) is directly used for linearization leading to the following expression:

$$\mathcal{L}[\delta W_c^{T,\,sticking}] = \int_{s_S} \left(\frac{dT_i^{el}}{dt}\delta\xi^i + T_i^{el}\mathcal{L}[\delta\xi^i] \right) ds_S =$$

$$= -\underbrace{\int_{s_S} \epsilon_T(\delta\mathbf{r}_S - \delta\boldsymbol{\rho}) \cdot a^{ij}\boldsymbol{\rho}_i \otimes \boldsymbol{\rho}_j(\mathbf{v}_S - \mathbf{v})ds_S}_{\mathcal{L}_{T,\,st}^m} \qquad (7.39)$$

$$-\underbrace{\int_{s_S} T_i^{el} \left[(\delta\mathbf{r}_S - \delta\boldsymbol{\rho}) \cdot a^{il}a^{jk}\boldsymbol{\rho}_k \otimes \boldsymbol{\rho}_l\mathbf{v}_j + \delta\boldsymbol{\rho}_{,j} \cdot a^{ik}a^{jl}\boldsymbol{\rho}_k \otimes \boldsymbol{\rho}_l(\mathbf{v}_S - \mathbf{v}) \right] ds_S}_{\mathcal{L}_{T,\,st}^r} \quad (7.40)$$

$$+\underbrace{\int_{s_S} T_i^{el}(\delta\mathbf{r}_S - \delta\boldsymbol{\rho}) \cdot h^{ij}\left(\boldsymbol{\rho}_j \otimes \mathbf{n} + \mathbf{n} \otimes \boldsymbol{\rho}_j \right)(\mathbf{v}_S - \mathbf{v})ds_S}_{\mathcal{L}_{T,\,st}^c} \quad (7.41)$$

For **the sliding case** the sliding traction T_i^{sl} in eqn. (7.18) is linearized taking into account the evolution equation (4.73). The full expression is, though fairly long, but can be characterized by parts:

$$\mathcal{L}[\delta W_c^{T,\,sliding}] = \int_{s_S} \left(\frac{dT_i^{sl}}{dt}\delta\xi^i + T_i^{sl}\mathcal{L}[\delta\xi^i] \right) ds_S =$$

$$= \underbrace{\int_{s_S} \frac{\mu\epsilon_N}{\|\mathbf{T}^{el}\|}(\delta\mathbf{r}_S - \delta\boldsymbol{\rho}) \cdot T_{el}^i\boldsymbol{\rho}_i \otimes \mathbf{n}(\mathbf{v}_S - \mathbf{v})ds_S}_{\mathcal{L}_{T,\,sl}^{m,\,N}} \qquad (7.42)$$

$$-\underbrace{\int_{s_S} \frac{\epsilon_T\mu|N|}{\|\mathbf{T}^{el}\|}(\delta\mathbf{r}_S - \delta\boldsymbol{\rho}) \cdot a^{ij}\boldsymbol{\rho}_i \otimes \boldsymbol{\rho}_j(\mathbf{v}_S - \mathbf{v})ds_S}_{\mathcal{L}_{T,\,sl}^{m,\,T}} \qquad (7.43)$$

$$+\underbrace{\int_{s_S} \frac{\epsilon_T\mu|N|}{\|\mathbf{T}^{el}\|^3}(\delta\mathbf{r}_S - \delta\boldsymbol{\rho}) \cdot T_{el}^iT_{el}^j\boldsymbol{\rho}_i \otimes \boldsymbol{\rho}_j(\mathbf{v}_S - \mathbf{v})ds_S}_{\mathcal{L}_{T,\,sl}^{m,\,|T|}} \qquad (7.44)$$

$$-\underbrace{\int_{s_S} T_i^{sl} \left[(\delta\mathbf{r}_S - \delta\boldsymbol{\rho}) \cdot a^{il}a^{jk}\boldsymbol{\rho}_k \otimes \boldsymbol{\rho}_l\mathbf{v}_j + \delta\boldsymbol{\rho}_{,j} \cdot a^{ik}a^{jl}\boldsymbol{\rho}_k \otimes \boldsymbol{\rho}_l(\mathbf{v}_S - \mathbf{v}) \right] ds_S}_{\mathcal{L}_{T,\,sl}^r} \quad (7.45)$$

$$+ \int_{s_S} T_i^{sl}(\delta\mathbf{r}_S - \delta\boldsymbol{\rho}) \cdot h^{ij} \left(\boldsymbol{\rho}_j \otimes \mathbf{n} + \mathbf{n} \otimes \boldsymbol{\rho}_j\right)(\mathbf{v}_S - \mathbf{v})ds_S \qquad (7.46)$$

$$\underbrace{\phantom{+ \int_{s_S} T_i^{sl}(\delta\mathbf{r}_S - \delta\boldsymbol{\rho}) \cdot h^{ij} \left(\boldsymbol{\rho}_j \otimes \mathbf{n} + \mathbf{n} \otimes \boldsymbol{\rho}_j\right)(\mathbf{v}_S - \mathbf{v})ds_S}}_{\mathcal{L}_{T,\,sl}^{c,\,sym}}$$

$$- \int_{s_S} \frac{\mu|N|}{\|\mathbf{T}^{el}\|^3}(\delta\mathbf{r}_S - \delta\boldsymbol{\rho}) \cdot T_{el}^i T_{el}^k T_{el}^n \Gamma_{n,kj} a^{jm} \boldsymbol{\rho}_i \otimes \boldsymbol{\rho}_j (\mathbf{v}_S - \mathbf{v})ds_S \quad (7.47)$$

$$\underbrace{\phantom{- \int_{s_S} \frac{\mu|N|}{\|\mathbf{T}^{el}\|^3}(\delta\mathbf{r}_S - \delta\boldsymbol{\rho}) \cdot T_{el}^i T_{el}^k T_{el}^n \Gamma_{n,kj} a^{jm} \boldsymbol{\rho}_i \otimes \boldsymbol{\rho}_j (\mathbf{v}_S - \mathbf{v})ds_S}}_{\mathcal{L}_{T,\,sl}^{c,\,ns1}}$$

$$+ \int_{s_S} \frac{\mu|N|}{\|\mathbf{T}^{el}\|^3}(\delta\mathbf{r}_S - \delta\boldsymbol{\rho}) \cdot T_{el}^i T_{el}^j T_{el}^k h_{jk} \boldsymbol{\rho}_i \otimes \mathbf{n}(\mathbf{v}_S - \mathbf{v})ds_S \qquad (7.48)$$

$$\underbrace{\phantom{+ \int_{s_S} \frac{\mu|N|}{\|\mathbf{T}^{el}\|^3}(\delta\mathbf{r}_S - \delta\boldsymbol{\rho}) \cdot T_{el}^i T_{el}^j T_{el}^k h_{jk} \boldsymbol{\rho}_i \otimes \mathbf{n}(\mathbf{v}_S - \mathbf{v})ds_S}}_{\mathcal{L}_{T,\,sl}^{c,\,ns2}}$$

The following notations are used here: $\|\mathbf{T}^{el}\| = \sqrt{T_i^{el} T_j^{el} a^{ij}}$, where all contravariant resp. covariant components are computed via the rising resp. lowering index rule $T_{el}^i = T_j^{el} a^{ij}$; $h^{ij} = h_{kl} a^{ik} a^{lj}$. Covariant components of the curvature tensor are computed as $h_{kl} = \left(\dfrac{\partial\boldsymbol{\rho}}{\partial\xi^k\xi^l} \cdot \mathbf{n}\right)$; the Christoffel symbols are given as $\Gamma_{n,kj} = \left(\dfrac{\partial\boldsymbol{\rho}}{\partial\xi^k\xi^j} \cdot \boldsymbol{\rho}_n\right)$. The matrix contains non-symmetric main parts $\mathcal{L}_{T,\,sl}^{m,\,N}$ in eqn. (7.42) and non-symmetric curvature parts $\mathcal{L}_{T,\,sl}^{c,\,ns1}$, $\mathcal{L}_{T,\,sl}^{c,\,ns2}$ in eqns. (7.47), (7.48).

All parts of the tangent matrix can be classified as main parts, or constitutive parts (marked as $(.)^m$), representing the constitutive law for contact interfaces; as rotational parts (marked as $(.)^r$) representing a stiffness due to the rotation of the master segment and as curvature parts (marked as $(.)^c$) representing a stiffness due to the curvature of the master segment.

8

Covariant description of contact interfaces considering anisotropy for adhesion and friction: formulation and analysis of the computational model[*]

Abstract

A covariant description for contact problems including anisotropy for both adhesion and sliding domains is proposed. The principle of maximum dissipation is used to obtain a computational model in the case of quasi-static motion. Various cases including curvilinear anisotropy on arbitrary surfaces illustrating the possibility to model machined surfaces are considered. The part is served to be a necessary preparation for further finite element implementations and numerical analysis.

Keywords

covariant description anisotropy contact adhesion Coulomb friction

8.1   Introduction

The majority of contact problems is solved under the assumption of smoothness of contact surfaces. However, some cases appear in prac-

---

[*]The chapter is published in [90]: A. Konyukhov, K. Schweizerhof *Covariant description of contact interfaces considering anisotropy for adhesion and friction: Part 1. Formulation and analysis of the computational model*, Computer Methods in Applied Mechanics and Engineering, **196**(1):103–117, 2006.

tice when it is impossible to neglect the roughness of the contact surfaces. Essentially two types can be distinguished: a) when a surface has randomly distributed asperities, and b) when asperities have algorithmic structure, e.g. the surface shows different macro properties in different directions.

Mechanical characteristics for the contact problem of the first type a) are obtained via statistically distributed asperities. Statistical analysis of a real rough surface and experimental aspects of its measurements have been developed in series of publications: Longuet-Higgins [116], Greenwood and Williamson [50], Whitehouse and Archard [179] and more recently Whitehouse and Phillips [181] and Greenwood [48]. A comparative analysis of these surface models is presented in McCool [122].

A statistical concept for the contact in the context of finite element methods was developed then in Zavarise et.al. [201], Wriggers and Zavarise [196], [195] for the non-frictional contact with normally distributed asperities. Buczkowski and Kleiber [23] considered first non-frictional contact with an isotropical statistical distribution of asperities, and then in [24] non-frictional contact with an anisotropic statistical distribution of asperities. The modeling of a contact surface with Bezier splines according to the statistical distribution of asperities was considered in Bandeira, Wriggers and Pimenta [9]. Various nonlinear friction models are considered in the monography of Wriggers [188].

A generalization of the isotropic macro characteristics is used to describe frictional contact problems of the second type b). Michalowski and Mroz [129] considered the sliding of a rigid block on an inclined surface and formulated various sliding rules for sliding displacements which depend on directions. Thus, an anisotropic friction model for sliding was introduced. In a purely theoretical description Zmitrowicz [210] developed the structure of the friction tensor for sliding forces based on the motion of an elementary block on an anisotropic surface and described its properties based on symmetry groups for the tensor. Curnier [32] presented a rate independent theory of anisotropic friction for contact interaction mentioning adhesion as a possible elastic part without any further development. Zmitrowicz [211] developed the structure of the sliding friction tensor according to a relative sliding velocity and introduced a classification of anisotropic surfaces based on the number of eigenvalues of the friction tensor. These cases were numerically illustrated for a material point on the anisotropic plane. He and Curnier [62] used the theory of tensor function representations to obtain the structure of the friction tensor for an arbitrary nonlinear case according to the relative sliding velocity and derived also thermodynamical restrictions for the friction tensor components. Mroz and Stupkiewicz [129] considered the structure of the friction tensor based on a statical model of interaction of springs located in a plane which has periodically inclined asperities.

Despite the extensive literature on finite element solutions for contact problems, there are only few publications on finite element models for anisotropic friction. Buczkowski and Kleiber [22], [24] created an interface element containing the orthotropic sliding law. The return-mapping scheme in a Cartesian coordinate system was then used to obtain the sliding displacements. The effect of orthotropy was interpreted considering small displacements for a flat punch on an elastic foundation. Hjiaj et. al. [68] formulated the problem via the bi-potential and applied Lagrangian multiplier methods to solve a problem with orthotropic friction considering also small displacements in Cartesian coordinates. Parametric quadratic programming was used in Zhang et. al. [203] to solve the almost identical problem. Recently, Jones and Papadopoulos [82] developed a finite element model for anisotropic friction, where the stick-slip condition is enforced via Lagrange multipliers.

The aforementioned publications include anisotropy only for the friction model and **do not assume any anisotropy for the elastic region, the adhesion.** In the current contribution we propose a general approach for the finite element solution of quasi-statical frictional contact problems including anisotropy for both adhesion and sliding assuming that contact surfaces in general possess an anisotropic structure for both, elastic and friction domains. This approach is based on the covariant description for contact problems which is applicable for an arbitrary geometry of contact surfaces and large displacements. Within the covariant description, as given in Konyukhov and Schweizerhof [86], [89] contact conditions are described on the tangent plane of the contact surface exploiting tensor analysis. Using a penalty regularization of Coulomb's friction law and the return-mapping algorithm leads to the evolution equations for contact friction. It becomes obvious, that the evolution equations are not only a regularization technique, but act also as the constitutive relation to model friction behavior for an adhesion domain. Keeping this idea in mind, the evolution equations are generalized for a more complex mechanical behavior exploiting tensor algebra on the tangent plane of the contact surface. The covariant description allows to formulate the main characteristics for surfaces with arbitrary geometry, e.g. the yield function is formulated via the friction tensor defined in surface metrics. Then both, anisotropy for adhesion resp. sticking and anisotropy for sliding are treated. Anisotropic resp. orthotropic tensors inherit their properties either from the spectral decomposition or, in the more general case, from arbitrary curvilinear coordinate systems defined on the surface. The last case has advantages in practical applications as e.g. for a homogenized average model of machined surfaces. This structure of tensors automatically satisfies all necessary theoretical restrictions developed earlier in [62] and [211]. Thus, a consistent model for anisotropic surfaces can be developed.

In order to define sliding forces as well as sliding displacements the principle of maximum dissipation is used. All models allow representative geometrical interpretations on the tangent plane. Via this principle, the sliding forces and sliding displacements are derived in the covariant form. The formulation in the covariant form easily allows to derive the consistently linearized equations, which are necessary for the iterative solution within a Newton type scheme, even for the case with nonlinear anisotropy in the reference Cartesian coordinate system.

The article is subdivided into two parts, where the current part is organized as follows: In the second section we recall all necessary details from the covariant description for the isotropic case. The generalization for the adhesive part and for the sliding part for the arbitrary geometry is developed in section 3. The orthotropic planar cases in Cartesian as well as in polar coordinates are considered as particular structures of tensors. As a specific case with curvilinear geometry spiral orthotropy for a cylinder is considered. The fourth section deals with the formulation of the friction problem as maximization problem for the energy dissipation function. Here also the geometrical interpretation of the derived model is discussed. The consistent linearization on the tangent plane, the finite element implementation and a discussion about the robustness of the developed approach on the basis of various numerical examples are included in the second part.

8.2 Basis of the covariant description

Several computational approaches for isotropic Coulomb frictional contact in context with finite element analysis have been developed in the literature. The general models – all using the elastoplastic analogy and the return-mapping algorithm for the penalty regularization of the friction law – are developed in Wriggers et. al. [194], Laursen and Simo [109], Peric and Owen [139], Parisch and Lübbing [138]. Reviews of contemporary contact models can be found in the monographs of Wriggers [188] and Laursen [106]. The covariant description was especially developed in Konyukhov and Schweizerhof [86], [89] to take advantage of the differential properties of contact surfaces. These derivations allow a straightforward geometrical interpretation of the characteristics for an iterative solution, such as regularization equations and tangent matrices. In the current section all important details of the covariant description necessary for a further generalization into the anisotropic domain are briefly outlined.

As starting basis of the covariant description, we introduce a spatial local coordinate system related to the master surface. This coordinate system is defined fitting the closest projection procedure. Let S be a slave point and C its projection on the surface, see Fig. 8.1. At point C we consider a coordinate system based on the following relation:

$$\mathbf{r}_s(\xi^1, \xi^2, \xi^3) = \boldsymbol{\rho}(\xi^1, \xi^2) + \mathbf{n}\xi^3. \tag{8.1}$$

The first two convective coordinates ξ^1, ξ^2 define properties of the surface and, therefore, are responsible for the tangential contact interaction. The third coordinate ξ^3 is the value of the penetration and is used to define the properties of the normal interaction. It is obtained after the aforementioned closest point procedure as projection on the third axis $\mathbf{n}$

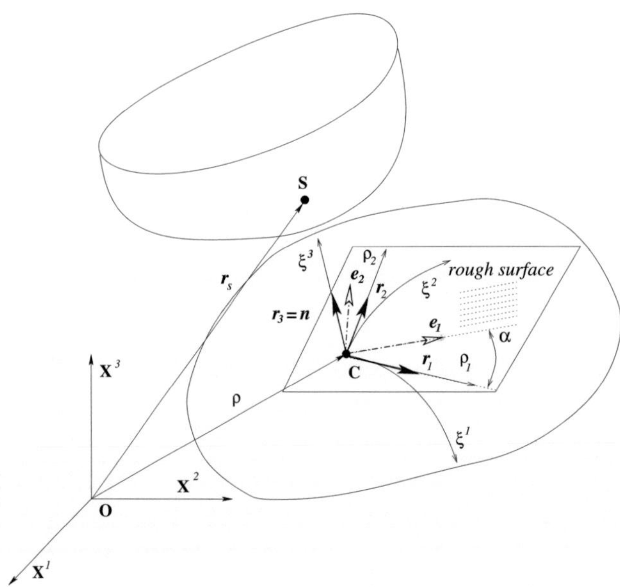

Figure 8.1: Contact between bodies. Definitions for closest point projection. Anisotropic contact surface.

in each iteration step:

$$\xi^3 = (\mathbf{r}_{\mathrm{s}} - \boldsymbol{\rho}) \cdot \mathbf{n}. \tag{8.2}$$

The basis vectors $\mathbf{r}_i$, $i = 1, 2, 3$ of the spatial coordinate system eqn. (8.1) are obtained via the basis surface vectors $\boldsymbol{\rho}_1$, $\boldsymbol{\rho}_2$, $\mathbf{n}$ as:

$$\mathbf{r}_i = \frac{\partial \mathbf{r}}{\partial \xi^i} = \boldsymbol{\rho}_i + \mathbf{n}_i \xi^3 = (a_i^k - h_i^k \xi^3)\boldsymbol{\rho}_k, \quad i, k = 1, 2, \quad \mathbf{r}_3 = \mathbf{n}, \tag{8.3}$$

where a_i^k are mixed components of the surface metric tensor and h_i^k are mixed components of the surface curvature tensor. A core of the covariant description is to consider contact dependencies in the 3D spatial system and to express them on the tangent plane, i.e. at $\xi^3 = 0$.

8.2.1 Convective velocities. Variation of relative displacements

Projections of the full time derivative of eqn. (8.1) to the local basis $\mathbf{r}_i$ considered at $\xi^3 = 0$ result in the convective velocities $\dot{\xi}^i$. The tangential components are defined as

$$\dot{\xi}^j = a^{ij}(\mathbf{v}_\mathrm{s} - \mathbf{v}) \cdot \boldsymbol{\rho}_i, \quad i,j = 1,2, \tag{8.4}$$

where $\mathbf{v}_\mathrm{s}$ is the velocity of the slave point S, and $\mathbf{v}$ is the velocity of the projection point C. The third component is given by the value in the normal direction

$$\dot{\xi}^3 = (\mathbf{v}_\mathrm{s} - \mathbf{v}) \cdot \mathbf{n}, \quad i = 1,2. \tag{8.5}$$

Considering the convective velocities also the variation of the relative displacements can be expressed in form of $\delta\xi^i$

$$\delta\mathbf{r}_\mathrm{s} - \delta\boldsymbol{\rho} = \delta\xi^i \boldsymbol{\rho}_i + \delta\xi^3 \mathbf{n}. \tag{8.6}$$

8.2.2 Evolution equations for contact tractions

The vector of contact tractions $\mathbf{R}$ is defined as a covariant vector and, therefore, is expressed via the contravariant basis vectors $\boldsymbol{\rho}^i$ and $\mathbf{n}$ as sum of the tangential and normal components

$$\mathbf{R} = \mathbf{T} + \mathbf{N} = T_i \boldsymbol{\rho}^i + N\mathbf{n}. \tag{8.7}$$

As is well known, the relations between two coordinates ξ^1, ξ^2 and the tangential force can be formulated in the differential form as so-called evolution equations. The penalty regularization process for the simple Coulomb friction law within the analogy to the rigid plasticity model leads to the following evolution equations for the trial tangential contact tractions T_i:

$$\frac{dT_i}{dt} = (-\epsilon_T a_{ij} + \Gamma^k_{ij} T_k)\dot{\xi}^j - h^k_i T_k \dot{\xi}^3, \tag{8.8}$$

where Γ^k_{ij} are the Christoffel symbols for the contact surface. Eqn. (8.8) serves to compute the trial tangent tractions. The final values are obtained via the return-mapping scheme, see [106], [188]. Equation (8.8) is a covariant scalar form of the proportionality condition between the full time derivative of the tangent traction vector $\mathbf{T}$ and the relative velocity vector $\dot{\xi}^i \boldsymbol{\rho}_i$ expressed on the tangent plane

$$\frac{d\mathbf{T}}{dt} = -\epsilon_T \dot{\xi}^i \boldsymbol{\rho}_i, \tag{8.9}$$

where a full time derivative $\frac{d\mathbf{T}}{dt}$ is taken in covariant form

$$\frac{d\mathbf{T}}{dt} = \frac{DT_i}{dt}\boldsymbol{\rho}^i, \qquad \frac{DT_i}{dt} \equiv \frac{dT_i}{dt} - \Gamma_{ij}^k T_k \dot{\xi}^j + h_i^k T_k \dot{\xi}^3, \qquad (8.10)$$

The regularization equation for the normal traction N satisfying the non-penetrability condition has the following form:

$$N = \epsilon_N H(-\xi^3)\xi^3, \qquad (8.11)$$

where a Heaviside function $H(-\xi^3)$ reflects the fact that N is not equal zero and is computed only when a penetration occurs, i.e. $\xi^3 < 0$.

The full time derivative $\dot{N}$ is then:

$$\dot{N} = -\epsilon_N H(-\xi^3)\dot{\xi}^3. \qquad (8.12)$$

8.2.2.1 Integration of evolution equations. Geometrical interpretation of the return-mapping scheme

As shown in Konyukhov and Schweizerhof [86], [89], within the contact description the curvature part can be omitted in numerous cases without loss of efficiency leading to considerable simplifications and a major gain in numerical effort. This is especially pronounced for such contact problems, where the development of sticking-sliding zones is important. In such cases we can e.g. simplify all equations, assuming constant metrics. The evolution equation is solved then numerically via the implicit backward scheme with n indicating the load step arising from a subdivision of the loads applied in sequential load steps.

$$T_i^{(n+1)} = T_i^{(n)} - \epsilon_T a_{ij}(\xi_{(n+1)}^j - \xi_{(n)}^j) = \qquad (8.13)$$

continuing recursively:

$$= T_i^{(n-1)} - \epsilon_T a_{ij}(\xi_{(n)}^j - \xi_{(n-1)}^j) - \epsilon_T a_{ij}(\xi_{(n+1)}^j - \xi_{(n)}^j)$$

$$= T_i^{(n-1)} - \epsilon_T a_{ij}(\xi_{(n+1)}^j - \xi_{(n-1)}^j) = \ldots$$

$$= T_i^{(0)} - \epsilon_T a_{ij}(\xi_{(n+1)}^j - \xi_{(0)}^j)$$

Assuming that the initial tangential forces are zero, $T_i^{(0)} = 0$, we get

$$T_i^{(n+1)} = -\epsilon_T a_{ij}\Delta\xi^j, \quad \text{with} \ \Delta\xi^j = (\xi_{(n+1)}^j - \xi_{(0)}^j). \qquad (8.14)$$

Eqn. (8.14) defines trial tractions T_i at load step $(n+1)$ at contact points on the tangent plane $\xi_{(n+1)}^1, \xi_{(n+1)}^2$. The absolute value of the tangent traction is computed as:

$$\|\mathbf{T}\|^2 = T_i^{(n+1)} T_j^{(n+1)} a^{ij} = \epsilon_T^2 a_{ij}(\xi_{(n+1)}^i - \xi_{(0)}^i)(\xi_{(n+1)}^j - \xi_{(0)}^j). \qquad (8.15)$$

With eqn. (8.15) the sticking zone in combination with the Coulomb law becomes then:

$$\|\mathbf{T}\|^2 \leq \mu|N|^2 \implies \epsilon_T^2 a_{ij}(\xi_{(n+1)}^i - \xi_{(0)}^i)(\xi_{(n+1)}^j - \xi_{(0)}^j) \leq \mu^2 N^2. \qquad (8.16)$$

Eqn. (8.16) describes a circle in a Cartesian coordinate system via the convective coordinates ξ_1, ξ_2. The circle is placed on the tangent plane with the center at $\xi_{(0)}^1, \xi_{(0)}^2$. The inner part of it defines the allowable elastic region for the projection of a slave point, the so-called adhesion domain. Thus, the geometrical interpretation of the solution of the evolution equation is a trajectory of the contact point which is allowed to be inside the circle in the case of sticking. If eqn. (8.11) is taken for regularization of the normal traction N, then we obtain a cone equation in the spatial coordinate system:

$$\epsilon_T^2 a_{ij}(\xi_{(n+1)}^i - \xi_{(0)}^i)(\xi_{(n+1)}^j - \xi_{(0)}^j) \leq \mu^2 \epsilon_N^2 (\xi^3)^2. \qquad (8.17)$$

This interpretation can be found in Krstolovic-Opara and Wriggers [101], where a so-called "frictional cone description" was proposed.

8.2.3 Weak form.

The work of the contact tractions $\mathbf{R}$ in eqn. (8.7) on the relative virtual displacement $\delta\mathbf{r}_s - \delta\boldsymbol{\rho}$ in eqn. (8.6) can be expressed on the contact surface as:

$$\delta W_c = \int_s \mathbf{R} \cdot (\delta\mathbf{r}_s - \delta\boldsymbol{\rho})ds = \int_s N\delta\xi^3 ds + \int_s T_j \delta\xi^j ds. \qquad (8.18)$$

The integral in eqn. (8.18) is computed on the slave surface ds, whereas all functions are defined on the master surface.

8.3 Generalization for complex contact interface laws

The regularization for a Coulomb type frictional law leads to a subdivision of the motion of the contact point on the master surface into reversible and irreversible parts. The first, reversible part appears due to the regularization and usually contains a penalty parameter. It describes the elastic tangent deformation, the so-called tangential adhesion, see Curnier [32]. The second, irreversible part is described by a flow rule due to a specific yield function. Both parts can be generalized for anisotropic domains by taking proper tensors and equations. In this section, we summarize all necessary equations in vector form convenient for the expansion into the anisotropic domain. For the representations of the anisotropic tensor we will use the representation based on the spectral decomposition in the simple case of constant orthotropy, and in the general case, the representation based on the tensor product of unit vectors of an arbitrary curvilinear surface coordinate system.

8.3.1 Vector form of the isotropic equations

The generalization is based upon the consideration of anisotropic tensors instead of isotropic tensors in the evolution equations. The structure of the anisotropic tensors for contact, if one or both contact surfaces have anisotropic structure is theoretically discussed by Zmitrowicz [210], [211] and also by He and Curnier [62]. Here we assume the described properties for tensors and will further discuss some restrictions for both the adhesion tensor and the friction tensor. All tensors are defined in the basis of the tangent plane of the contact master surface.

8.3.1.1 Elastic part of the contact deformation

To start the development, we reorganize the evolution equations (8.8) and (8.12), describing in fact the elastic reversible part of the deformation in vector form in the local surface coordinate system.

$$\frac{d\mathbf{R}}{dt} = \hat{\mathbf{E}}\mathbf{v}^r, \tag{8.19}$$

where $\mathbf{R}$ is a traction vector acting on a slave point S; $\mathbf{v}^r = \mathbf{v}_s - \mathbf{v} = \boldsymbol{\rho}_i\dot{\xi}^i + \mathbf{n}\dot{\xi}^3$ is a relative velocity of a slave point and $\hat{\mathbf{E}}$ is an isotropic tensor of penalty parameters:

$$\hat{\mathbf{E}} = \begin{bmatrix} -\epsilon_T & 0 & 0 \\ 0 & -\epsilon_T & 0 \\ 0 & 0 & -\epsilon_N \end{bmatrix}. \tag{8.20}$$

Eqn. (8.19) describes the force-displacement relationship in a rate form for the reversible elastic part of the contact interaction. The irreversible part in the case of a simple Coulomb friction law is correlated to the rigid plasticity model.

8.3.1.2 Yield function for the isotropic Coulomb friction law

Remembering, that the scalar product is computed on the surface via the metric tensor components a^{ij}, we can define a yield function for the Coulomb friction law as

$$\Phi := \frac{\sqrt{\mathbf{T} \cdot \mathbf{T}}}{\mu|N|} - 1 \equiv \frac{\sqrt{T_iT_ja^{ij}}}{\mu|N|} - 1. \tag{8.21}$$

The sticking and sliding zones are now defined by the rule:

$$\Phi \leq 0 \text{ means } \textit{sticking}; \quad \Phi > 0 \text{ means } \textit{sliding}. \tag{8.22}$$

Irreversible parts including the sliding forces etc. can be obtained via the flow rule.

8.3.2 General interface model

In order to take into account a diversity of linear mechanical models including viscoelasticity etc., we can generalize eqn. (8.19) as follows:

$$\frac{d\mathbf{R}}{dt} + \hat{\mathbf{A}}\mathbf{R} = \hat{\mathbf{B}}\mathbf{v}^r + \hat{\mathbf{C}}\Delta\boldsymbol{\xi}, \tag{8.23}$$

where $\hat{\mathbf{A}}$, $\hat{\mathbf{B}}$, $\hat{\mathbf{C}}$ are tensors defined in the local surface coordinate system and $\Delta\boldsymbol{\xi}$ is a vector of the relative displacements

$$\Delta\boldsymbol{\xi} = \Delta\xi^i\boldsymbol{\rho}_i + \Delta\xi^3\mathbf{n}, \quad \Delta\xi^i = \xi^i - \xi^i_{(0)}, \quad i = 1, 2, 3. \tag{8.24}$$

A point with convective coordinates $\xi^i_{(0)}$ indicates an initial position where the contact traction vector $\mathbf{R}$ equals zero, i.e.

$$\mathbf{R}\big|_{\xi^i=\xi^i_{(0)}} = \mathbf{0}. \tag{8.25}$$

The equation for the elastic region in the form of eqn. (8.23) covers various viscoelastic models such as Maxwell and Kelvin models in arbitrary anisotropic forms including adhesion. A generalization of the isotropic Coulomb friction model according to Maxwell and Kelvin viscoelastic models was considered in Araki and Hjelmstad [5]. A rate-dependent model with orthotropic friction coefficients in Cartesian coordinate system was considered in Oancea and Laursen [131].

Assuming the decoupling of the third normal coordinate ξ^3, we can rewrite eqn. (8.23) for the surface components in the form:

$$\frac{d\mathbf{T}}{dt} + \mathbf{AT} = \mathbf{B}\mathbf{v}^r + \mathbf{C}\Delta\boldsymbol{\xi}, \tag{8.26}$$

where $\mathbf{A}$, $\mathbf{B}$, $\mathbf{C}$ are tensors defined on the tangent plane, $\mathbf{T}$ is a tangent force vector, $\mathbf{v}^r = \mathbf{v}_s - \mathbf{v}$ is a relative tangent velocity vector expressed on the tangent plane, $\Delta\boldsymbol{\xi} = \Delta\xi^i\boldsymbol{\rho}^i$ is a relative tangent displacement vector and $\dfrac{d}{dt}$ is a full time derivative in the covariant form on the contact surface.

The third equation for the normal force N and for the penetration ξ^3, is treated separately

$$\dot{N} + aN = b\dot{\xi}^3 + c\Delta\xi^3. \tag{8.27}$$

For the further development we consider only a rate independent motion, i.e. assume a specific structure of the evolution equations (8.23) and (8.27) excluding the direct time dependency of the contact tractions.

8.3.2.1 Anisotropic evolution equations. Rate-independent model

Considering a case of rate-independent motions by taking $\mathbf{A} = 0$, $\mathbf{C} = 0$ in eqn. (8.26) and $a = 0$, $c = 0$ in eqn. (8.27) we define anisotropy for an elastic part of the contact conditions. Therefore, from eqn. (8.26) the following rate forms remain:

$$\frac{d\mathbf{T}}{dt} = \mathbf{B}(\mathbf{v}_{\mathrm{s}} - \mathbf{v}). \tag{8.28}$$

Expressing this in the tangent plane by coordinates, we get the following evolution equation

$$\frac{\partial T_i}{\partial t} + \nabla_j T_i \dot{\xi}^j = b_{ij}\dot{\xi}^j, \quad j = 1, 2. \tag{8.29}$$

The evolution equations (8.28) resp. (8.29) describe the fact that the reaction tangential forces are acting not in the opposite direction to the velocity vector, but in the direction defined by tensor $\mathbf{B}$. In other words, if a force $\mathbf{T}$ is acting on point C on the surface, then this point is moved in a somewhat different direction defined by the angle β, see Fig. 8.2, but, in general, not in the direction of the force.

Remark.
The evolution equations (8.28) describe the elastic deformations of contact interaction in the rate form. This elastic tangential deformation is known as tangential adhesion, see Curnier [32]. Thus, we call a tensor $\mathbf{B}$ **the elastic adhesion tensor**, or simply **the adhesion tensor**.

Remark.
The mechanical restriction for the elastic force $\mathbf{T}$ to act in opposite direction to the relative velocity can be formulated in an energy sense according to the thermodynamical restriction: the power of the elastic force $\mathbf{T}$ must be negative. This requires that **the adhesion tensor taken with a minus sign** $-\mathbf{B}$ is a positively defined tensor.

Since we are working with a decoupled model regarding the normal and tangential contact interactions, the evolution equation for the normal force N and thus the parameter b in eqn. (8.27) is kept in the form given by eqn. (8.12).

8.3.3 Anisotropic yield function

Several theoretical approaches are known in literature for formulations for the yield function and for the sliding rule. Before presenting a particular structure of the tensor we will briefly review these approaches as well as restrictions for the tensor which are known in literature.

8.3.3.1 Various approaches for formulations of yield criteria and sliding rules

In a first publication, Michalowski and Mroz [127] proposed to distinguish functions for limit criteria and for sliding, introducing the so-called associated and non-associated sliding rules. These functions were built by analogy looking at the solution for the sliding of a rigid block on an inclined surface.

Zmitrowicz [210] introduced the friction tensor $\mathbf{F}$ into the originally isotropic Coulomb criteria (8.21) and described its properties based on the groups of material symmetry. The sliding forces were formulated directly without correspondence to the yield criteria. It was also obtained that *"There is no restrictions for the arbitrary anisotropic friction tensor except its positive definition, but the orthotropic friction tensor is symmetric"*. The superposition of two friction tensors when two anisotropic surfaces are in contact was also discussed. In a later publication Zmitrowicz [211] assumed that the sliding force nonlinearly depends on the relative velocity introducing in addition a 4th-order friction tensor.

He and Curnier [62] applied the theory of tensor function representations – the fundamental aspects of this theory see Zheng [204] – to obtain the general irreducible structure for the nonlinear friction tensor in the case of friction between two orthotropic surfaces. The structure at the contact point is defined by [62] as follows:

- $\mathbf{m}$ is a unit vector of the preferable direction for the first surface,

- $\mathbf{k}$ is a unit vector of the preferable direction for the second surface,

- $\mathbf{u}$ is a unit vector of the relative sliding velocity.

Then the sliding force $\mathbf{T}$ is defined as

$$\mathbf{T} = N\left[\,\alpha_1(I_1, I_2)\mathbf{E} + \alpha_2(I_1, I_2)(\mathbf{m} \otimes \mathbf{m}) + \alpha_3(I_1, I_2)(\mathbf{k} \otimes \mathbf{k})\,\right]\mathbf{u}\,, \quad (8.30)$$

where N is a normal force, $\mathbf{E}$ is a unit tensor, α_i are scalar functions of the following invariants: $I_1 = \mathbf{u} \cdot (\mathbf{m} \otimes \mathbf{m})\mathbf{u}$, $I_2 = \mathbf{u} \cdot (\mathbf{k} \otimes \mathbf{k})\mathbf{u}$. Also the symmetric properties of the friction tensor for orthotropic surfaces were derived and it was shown, that a yield function is a direct consequence of the rate-independence condition. Other important results were *"the thermodynamical restrictions for the friction tensor leading to its positivity and the formulation of the convex energy dissipation function"*.

8.3.3.2 Coulomb type yield functions

Summarizing the results of the previous developments, we will later use a generalization of the isotropic yield criterion in eqn. (8.21) by replacing the scaled metric components $a^{ij}/\mu^2 N^2$ with tensor components f^{ij}, assuming the a-priori necessary properties as discussed above. Thus, we can obtain:

$$\Phi_N = \sqrt{f_N^{ij} T_i T_j} - 1 = \sqrt{(\mathbf{T} \cdot \mathbf{F}_N \mathbf{T})} - 1. \tag{8.31}$$

The standard assumption of *proportionality of the sliding force* $\mathbf{T}^{sl}$ *to the normal traction* N, the so-called Coulomb's form, see Curnier [32], leads to

$$\mathbf{F}_N = \frac{\mathbf{F}}{N^2}, \tag{8.32}$$

and the yield function can then be written as:

$$\Phi = \sqrt{f^{ij} T_i T_j} - |N| = \sqrt{\mathbf{T} \cdot \mathbf{F} \mathbf{T}} - |N|. \tag{8.33}$$

The sliding criteria (8.21-8.22) become then

$$\text{if } \Phi \leq 0 \text{ then } sticking, \tag{8.34}$$

describing a contact point inside the adhesion domain,

$$\text{if } \Phi > 0 \text{ then } sliding, \tag{8.35}$$

describing a contact point outside the adhesion domain.

The tensor $\mathbf{F}$ is called **friction tensor** and is defined by its components f^{ij} in the surface tensor basis as

$$\mathbf{F} = f^{ij}\boldsymbol{\rho}_i \otimes \boldsymbol{\rho}_j. \tag{8.36}$$

Comparing with the isotropic function, we introduce anisotropic friction coefficients:

$$f^{ij} = \frac{a^{ij}}{\mu_{ij}^2}, \quad i,j = 1,2, \tag{8.37}$$

where μ_{ij} are coefficients of friction.

8.3.4 Tensor representations for anisotropy

Here we consider particular structures for the anisotropic tensors for the evolution equations (8.28) as well as for the yield function (8.33), automatically satisfying the necessary restrictions mentioned in the previous sections. We start with the simplest case – a constant orthotropy on the plane. In this case all tensors are symmetrical ones. The more general anisotropic case can be defined then in an arbitrary coordinate system by setting different properties along the coordinate lines.

8.3.4.1 Spectral representation of the tensor – constant orthotropy in the plane

As mentioned above, constant orthotropy in the plane is described by constant symmetric tensors for which a spectral representation is chosen. Any symmetric positive tensor $\mathbf{A}$ can be decomposed as:

$$\mathbf{A} = \mathbf{Q}\mathbf{\Lambda}\mathbf{Q}^T, \tag{8.38}$$

where $\mathbf{\Lambda}$ is defined in the mixed tensorial basis as a diagonal matrix of eigenvalues:

$$\mathbf{\Lambda} = \begin{bmatrix} \lambda_1 & 0 \\ 0 & \lambda_2 \end{bmatrix}; \tag{8.39}$$

and $\mathbf{Q}$ is an orthogonal tensor. Since it describes the rotation on the tangent plane between the main axes $\mathbf{e}_i$ and the axes ξ^i, see Fig. 8.2, it becomes

$$\mathbf{Q}_\alpha = \begin{bmatrix} \cos\alpha & -\sin\alpha \\ \sin\alpha & \cos\alpha \end{bmatrix}. \tag{8.40}$$

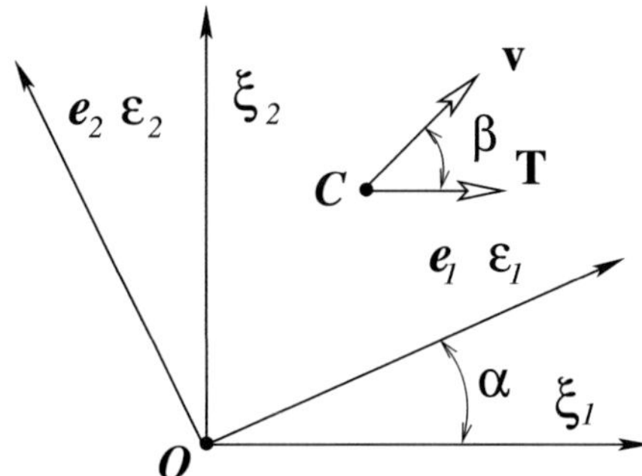

Figure 8.2: Main orthotropy axes and local surface coordinate system.

The main axes are hereby defined as orthotropy axes. Then the spectral representation of $\mathbf{A}$ is obtained as

$$\mathbf{A} = \mathbf{Q}_\alpha \boldsymbol{\Lambda} \mathbf{Q}_\alpha^T = [A_j^i] = \begin{bmatrix} \lambda_1 \cos^2 \alpha + \lambda_2 \sin^2 \alpha & (\lambda_1 - \lambda_2) \sin \alpha \cos \alpha \\ (\lambda_1 - \lambda_2) \sin \alpha \cos \alpha & \lambda_1 \sin^2 \alpha + \lambda_2 \cos^2 \alpha \end{bmatrix} . \quad (8.41)$$

Taking into account, that the model for constant orthotropic friction contains the tensor in the evolution equations (8.28) and in the yield function (8.33) in mixed form which allows the spectral decomposition given in eqn. (8.41), the following derivations are possible:

First we focus on **the evolution equation** given by eqn. (8.29), where the mixed components are introduced via the spectral decomposition (8.41). Taking into account that the adhesion tensor $\mathbf{B}$ (see Remark 8.3.2.1 in Sec. 8.3.2.1) is negative, we introduce positive values $\varepsilon_i = -\lambda_i > 0$ and an orthotropy angle α. Thus, after a transformation into a local coordinate system the following matrix description for the tensor is obtained:

$$\mathbf{B} = [b_j^i] = - \begin{bmatrix} \varepsilon_1 \cos^2 \alpha + \varepsilon_2 \sin^2 \alpha & (\varepsilon_1 - \varepsilon_2) \sin \alpha \cos \alpha \\ (\varepsilon_1 - \varepsilon_2) \sin \alpha \cos \alpha & \varepsilon_1 \sin^2 \alpha + \varepsilon_2 \cos^2 \alpha \end{bmatrix} . \quad (8.42)$$

Matrix $\mathbf{B}$ in eqn. (8.42) describes the orthotropic properties in the reversible part of the tangential interaction, where the parameters ε_1, ε_2 can be seen as orthotropic moduli of the tangential adhesion. Assuming $\varepsilon_1 = \varepsilon_2$ leads to isotropic behavior and then the isotropic evolution equation is recovered.

Second, **the yield function** for the constant orthotropic friction is defined via the orthotropic friction tensor $\mathbf{F}$ which allows a spectral decomposition. After introducing the angle β between the local coordinate axes and the orthotropy axes, and eigenvalues as $\lambda_i = 1/\mu_i^2$, the following matrix $[f_k^i]$ is obtained:

$$\mathbf{F} = [f_k^i] = \begin{bmatrix} \dfrac{1}{\mu_1^2}\cos^2\beta + \dfrac{1}{\mu_2^2}\sin^2\beta & (\dfrac{1}{\mu_1^2} - \dfrac{1}{\mu_2^2})\sin\beta\cos\beta \\[2em] (\dfrac{1}{\mu_1^2} - \dfrac{1}{\mu_2^2})\sin\beta\cos\beta & \dfrac{1}{\mu_1^2}\sin^2\beta + \dfrac{1}{\mu_2^2}\cos^2\beta \end{bmatrix} . \tag{8.43}$$

The restriction of positivity for the friction tensor $\mathbf{F}$ leads to the known positive orthotropic friction coefficients $\mu_i > 0$.

8.3.4.2 Structure of the anisotropic tensor inherited from an arbitrary surface curvilinear coordinate system

We assume that on the surface defined via the Gaussian coordinates ξ^1, ξ^2 as

$$\boldsymbol{\rho} = \boldsymbol{\rho}(\xi^1, \xi^2), \quad \boldsymbol{\rho} = \left\{ \begin{array}{c} x_1(\xi^1, \xi^2) \\ x_2(\xi^1, \xi^2) \\ x_3(\xi^1, \xi^2) \end{array} \right\} \tag{8.44}$$

another Gaussian coordinate system is defined. Thus, the Cartesian coordinates of the same surface x_i are defined by Gaussian convective coordinates α^1, α^2:

$$x_i = \phi_i(\alpha^1, \alpha^2), \quad i = 1, 2, 3. \tag{8.45}$$

One can say, that we have re-parameterization of the surface

$$\alpha^i = \alpha^i(\xi^1, \xi^2), \quad i = 1, 2, \tag{8.46}$$

or, in another words, eqn. (8.46) defines a new coordinate system on the same surface.

Arbitrary anisotropic properties of a surface can be defined as different characteristics along these coordinate lines. The unit tangent vectors along the coordinate lines are then defined as:

$$\mathbf{e}_i = \frac{\mathbf{r}_i}{\sqrt{g_{ii}}}, \quad i = 1, 2 \quad \text{no summation over } i \tag{8.47}$$

where $\mathbf{r}_i = \dfrac{\partial \mathbf{r}}{\partial \alpha_i}$ are the basis vectors, and $g_{ii} = \mathbf{r}_i \cdot \mathbf{r}_i$ are diagonal coefficients of the covariant metrics tensor.

The tensor of anisotropy $\mathbf{A}$ can then be defined via the unit tensor basis as:

$$\mathbf{A} := \lambda_1 \mathbf{e}_1 \otimes \mathbf{e}_1 + \lambda_2 \mathbf{e}_2 \otimes \mathbf{e}_2 = \lambda_1 \frac{\mathbf{r}_1 \otimes \mathbf{r}_1}{g_{11}} + \lambda_2 \frac{\mathbf{r}_2 \otimes \mathbf{r}_2}{g_{22}} = \lambda_i \frac{\mathbf{r}_i \otimes \mathbf{r}_i}{g_{ii}}. \tag{8.48}$$

Remark.
From now on and further in the last representation in eqn. (8.48) the summation convention is implied also for the sum over i-index, though the index i is repeated four times.

It is obvious, that the tensor structure prescribed above, see eqn. (8.30), is preserved. As a next step a transformation into the surface basis $\rho_i(\xi^1, \xi^2)$ is necessary for the evolution equation as well as for the yield equation.

Remark.
The tensor in eqn. (8.48) is defined, in general, via the coordinates α^1, α^2. We introduce a notation $\mathbf{A}^C$, if the tensor $\mathbf{A}$ is defined in the global reference Cartesian coordinates, and will keep the notation $\mathbf{A}$ if the tensor is defined after the tensor transformation to the surface coordinate system ξ^1, ξ^2, eqn. (8.44).

The evolution equation is transformed as follows. Assuming the stiffnesses along the coordinate lines as $\lambda_i = -\epsilon_i$, we obtain

$$\mathbf{B}^C := -\left(\epsilon_1 \frac{\mathbf{r}_1 \otimes \mathbf{r}_1}{g_{11}} + \epsilon_2 \frac{\mathbf{r}_2 \otimes \mathbf{r}_2}{g_{22}} \right). \tag{8.49}$$

The tensor $\mathbf{B}^C$ is a Cartesian tensor. Some steps are necessary to transform it to surface coordinates ξ^1, ξ^2. Substitution into the right hand side of the evolution equation (8.28) leads to:

$$\mathbf{B}^{\mathrm{C}}(\mathbf{v}_{\mathrm{s}} - \mathbf{v}) := -\epsilon_i \frac{\mathbf{r}_i \otimes \mathbf{r}_i}{g_{ii}} \cdot \boldsymbol{\rho}_j \dot{\xi}^j = -\frac{\epsilon_i(\mathbf{r}_i \cdot \boldsymbol{\rho}_j)\dot{\xi}^j}{g_{ii}}\mathbf{r}_i. \tag{8.50}$$

The dot product of the evolution equation (8.28) with $\boldsymbol{\rho}_k$, taking into account eqn. (8.50), leads to the equation in components according to the surface metrics:

$$\frac{dT_k}{dt} = -\frac{\epsilon_i(\mathbf{r}_i \cdot \boldsymbol{\rho}_k)(\mathbf{r}_i \cdot \boldsymbol{\rho}_j)}{g_{ii}}\dot{\xi}^j. \tag{8.51}$$

For the further implementation into a finite element code, it is more appropriate to introduce a tensor decomposition:

$$\frac{dT_k}{dt} = -\frac{\epsilon_i \mathbf{r}_i \otimes \mathbf{r}_i}{g_{ii}} : (\boldsymbol{\rho}_k \otimes \boldsymbol{\rho}_j)\dot{\xi}^j, \tag{8.52}$$

where the components of the first tensor $\mathbf{B}^{\mathrm{C}} = -\dfrac{\epsilon_i \mathbf{r}_i \otimes \mathbf{r}_i}{g_{ii}}$ can be computed in a Cartesian coordinate system separately for the surface as:

$$b_{ln}^{\mathrm{C}} = -\frac{\epsilon_i}{g_{ii}}\frac{\partial \phi_l}{\partial \alpha_i}\frac{\partial \phi_n}{\partial \alpha_i}. \tag{8.53}$$

After this the evolution eqn. (8.52) is defined as:

$$\frac{dT_k}{dt} = b_{ln}^{\mathrm{C}}\frac{\partial x_l}{\partial \xi^k}\frac{\partial x_n}{\partial \xi^j}\dot{\xi}^j. \tag{8.54}$$

From the last equation the covariant components of the tensor $\mathbf{B}$ in the global surface basis $\boldsymbol{\rho}_1, \boldsymbol{\rho}_2$ are obtained as tensor transformations, i.e.

$$b_{ij} = b_{ln}^{\mathrm{C}}\frac{\partial x_l}{\partial \xi^i}\frac{\partial x_n}{\partial \xi^j}. \tag{8.55}$$

Introducing the friction coefficients $\lambda_i = 1/\mu_i^2$ *into the yield function* and then applying analogous tensor operations as for the evolution equation lead to the following form:

$$\Phi = \sqrt{\frac{\mathbf{r}_i \otimes \mathbf{r}_i}{\mu_i^2 g_{ii}} : (T_k T_l \boldsymbol{\rho}^k \otimes \boldsymbol{\rho}^l)} - |N|, \tag{8.56}$$

from which a similar definition of covariant components for the friction tensor are obtained:

$$f_{kl} = \frac{\mathbf{r}_i \otimes \mathbf{r}_i}{\mu_i^2 g_{ii}} : (\boldsymbol{\rho}^k \otimes \boldsymbol{\rho}^l).$$ (8.57)

In the following we consider particular structures for covariant components in the Cartesian coordinate system b_{ij}^C in eqn. (8.53) as well as for b_{ij} in eqn. (8.55) in the local surface coordinate system for the anisotropic plane, for polar orthotropy on a plane and for spiral orthotropy on a cylinder.

8.3.4.3 Anisotropic plane. Structure of the $\mathbf{B}^C$ and $\mathbf{B}$ tensors in Cartesian coordinates

On the plane $x_3 = 0$, we consider anisotropic properties defined by two unit vectors $\mathbf{r}_1$ and $\mathbf{r}_2$:

$$\mathbf{r}_1 = \left\{ \begin{array}{c} \cos\alpha \\ \sin\alpha \\ 0 \end{array} \right\}, \quad \mathbf{r}_2 = \left\{ \begin{array}{c} \cos\beta \\ \sin\beta \\ 0 \end{array} \right\}.$$ (8.58)

The Cartesian components of the adhesion tensor b_{ln}^C are obtained as:

$$\mathbf{B}^C = - \begin{bmatrix} \varepsilon_1 \cos^2\alpha + \varepsilon_2 \cos^2\beta & \varepsilon_1 \sin\alpha\cos\alpha + \varepsilon_2 \sin\beta\cos\beta & 0 \\ \varepsilon_1 \sin\alpha\cos\alpha + \varepsilon_2 \sin\beta\cos\beta & \varepsilon_1 \sin^2\alpha + \varepsilon_2 \sin^2\beta & 0 \\ 0 & 0 & 0 \end{bmatrix}.$$ (8.59)

For an analysis, the approximation of the surface is necessary in order to obtain the structure of the tensor in surface coordinates ξ^1, ξ^2 in eqn. (8.55). If the latter coincide with the global Cartesian coordinates $\xi^i = x_i$, we obtain

$$\mathbf{B} = - \begin{bmatrix} \varepsilon_1 \cos^2\alpha + \varepsilon_2 \cos^2\beta & \varepsilon_1 \sin\alpha\cos\alpha + \varepsilon_2 \sin\beta\cos\beta \\ \varepsilon_1 \sin\alpha\cos\alpha + \varepsilon_2 \sin\beta\cos\beta & \varepsilon_1 \sin^2\alpha + \varepsilon_2 \sin^2\beta \end{bmatrix}.$$ (8.60)

It is obvious, that isotropy is no longer recovered simply by taking $\varepsilon_1 = \varepsilon_2$. However, if we take the unit vectors to be orthogonal, i.e. $\beta = \pi/2 + \alpha$, the orthotropic matrix obtained previously via the spectral representation in eqn. (8.42) is recovered.

8.3.4.4 Orthotropic surface in polar coordinates. Structure of the B^C and B tensors in Cartesian coordinates

As an example with curvilinear orthotropy, a plane with orthotropic properties in polar coordinates, see Fig. 8.3, is considered. These properties are defined by the elastic constants $\varepsilon_r, \varepsilon_\phi$ acting along the coordinate lines. The definition in eqn. (8.45) is then a transformation to polar coordinates:

$$x = r\cos\phi, \quad y = r\sin\phi, \quad z = 0. \tag{8.61}$$

The Cartesian components of the orthotropic tensor b_{kn}^C are then defined by the following matrix:

$$\mathbf{B}^C = -\begin{bmatrix} \varepsilon_r \cos^2\phi + \varepsilon_\phi \sin^2\phi & (\varepsilon_r - \varepsilon_\phi)\sin\phi\cos\phi & 0 \\ (\varepsilon_r - \varepsilon_\phi)\sin\phi\cos\phi & \varepsilon_r \sin^2\phi + \varepsilon_\phi \cos^2\phi & 0 \\ 0 & 0 & 0 \end{bmatrix}. \tag{8.62}$$

Applying the inverse mapping from polar coordinates to Cartesian coordinates in eqn. (8.61) (see also eqn. (8.55), we obtain:

$$\mathbf{B} = -\frac{1}{x^2 + y^2}\begin{bmatrix} \varepsilon_r x^2 + \varepsilon_\phi y^2 & (\varepsilon_r - \varepsilon_\phi)xy \\ (\varepsilon_r - \varepsilon_\phi)xy & \varepsilon_r y^2 + \varepsilon_\phi x^2 \end{bmatrix}. \tag{8.63}$$

The adhesion tensor for polar orthotropy in eqn. (8.63) contains a typical example of nonlinear surface properties in the reference Cartesian coordinate system.

8.3.4.5 Spiral orthotropy on a cylindrical surface. Structure of the B^C and B tensors in cylindrical coordinates

As a more complex case, we consider a circular cylinder and the surface orthotropy resulting from spiral coordinate lines on the cylinder. This example can be practically interesting to model e.g. screw connections. First, we define a rigid cylinder with a surface described by the following

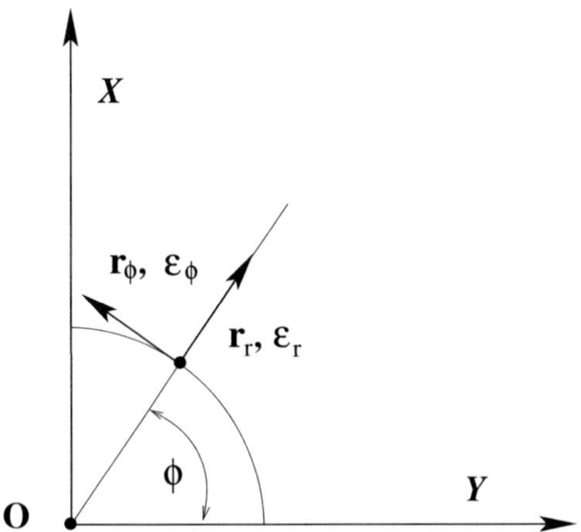

Figure 8.3: Polar orthotropic surface.

parameterization $\boldsymbol{\rho}(\alpha, z)$:

$$\boldsymbol{\rho} = \left\{ \begin{array}{c} R\cos\alpha \\ R\sin\alpha \\ z \end{array} \right\}. \tag{8.64}$$

The necessary surface characteristics are the tangent vectors and the normal vector

$$\boldsymbol{\rho}_1 = \left\{ \begin{array}{c} -R\sin\alpha \\ R\cos\alpha \\ 0 \end{array} \right\}, \quad \boldsymbol{\rho}_2 = \left\{ \begin{array}{c} 0 \\ 0 \\ 1 \end{array} \right\}, \quad \mathbf{n} = \left\{ \begin{array}{c} \cos\alpha \\ \sin\alpha \\ 0 \end{array} \right\}, \tag{8.65}$$

resulting in a covariant surface metrics tensor

$$[a_{ij}] = \left[\begin{array}{cc} (\boldsymbol{\rho}_1 \cdot \boldsymbol{\rho}_1) & (\boldsymbol{\rho}_1 \cdot \boldsymbol{\rho}_2) \\ (\boldsymbol{\rho}_2 \cdot \boldsymbol{\rho}_1) & (\boldsymbol{\rho}_2 \cdot \boldsymbol{\rho}_2) \end{array} \right] = \left[\begin{array}{cc} R^2 & 0 \\ 0 & 1 \end{array} \right]. \tag{8.66}$$

Orthotropic properties of the surface are obtained as follows. The equation for a family of cylindrical spiral lines on the cylinder (8.64) is, see also the geometry given in Fig. 8.4:

$$\mathbf{r} = \left\{ \begin{array}{c} R\cos\alpha \\ R\sin\alpha \\ \dfrac{H}{2\pi}\alpha + const \end{array} \right\}. \tag{8.67}$$

The first tangent vector $\mathbf{r}_1$ along the spiral line necessary for the tensor representation (8.49) becomes:

$$\mathbf{r}_1 = \frac{\partial \mathbf{r}}{\partial \alpha} = \left\{ \begin{array}{c} -R\sin\alpha \\ R\cos\alpha \\ \dfrac{H}{2\pi} \end{array} \right\}. \tag{8.68}$$

The second tangent vector $\mathbf{r}_2$ is defined to be orthogonal to the first vector $\mathbf{r}_1$ and to the normal on the cylinder surface, see eqn. (8.65.3) and Fig. 8.4:

$$\mathbf{r}_2 = [\mathbf{n} \times \mathbf{r}_1] = \left\{ \begin{array}{c} \dfrac{H}{2\pi}\sin\alpha \\ -\dfrac{H}{2\pi}\cos\alpha \\ R \end{array} \right\}. \tag{8.69}$$

With this expression an equation for the line orthogonal to the main spiral line (8.67), see line **AC** in Fig. 8.4, can be found from the condition that the integrated and scaled tangent vector $\mathbf{r}_2$ must belong to the cylinder surface

$$A\int \mathbf{r}_2 d\alpha \in cylinder \implies A = \frac{2\pi R}{H}, \tag{8.70}$$

which leads to the following definition of a vector $\hat{\mathbf{r}}$, orthogonal to $\mathbf{r}$:

$$\hat{\mathbf{r}} = \left\{ \begin{array}{c} R\cos\alpha \\ R\sin\alpha \\ \dfrac{2\pi R^2}{H}\alpha + const \end{array} \right\}. \tag{8.71}$$

Thus, orthotropic properties are inherited from the orthogonal spiral net on the cylinder via eqns. (8.67) and (8.71).

Remark.

For further analyses, one can define from eqns. (8.67) and (8.71) the distances between two adjacent threads as H for the main spiral and $\hat{H} = \dfrac{(2\pi R)^2}{H}$ for the orthogonal spiral **AC**, see Fig. 8.4.

The covariant components of the metric tensor g_{ij} are defined as

$$[g_{ij}] = \begin{bmatrix} R^2 + \left(\dfrac{H}{2\pi}\right)^2 & 0 \\ 0 & R^2 + \left(\dfrac{H}{2\pi}\right)^2 \end{bmatrix}. \tag{8.72}$$

The tensor of orthotropy $\mathbf{B}^\alpha$ in the Cartesian basis, see eqn. (8.49), becomes then:

$$\mathbf{B}^C = -\epsilon_1 \frac{\mathbf{r}_1 \otimes \mathbf{r}_1}{g_{11}} - \epsilon_2 \frac{\mathbf{r}_2 \otimes \mathbf{r}_2}{g_{22}} = \tag{8.73}$$

$$= -\frac{1}{R^2 + \left(\dfrac{H}{2\pi}\right)^2} \begin{bmatrix} g_\varepsilon \sin^2\alpha & -g_\varepsilon \sin\alpha\cos\alpha & -(\varepsilon_1 - \varepsilon_2)\dfrac{RH}{2\pi}\sin\alpha \\ -g_\varepsilon \sin\alpha\cos\alpha & g_\varepsilon \cos^2\alpha & (\varepsilon_1 - \varepsilon_2)\dfrac{RH}{2\pi}\cos\alpha \\ -(\varepsilon_1 - \varepsilon_2)\dfrac{RH}{2\pi}\sin\alpha & (\varepsilon_1 - \varepsilon_2)\dfrac{RH}{2\pi}\cos\alpha & \varepsilon_1\left(\dfrac{H}{2\pi}\right)^2 + \varepsilon_2 R^2 \end{bmatrix},$$

with

$$g_\varepsilon = \varepsilon_1 R^2 + \varepsilon_2 \left(\frac{H}{2\pi}\right)^2. \tag{8.74}$$

The backward tensor transformation (8.55) with the matrix $\mathbf{H}$ defined via the cylindrical coordinates

$$\mathbf{H} = \left[\frac{\partial x_i}{\partial \xi^j}\right] = \left[\frac{\partial \boldsymbol{\rho}}{\partial \boldsymbol{\xi}}\right] = \begin{bmatrix} -R\sin\alpha & 0 \\ R\cos\alpha & 0 \\ 0 & 1 \end{bmatrix} \tag{8.75}$$

gives us the covariant components b_{ij} of the tensor $\mathbf{B}$ in the contravariant surface basis ρ^1, ρ^2:

$$\mathbf{B} = \mathbf{H}^T \mathbf{B}^C \mathbf{H} = -\cfrac{1}{R^2 + \left(\cfrac{H}{2\pi}\right)^2} \begin{bmatrix} g_\varepsilon R^2 & (\varepsilon_1 - \varepsilon_2)\dfrac{R^2 H}{2\pi} \\[2ex] (\varepsilon_1 - \varepsilon_2)\dfrac{R^2 H}{2\pi} & \varepsilon_1 \left(\dfrac{H}{2\pi}\right)^2 + \varepsilon_2 R^2 \end{bmatrix}.$$

(8.76)

Matrix $\mathbf{B}$ in eqn. (8.76) represents the constant spiral orthotropy for the cylindrical surface and, therefore, is a generalization of the constant plane orthotropy in eqn. (8.42) for the case of a cylindrical geometry.

Remark.

With the assumption of isotropy $\varepsilon_1 = \varepsilon_2 = \varepsilon$ the unit matrix is recovered only in mixed components. In covariant components we obtain

$$\mathbf{B}_{\varepsilon_1=\varepsilon_2=\varepsilon} = [b_{ij}]_{\varepsilon_1=\varepsilon_2=\varepsilon} = -\varepsilon \begin{bmatrix} R^2 & 0 \\ 0 & 1 \end{bmatrix}.$$

(8.77)

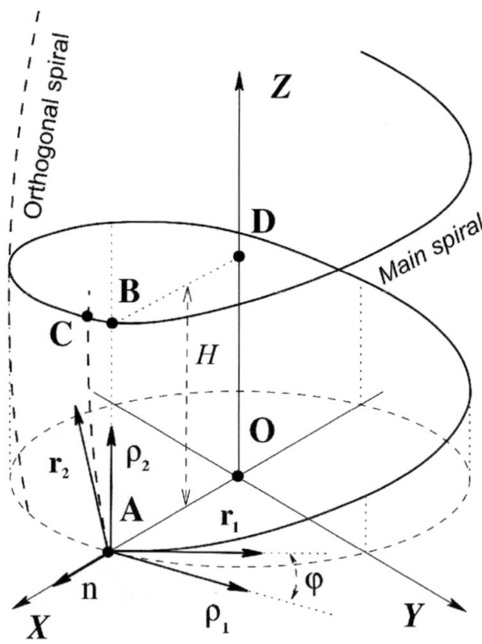

Figure 8.4: Spiral orthotropy on the cylindrical surface.

8.4 Derivation of the frictional contact problem via the principle of maximum dissipation

The principle of maximum dissipation is known in plasticity for the formulation of the necessary characteristics, such as plastic strains etc.. The application of this principle for the construction of computational algorithms in linear and nonlinear isotropic plasticity was developed in Simo and Hughes [160]. He and Curnier [62] formulated the dissipation function for the anisotropic function and investigated its extremal properties applying convex analysis. The correspondence between the dissipation function and the sliding rule was shown. Here we will also formulate the frictional problem as an extremal problem in a continuous form, and then applying the finite difference scheme in an incremental form. Afterwards, the return-mapping algorithm is applied to obtain all characteristics for sliding, such as a sliding force $\mathbf{T}^{sl}$ and a sliding displacement vector $\Delta\boldsymbol{\xi}^{sl}$. These variables can be viewed as a pair of conjugate variables for the energy dissipation function, which allows to define them separately.

8.4.1 Continuous formulation

According to the elastic-plastic analogy, a frictional contact problem via the energy dissipation function can be formulated as follows:

a) The relative velocity vector of the contact point is decomposed additively into an elastic part $\mathbf{v}^{el}$ and a sliding part $\mathbf{v}^{sl}$

$$\mathbf{v}^{r} = \mathbf{v}^{el} + \mathbf{v}^{sl}. \tag{8.78}$$

b) The elastic part $\mathbf{v}^{el}$ is responsible for reversible deformations (adhesion) and satisfies the evolution equations (8.28):

$$\frac{d\mathbf{T}}{dt} = \mathbf{B}\mathbf{v}^{el}. \tag{8.79}$$

c) The tangential force $\mathbf{T}$ must satisfy the following inequalities defined via the yield function eqn. (8.33), which in tensor form can be written as:

$$\Phi := \sqrt{f^{ij}T_iT_j} - |N| = \sqrt{\mathbf{T} \cdot \mathbf{FT}} - |N| : \qquad (8.80)$$

- if $\Phi < 0$ then the contact point is inside the elastic domain and $\mathbf{T} = \mathbf{T}^{el}$ is an elastic force,
- if $\Phi = 0$ then the contact point is sliding and $\mathbf{T} = \mathbf{T}^{sl}$ is a sliding force.

d) The power of the sliding forces, described by the energy dissipation function D achieves its maximum:

$$D := \dot{\xi}^i_{sl}T^{sl}_i = \mathbf{v}^{sl} \cdot \mathbf{T}^{sl}, \quad D \longrightarrow \max. \qquad (8.81)$$

For the convenient application of standard methods of convex analysis [119], [19], [46] we transform the problem (8.81) into a minimization problem

$$D_{min} := -\dot{\xi}^i_{sl}T^{sl}_i = -\mathbf{v}^{sl} \cdot \mathbf{T}^{sl}, \quad D_{min} \longrightarrow \min. \qquad (8.82)$$

The principle of maximum dissipation requires that the plastic dissipation function D subjected to the inequality conditions (8.80) achieves a maximum. A system of ordinary equations (8.78) - (8.79) is defined in convective surface coordinates ξ^1, ξ^2, identifying a contact point on the surface.

8.4.2 Incremental formulation

The application of the backward Euler scheme to the continuous problem a)-d) described above, namely to a system of ordinary differential equations (8.78)-(8.79) with an additional extremal condition (8.82) leads to an incremental formulation. Here we investigate only quasi-static contact problems, therefore we can take $\Delta t = 1$. The return-mapping scheme – for plasticity see [160] and among the first applications for contact problems see [45] – is applied to obtain the real sliding force and sliding displacements: the trial tangential force $\mathbf{T}^{tr}$ is assumed to be elastic and can be computed from the incremental solution. Thus, the following incremental formulation for the trial tangential force $\mathbf{T}^{tr}$ is found:

i) The full displacement vector $\Delta\boldsymbol{\xi} = \boldsymbol{\xi}^{(n+1)} - \boldsymbol{\xi}^{(n)}$ is decomposed additively into an elastic increment $\Delta\boldsymbol{\xi}^{el}$ and into a sliding increment $\Delta\boldsymbol{\xi}^{sl}$:

$$\Delta\boldsymbol{\xi} = \Delta\boldsymbol{\xi}^{el} + \Delta\boldsymbol{\xi}^{sl}, \tag{8.83}$$

where both vectors are defined in the surface metrics, namely,

$$\Delta\boldsymbol{\xi} := \Delta\xi^i \boldsymbol{\rho}_i = (\xi^i_{(n+1)} - \xi^i_{(n)})\boldsymbol{\rho}_i. \tag{8.84}$$

ii) The trial force $\mathbf{T}^{tr}_{(n+1)}$ is computed via the incremental evolution equations:

$$\mathbf{T}^{tr}_{(n+1)} - \mathbf{T}_{(n)} = \mathbf{B}^{(n+1)}(\boldsymbol{\xi}^{el}_{(n+1)} - \boldsymbol{\xi}^{el}_{(n)}). \tag{8.85}$$

iii) In order to decide whether the trial force $\mathbf{T}^{tr}$ is a sliding force $\mathbf{T}^{sl}$ or a sticking force $\mathbf{T}^{st}$ the yield condition is checked in each load step:

$$\begin{aligned}
\Phi^{tr} : &= \sqrt{\mathbf{T}^{tr}_{(n+1)} \cdot \mathbf{F}^{(n+1)}\mathbf{T}^{tr}_{(n+1)}} - |N_{(n+1)}| \\
&= \sqrt{f^{ij}T^{tr}_{i\,(n+1)}T^{tr}_{j\,(n+1)}} - |N_{(n+1)}|,
\end{aligned} \tag{8.86}$$

- If $\Phi^{tr} < 0$ then the trial force is a real sticking force $\mathbf{T} = \mathbf{T}^{tr}$.
- If $\Phi^{tr} \geq 0$ then the sliding force must be obtained via the maximum of the energy dissipation function given in the incremental form.

iv) The incremental analog of the continuous formulation eqn. (8.82) is then:

$$D^{(n+1)}_{min} := -\Delta\boldsymbol{\xi}^{sl} \cdot \mathbf{T}^{sl}_{(n+1)} = -\Delta\xi^i_{sl}T^{sl}_{i\,(n+1)}, \quad D^{(n+1)}_{min} \longrightarrow \min. \tag{8.87}$$

For large sliding problems especially with reversible loading, it is necessary to define both, the sliding force and the sliding displacements.

Taking the sliding distance $\Delta\xi_{sl}$ as an independent variable we can determine the sliding force $\mathbf{T}^{sl}$ via the minimum of the function $D_{min}^{(n+1)}$ in eqn. (8.87). The expression for the sliding force will be used for numerical computation within each load step during an iterative Newton solution scheme. The sliding distance, in due course, is defined after convergence is achieved in the load step and is computed via the consistency condition.

A Lagrange function for the constraint minimization problem is given as:

$$
\mathcal{L}_{(n+1)} \quad := \quad -\Delta\boldsymbol{\xi}^{sl} \cdot \mathbf{T}^{sl}_{(n+1)} + \lambda \left(\sqrt{\mathbf{T}^{tr}_{(n+1)} \cdot \mathbf{F}_{(n+1)} \mathbf{T}^{tr}_{(n+1)}} - |N_{(n+1)}| \right),
$$
$$
\mathcal{L}_{(n+1)} \quad \longrightarrow \quad \min, \tag{8.88}
$$

where the complementary Kuhn-Tucker conditions (see [119], [19]) are given as:

$$
\Phi := \sqrt{\mathbf{T}^{tr}_{(n+1)} \cdot \mathbf{F}_{(n+1)} \mathbf{T}^{tr}_{(n+1)}} - |N_{(n+1)}| \leq 0, \quad \lambda \geq 0, \quad \lambda\Phi_{(n+1)} = 0. \tag{8.89}
$$

For the next transformations a gradient of the yield function is necessary:

$$
\frac{\partial\Phi_{(n+1)}}{\partial\mathbf{T}^{tr}_{(n+1)}} = \frac{\mathbf{F}_{(n+1)}\mathbf{T}^{tr}_{(n+1)}}{\sqrt{\mathbf{T}^{tr}_{(n+1)} \cdot \mathbf{F}_{(n+1)}\mathbf{T}^{tr}_{(n+1)}}}, \tag{8.90}
$$

as well as a derivative of the trial force at load step $(n+1)$, computed via the chain rule, see eqns. (8.85) and (8.83):

$$
\frac{\partial\mathbf{T}^{tr}_{(n+1)}}{\partial\Delta\boldsymbol{\xi}^{sl}} = -\mathbf{B}_{(n+1)}. \tag{8.91}
$$

In the following sections the subscript $(n+1)$ will be omitted everywhere for simplicity reasons.

8.4.2.1 Derivation of the sliding force $\mathbf{T}^{sl}$

In order to obtain the sliding force $\mathbf{T}^{sl}$, the sliding incremental displacement $\Delta\xi^{sl}$ is taken as an independent variable. A formal application of convex analysis, see [119], [19], to the Lagrange function (8.88) gives us the necessary condition of the minimum:

$$\frac{\partial \mathcal{L}}{\partial \Delta \xi^{sl}} = 0,$$

$$(8.92)$$

leading to the definition of the sliding force as:

$$\mathbf{T}^{sl} = \lambda \frac{\partial \Phi}{\partial \Delta \boldsymbol{\xi}^{sl}}.$$

$$(8.93)$$

Exploiting the chain rule and using eqns. (8.90), (8.91), we obtain:

$$\mathbf{T}^{sl} = \lambda \frac{\partial \Phi}{\partial \mathbf{T}^{tr}} \frac{\partial \mathbf{T}^{tr}}{\partial \Delta \boldsymbol{\xi}^{sl}} = -\lambda \mathbf{B} \frac{\mathbf{FT}^{tr}}{\sqrt{\mathbf{T}^{tr} \cdot \mathbf{FT}^{tr}}}.$$

$$(8.94)$$

For the current sliding case, i.e. when $\lambda > 0$, we have to satisfy the Kuhn-Tucker condition $\Phi = 0$, in order to define λ, i.e. substitute $\mathbf{T}^{sl}$ in the yield function $\Phi = \sqrt{\mathbf{T}^{sl} \cdot \mathbf{FT}^{sl}} - |N|$. This leads to the following equation for λ:

$$\lambda = |N| \sqrt{\frac{\mathbf{T}^{tr} \cdot \mathbf{FT}^{tr}}{\mathbf{BFT}^{tr} \cdot \mathbf{FBFT}^{tr}}},$$

$$(8.95)$$

where the positive λ is taken due to the second Kuhn-Tacker condition in eqn. (8.89). Thus, the sliding force $\mathbf{T}^{sl}$ in eqn. (8.94) is defined as

$$\mathbf{T}^{sl} = -\frac{\mathbf{BFT}^{tr}}{\sqrt{\mathbf{BFT}^{tr} \cdot \mathbf{FBFT}^{tr}}} |N|.$$

$$(8.96)$$

We now introduce an auxiliary vector

$$\hat{\mathbf{T}} = \mathbf{BFT}^{tr}$$

$$(8.97)$$

to compute the sliding force $\mathbf{T}^{sl}$. The covariant components of the auxiliary vector are defined via various components of the tensor $\mathbf{B}$ and $\mathbf{F}$ as:

$$\hat{T}_i = b_{ij} f^{jk} T_k^{tr} = b_{ij} f_{ln} a^{jl} a^{kn} T_k^{tr} = b_i^j f_j^k T_k^{tr}.$$

$$(8.98)$$

Then the covariant components of the sliding vector (8.96) can be computed as

$$\mathbf{T}^{sl} = -\frac{\hat{\mathbf{T}}}{\sqrt{\hat{\mathbf{T}} \cdot \mathbf{F}\hat{\mathbf{T}}}} \implies T_i^{sl} = -\frac{\hat{T}_i}{\sqrt{\hat{T}_k \hat{T}_l f^{kl}}}$$

$$(8.99)$$

The isotropic case is recovered from eqn. (8.96) by taking $\mathbf{B} = -\epsilon_T \mathbf{E}$ and $\mathbf{F} = \mathbf{E}/\mu^2$ to

$$\mathbf{T}^{sl} = \mu|N|\frac{\mathbf{T}^{tr}}{\sqrt{\mathbf{T}^{tr} \cdot \mathbf{T}^{tr}}}. \tag{8.100}$$

8.4.3 Specification of initial conditions for the return-mapping scheme

Initial conditions are necessary for the incremental solution of eqns.(8.83) and (8.85) formulated as Cauchy problem for a system of ordinary differential equations. These initial conditions can be defined assuming that the initial configuration is unstressed, thus with zero external loading:

$$\mathbf{T} = 0, \ N = 0 \ \implies \ \boldsymbol{\xi} = \boldsymbol{\xi}_{(0)}, \ \boldsymbol{\xi}^{sl}_{(0)} = 0 \ \text{ for } \ \xi^3 \leq 0. \tag{8.101}$$

The conditions are formulated for all points which are in contact at the initial configuration, i.e. for all points satisfying $\xi^3 \leq 0$. The vector $\boldsymbol{\xi}_{(0)}$ is obtained via a projection procedure, and defines the center of the ellipse for the adhesion domain, see Fig. 8.5. The additional initialization of the sliding displacement and the update procedure will be discussed after the geometrical interpretation of the solution process.

8.4.4 Derivation of the sliding incremental displacement $\Delta \xi^{sl}$ and update scheme for the history variables

We consider here a "step-by-step" scheme for the case with nonlinear tensors $\mathbf{B}$ and $\mathbf{F}$ concentrating on computational aspects for the numerical implementation. For the nonlinear case let us assume that the first converged load step is elastic, i.e. $\Phi < 0$ in eqn. (8.89), while for the second load step the sliding condition is achieved, i.e. $\Phi = 0$ and the load step was computed with the sliding force $\mathbf{T}^{sl}$. Let $\boldsymbol{\xi}_{(1)}$ and $\boldsymbol{\xi}_{(2)}$ are convective coordinates of a contact point after the corresponding first and the second load steps. The trial force for the second load step $\mathbf{T}^{tr}_{(2)}$ is then computed as in an Euler backward scheme:

$$\mathbf{T}^{tr}_{(2)} = \mathbf{T}_{(1)} + \mathbf{B}^{(2)}(\boldsymbol{\xi}_{(2)} - \boldsymbol{\xi}_{(1)}), \tag{8.102}$$

where for the first load step a force $\mathbf{T}_{(1)}$ is computed taking into account the initial conditions in eqn. (8.101)

$$\mathbf{T}_{(1)} = \mathbf{B}^{(1)}(\boldsymbol{\xi}_{(1)} - \boldsymbol{\xi}_{(0)}). \tag{8.103}$$

The value of the sliding displacement $\boldsymbol{\xi}^{sl}_{(2)}$ and resp. the elastic part $\boldsymbol{\xi}^{el}_{(2)}$ are defined via the strict fulfillment of the Kuhn-Tucker condition $\Phi_{(2)} = 0$ for the elastic part $\boldsymbol{\xi}^{el}_{(2)}$:

$$\Phi_{(2)} = \Phi_{(2)}(\boldsymbol{\xi}, \boldsymbol{\xi}^{sl}) =$$

$$\sqrt{(\mathbf{T}_{(1)} + \mathbf{B}^{(2)}(\boldsymbol{\xi}^{el}_{(2)} - \boldsymbol{\xi}_{(1)})) \cdot \mathbf{F}^{(2)}(\mathbf{T}_{(1)} + \mathbf{B}^{(2)}(\boldsymbol{\xi}^{el}_{(2)} - \boldsymbol{\xi}_{(1)}))} - |N|, \tag{8.104}$$

where the full displacement vector $\boldsymbol{\xi}_{(2)}$ after the converged second load step can be decomposed as:

$$\boldsymbol{\xi}_{(2)} = \boldsymbol{\xi}^{el}_{(2)} + \boldsymbol{\xi}^{sl}_{(2)}. \tag{8.105}$$

The consistency condition (see [119], [19]) for the constraint function $\Phi_{(2)}$ leads to an additional equation allowing to determine the direction of the sliding displacement:

$$\dot{\Phi} = \frac{\partial \Phi}{\partial \boldsymbol{\xi}_{(2)}} \cdot \frac{d\boldsymbol{\xi}_{(2)}}{dt} + \frac{\partial \Phi}{\partial \boldsymbol{\xi}^{sl}_{(2)}} \cdot \frac{d\boldsymbol{\xi}^{sl}_{(2)}}{dt} = 0. \tag{8.106}$$

Continuing with the chain rule we obtain:

$$\frac{\partial \Phi}{\partial \boldsymbol{\xi}^{el}_{(2)}} \frac{\partial \boldsymbol{\xi}^{el}_{(2)}}{\partial \boldsymbol{\xi}_{(2)}} \cdot \frac{d\boldsymbol{\xi}_{(2)}}{dt} + \frac{\partial \Phi}{\partial \boldsymbol{\xi}^{el}_{(2)}} \frac{\partial \boldsymbol{\xi}^{el}_{(2)}}{\partial \boldsymbol{\xi}^{sl}_{(2)}} \cdot \frac{d\boldsymbol{\xi}^{sl}_{(2)}}{dt} = \frac{\partial \Phi}{\partial \boldsymbol{\xi}^{el}_{(2)}} \cdot \left(\frac{d\boldsymbol{\xi}_{(2)}}{dt} - \frac{d\boldsymbol{\xi}^{sl}_{(2)}}{dt} \right) = 0. \tag{8.107}$$

From the last equation, we can obtain the following condition

$$\frac{d\boldsymbol{\xi}^{sl}_{(2)}}{dt} = \frac{d\boldsymbol{\xi}_{(2)}}{dt}. \tag{8.108}$$

Then the sliding displacement update vector $\boldsymbol{\xi}^{sl}_{(2)}$ can be defined in the direction of the last converged full displacement vector $\boldsymbol{\xi}_{(2)}$. For the computation of the nonlinear case this direction can be approximately taken as

$$\mathbf{e} = \frac{\boldsymbol{\xi}_{(2)} - \boldsymbol{\xi}_{(1)}}{|\boldsymbol{\xi}_{(2)} - \boldsymbol{\xi}_{(1)}|} \qquad (8.109)$$

leading to the sliding displacement

$$\boldsymbol{\xi}_{(2)}^{sl} = \lambda \mathbf{e}. \qquad (8.110)$$

Parameter λ defining the length of the vector is obtained from the Kuhn-Tucker condition for the function $\Phi_{(2)}$ in eqn. (8.104). This leads to the following algebraic equation:

$$\lambda^2 \left(\mathbf{B}^{(2)}\mathbf{e} \cdot \mathbf{F}^{(2)}\mathbf{B}^{(2)}\mathbf{e} \right) - 2\lambda \left(\mathbf{B}^{(2)}\mathbf{e} \cdot \mathbf{F}^{(2)}\mathbf{T}_{(2)} \right) \quad +$$

$$+ \mathbf{T}_{(2)} \cdot \mathbf{F}^{(2)}\mathbf{T}_{(2)} - N^2 = 0, \qquad (8.111)$$

where the positive root $\lambda > 0$, minimizing globally the function in eqn. (8.87), should be taken.

8.4.5 Computational aspects for further implementation considering nonlinear and constant tensors

For the further implementations we consider the following cases: a) with nonlinear tensors for large displacement problems; b) with constant tensors, i.e. the case of constant orthotropy.

8.4.5.1 A case with nonlinear tensors for large displacement problems

Within the adjustment of the sliding force the strict execution of the backward scheme in eqn. (8.85) leads to the necessity to store as history variables in addition to $\boldsymbol{\xi}_{(n)}^{el}$ also updated sliding variables $\boldsymbol{\xi}_{(n)}^{sl}$ at load step (n), which is computationally rather expensive. However, the numerical experience from some cases even with non-constant $\mathbf{B}$, e.g. for polar orthotropy, shows that it may be mostly sufficient to use a simplified scheme, which is identical with the backward scheme computed with the updated matrix $\mathbf{B}^{(n+1)}$ at load step $(n + 1)$, namely, with the following finite difference scheme:

$$\mathbf{T}_{(n+1)}^{tr} = \mathbf{T}_{(n)} + \mathbf{B}^{(n+1)}(\boldsymbol{\xi}_{(n+1)} - \boldsymbol{\xi}_{(n)}), \qquad (8.112)$$

where $\boldsymbol{\xi}_{(n)}$ is a displacement vector from the converged load step (n). For large displacement problems the elastic part $\boldsymbol{\xi}^{el}$ can be neglected leading to the result that the update sliding vector $\boldsymbol{\xi}^{sl}_{(n)}$ is equal to the displacement vector $\boldsymbol{\xi}_{(n)}$ in the last converged load step. Computations show that the scheme in eqn. (8.112) is robust and requires only $\mathbf{T}_{(n)}$ and $\boldsymbol{\xi}_{(n)}$ as history variables.

8.4.5.2 A case with constant orthotropy

For the case with a constant tensor $\mathbf{B}$, we can proceed recursively transforming eqn. (8.85) similar to the isotropic case, see eqn. (8.13), and obtain

$$\mathbf{T}^{tr}_{(n+1)} = \mathbf{T}_{(n)} + \mathbf{B}\Delta\boldsymbol{\xi}^{el} = \ldots = \mathbf{B}(\Delta\boldsymbol{\xi}_{(n+1)} - \Delta\boldsymbol{\xi}^{sl}), \qquad (8.113)$$

$$\Delta\boldsymbol{\xi}_{(n+1)} = \boldsymbol{\xi}_{(n+1)} - \boldsymbol{\xi}_{(0)},$$

where $\boldsymbol{\xi}_{(0)}$ is a center of the elliptical adhesion domain, for which a geometrical interpretation is given later.

For the case with constant tensors the update algorithm in Sect. 8.4.4 leads to the exact definition of the sliding displacement as:

$$\Delta\boldsymbol{\xi}^{sl} = \boldsymbol{\xi} - \boldsymbol{\xi}^{(0)} - |N|\frac{\boldsymbol{\xi} - \boldsymbol{\xi}^{(0)}}{\sqrt{(\boldsymbol{\xi} - \boldsymbol{\xi}^{(0)}) \cdot \mathbf{BFB}(\boldsymbol{\xi} - \boldsymbol{\xi}^{(0)})}}, \qquad (8.114)$$

where $\boldsymbol{\xi}$ is the full displacement vector after the converged load step. A geometrical interpretation of this scheme is discussed in the next section.

8.4.6 Geometrical interpretation of the solution process

A simple geometrical interpretation for the solution can be given for a plane surface with constant orthotropy in the case of both, elastic sticking and sliding behavior. Interpretations can be formulated in the trial force space T^{tr}_1, T^{tr}_2 and on the tangent plane ξ^1, ξ^2 where we assume Cartesian metrics, i.e. $a_{ij} = \delta_{ij}$. In the trial force space the yield function in eqn. (8.86) represents an ellipse due to the positivity of the friction tensor $\mathbf{F}$:

$$\mathbf{T}^{tr} \cdot \mathbf{F}\mathbf{T}^{tr} = N^2. \tag{8.115}$$

In order to obtain the interpretation on the tangent plane, the incremental evolution equation (8.113) is used. Then, we also obtain an ellipse on the tangent plane

$$\mathbf{B}\Delta\boldsymbol{\xi} \cdot \mathbf{F}\mathbf{B}\Delta\boldsymbol{\xi} = N^2. \tag{8.116}$$

In order to derive further characteristics, the symmetrical tensors $\mathbf{F}$ and $\mathbf{B}$ are expressed via the spectral representation (8.41):

$$\mathbf{B} = \mathbf{Q}_\alpha \mathbf{D}_B \mathbf{Q}_\alpha^T, \quad \mathbf{F} = \mathbf{Q}_\beta \mathbf{D}_F \mathbf{Q}_\beta^T, \tag{8.117}$$

where $\mathbf{Q}_\alpha$ and $\mathbf{Q}_\beta$ are rotational matrices with corresponding angles α and β, see eqn. (8.40), and $\mathbf{D}_B$, $\mathbf{D}_F$ are diagonal matrices:

$$\mathbf{D}_B = -\begin{bmatrix} \varepsilon_1 & 0 \\ 0 & \varepsilon_2 \end{bmatrix}, \quad \mathbf{D}_F = \begin{bmatrix} \dfrac{1}{\mu_1^2} & 0 \\ 0 & \dfrac{1}{\mu_2^2} \end{bmatrix}. \tag{8.118}$$

Spectral representations leads to the canonical form of ellipses. Thus, eqn. (8.115) becomes then

$$\mathbf{Q}_\beta^T \mathbf{T}^{tr} \cdot \begin{bmatrix} \dfrac{1}{\mu_1^2 N^2} & 0 \\ 0 & \dfrac{1}{\mu_2^2 N^2} \end{bmatrix} \mathbf{Q}_\beta^T \mathbf{T}^{tr} = 1, \tag{8.119}$$

describing a central ellipse inclined with angle β in the force plane T_1^{tr}, T_2^{tr} with the main axes $\mu_i N$.

Eqn. (8.116) describes a domain with orthotropic elastic properties, where the contact point S is attracted by the center (adhesion domain). Its transformation according to the representation (8.117) and (8.118) gives the following canonical equation:

$$(\mathbf{Q}_\alpha \mathbf{Q}_\alpha \mathbf{Q}_\beta^T)^T \Delta\boldsymbol{\xi} \cdot \begin{bmatrix} \dfrac{1}{(\mu_1 N/\varepsilon_1)^2} & 0 \\ 0 & \dfrac{1}{(\mu_2 N/\varepsilon_2)^2} \end{bmatrix} (\mathbf{Q}_\alpha \mathbf{Q}_\alpha \mathbf{Q}_\beta^T)^T \Delta\boldsymbol{\xi} = 1, \tag{8.120}$$

$$\Delta\boldsymbol{\xi} = \boldsymbol{\xi}_{(n+1)} - \boldsymbol{\xi}_{(0)}$$

describing an ellipse inclined by the matrix $\mathbf{Q}_\alpha \mathbf{Q}_\alpha \mathbf{Q}_\beta^T$. The ellipse center is shifted by the distance $\boldsymbol{\xi}_{(0)}$ on the tangent plane ξ^1, ξ^2, see Fig. 8.5. The lengths of the main axes of the ellipse are a resp. $b = \mu_i N/\varepsilon_i$. The inclination angle becomes $\phi = 2\alpha - \beta$, which is verified from the matrix:

$$\mathbf{Q}_\alpha \mathbf{Q}_\alpha \mathbf{Q}_\beta^T = \begin{bmatrix} \cos(2\alpha - \beta) & -\sin(2\alpha - \beta) \\ \sin(2\alpha - \beta) & \cos(2\alpha - \beta) \end{bmatrix} \tag{8.121}$$

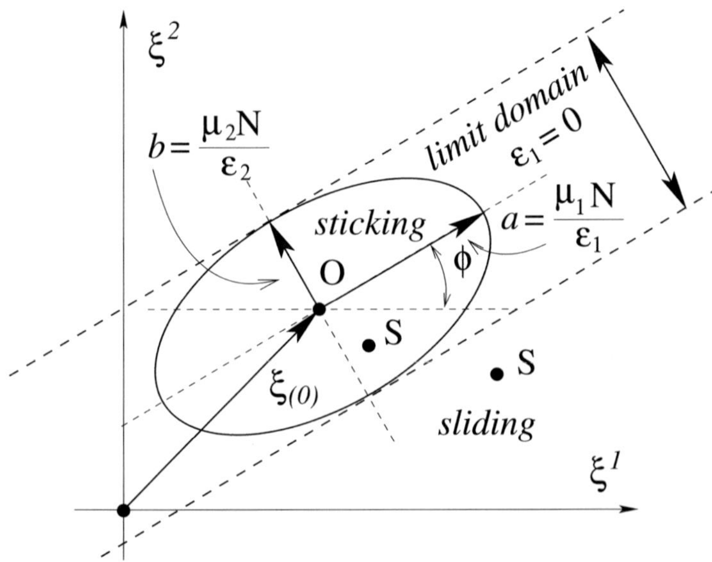

Figure 8.5: Allowable elastic region (adhesion domain).

In the numerical examples, we will also investigate a case with non-linear orthotropy with $\varepsilon_1 = 0$. In this limit case with $\lim a = \infty$ the ellipse degenerates into an infinite strip of width $2b = 2\dfrac{\mu_2 N}{\varepsilon_2}$, see Fig. 8.5. The properties inside the strip defined by eqn. (8.113) are elastic, but the motion along the strip causes the corresponding elastic force to be zero $T_a = 0$.

Thus, *a geometrical interpretation of the solution* is as follows. The ellipse describes an elastic domain with orthotropic properties obtained by the incremental evolution equation (8.113). The sticking condition is fulfilled when a contact point S remains inside the ellipse. If a contact

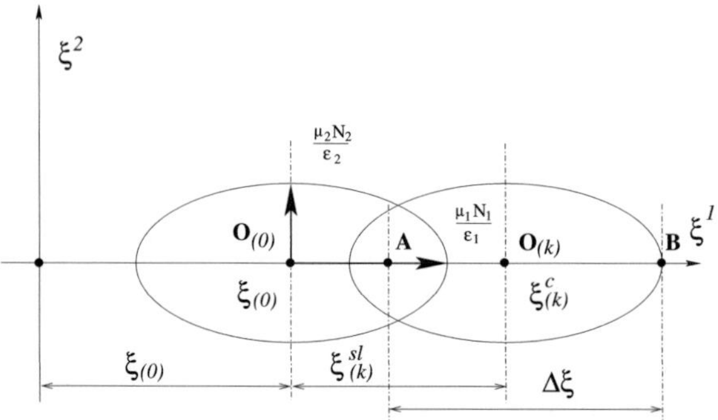

Figure 8.6: Update scheme. Particular case.

point S appears outside of the ellipse, then this point is sliding, i.e. the acting force is computed via eqn. (8.99), see Fig. 8.5. Due to the condition $\Phi = 0$ this point is on the boundary of the elliptic region, which leads to a shifting of the ellipse in order to define the forces in the next load step correctly. This shifting is originated by the sliding vector $\Delta\boldsymbol{\xi}^{sl}$ which leads to an update of the ellipse center and which is computed via the update scheme in eqn. (8.114). The ellipse center is an attraction point for the domain with anisotropic elastic central forces defined by the evolution equation (8.113). As long as the contact point S is inside of the adhesion domain, the sliding displacements as well as the sliding forces have not to be computed. This is, the so-called, "sticking zone". Let k be the number of the load step, when sliding is detected the first time, i.e. the contact point has been moved outside the adhesion domain. Then, the sliding displacement $\boldsymbol{\xi}^{sl}_{(k)}$ is computed via eqn. (8.114). Now, the trial procedure in eqn. (8.113) is considered in the next load step $(k+1)$. Since the trial procedure is a computation in the elastic region, the sliding displacement $\boldsymbol{\xi}^{sl}_{(k+1)}$ is assumed to be zero.

$$
\begin{aligned}
\mathbf{T}^{tr}_{(k+1)} &= \mathbf{B}(\Delta\boldsymbol{\xi}_{(k+1)} - \Delta\boldsymbol{\xi}^{sl}_{(k+1)}) = \\
&= \mathbf{B}(\boldsymbol{\xi}_{(k+1)} - \boldsymbol{\xi}_{(0)} - \boldsymbol{\xi}^{sl}_{(k)}) = \\
&= \mathbf{B}(\boldsymbol{\xi}_{(k+1)} - \underbrace{(\boldsymbol{\xi}_{(0)} + \boldsymbol{\xi}^{sl}_{(k)})}_{\boldsymbol{\xi}^{c}_{(k)}}).
\end{aligned}
\tag{8.122}
$$

Vector $\boldsymbol{\xi}^{c}_{(k)}$ defines the update scheme for the sliding displacements and allows to describe the shift of the ellipse center. The incremental evolution equation (8.113) is corrected then in accordance to this update scheme as

$$\mathbf{T}^{tr}_{(n+1)} = \mathbf{B}(\Delta\boldsymbol{\xi}_{(n+1)} - \boldsymbol{\xi}^c_n),$$
$$\text{with} \qquad \boldsymbol{\xi}^c_n = \boldsymbol{\xi}^c_{n-1} + \boldsymbol{\xi}^{sl}_n = ... = \boldsymbol{\xi}^0 + \boldsymbol{\xi}^c_k + ... + \boldsymbol{\xi}^{sl}_n. \qquad (8.123)$$

Now a particular case is considered, when the orthotropy axes coincide with the Cartesian coordinate axes (i.e. $\mathbf{Q}_\alpha = \mathbf{Q}_\beta = \mathbf{E}$) and a contact point being at the position **A** during the load step (k) moved to the position **B** during the load step $(k+1)$ along the ξ^1 axis with the distance $\Delta\xi$, see Fig. 8.6. In this case, the trial force is obtained as $T^{tr}_1 = -\varepsilon_1\Delta\xi$, and the following sliding displacement $\Delta\xi^{sl}$ in eqn. (8.114) becomes then

$$\Delta\xi^{sl} = \Delta\xi - |N|\frac{\Delta\xi\mu_1}{\sqrt{(\Delta\xi\varepsilon_1)^2}} = \frac{T^{tr}_1}{\varepsilon_1|T^{tr}_1|}(-|T^{tr}_1| + \mu_1|N|). \qquad (8.124)$$

From Fig. 8.6 it becomes obvious that the sliding displacement $\Delta\xi^{sl}$ is updating the position of the ellipse center O. The last result (8.124) can already be found for the isotropic case in Wriggers and Krstulovic-Opara [190] and [102]. In addition, the analogy between the geometrical interpretations of friction and plasticity with kinematical hardening, as described in Simo and Hughes [160] becomes obvious.

8.5 Conclusion

In this contribution a model for anisotropic surfaces including both anisotropy for adhesion and anisotropy for friction domains has been developed. The principle of maximum dissipation is applied to derive all necessary parameters. The problem is formulated in a covariant form in the surface coordinate system. Various types of anisotropy based either on the spectral decomposition, or inherited from the arbitrary curvilinear coordinate system are considered. As an example, the adhesion tensor and the friction tensor are derived for the polar orthotropy on a plane and the spiral orthotropy on a cylinder. A special attention is paid to the geometrical interpretation of the solution process. The current consideration is the necessary step for an iterative Newton type solution within the finite element method. The subsequent linearization procedure, details of the finite element implementation and numerical examples will be considered in the second part.

9

Covariant description of contact interfaces considering anisotropy for adhesion and friction: linearization, finite element implementation and numerical analysis of the model*

Abstract

A covariant description for contact problems including anisotropy for both adhesion and sliding domains is proposed. The principle of maximum dissipation is used to obtain a computational model in the case of quasi-static motions. This second part is first focusing on the linearization of the nonlinear equations necessary for the solution process. Then the finite element implementation for various contact elements is developed. In addition, a mechanical interpretation via a rheological model is discussed. Finally, different cases including curvilinear anisotropy on arbitrary surfaces are considered. The numerical examples are chosen to show the influence of the orthotropy type on the development of the sticking-sliding zone as well as on the kinematical behavior of the contact bodies.

Keywords

covariant description anisotropy contact adhesion Coulomb friction linearization FE discretization geometrical isotropy

---

*The chapter is published in [91]: A. Konyukhov, K. Schweizerhof *Covariant description of contact interfaces considering anisotropy for adhesion and friction. Part 2. Linearization, finite element implementation and numerical analysis of the model*, Computer Methods in Applied Mechanics and Engineering, **196**(3):289–303, 2006.

9.1 Introduction

In the first part of this contribution contact problems with surfaces possessing anisotropic structure have been formulated via the principle of maximum dissipation in a continuous as well as in an incremental form. The model includes both anisotropy for friction and anisotropy for adhesion. An iterative solution, e.g. of a Newton's type, is required for the solution of the nonlinear contact problem. Thus, we consider in this part the derivation of the necessary consistent tangent matrices for the return-mapping scheme.

The finite implementation of an anisotropic friction law is briefly discussed in Montmitonnet and Hasquin [128] with an application to hot rolling processes, and presented in details in Alart and Heege [4]. A symbolic computation software has been exploited to derive the corresponding tangent matrices in [4]. In the current publication, particular attention is paid to the derivation of tangent matrices in a covariant form allowing the straightforward implementation into a finite element code even for arbitrary curved contact surfaces possessing anisotropic properties for both friction and adhesion.

A rheological model is discussed as a simple mechanical interpretation of the continuous constitutive model. In addition details of finite element implementations for various types of finite element approximations are presented. The set of numerical examples is chosen to show the influence of the orthotropy type on the development of the sticking-sliding zone as well as on the kinematical behavior of the contact bodies. Thus, constant orthotropy is classified by the eigenvalue ratio of the corresponding tensor defining the adhesion region. These cases are thoroughly investigated by the development of the sticking-sliding zone. It will be shown that a specific combination of both, orthotropy for adhesion and orthotropy for friction, can lead to the so-called geometrical isotropy, when the contact bodies show kinematically isotropic behavior. Finally, it is demonstrated that various kinematical properties of arbitrarily curved contact surfaces can be modeled by means of the adhesion tensor.

9.2 Consistent linearization for a Newton type solution

The full contact integral can be split into parts for the normal and the tangential directions:

$$\delta W_c = \underbrace{\int_s N\delta\xi^3 ds}_{\delta W_c^N} + \underbrace{\int_s T_j \delta\xi^j ds}_{\delta W_c^T}, \tag{9.1}$$

therefore, the linearization procedure for a Newton type solution will lead to a normal part and to a tangential part of the tangent matrix. The algorithmic aspects of the linearization include the following operations:

a) linearization of the convective variations, $\delta\xi^i$, $i = 1, 2, 3$

b) linearization of contact traction N and tangential traction T_i taking the return-mapping scheme properly into account.

In order to keep the information as brief as possible, we focus on the specifications for the anisotropic part and refer to previous derivations wherever possible. For the nomenclature we urge the reader to check the first part of the contribution [90].

9.2.1 Linearization of the variations $\delta\xi^i$

Since the spatial coordinate system is chosen according to the surface geometry, the variational expressions are linearized separately for the tangential variations $\delta\xi^i$, $i = 1, 2$ and for the normal variation $\delta\xi^3$. For details we refer to the derivations already given in [86], [89].

9.2.1.1 Linearization of the normal variation $\delta\xi^3$

$$\frac{d}{dt}\delta\xi^3 = -\left(\delta\boldsymbol{\rho}_{,j} \cdot a^{ij}(\mathbf{n}\otimes\boldsymbol{\rho}_i)(\mathbf{v}_s - \mathbf{v}) + (\delta\mathbf{r}_s - \delta\boldsymbol{\rho})\cdot a^{ij}(\boldsymbol{\rho}_j\otimes\mathbf{n})\mathbf{v}_i\right)$$
$$-(\delta\mathbf{r}_s - \delta\boldsymbol{\rho})\cdot h^{ij}(\boldsymbol{\rho}_i\otimes\boldsymbol{\rho}_j)(\mathbf{v}_s - \mathbf{v}). \tag{9.2}$$

9.2.1.2 Linearization of the tangential variations $\delta\xi^i$, $i = 1,2$

$$
\begin{aligned}
\frac{d}{dt}\delta\xi^i =\;
&- \left((\delta\mathbf{r}_{\mathrm{s}} - \delta\boldsymbol{\rho}) \cdot a^{il} a^{jk}\, \boldsymbol{\rho}_k \otimes \boldsymbol{\rho}_l\, \mathbf{v}_j + \delta\boldsymbol{\rho}_j \cdot a^{ik} a^{jl}\, \boldsymbol{\rho}_k \otimes \boldsymbol{\rho}_l\, (\mathbf{v}_{\mathrm{s}} - \mathbf{v}) \right) \\
&+ (\delta\mathbf{r}_{\mathrm{s}} - \delta\boldsymbol{\rho}) \cdot h^{ij} \left(\boldsymbol{\rho}_j \otimes \mathbf{n} + \mathbf{n} \otimes \boldsymbol{\rho}_j \right) (\mathbf{v}_{\mathrm{s}} - \mathbf{v}) \\
&+ h^i_n \dot{\xi}^3 \delta\xi^n - \Gamma^i_{kj}\dot{\xi}^j \delta\xi^k.
\end{aligned}
\tag{9.3}
$$

9.2.2 Linearization of the contact tractions

The derivative of the normal traction N is written as

$$
\dot{N} = -\epsilon_N \dot{\xi}^3.
\tag{9.4}
$$

For the linearization of the tangent traction of the reversible part, we recall the evolution equations from Part 1. For the linearization the tangent traction $\mathbf{T}$ has to be considered in the covariant form

$$
\frac{d\mathbf{T}}{dt} = \mathbf{B}(\mathbf{v}_{\mathrm{s}} - \mathbf{v}),
\tag{9.5}
$$

leading to the component form in the surface metrics as

$$
\frac{\partial T_i}{\partial t} + \nabla_j T_i \dot{\xi}^j = b_{ij}\dot{\xi}^j, \quad j = 1,2,
\tag{9.6}
$$

where the adhesion tensor $\mathbf{B} = b_{ij}\boldsymbol{\rho}^i \otimes \boldsymbol{\rho}^j$ is defined in the surface metrics.

Remark.
For a consistent linearization we adopt the assumption that all terms describing the curvature properties of the master surface, i.e. including the second derivatives with respect to convective coordinates, can be neglected based on the numerical investigations in [86] and [89]. This allows to reduce the size of various expressions considerably.

9.2.3 Linearization of the normal part δW_c^N

According to **Remark 9.2.2** we write the result given in [86] without the curvature term:

$$D(\delta W_c^N) = \int_s \dot{N}\delta\xi^3 ds + \int_s N\frac{d}{dt}\delta\xi^3 ds$$

$$= -\int_s \epsilon_N (\delta\mathbf{r}_s - \delta\boldsymbol{\rho}) \cdot (\mathbf{n} \otimes \mathbf{n})(\mathbf{v}_s - \mathbf{v})ds \tag{9.7}$$

$$-\int_s \epsilon_N\xi^3 \left(\delta\boldsymbol{\rho}_{,j} \cdot a^{ij}(\mathbf{n} \otimes \boldsymbol{\rho}_i)(\mathbf{v}_s - \mathbf{v}) + (\delta\mathbf{r}_s - \delta\boldsymbol{\rho}) \cdot a^{ij}(\boldsymbol{\rho}_j \otimes \mathbf{n})\mathbf{v}_{,i} \right) ds.$$

Here, for the first term (the main part) the evolution equation (9.4) together with the representation of $\delta\xi^3$ by the geometry of the surface is used. For the second term (the rotational part) the regularization for the normal traction together with the linearization of the variation $\delta\xi^3$ in eqn. (9.2) is taken into account.

9.2.4 Linearization of the tangential part δW_c^T

The tangential part of the contact integral δW_c^T is considered taking into account the anisotropic evolution equations and the return mapping algorithm. The cases of sticking and sliding have to be treated separately.

9.2.4.1 The sticking case

Sticking is identified when the trial contact tractions T_i^{tr} computed at load step $(n + 1)$ satisfy the conditions imposed by Coulomb's law :

$$\Phi := \sqrt{f^{ij}T_j^{tr}T_i^{tr}} - N < 0. \tag{9.8}$$

In this case, the real tractions are identical to the trial ones $T_i = T_i^{tr}$, therefore, the linearized traction terms are obtained from the evolution equation in (9.6) directly. For the convective velocities $\delta\xi^i$, the linearized equations (9.3) can be used.

$$D_v(\delta W_c^T) = \int_s \left(\frac{dT_i}{dt}\delta\xi^i + T_i\frac{d\delta\xi^i}{dt} \right) ds = \tag{9.9}$$

$$= \int_s (\delta \mathbf{r}_s - \delta \boldsymbol{\rho}) \cdot \mathbf{B}(\mathbf{v}_s - \mathbf{v})ds - \tag{9.10a}$$

$$-\int_s T_i \left((\delta \mathbf{r}_s - \delta \boldsymbol{\rho}) \cdot a^{il} a^{jk} \, \boldsymbol{\rho}_k \otimes \boldsymbol{\rho}_l \, \mathbf{v}_j + \delta \boldsymbol{\rho}_{,j} \cdot a^{ik} a^{jl} \, \boldsymbol{\rho}_k \otimes \boldsymbol{\rho}_l \, (\mathbf{v}_s - \mathbf{v}) \right) ds, \tag{9.10b}$$

or component-wise

$$= \int_s (\delta \mathbf{r}_s - \delta \boldsymbol{\rho}) \cdot b^{ij} \boldsymbol{\rho}_i \otimes \boldsymbol{\rho}_j (\mathbf{v}_s - \mathbf{v})ds - \tag{9.11a}$$

$$-\int_s T_i \left((\delta \mathbf{r}_s - \delta \boldsymbol{\rho}) \cdot a^{il} a^{jk} \, \boldsymbol{\rho}_k \otimes \boldsymbol{\rho}_l \, \mathbf{v}_j + \delta \boldsymbol{\rho}_{,j} \cdot a^{ik} a^{jl} \, \boldsymbol{\rho}_k \otimes \boldsymbol{\rho}_l \, (\mathbf{v}_s - \mathbf{v}) \right) ds. \tag{9.11b}$$

The matrices included in this integral obviously preserve symmetry.

9.2.4.2   The sliding case

The sliding case is identified if the inequality in eqn. (9.8) is not satisfied. Then the sliding force $\mathbf{T}^{sl}$ is derived from the principle of the maximum dissipation (see Part 1) as

$$\mathbf{T}^{sl} = -\frac{\mathbf{BFT}^{tr}}{\sqrt{\mathbf{BFT}^{tr} \cdot \mathbf{FBFT}^{tr}}} |N| = -\frac{\hat{\mathbf{T}} |N|}{\Psi}, \tag{9.12}$$

or component-wise as

$$T_i^{sl} = -\frac{b_{ij} f^{jk} T_k^{tr}}{b_{ij} b_{lm} f^{jk} f^{il} f^{mn} T_k^{tr} T_n^{tr}} \tag{9.13}$$

In eqn. (9.12) an auxiliary force $\hat{\mathbf{T}}$ – allowing some reductions in the following expressions – is introduced as

$$\hat{\mathbf{T}} = \mathbf{BFT}^{tr}, \tag{9.14}$$

or component-wise as

$$\hat{T}_i = b_{ij} f^{jk} T_k^{tr} = b_{ij} f_{ln} a^{jl} a^{kn} T_k^{tr} = b_i^j f_j^k T_k^{tr}. \tag{9.15}$$

and a function Ψ as

$$\Psi := \sqrt{\mathbf{BFT}^{tr} \cdot \mathbf{FBFT}^{tr}} = \sqrt{\hat{\mathbf{T}}^{tr} \cdot \mathbf{F}\hat{\mathbf{T}}^{tr}}. \tag{9.16}$$

The derivative of the sliding force $\mathbf{T}^{sl}$ is computed according to the chain rule as:

$$\frac{d\mathbf{T}^{sl}}{dt} = \frac{d}{dt}\left(-\frac{\hat{\mathbf{T}}|N|}{\Psi}\right) = -\left(\frac{d|N|}{dt}\frac{\hat{\mathbf{T}}}{\Psi} + \frac{|N|}{\Psi}\frac{d\hat{\mathbf{T}}}{dt} - \frac{|N|\hat{\mathbf{T}}}{\Psi^2}\frac{\partial\Psi}{\partial\hat{\mathbf{T}}}\cdot\frac{d\hat{\mathbf{T}}}{dt}\right)$$

$$= -\left(\frac{d|N|}{dt}\frac{\hat{\mathbf{T}}}{\Psi} + \frac{|N|}{\Psi}\left[\frac{d\hat{\mathbf{T}}}{dt} - \hat{\mathbf{T}}\frac{\mathbf{F}\hat{\mathbf{T}}}{\Psi^2}\cdot\frac{d\hat{\mathbf{T}}}{dt}\right]\right) \tag{9.17}$$

The evolution equation (9.4) for the normal traction is used for the linearization of $|N|$. The auxiliary force $\hat{\mathbf{T}}$ defined in eqn. (9.14) is linearized according to the chain rule

$$\frac{d\hat{\mathbf{T}}}{dt} = \frac{\partial\hat{\mathbf{T}}}{\partial\mathbf{T}^{tr}}\frac{d\mathbf{T}^{tr}}{dt} = \mathbf{BFB}(\mathbf{v}^s - \mathbf{v}), \tag{9.18}$$

where for the linearization of the trial traction $\mathbf{T}^{tr}$ the evolution equation (9.5) is used directly. It is an interesting fact that due to the tensor representation the linearization is valid even in the case of arbitrarily varying surface tensors $\mathbf{B}$ and $\mathbf{F}$. Eqn. (9.17) is then transformed into

$$\frac{d\mathbf{T}^{sl}}{dt} = \epsilon_N \dot{\xi}^3 \frac{\hat{\mathbf{T}}}{\Psi} - |N|\left[\frac{\mathbf{BFB}}{\Psi} - \frac{\hat{\mathbf{T}}\otimes(\mathbf{BFB})^T\mathbf{F}\hat{\mathbf{T}}}{\Psi^3}\right](\mathbf{v}^s - \mathbf{v}). \tag{9.19}$$

Eqn. (9.19) is used for the further linearization. After some transformations the following expression in components is obtained for the tangential part of the contact integral

$$D_v(\delta W_c^T) = \int_s \left(\frac{dT_i}{dt}\delta\xi^i + T_i\frac{d\delta\xi^i}{dt}\right) ds = \tag{9.20}$$

in tensor form denoted by $(\ldots\_t)$

$$= \int_s \left((\delta\mathbf{r}_s - \delta\boldsymbol{\rho}) \cdot \frac{\epsilon_N\hat{\mathbf{T}}\otimes\mathbf{n}}{\Psi}(\mathbf{v}_s - \mathbf{v})\right) ds \tag{9.21a}$$

$$-\int_s \left((\delta \mathbf{r}_s - \delta \boldsymbol{\rho}) \cdot \frac{|N|\,\mathbf{BFB}}{\Psi}(\mathbf{v}_s - \mathbf{v}) \right) ds \qquad (9.21\text{b})$$

$$+\int_s \left((\delta \mathbf{r}_s - \delta \boldsymbol{\rho}) \cdot \frac{|N|\,\hat{\mathbf{T}} \otimes (\mathbf{BFB})^T \mathbf{F}\hat{\mathbf{T}}}{\Psi^3}(\mathbf{v}_s - \mathbf{v}) \right) ds \qquad (9.21\text{c})$$

$$-\int_s T_i^{sl} \left[(\delta \mathbf{r}_s - \delta \boldsymbol{\rho}) \cdot a^{il} a^{jk}\, \boldsymbol{\rho}_k \otimes \boldsymbol{\rho}_l\, \mathbf{v}_j + \delta \boldsymbol{\rho}_{,j} \cdot a^{ik} a^{jl}\, \boldsymbol{\rho}_k \otimes \boldsymbol{\rho}_l\, (\mathbf{v}_s - \mathbf{v}) \right] ds., \qquad (9.21\text{d})$$

or component-wise denoted by $(..._c)$

$$=\int_s \left((\delta \mathbf{r}_s - \delta \boldsymbol{\rho}) \cdot \frac{\epsilon_N \hat{T}_i a^{ij}}{\Psi} \boldsymbol{\rho}_j \otimes \mathbf{n}(\mathbf{v}_s - \mathbf{v}) \right) ds \qquad (9.22\text{a})$$

$$-\int_s \left((\delta \mathbf{r}_s - \delta \boldsymbol{\rho}) \cdot \frac{|N|\, b_i^k f_j^i b^{jl}}{\Psi} \boldsymbol{\rho}_k \otimes \boldsymbol{\rho}_l(\mathbf{v}_s - \mathbf{v}) \right) ds \qquad (9.22\text{b})$$

$$+\int_s \left((\delta \mathbf{r}_s - \delta \boldsymbol{\rho}) \cdot \frac{|N|\, b_j^i f_m^j b^{ml} f_i^q \hat{T}_q \hat{T}_n a^{nk}}{\Psi^3} \boldsymbol{\rho}_k \otimes \boldsymbol{\rho}_l(\mathbf{v}_s - \mathbf{v}) \right) ds \qquad (9.22\text{c})$$

$$-\int_s T_i^{sl} \left[(\delta \mathbf{r}_s - \delta \boldsymbol{\rho}) \cdot a^{il} a^{jk}\, \boldsymbol{\rho}_k \otimes \boldsymbol{\rho}_l\, \mathbf{v}_j + \delta \boldsymbol{\rho}_{,j} \cdot a^{ik} a^{jl}\, \boldsymbol{\rho}_k \otimes \boldsymbol{\rho}_l\, (\mathbf{v}_s - \mathbf{v}) \right] ds. \qquad (9.22\text{d})$$

Here, the components of the sliding force T_i^{sl} are computed via eqn. (9.13), and the components of the auxiliary vector $\hat{T}_i$ via eqn. (9.15). It becomes obvious, that anisotropy leads to the loss of symmetry of part (9.21c) and (9.22c), by looking at the non-diagonal components of $\hat{\mathbf{T}} \otimes \mathbf{A}\hat{\mathbf{T}}$, where

$$\mathbf{A} = (\mathbf{BFB})^T \mathbf{F} \qquad (9.23)$$

then we have

$$\mathbf{C} = \hat{\mathbf{T}} \otimes \mathbf{A}\hat{\mathbf{T}} \implies \begin{cases} c_{12} &= \hat{T}_1 A_2^1 \hat{T}_1 + \hat{T}_1 A_2^2 \hat{T}_2 \\ c_{21} &= \hat{T}_2 A_1^1 \hat{T}_1 + \hat{T}_2 A_1^2 \hat{T}_2 \end{cases} . \qquad (9.24)$$

Symmetry in eqn. (9.24) is recovered only in the isotropic case, i.e. if

$A^i_j = \delta^i_j$. In the case of isotropy we have $f^i_j = \dfrac{\delta^i_j}{\mu^2}$ and $b^i_j = -\epsilon_T \delta^i_j$ leading to $\Psi = \dfrac{\|\mathbf{T}\|}{\mu^3}$ and $\hat{\mathbf{T}} = -\dfrac{\epsilon_T}{\mu^2} T^{tr}$.

Summarizing we obtain, that the tangent matrix in eqn. (9.20) consists of the standard constitutive non-symmetric part (9.21a), a constitutive symmetric part (9.21b), and a constitutive non-symmetric part (9.21c), which is symmetric only in the isotropic case, and, finally, the standard symmetric rotational part (9.21d).

Remark.
The component-wise formulas in eqns. (9.22a, 9.22b, 9.22c, 9.22d) show the possible representations of the corresponding tensor formulas in eqns. (9.21a, 9.21b, 9.21c, 9.21d). It obvious that variations within the sequence of covariant and contravariant components are possible.

9.3 Finite element implementation

In this section we will discuss details of the finite element implementation and necessary definitions for the proposed model. In particular, the anisotropic structure has to be defined on the whole contact surface. This leads to additional difficulties concerning a unique description for the whole surface and not only an approximation for the corresponding contact elements. Therefore, we start with the simplest contact element defining contact with an anisotropic rigid surface a so-called a point-to-analytical surface contact element. For this a node of a FE-mesh as well as an integration point of an element can be taken as the contact point. In this case, the contact point itself can be seen as containing history variables. If the anisotropic surface is deformable then the node-to-segment strategy has to be applied. In this case the contact segments store the history variables of passing nodes. As a more general case, the re-parameterization of the complete contact surface in the case of contact of two deformable bodies is discussed.

9.3.1 Point-to-analytical surface contact element. Linear surface approximation of a deformable body

The contact of a body meshed with bilinear finite elements – and thus bilinear contact surface elements – with a rigid anisotropic surface is one of the simplest cases to define a contact element. In this case, a node of the FE meshed surface is taken as a contact point, while all necessary anisotropic tensors are defined on the rigid surface which can geometrically be described by analytical functions. The corresponding geometrical characteristics as normal vector $\mathbf{n}$ and tangent vector $\boldsymbol{\rho}_i$ are then directly given by the analytical surface description. All integrals for the tangent matrix as well as for the residuum are defined for one nodal point. Since the anisotropic surface is rigid, all rotational parts can be set to zero. Thus, we obtain the following matrices for the contact contributions.

9.3.1.1 Matrix for the normal part

$$\mathbf{K}_N = -\epsilon_N \mathbf{n} \otimes \mathbf{n}. \tag{9.25}$$

9.3.1.2 Matrix for the tangential part. Sticking case

$$\mathbf{K}_T^{stick} = b^{ij} \boldsymbol{\rho}_i \otimes \boldsymbol{\rho}_j. \tag{9.26}$$

9.3.1.3 Matrix for the tangential part. Sliding case

$$\mathbf{K}_T^{slide} = \frac{\epsilon_N \hat{T}_i a^{ij}}{\Psi} \boldsymbol{\rho}_j \otimes \mathbf{n} \; - \; \frac{b_i^j f_j^n b_n^l a^{ik} |N|}{\Psi} \boldsymbol{\rho}_k \otimes \boldsymbol{\rho}_l \tag{9.27}$$
$$+ \; \frac{f_i^q b_j^i f_m^j b^{ml} \hat{T}_q \hat{T}_n a^{nk} |N|}{\Psi^3} \boldsymbol{\rho}_k \otimes \boldsymbol{\rho}_l.$$

All matrices contain only constitutive parts and belong to the corresponding nodes. The contact node then owns also the necessary history variables from the previous converged step (n): convective coordinates $\xi_{(n)}^1, \xi_{(n)}^2$ and tangential contact forces $T_1^{(n)}, T_2^{(n)}$.

9.3.2 Point-to-analytical surface contact element. Arbitrary surface approximation of the deformable body

If a body is meshed with finite elements of higher order of approximation for the surface – leading to a contact element of the same high order – then an integration point of the FE surface elements has to be taken as a contact point. In this case, the vector $\mathbf{r}_s$ describes the analytical surface and the vector ρ is computed from the finite element mesh. Algorithmic aspects of contact problems with a surface described by analytical functions are discussed in [57]. The algorithmic discretization of the tangent matrix, presented in Sect. 9.2.2, is obtained as follows. Let $\mathbf{u}_e$ be the nodal displacement vector taken from the finite element discretization as

$$\mathbf{u}_e^T = \{u_1^{(1)}, u_2^{(1)}, u_3^{(1)}, u_1^{(2)}, u_2^{(2)}, u_3^{(2)}, ..., u_1^{(n)}, u_2^{(n)}, u_3^{(n)}\}^T, \tag{9.28}$$

where (n) is a number of nodal points of the contact surface element. Assuming that the approximation is performed with (n) shape functions, only a single position matrix $\mathbf{A}$ is necessary for discretization of all contact contributions.

$$\mathbf{A} = \begin{bmatrix} N_1 & 0 & 0 & N_2 & 0 & 0 & ... & ... & ... & N_{(n)} & 0 & 0 \\ 0 & N_1 & 0 & 0 & N_2 & 0 & ... & ... & ... & 0 & N_{(n)} & 0 \\ 0 & 0 & N_1 & 0 & 0 & N_2 & ... & ... & ... & 0 & 0 & N_{(n)} \end{bmatrix}. \tag{9.29}$$

The contact matrix for the normal part in eqn. (9.7) is then obtained as:

$$\mathbf{K}_N = -\int_s \epsilon_N \mathbf{A}^T \mathbf{n} \otimes \mathbf{n} \mathbf{A} ds = \tag{9.30}$$

$$= -\sum_{I,J=1}^{N_p} \left(\epsilon_N \mathbf{A}^T \mathbf{n} \otimes \mathbf{n} \mathbf{A} W_I W_J \det \mathbf{J}(\xi_I^1, \xi_J^2) \right),$$

where N_P is the number of integration points and W_I, $I = 1, 2, ..., N_P$ are weights of the chosen quadrature formula. The determinant of the Jacobian $\det \mathbf{J}(\xi_I^1, \xi_J^2)$ is computed for the surface segment – the contact segment – of the body. For each integration point a set of history variables from the previous converged step (n) must be stored: convective coordinates $\xi_{(n)}^1, \xi_{(n)}^2$ and tangential contact forces $T_1^{(n)}, T_2^{(n)}$. The tangent matrices for the tangential part as defined in Sect. 9.2.4.1 and 9.2.4.2 can be derived in similar fashion.

Remark.

For this specific case all geometrical characteristics such as normal vector $\mathbf{n}$, tangent vector ρ_i as well as the anisotropic tensors are taken from the rigid surface which is defined by analytical functions.

9.3.3 Node-to-segment approach. Deformable anisotropic contact surface

If a deformable body has a surface with anisotropic properties, then the node-to-segment approach can be applied. In this case, the corresponding contact segments covering the anisotropic surface are taken as master segments. A nodal displacement vector contains then an additional $(n+1)$ slave node besides the first (n) nodes from the master segment:

$$\mathbf{u}_e^T = \{u_1^{(1)}, u_2^{(1)}, u_3^{(1)}, \ldots, u_1^{(n)}, u_2^{(n)}, u_3^{(n)}, u_1^{(n+1)}, u_2^{(n+1)}, u_3^{(n+1)}\}^T. \quad (9.31)$$

The position matrix $\mathbf{A}$ is modified as

$$\mathbf{A} = \begin{bmatrix} N_1 & 0 & 0 & N_2 & 0 & 0 & \ldots & N_{(n)} & 0 & 0 & N_{(n+1)} & 0 & 0 \\ 0 & N_1 & 0 & 0 & N_2 & 0 & \ldots & 0 & N_{(n)} & 0 & 0 & N_{(n+1)} & 0 \\ 0 & 0 & N_1 & 0 & 0 & N_2 & \ldots & 0 & 0 & N_{(n)} & 0 & 0 & N_{(n+1)} \end{bmatrix}. \quad (9.32)$$

The components of the tangent matrices (normal and tangent vectors etc.) are computed in the projection point of the master segment. The structure is again algorithmic, e.g. a part for the normal contact has the form:

$$\mathbf{K}_N = -\int_s \epsilon_N \mathbf{A}^T \mathbf{n} \otimes \mathbf{n} \mathbf{A} ds = \quad (9.33)$$

$$= -\epsilon_N \mathbf{A}^T \mathbf{n} \otimes \mathbf{n} \mathbf{A}.$$

Here, the segment contains the aforementioned history variables.

9.3.3.1 Mapping of anisotropic properties from the surface to a contact segment

It is expected, that the anisotropic properties are defined for the complete surface, not only for a segment. Thus, the main problem is, how to properly transfer the anisotropic properties from the surface to the contact segment. In the case of a simple curvilinear rectangular patch this can be organized as follows. Let s^1, s^2 are convective coordinates defining the parameterization of the patch, see Fig. 9.1, with $0 \leq s^1, s^2 \leq 1$. The anisotropic properties are determined then via the tensor basis $e_1(s^1, s^2) \otimes e_2(s^1, s^2)$. The regular numbering $i = 1, ..., m$ and $j = 1, ..., n$ is introduced according to the mapped mesh on the patch, see Fig. 9.2. Therefore, a direct mapping of the convective coordinate on the element ξ^1 can be defined as

$$\left. \begin{aligned} \xi^1 = -1 \quad &\longrightarrow \quad s^1 = \frac{j-1}{n} \\ \xi^1 = 1 \quad &\longrightarrow \quad s^1 = \frac{j}{n} \end{aligned} \right\} \implies s^1 = \frac{2j - 1 + \xi^1}{2n}. \tag{9.34}$$

According to the introduced direct transformation, the backward trans-

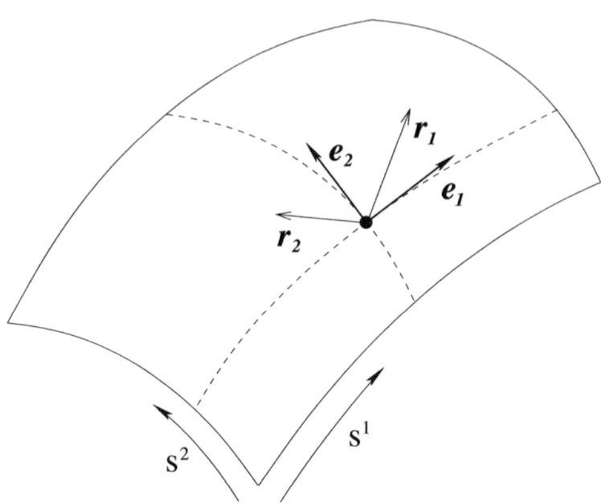

Figure 9.1: Curvilinear rectangular patch.

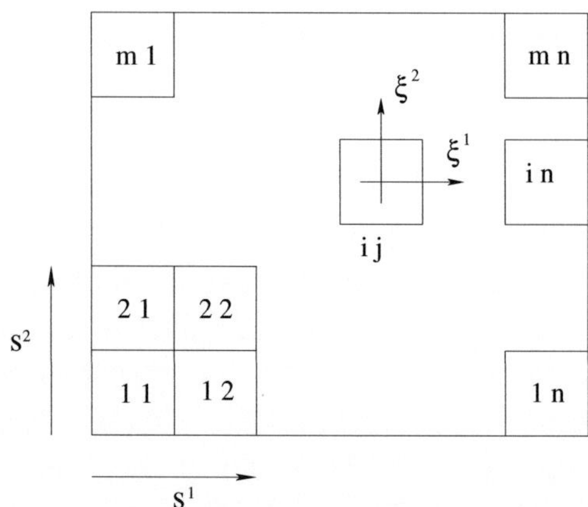

Figure 9.2: Curvilinear rectangular patch. Mapping scheme.

formation is defined according to the following algorithm:

$$do\ j = 1, n$$

$$if\ \frac{j-1}{n} \leq s^1 \leq \frac{j}{n}$$

$$then\ k = j$$

$$\xi^1 = 2ns^1 - 2k + 1$$

$$endif$$

$$enddo$$

(9.35)

The second coordinate s^2 is analogously computed.

9.4 Numerical examples

In this section we present numerical examples illustrating several types of the orthotropy. As known, the orthotropic frictional properties of the surface leads to changing of kinematical behavior of the contact bodies, see [22], [24], [68], [82], [212], [213], [214], [203], therefore, the set of numerical examples is chosen to illustrate particular kinematical effects which appear due to presence of anisotropy for adhesion and friction. In the first example we chose constant orthotropy on a plane represented by a tensor with spectral decomposition. This model possesses a simple mechanical interpretation a so-called rheological model. Constant orthotropy is thoroughly investigated for the case with small displacements in order to show the development and the distribution of the sticking-sliding zone for different types of orthotropy. These cases are considered for the start of sliding as well as for large sliding deformations in order to consider the trajectories of a block for different types of orthotropy. Then a large displacement problem for a plane with polar orthotropy is taken as an example for curvilinear orthotropy on the plane. In order to show the robustness of the developed approach for curvilinear surfaces, kinematical effects of a bolt connection are modeled with spiral orthotropy defined on a cylinder.

9.4.1 Rheological model of the orthotropic adhesion-friction problem

As is well known [160], the return-mapping scheme used for the model of elasto-plasticity can be interpreted via a one-dimensional spring-sliding system. A generalization of this model into 2D anisotropy is a point on the plane with a two spring-two slider system with different properties: $\varepsilon_1, \varepsilon_2$ as stiffnesses of the springs and μ_1, μ_2 as coefficients of friction for the sliding devices, see Fig. 9.3. A constant force $\mathbf{F}$ is applied to the point at an angle γ. It becomes obvious, that after transformation of the coordinate system in such a way that e.g. the X-axis coincides with the direction a force $\mathbf{F}$, the problem exactly corresponds to the constant orthotropy on the plane. The latter is given by a tensor with the spectral decomposition in the case of coinciding orthotropy angles for adhesion α and for friction β, see Part 1. The trajectory of the point is then a straight line inclined with an angle φ, the value of which depends on the ratio of the eigenvalues defining the adhesion ellipse, see the geometrical interpretation in Part 1:

$$r_\lambda = \frac{\lambda_1}{\lambda_2} = \frac{\varepsilon_1}{\mu_1} \cdot \frac{\mu_2}{\varepsilon_2} \tag{9.36}$$

In forthcoming computations, we will show, that it is possible to represent geometrically isotropic behavior, in such a way that the trajectory of the

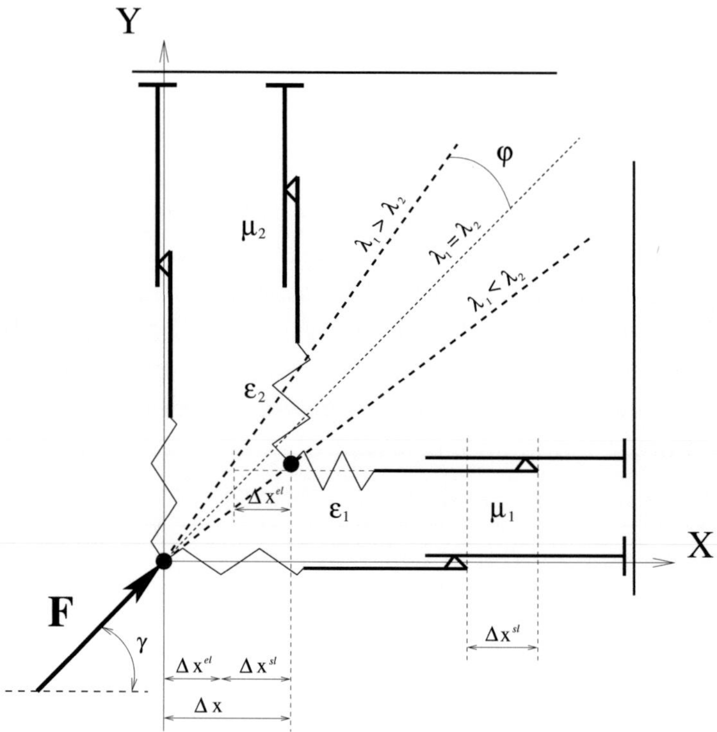

Figure 9.3: Mechanical interpretation of orthotropic friction − orthotropic adhesion model. A material point on a plane with two spring-slider systems driven by a tangential force.

point is coinciding with the direction of the force $\mathbf{F}$, though the properties of this contact surface remain orthotropic.

9.4.2   Linear constant orthotropy on the plane

In order to investigate the properties of the proposed model, a rectangular block is considered on an orthotropic plane. The dimensions of the block are $10 \times 10 \times 4$ with linear elastic properties: Young's modulus $E = 2.10 \cdot 10^4$ and Poisson ratio $\nu = 0.3$, assumed within a consistent dimension system. The block (see Fig. 9.4) is located on the XOY plane and loaded by prescribing displacements on the upper surface in two steps: 1) vertical loading with $w = 1.0 \cdot 10^{-2}$, 2) incremental loading with Δu along the X axis. Contact with regard to the constant orthotropic model with the adhesion tensor

$$\mathbf{B} = [b^i_j] = - \begin{bmatrix} \varepsilon_1 \cos^2 \alpha + \varepsilon_2 \sin^2 \alpha & (\varepsilon_1 - \varepsilon_2) \sin \alpha \cos \alpha \\ (\varepsilon_1 - \varepsilon_2) \sin \alpha \cos \alpha & \varepsilon_1 \sin^2 \alpha + \varepsilon_2 \cos^2 \alpha \end{bmatrix}, \qquad (9.37)$$

and the friction tensor

$$\mathbf{F} = [f^i_k] = \begin{bmatrix} \dfrac{1}{\mu_1^2} \cos^2 \beta + \dfrac{1}{\mu_2^2} \sin^2 \beta & (\dfrac{1}{\mu_1^2} - \dfrac{1}{\mu_2^2}) \sin \beta \cos \beta \\[3ex] (\dfrac{1}{\mu_1^2} - \dfrac{1}{\mu_2^2}) \sin \beta \cos \beta & \dfrac{1}{\mu_1^2} \sin^2 \beta + \dfrac{1}{\mu_2^2} \cos^2 \beta \end{bmatrix} \qquad (9.38)$$

is specified between the plane and the block. To compare both approaches contact is modeled with a point-to-analytical surface contact element as well as with a node-to-segment approach. The rigid plane is taken as master segment within the latter approach. As an example coincident orthotropy angles $\alpha = \beta$ are chosen. First, we will investigate the development of a sticking-sliding zone for small displacements for various cases of the ratio r_λ, see eqn. (9.36). Afterwards, we will consider the large displacement problem and investigate the trajectories of the block depending on the surface properties.

9.4.2.1 Small displacement problem.
Development of the sticking-sliding zone

In order to investigate the development of the sticking-sliding zone small displacement increments with $\Delta u = 1.0 \cdot 10^{-4}$ along the X axis are applied. The following cases are considered:

1. Isotropic case; both, the friction and the adhesion tensors are isotropic with: penalty parameter for the normal traction $\varepsilon_N = 2.1 \cdot 10^5$, parameters of the adhesion tensor $\varepsilon_1 = \varepsilon_2 = 2.1 \cdot 10^5$, parameters of the friction tensor $\mu_1 = \mu_2 = 0.3$. Orthotropy angles: $\alpha = \beta = 0^o$.

2. Geometrically isotropic case with: penalty parameter for the normal traction $\varepsilon_N = 2.1 \cdot 10^5$, parameters of the adhesion tensor $\varepsilon_1 = 3.0 \cdot 10^5$, $\varepsilon_2 = 2.0 \cdot 10^5$, parameters of the friction tensor $\mu_1 = 0.3$, $\mu_2 = 0.2$. Orthotropy angles: $\alpha = \beta = 45^o$. This case leads to

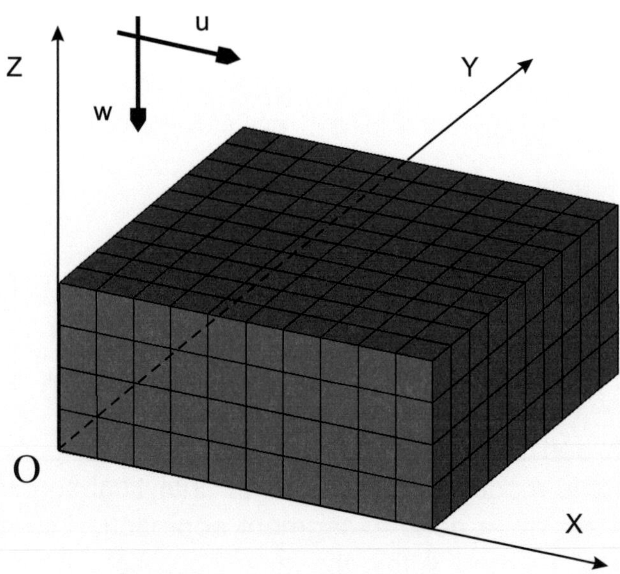

Figure 9.4: Geometry and loading of the rectangular in plane block.

the ratio of eigenvalues eqn. (9.36) $r_\lambda = 1$ and, thus, to a circular adhesion region.

3. Orthotropic case; orthotropic adhesion, isotropic friction with: penalty parameter for the normal traction $\varepsilon_N = 2.1 \cdot 10^5$, parameters of the adhesion tensor $\varepsilon_1 = 2.0 \cdot 10^5$, $\varepsilon_2 = 3.0 \cdot 10^5$, parameters of the friction tensor $\mu_1 = \mu_2 = 0.3$. Orthotropy angles: $\alpha = \beta = 45^o$. The eigenvalue ratio is then $r_\lambda = 2/3$.

4. Orthotropic case; isotropic adhesion, orthotropic friction with: penalty parameter for the normal traction $\varepsilon_N = 2.1 \cdot 10^5$, parameters of the adhesion tensor $\varepsilon_1 = \varepsilon_2 = 2.1 \cdot 10^5$, parameters of the friction tensor $\mu_1 = 0.3$, $\mu_2 = 0.2$. Orthotropy angles: $\alpha = \beta = 45^o$. The eigenvalue ratio is then $r_\lambda = 2/3$.

Isotropic case (1). The results are depicted in the diagram in Figure 9.6 showing the developed sticking area (in grey color) for the applied horizontal displacements on the lower contact surface in several states from the top view. This area is identified by sticking nodes on the lower contact surface; these nodes are inside the adhesion ellipse. For

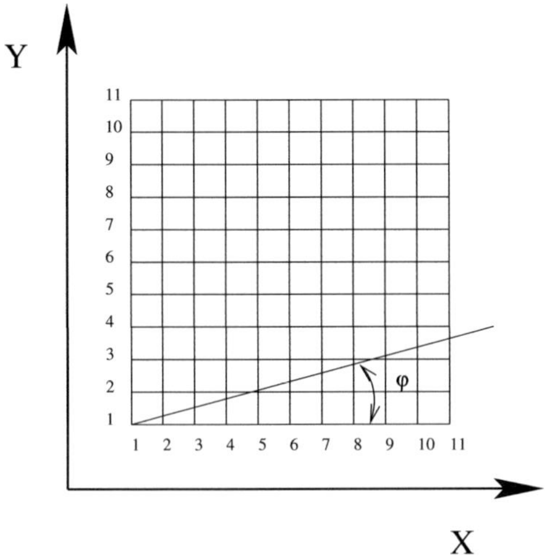

Figure 9.5: Numbering of nodes on the lower surface and direction φ of the development of a sliding zone.

the results of the investigations we use the node numbering and the inclination angle φ for the block as given in Fig. 9.5. One can observe that the edge nodes are sliding from the beginning, i.e. when only vertical loading is applied. This effect is due to the singularity of stresses on the edges known from the analytical solution for a rigid punch problem, see e.g. Johnson [77]. The symmetrical sticking region is vanishing and the block begins to slide fully, once the applied horizontal displacements are beyond the value $u = 10.0 \cdot 10^{-3}$.

It is also interesting, especially for the forthcoming orthotropic cases, to observe the beginning of full sliding. Both the initial configuration and the scaled deformed configuration from the bottom view are depicted in Fig. 9.7 for loading $u = 9.0 \cdot 10^{-3}$ (the block begins to slide partially) as well as for loading $u = 20.0 \cdot 10^{-3}$ (fully developed sliding of the block). The deformation is symmetric as the horizontal axis and the current horizontal symmetry axis are moving along the reference axis.

Geometrically isotropic case (2). The results are depicted in the diagrams in Figure 9.8 showing the development of the sticking area. It is interesting to observe that the initial sticking area $u = 0.0$ in this case is symmetric along the main orthotropy axes which are turned according to the references coordinate system by the angles $\alpha = \beta = 45^o$.

We recall the geometrical interpretation to explain this phenomena. The adhesion region becomes a circle in this particular example, see Fig. 9.10. The elastic properties inside the circle are orthotropic according to the computation of the trial force as $\mathbf{T}^{tr} = \mathbf{B}\Delta\mathbf{x}$ with the adhesion tensor computed with $\alpha = 45^o$, see eqn. (9.37):

$$\mathbf{B} = [b^i_j] = -\frac{1}{2}\begin{bmatrix} \varepsilon_1 + \varepsilon_2 & \varepsilon_1 - \varepsilon_2 \\ \varepsilon_1 - \varepsilon_2 & \varepsilon_1 + \varepsilon_2 \end{bmatrix} = -\frac{10^5}{2}\begin{bmatrix} 5 & 1 \\ 1 & 5 \end{bmatrix}. \tag{9.39}$$

Since, in the current example the stiffness in the second direction is less then in the first one $\varepsilon_2 < \varepsilon_1$, all points $\mathbf{A}^{el}$ tend to reach the boundary of the adhesion circle in the direction of the ξ^2-axis. This can explain the asymmetric behavior of the initial sticking area in Fig. 9.8 (a). The direction of the sliding force $\mathbf{T}^{sl}$, which is starting to act from the boundary of the circle, is defined by the matrix $\mathbf{BFB}$, see eqn. (9.12), which in the current example becomes:

$$\mathbf{BFB} = \frac{1}{2}\begin{bmatrix} \dfrac{\varepsilon_1^2}{\mu_1^2} + \dfrac{\varepsilon_2^2}{\mu_2^2} & \dfrac{\varepsilon_1^2}{\mu_1^2} - \dfrac{\varepsilon_2^2}{\mu_2^2} \\ \dfrac{\varepsilon_1^2}{\mu_1^2} - \dfrac{\varepsilon_2^2}{\mu_2^2} & \dfrac{\varepsilon_1^2}{\mu_1^2} + \dfrac{\varepsilon_2^2}{\mu_2^2} \end{bmatrix} = \begin{bmatrix} 10^6 & 0 \\ 0 & 10^6 \end{bmatrix}. \tag{9.40}$$

This recovered isotropic behavior of the sliding force $\mathbf{T}^{sl}$ is depicted in Fig. 9.10. This effect is depicted for the developed sliding in Fig. 9.9, where the horizontal axis of symmetry of the deformed body is moving along the reference axis.

For completeness, we would like to present the results showing the convergence of the sticking zone with regard to: 1) adhesion parameters; 2) mesh density. Thus, Fig. 9.11 represents the comparison of the initial sticking zones computed first for the set of parameters $\varepsilon_N = 2.1 \cdot 10^3$, $\varepsilon_1 = 3.0 \cdot 10^3$, $\varepsilon_2 = 2.0 \cdot 10^3$, then for the same set, but instead of 10^3 scaled sequentially by the factors 10^4, 10^5 and 10^6. For the latter computation the vertical loading has been provided in 10 load steps. The results serve also to show the convergence of the results with increasing penalty parameters.

The influence of the mesh density on the sticking zone for the case of parameters $\varepsilon_N = 2.1\cdot10^5$, $\varepsilon_1 = 3.0\cdot10^5$, $\varepsilon_2 = 2.0\cdot10^5$ is shown in Fig. 9.12. The uniform mesh was varied as $10 \times 10 \times 4$, $16 \times 16 \times 6$, $20 \times 20 \times 8$ and $32 \times 32 \times 8$ respectively, the third number always representing the number of elements in thickness direction.

Orthotropic case (3). The results are depicted in the diagrams in Fig. 9.13 showing the development of a closed sticking area. The orthotropy of the adhesion region results in the sticking region being turned by the angle 45°. The development of a sliding zone starts at the upper right corner and continues unsymmetrically leading further to a parallel shifting of the block and a straight trajectory inclined at an angle φ (see definition at Fig. 9.5) in the case of large displacements. This effect is depicted in Fig. 9.14, a bottom view.

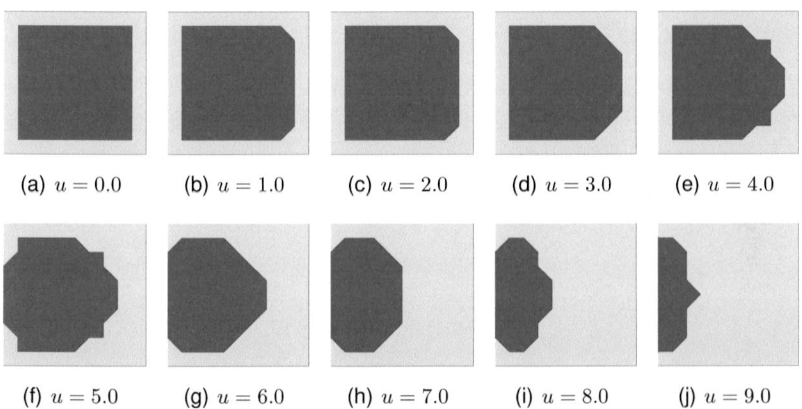

(a) $u = 0.0$ (b) $u = 1.0$ (c) $u = 2.0$ (d) $u = 3.0$ (e) $u = 4.0$

(f) $u = 5.0$ (g) $u = 6.0$ (h) $u = 7.0$ (i) $u = 8.0$ (j) $u = 9.0$

Figure 9.6: Isotropic case (1). Development (degeneration) of sticking zone for several displacement states. Horizontally applied displacement u pointing into the right direction (u is scaled by 10^{-3}).

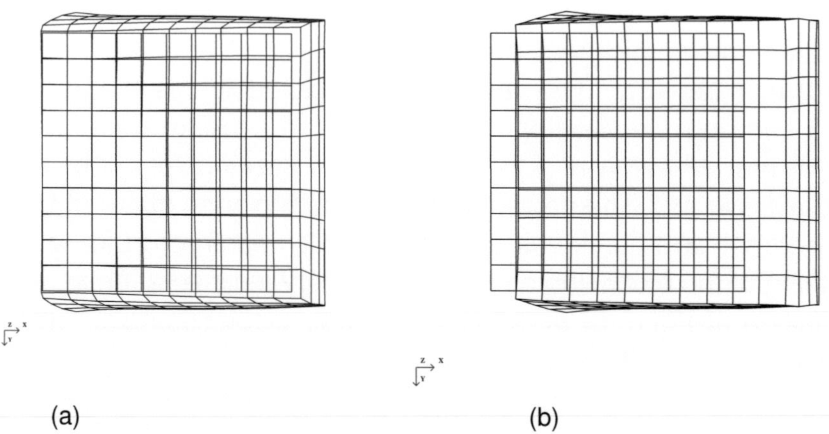

(a)   (b)

Figure 9.7: Isotropic case (1). Deformed and initial configuration. Bottom view. Displacements scaled by factor 150. Applied horizontal displacements on the upper surface: (a) $u = 9.0 \cdot 10^{-3}$ (start of sliding); (b) $u = 20.0 \cdot 10^{-3}$ (fully developed sliding).

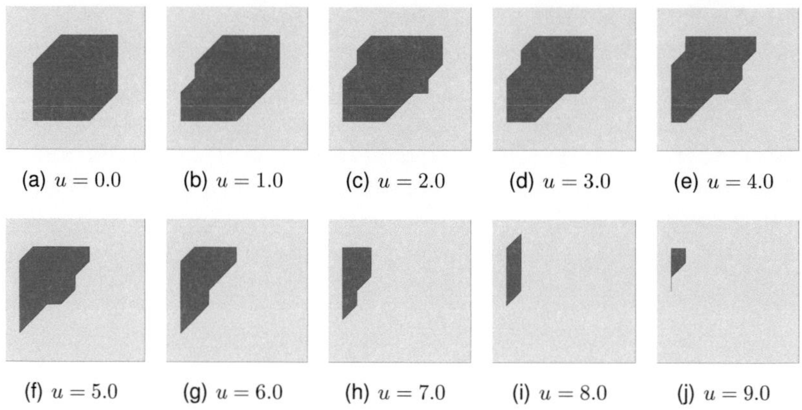

(a) $u = 0.0$ (b) $u = 1.0$ (c) $u = 2.0$ (d) $u = 3.0$ (e) $u = 4.0$

(f) $u = 5.0$ (g) $u = 6.0$ (h) $u = 7.0$ (i) $u = 8.0$ (j) $u = 9.0$

Figure 9.8: Geometrically isotropic case (2). Development of the sticking zone for various displacement states. Horizontally applied displacement u pointing into the right direction (u is scaled by 10^{-3}).

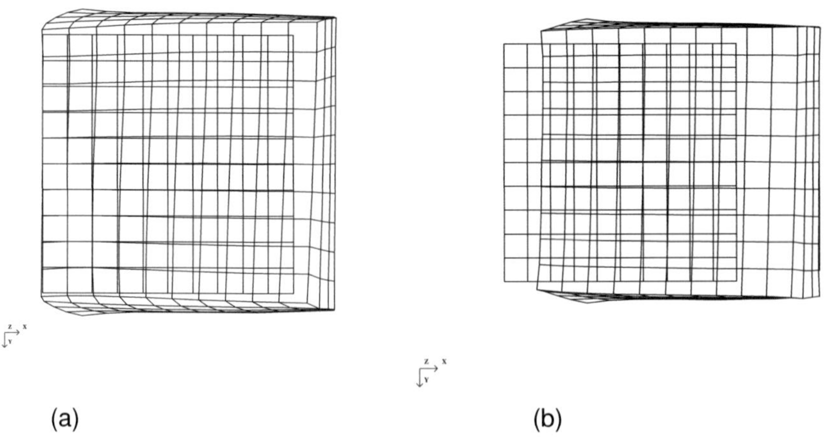

(a)   (b)

Figure 9.9: Geometrically isotropic case (2). Deformed and initial configuration. Bottom view. Displacements scaled by factor 150. Applied horizontal displacements on the upper surface: (a) $u = 9.0 \cdot 10^{-3}$ (start of sliding); (b) $u = 20.0 \cdot 10^{-3}$ (fully developed sliding).

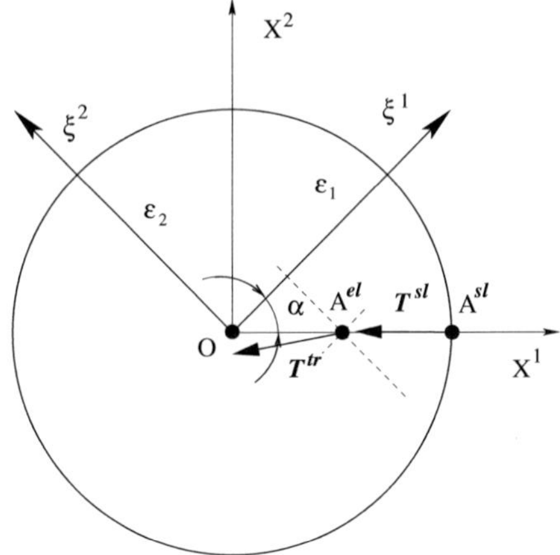

Figure 9.10: Adhesion region and direction of an elastic trial force and a sliding force for specially chosen parameters leading to the geometrically isotropic case (2). $\varepsilon_1 < \varepsilon_2$, $\alpha = 45^o$.

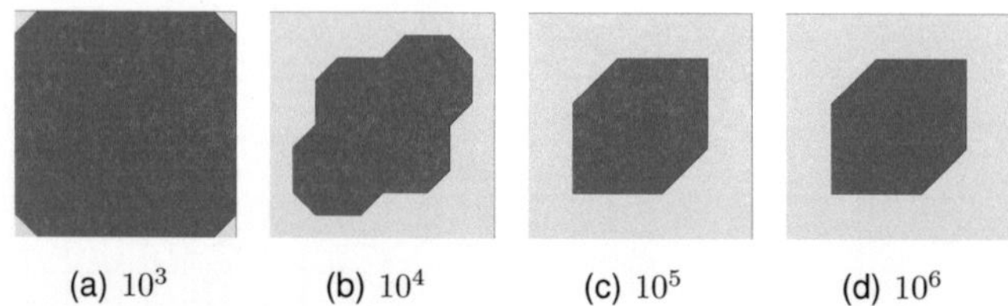

(a) 10^3 (b) 10^4 (c) 10^5 (d) 10^6

Figure 9.11: Geometrically isotropic case (2). Initial sticking zone. Variation of penalty and adhesion tensor parameters. (a) scale factor 10^3, one load step. (b) scale factor 10^4, one load step. (c) scale factor 10^5, one load step. (d) scale factor 10^6, 10 load steps.

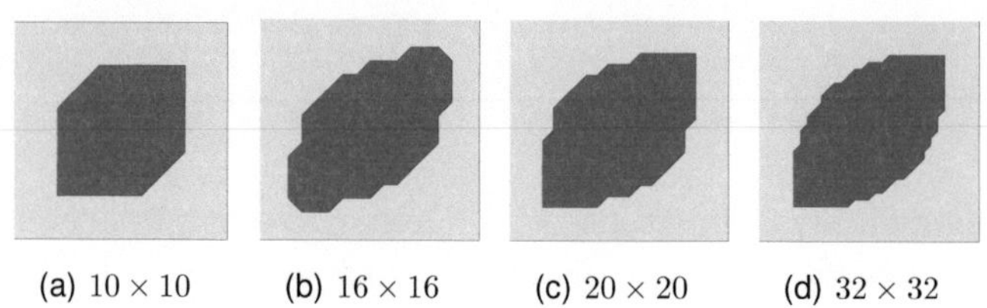

(a) 10×10 (b) 16×16 (c) 20×20 (d) 32×32

Figure 9.12: Geometrically isotropic case (2). Initial sticking zone. Variation of mesh density. (a) 10×10 elements in plane. (b) 16×16 elements in plane. (c) 20×20 elements in plane. (d) 32×32 elements in plane.

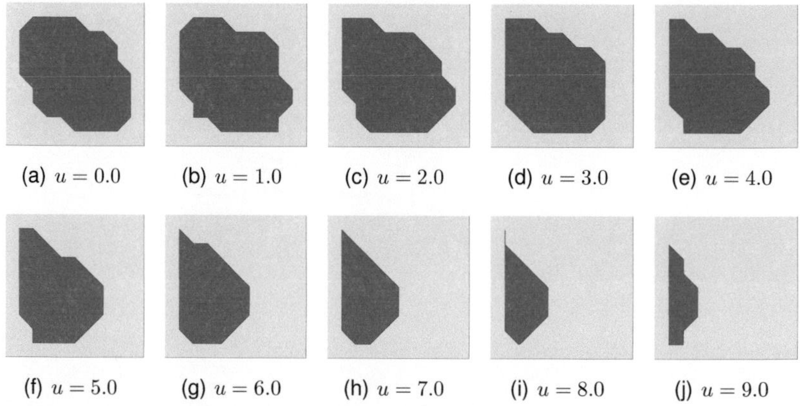

(a) $u = 0.0$ (b) $u = 1.0$ (c) $u = 2.0$ (d) $u = 3.0$ (e) $u = 4.0$

(f) $u = 5.0$ (g) $u = 6.0$ (h) $u = 7.0$ (i) $u = 8.0$ (j) $u = 9.0$

Figure 9.13: Orthotropic case (3). Development (degeneration) of sticking zone for several displacement states. Horizontally applied displacement u pointing into the right direction (u is scaled by 10^{-3}).

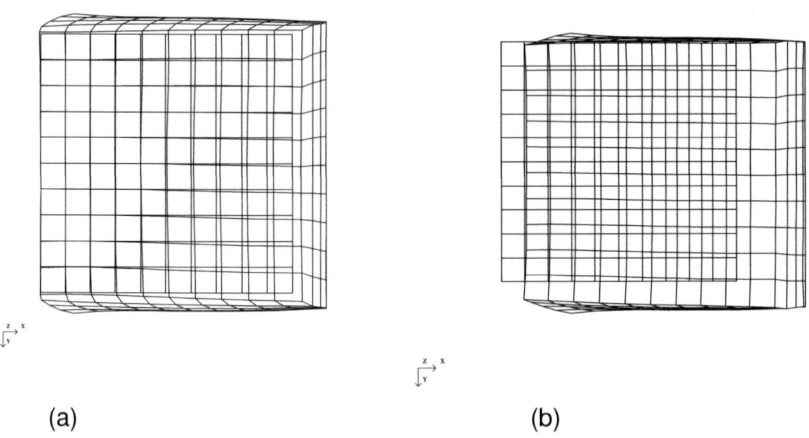

(a)   (b)

Figure 9.14: Orthotropic case (3). Deformed and undeformed configuration. Bottom view. Displacements scaled by factor 150. Applied horizontal displacements on the upper surface: (a) $u = 9.0 \cdot 10^{-3}$ (start of sliding); (b) $u = 20.0 \cdot 10^{-3}$ (fully developed sliding).

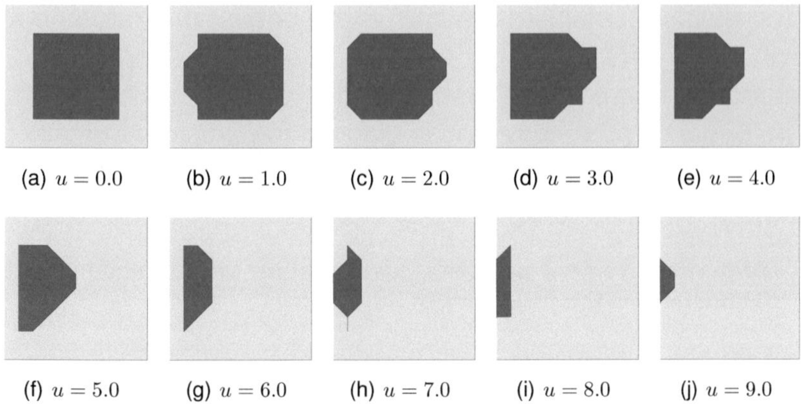

(a) $u = 0.0$ (b) $u = 1.0$ (c) $u = 2.0$ (d) $u = 3.0$ (e) $u = 4.0$

(f) $u = 5.0$ (g) $u = 6.0$ (h) $u = 7.0$ (i) $u = 8.0$ (j) $u = 9.0$

Figure 9.15: Orthotropic case(4). Development (degeneration) of sticking zone for several displacement states. Horizontally applied displacement u pointing into the right direction (u is scaled by 10^{-3}).

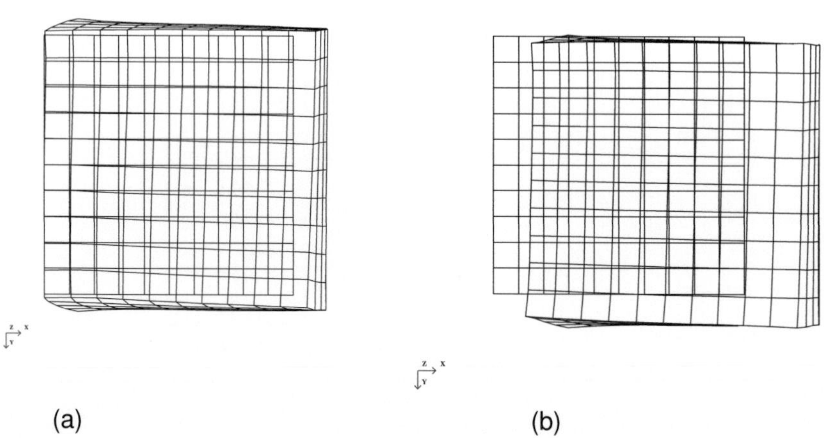

(a)   (b)

Figure 9.16: Orthotropic case (4). Deformed and undeformed configuration. Bottom view. Displacements scaled by factor 150. Applied horizontal displacements on the upper surface: (a) $u = 9.0 \cdot 10^{-3}$ (start of sliding); (b) $u = 20.0 \cdot 10^{-3}$ (fully developed sliding).

Orthotropic case (4). The results are depicted in the diagrams in Fig. 9.15 showing the development of a closed sticking area. Now, isotropy for the adhesion region together with a small difference between the friction coefficients results in the obtained symmetric region. As known, large differences between the coefficients lead to an unsymmetrical region, see e.g. the results for the classical orthotropic friction model in [68]. The sticking region is diminishing unsymmetrically resulting globally in a shifting of the block, see Fig. 9.16.

9.4.2.2 Large displacement problem. Investigation on influence of adhesion parameters on the trajectory of the block

The goal of this analysis is to show that with properly chosen parameters for the adhesion tensor it is possible to achieve a predefined motion on the surface. Namely, if the surface is uniformly orthotropic with a given orthotropy angle α then it is possible to prescribe a straight trajectory of the block with an angle $\varphi \approx \alpha$, see Fig. 9.5, keeping the driving force at a certain level. Moreover, it is possible to obtain even the geometrically isotropic case, when a trajectory is a straight line coinciding with the direction of the force, though the friction tensor remains orthotropic leading to different global forces in the different global directions.

In the numerical examples the penalty parameter for the normal traction is kept to $\varepsilon_N = 2.1 \cdot 10^4$ for all cases; the orthotropy angles are given as $\alpha = \beta = 45^o$. The horizontal incremental displacements are taken as $\Delta u = 5.0 \cdot 10^{-2}$ in order to reach a sliding state of the block from the beginning. The other parameters are varied to achieve different cases as presented in Table 9.1.

Case	Adhesion tensor	Friction tensor	Eigenvalue ratio	Resulting angle φ
1	$\varepsilon_1 = 3.0 \cdot 10^3$ $\varepsilon_2 = 0.0$	$\mu_1 = \mu_2 = 0.3$	∞	-44.95^o
2	$\varepsilon_1 = 3.0 \cdot 10^3$ $\varepsilon_2 = 2.0 \cdot 10^3$	$\mu_1 = \mu_2 = 0.3$	$3/2$	-21.01^o
3	$\varepsilon_1 = 3.0 \cdot 10^3$ $\varepsilon_2 = 2.0 \cdot 10^3$	$\mu_1 = 0.3$ $\mu_2 = 0.2$	1 geom. isotropy	0°
4	$\varepsilon_1 = 2.0 \cdot 10^3$ $\varepsilon_2 = 3.0 \cdot 10^3$	$\mu_1 = \mu_2 = 0.3$	$2/3$	21.01^o
5	$\varepsilon_1 = 0.0$ $\varepsilon_2 = 3.0 \cdot 10^3$	$\mu_1 = \mu_2 = 0.3$	0	44.95^o

Table 9.1: Sliding of a block on a plane. Large displacement problem. Variation of orthotropy. Computed inclination angle of a sliding block φ.

The trajectories of the block for all cases are depicted in Fig. 9.17. It becomes obvious that the block tends to move into the direction of the eigenvector with smaller eigenvalue λ. E.g. if $r_\lambda = \varepsilon_1/\varepsilon_2 = 0$, then the trajectory is inclined at the angle of orthotropy $\varphi \approx \alpha = 45^o$ and vice versa if $r_\lambda = \varepsilon_1/\varepsilon_2 = \infty$, i.e. $\varepsilon_2 = 0$, then the block is inclined at angle $\varphi \approx -45^o$. The other parameter variations lead to different trajectories with angles $-45^o < \varphi < 45^o$. As a particular example, the geometrically isotropic case is recovered for the ratio $r_\lambda = 1$ leading to a circular adhesion region. In this particular case the block is moving into the direction of the applied force.

9.4.3 Polar orthotropy on a plane. Large displacement problem

A more complex orthotropy is given by a polar orthotropy on a plane, see Part 1. As in the previous example, we will show that it is possible to

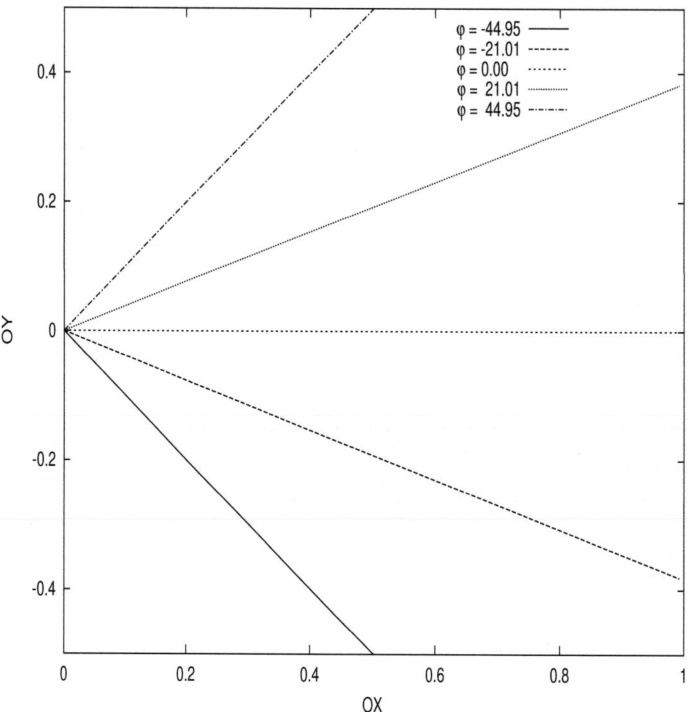

Figure 9.17: Trajectories of the block and inclination angle φ for various cases of orthotropy with the eigenvalue ratios: $r_\lambda = \infty \Longrightarrow \varphi = -44.95^\circ$; $r_\lambda = 3/2 \Longrightarrow \varphi = -21.01^\circ$; $r_\lambda = 1.0 \Longrightarrow \varphi = 0.00^\circ$; $r_\lambda = 2/3 \Longrightarrow \varphi = 21.01^\circ$; $r_\lambda = 0.0 \Longrightarrow \varphi = 44.95^\circ$.

define the orthotropic structure of a plane by using the adhesion tensor. An elastic block with dimensions $1 \times 1 \times 0.25$ and mesh $4 \times 4 \times 1$ is sitting on a rigid block, see Fig. 9.18. Linear elastic material is assumed within a consistent dimension system: Young's modulus $E = 2.10 \cdot 10^4$; Poisson ratio $\nu = 0.3$. The loading is applied sequentially by prescribing displacements on the upper surface in $(1 + n)$ steps: 1) vertical loading with $w = 1.0 \cdot 10^{-2}$, 2) n steps with horizontal displacement increments $\Delta u = 1.0 \cdot 10^{-2}$ along the X axis. The structure of the adhesion tensor is as follows, see the derivation in Part 1:

$$\mathbf{B} = -\frac{1}{x^2 + y^2} \begin{bmatrix} \varepsilon_r x^2 + \varepsilon_\varphi y^2 & (\varepsilon_r - \varepsilon_\varphi)xy \\ (\varepsilon_r - \varepsilon_\varphi)xy & \varepsilon_r y^2 + \varepsilon_\varphi x^2 \end{bmatrix}. \tag{9.41}$$

The chosen contact surface parameters are: penalty parameter for the normal traction $\varepsilon_N = 2.1 \cdot 10^5$; isotropic friction tensor with $\mu_1 = \mu_2 = 0.2$. Adhesion tensor with cases: a) $\varepsilon_r = 100$, $\varepsilon_\varphi = 1000$, b) $\varepsilon_r = 2000$, $\varepsilon_\varphi = 5000$, c) $\varepsilon_r = 1000$, $\varepsilon_\varphi = 1000$, d) $\varepsilon_r = 5000$, $\varepsilon_\varphi = 2000$, e) $\varepsilon_r = 1000$, $\varepsilon_\varphi = 100$, f) $\varepsilon_r = 1000$, $\varepsilon_\varphi = 0.0$.

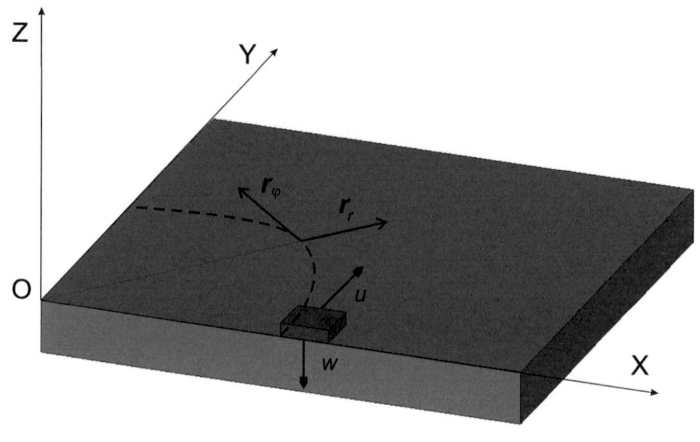

Figure 9.18: Geometry and loading for the case of polar orthotropy.

In Fig. 9.19 the sequence of the motions for all cases leading to different ratios $r_\varepsilon = \varepsilon_\varphi / \varepsilon_r$ is depicted. It is obvious, that the desired circular motion can be achieved by prescribing the corresponding eigenvalue to a small value, e.g. $r_\varepsilon = 0.0$ in the current example. The last result can be derived also analytically from the analysis of global motion of a block, see proof in **Appendix**.

Remark.

The trajectory of the block depends on the ratio $\lambda_\varphi / \lambda_r$. However, the numerical computations show that it is possible to prescribe the desired trajectory by only controlling the adhesion tensor parameters. Attempts to achieve the desired trajectory controlling the friction tensor parameters, e.g. taking $\mu_r \approx 0$ for the case (f), lead to disconvergence. A straightforward conclusion is that the geometrical structure of the surface in the sense of the desired trajectory can be defined via the ad-

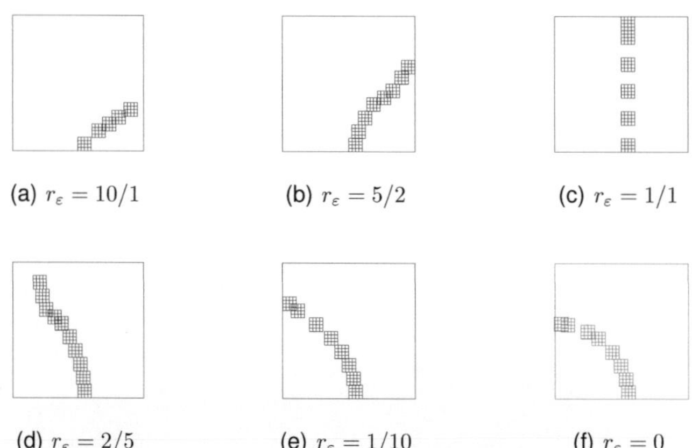

(a) $r_\varepsilon = 10/1$ (b) $r_\varepsilon = 5/2$ (c) $r_\varepsilon = 1/1$

(d) $r_\varepsilon = 2/5$ (e) $r_\varepsilon = 1/10$ (f) $r_\varepsilon = 0$

Figure 9.19: Motion of the block in the case of polar orthotropy on the plane. Varying adhesion tensor parameters: $r_\varepsilon = \varepsilon_\varphi/\varepsilon_r = 10/1,\ 5/2,\ 1,\ 2/5,\ 1/10,\ 0$; isotropic friction tensor parameters: $\mu_1 = \mu_2 = 0.2$. Loading by prescribed vertical displacement w and incremental y-displacement Δu.

hesion tensor, while other mechanical characteristics such as the measured global forces are defined via the friction tensor.

9.4.4 Spiral orthotropy on the cylinder

Another rather complex kinematical behavior of a curved contact surface, e.g. a machined surface of a bolt can be described by means of controlling the adhesion tensor parameters. The model of spiral orthotropy on the cylinder developed in Part 1 allows to describe the kinematical behavior of a bolt connection with a rather coarse mesh. In order to show this, we consider a finite element model of the bolt, see Fig. 9.20.

The bolt is modeled with linear finite elements with elastic properties: $E = 2.1 \cdot 10^4$, $\nu = 0.3$. Contact is modeled with the point-to-analytical surface contact element and specified on the cylindrical surface of the bolt, i.e. each node on the bolt surface is a contact node. The important dimensions of the example are the radius of a cylinder $R = 3.0$ and the distance H between threads of a spiral line $H = 3.3333$. The central axis is constrained to move along the OZ-axis and a rotation with an angle increment $\Delta\varphi = 1^o$ is applied to the head of the bolt. In order to

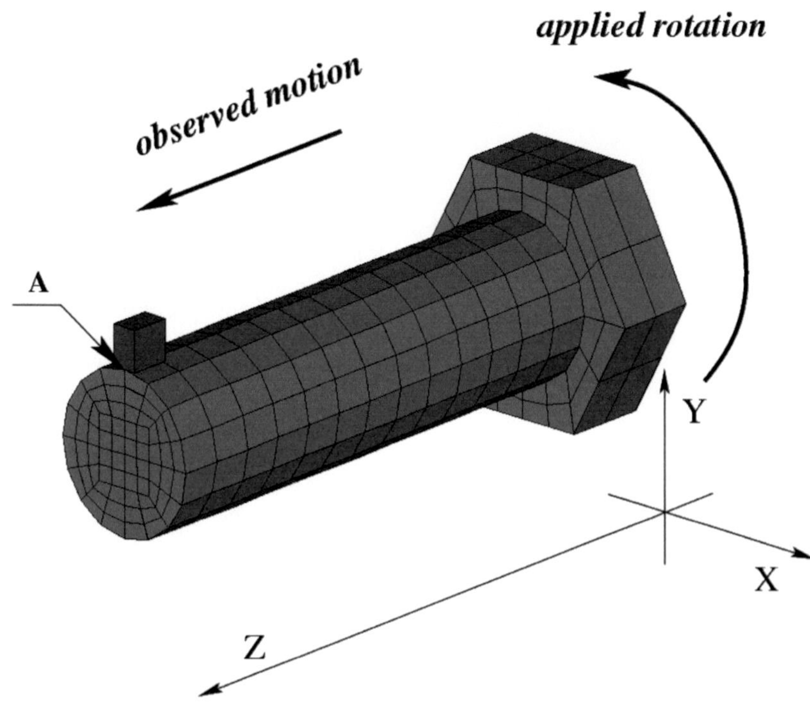

Figure 9.20: Finite element model of a bolt connection. Outer surface is rigid and is illustrated with one element only.

supply contact with a rigid external cylindrical surface an initial penetration is specified as $1.0 \cdot 10^{-4}$ with a normal penalty parameter $\varepsilon_N = 10^3$. The friction tensor is chosen to be isotropic with $\mu_1 = \mu_2 = 0.01$. The adhesion tensor parameters $\varepsilon_1, \varepsilon_2$ are chosen according to the following tensor representation (see derivation in Part 1)

$$
\mathbf{B} = b_{ij}\boldsymbol{\rho}^i \otimes \boldsymbol{\rho}^j = -\frac{1}{R^2 + \left(\dfrac{H}{2\pi}\right)^2}
\begin{bmatrix}
g_\varepsilon R^2 & (\varepsilon_1 - \varepsilon_2)\dfrac{R^2 H}{2\pi} \\[2ex]
(\varepsilon_1 - \varepsilon_2)\dfrac{R^2 H}{2\pi} & \varepsilon_1 \left(\dfrac{H}{2\pi}\right)^2 + \varepsilon_2 R^2
\end{bmatrix}
\tag{9.42}
$$

with $g_\varepsilon = \varepsilon_1 R^2 + \varepsilon_2 \left(\dfrac{H}{2\pi}\right)^2$.

The adhesion tensor parameters are varying according to Table 9.2.

Case	Adhesion tensor	Ratio r_ε	Resulting distance h
1	$\varepsilon_1 = 0.0$ $\varepsilon_2 = 10.0$	0.0	$1.000\,H$
2	$\varepsilon_1 = 5.0$ $\varepsilon_2 = 10.0$	$1/2$	$0.742\,H$
3	$\varepsilon_1 = 7.5$ $\varepsilon_2 = 10.0$	$3/4$	$0.430\,H$
4	$\varepsilon_1 = 10.0$ $\varepsilon_2 = 10.0$	1.0 geom. isotropy	$0.000\,H$
5	$\varepsilon_1 = 10.0$ $\varepsilon_2 = 5.0$	$2/1$	$-0.487\,H$
6	$\varepsilon_1 = 10.0$ $\varepsilon_2 = 7.5$	$4/3$	$-2.758\,H$
7	$\varepsilon_1 = 10.0$ $\varepsilon_2 = 0.0$	∞	$-31.98\,H = \hat{H}$

Table 9.2: Rotation of a bolt. Variation of parameters of the adhesion tensor. Resulting distance of a longitudinal motion h after rotation of a bolt by 360°.

The results of the analyses are depicted in Fig. 9.21 showing the computed longitudinal motion of the bolt along the OZ-axis vs. the applied rotation angle. The resulting distance of a longitudinal motion h after the rotation of a bolt by 360° is also presented in Table 9.2. Case (1) with ratio $r_\varepsilon = 0$ leads to a pure forward motion according to the motion along the main spiral line, see Part 1. The bolt moves forward at the distance H while the bolt is rotating at the full angle 360°. The geometrically isotropic case (4) does not lead to any longitudinal motion. For ratios $r_\varepsilon > 1$ a backward motion is obtained. Finally, the case (7) with ratio $r_\varepsilon = \infty$ (not shown in Fig. 9.21) leads to a motion along the orthogonal spiral, i.e. the bolt moves backwards at the distance $\hat{H} = \dfrac{(2\pi R)^2}{H}$, see **Remark 4** Sect. 3 in Part 1.

Remark.

This example is only chosen to illustrate the possibility to describe machined surface from a kinematical point of view. The applicability of the proposed model to stress analysis of the bolt connection requires, certainly, a more sophisticated analysis.

Remark.

The presented technique has been implemented into the FEAP-MeKa code [172]. The unsymmetric solver is based on a standard LU-decomposition combined with an iterative Newton scheme.

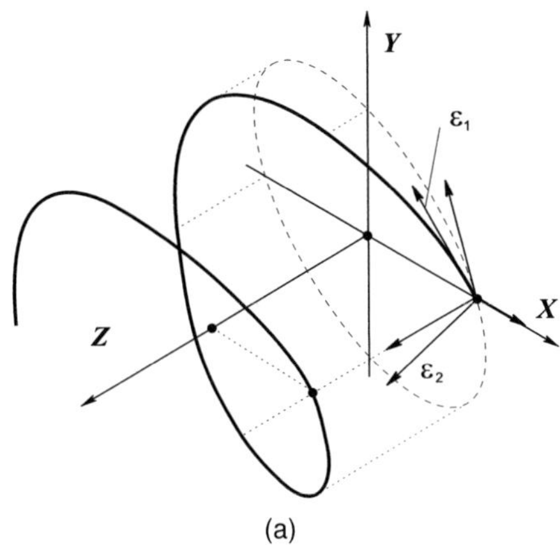

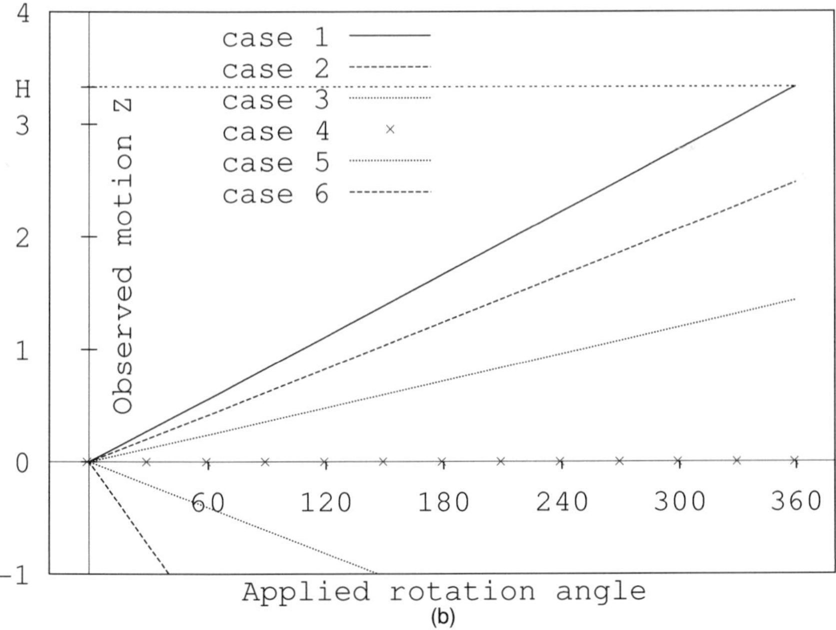

Figure 9.21: a) Spiral orthotropy on a cylinder. b) Observed longitudinal motion vs. applied rotation. Variation of ratio $r_\varepsilon = \varepsilon_1/\varepsilon_2$. Loading cases according to Table 9.2.

9.5 Conclusions

In the current part the numerical analysis of the model of contact interfaces including anisotropy for adhesion and friction developed in Part 1 is presented. The linearization necessary for the numerical iterative solution is considered in a covariant form in the metrics of a tangent plane. The covariant form makes the application of the derived scheme to contact problems with arbitrary curved surfaces possible. A numerical implementation into finite element codes is also considered leading to a family of contact elements based on a) a point-to-analytical approach for both linear and curvilinear approximations of the contact body and b) a node-to-segment approach. In addition, a rheological model based on a simple mechanical interpretation of the model is discussed.

The proposed model for contact interfaces is analyzed for constant orthotropy on a plane, for polar orthotropy on a plane and for spiral orthotropy on a cylinder. The reliability of the numerical results is controlled by checking the convergence with increasing the adhesion parameters and by remeshing. A classification of orthotropy based on the ratio of eigenvalues for the corresponding orthotropic tensor is proposed and the kinematical effects for the classified cases are numerically investigated. In particular, as a specific case of complex anisotropy the geometrically isotropic kinematic behavior of the contact bodies can be found despite the presence of both, anisotropy for adhesion and friction. The possibility of modeling machined surfaces with the help of the adhesion tensor is shown by numerical examples for polar orthotropy and for the model of a bolt connection.

The presented approach shows an algorithmic inclusion in covariant form only for linear elastic adhesion and for an associated Coloumb's friction law. Any combination of elastic, visco-elastic or nonlinear laws for the adhesion region together with non-associative or associative friction laws can be derived in straightforward covariant manner based on the metrics of the contact surfaces.

9.6 Acknowledgments

We thank the DFG for the support given by grant SCHW 307/18-2. We also thank Prof. Vielsack, Institut für Mechanik, Universität Karlsruhe for many creative discussions, in particular, concerning the rheological model for the contact interfaces including adhesion and friction.

9.7 APPENDIX. Recovering a circular motion for polar orthotropy

A family of curves in the case of polar orthotropy with the parameter $\varepsilon_\varphi = 0$ for the problem considered in Sect. 9.4.3 is a family of circles.

In order to prove this statement we consider the equilibrium equations for the quasi-statical sliding of a block, see Fig. 9.18, Sect. 9.4.3.

$$\begin{aligned} T_1 &= 0 \\ T_2 + F &= 0. \end{aligned} \tag{9.43}$$

Here a force F is associated with applied displacements along OY axis and T_1, T_2 are components of a sliding force $\mathbf{T}^{sl}$ defined in eqn. (9.12). Following the incremental displacement loading process and the definition of the sliding force via the trial force, see also eqn. (9.14) we can write:

$$\mathbf{T}^{sl} = -|N|\frac{\mathbf{BFB}}{\Psi}\left\{ \begin{array}{c} \Delta u \\ \Delta v \end{array} \right\}, \tag{9.44}$$

where Δu is a prescribed displacement component and Δv is a computed component from the equilibrium equation. Then from the first equilibrium equation (9.43) we obtain:

$$\mathbf{BFB}_{11}\Delta u + \mathbf{BFB}_{12}\Delta v = 0. \tag{9.45}$$

The limit of the ratio of the displacement increments when Δu goes to zero leads to the exact definition of the derivative:

$$\lim_{\Delta u \to 0}\frac{\Delta v}{\Delta u} = y' = -\frac{\mathbf{BFB}_{11}}{\mathbf{BFB}_{12}}. \tag{9.46}$$

Now, we specify the particular case with an isotropic friction tensor $\mathbf{F} = \mathbf{E}/\mu^2$ and $\varepsilon_\varphi = 0$ in the adhesion tensor $\mathbf{B}$ in eqn. (9.41) leading to

$$\mathbf{B} = -\frac{\varepsilon_r}{x^2 + y^2} \begin{bmatrix} x^2 & xy \\ xy & y^2 \end{bmatrix}. \tag{9.47}$$

Inserting the corresponding components from eqn. (9.47) into eqn. (9.46) we obtain an ordinary differential equation (ODE) describing a family of curves (trajectory of the ODE):

$$y' = -\frac{x}{y}. \tag{9.48}$$

The integration of this differential equation leads to

$$y^2 + x^2 = const, \tag{9.49}$$

which describes a family of circles.

10

Symmetrization of various friction models based on an Augmented Lagrangian approach[*]

Abstract

The standard implementation of the classical Coulomb friction model together with the Newton iterative method for the finite element method leads to non-symmetric tangent matrices for sliding zones of contact surfaces. This fact is known in literature as consequence of the non-associativity of the friction law. Considering anisotropic models for friction, especially including coupling of adhesion and friction, leads to additional non-symmetries due to anisotropy. Since, non-symmetry of matrices is a non-desirable feature of most engineering problems, various proposals for symmetrization are known in computational mechanics. A further suggestion is made in this contribution. The covariant approach for both isotropic and anisotropic frictional contact problems leads to a very simple structure of the tangent matrices. This allows to obtain very robust tangent matrices within the symmetrized Augmented Lagrangian method. In the current contribution, the nested Uzawa algorithm is applied for symmetrization within the Augmented Lagrangian approach for an anisotropic friction model including adhesion and friction. The numerical examples show the good convergence behavior for various problems such as small and large sliding problems.

---

[*]The chapter is published in [94]: A. Konyukhov, K. Schweizerhof *Symmetrization of various friction models based on an Augmented Lagrangian approach*, In *IUTAM Symposium on Computational Contact Mechanics*, U.Nackenhorst, P. Wriggers eds., IUTAM Bookseries. Springer, pp. 97–111, 2007.

10.1 Introduction

The penalty method for frictional contact problems [188], [106] currently is among the most popular schemes in finite element packages, leads to the satisfaction of the contact constraints, such as non-penetration and sticking conditions, only within a certain tolerance. This tolerance is defined by the penalty parameters for both normal and tangential direction. As is known, the classical method of Lagrangian multipliers leads to an exact satisfaction of contact constraints, however, one should take care of the number of multipliers due to the often overstiff behavior of contact interfaces, see e.g. in [57]. Additional degrees of freedoms for the contact tractions are often mentioned among the disadvantages of this method. Recently, various combinations of the Mortar method have been developed in e.g. [150], [151], [40] allowing to overcome overstiff behavior especially in the case where contact tractions are computed point-wise at integration points. In addition, good results for the patch-test have been shown also in [57] exploiting the, so-called, segment-to-segment approach, coinciding with the Mortar method with penalty descriptions of the contact traction. So-called dual Lagrange multipliers have been developed for non-frictional contact in Wohlmuth [186] allowing to condense degrees of freedom for contact traction. Using this approach, the frictional constraints should be carefully treated as a sequence of the Tresca type friction model, see [69]. This approach is similar to the Augmented Lagrangian approach allowing to satisfy the contact constraints for non-penetration and sticking within the nested algorithm. The method is described theoretically in Bertsekas [12] and Fortin and Glowinski [42]. Pietrzak and Curnier [144] developed the Augmented Lagrangian approach for frictional contact including the corresponding saddle point functional. Laursen and Simo [109] proposed a symmetrization procedure based on the nested Uzawa algorithm. This approach is based on symmetrization of the corresponding tangent matrices for the solution of the equilibrium equations, while the contact constraints have been still iteratively satisfied within the external loop with a specified tolerance.

In the current contribution, we will describe the symmetric Augmented Lagrangian method for the coupled anisotropic contact model including anisotropy for friction and adhesion as developed in Konyukhov and Schweizerhof [90], [91]. The geometrical structure of the corresponding tangent matrices allows us to construct very simple symmetric tangent matrices for the anisotropic case. The isotropic case can then be defined as a reduction of the anisotropic case. Good convergence rates are illustrated in the numerical examples.

10.2 Covariant description of the coupled anisotropic friction model

We shortly present here the main details of the coupled interface contact model including anisotropy for adhesion and friction and refer to [90], [91] for further details and corresponding derivations.

At the beginning, a local surface coordinate system is introduced as

$$\mathbf{r}_\mathrm{s}(\xi^1, \xi^2, \xi^3) = \boldsymbol{\rho}(\xi^1, \xi^2) + \mathbf{n}\xi^3, \tag{10.1}$$

where ξ^1, ξ^2 are two convective coordinates and responsible for the tangential contact interaction. The third coordinate ξ^3 is the value of the penetration and is used to define the properties of the normal interaction

$$\xi^3 = (\mathbf{r}_\mathrm{s} - \boldsymbol{\rho}) \cdot \mathbf{n}. \tag{10.2}$$

The vector of contact tractions $\mathbf{R}$ is defined as a covariant vector and, therefore, is expressed via the contravariant basis vectors $\boldsymbol{\rho}^i$ and $\mathbf{n}$ in the coordinate system (10.1) as sum of the tangential and normal components

$$\mathbf{R} = \mathbf{T} + \mathbf{N} = T_i\boldsymbol{\rho}^i + N\mathbf{n}. \tag{10.3}$$

For the generalization into anisotropy for adhesion, the tangential traction vector $T_i\boldsymbol{\rho}^i$ and the normal traction vector $N\mathbf{n}$ are assumed to be decoupled. Therefore, the generalized constitutive equations for tangential tractions are taken in rate form as

$$\frac{d\mathbf{T}}{dt} = \mathbf{B}(\mathbf{v}_\mathrm{s} - \mathbf{v}), \tag{10.4}$$

where B is the anisotropic adhesion tensor. The covariant differentiation operations are involved throughout whenever the rate form is given.

The constitutive equation for normal traction is given in closed form and can be viewed as a simple penalty regularization procedure for this normal traction N

$$N = \epsilon_N \xi^3, \tag{10.5}$$

where ϵ_N is a parameter of normal compliance, or a penalty parameter.

The anisotropy for friction is chosen as a model of Coulomb type involving proportionality of the frictional force to the normal traction N. The corresponding yield function is written as

$$\Phi = \sqrt{f^{ij}T_iT_j} - |N| = \sqrt{\mathbf{T} \cdot \mathbf{FT}} - |N|, \tag{10.6}$$

where $\mathbf{F} = f^{ij}\boldsymbol{\rho}_i\boldsymbol{\rho}_j$ is the anisotropic friction tensor.

10.2.1 Incremental formulation of the coupled anisotropic model

Though, initially the model is formulated in the continuous rate form, it has been transformed into incremental form for the final computational model via the application of the backward Euler scheme.

i) The full displacement vector $\Delta\boldsymbol{\xi} = \boldsymbol{\xi}^{(n+1)} - \boldsymbol{\xi}^{(n)}$ is decomposed additively into an elastic increment $\Delta\boldsymbol{\xi}^{el}$ and into a sliding increment $\Delta\boldsymbol{\xi}^{sl}$:

$$\Delta\boldsymbol{\xi} = \Delta\boldsymbol{\xi}^{el} + \Delta\boldsymbol{\xi}^{sl}, \tag{10.7}$$

where both vectors are defined in the surface metrics, namely,

$$\Delta\boldsymbol{\xi} := \Delta\xi^i\boldsymbol{\rho}_i = (\xi^i_{(n+1)} - \xi^i_{(n)})\boldsymbol{\rho}_i. \tag{10.8}$$

ii) The trial elastic, or adhesion force $\mathbf{T}^{tr}_{(n+1)}$ is computed via the incremental evolution equations:

$$\mathbf{T}^{tr}_{(n+1)} - \mathbf{T}_{(n)} = \mathbf{B}^{(n+1)}(\boldsymbol{\xi}^{el}_{(n+1)} - \boldsymbol{\xi}^{el}_{(n)}). \tag{10.9}$$

iii) The final result if the tangential traction $\mathbf{T}$ is the elastic one (belongs to the adhesion region), or the plastic one if provided by the yield condition of the Coulomb type in each load step becomes:

$$
\begin{aligned}
\Phi^{tr} : &= \sqrt{\mathbf{T}^{tr}_{(n+1)} \cdot \mathbf{F}^{(n+1)} \mathbf{T}^{tr}_{(n+1)}} - |N_{(n+1)}| \\
&= \sqrt{f^{ij} T^{tr}_{i\,(n+1)} T^{tr}_{j\,(n+1)}} - |N_{(n+1)}|,
\end{aligned}
\qquad (10.10)
$$

- If $\Phi^{tr} < 0$ then the trial force is a real sticking force $\mathbf{T} = \mathbf{T}^{tr}$.

- If $\Phi^{tr} \geq 0$ then the sliding force must be obtained via the maximum of the energy dissipation function given in the incremental form.

iv) All contact parameters such as sliding traction and sliding distance should be derived via the principle of the maximum dissipation

$$
D^{(n+1)} := \Delta\boldsymbol{\xi}^{sl} \cdot \mathbf{T}^{sl}_{(n+1)} = \Delta\xi^i_{sl} T^{sl}_{i\,(n+1)}, \quad D^{(n+1)} \longrightarrow \max. \qquad (10.11)
$$

Using the necessary optimization conditions for the functional $D^{(n+1)}$ together with the Kuhn-Tucker conditions from *iii)* the closed form for the sliding force $\mathbf{T}^{sl}$ is obtained as:

$$
\mathbf{T}^{sl} = -\frac{\mathbf{BFT}^{tr}}{\sqrt{\mathbf{BFT}^{tr} \cdot \mathbf{FBFT}^{tr}}} |N|. \qquad (10.12)
$$

10.3 Linearization process and structure of matrices

Since the frictional problem is nonlinear, an iterative solution based on a Newton scheme should be applied. The important part of the implementation is then a consistent tangent matrix which differs for sticking and sliding according to the return-mapping scheme. We present here the results with particular focus on the structure of the matrices especially their symmetry. According to numerical experiences reported in [86] and [89]. we will also exclude the curvature parts of the matrices.

10.3.1 Linearization of the normal part δW_c^N

We denote $D(f)$ as a linearization operator acting on a functional f in the covariant form. Thus, linearization of the virtual work of the contact normal traction N is given as:

$$D(\delta W_c^N) = D(\int_s N\delta\xi^3 ds) = -\underbrace{\int_s \epsilon_N(\delta\mathbf{r}_s - \delta\boldsymbol{\rho}) \cdot (\mathbf{n} \otimes \mathbf{n})(\mathbf{v}_s - \mathbf{v})ds}_{\text{leads to } \mathbf{K}_N^m} \quad (10.13a)$$

$$-\underbrace{\int_s \epsilon_N\xi^3 \left(\delta\boldsymbol{\rho}_{,j} \cdot a^{ij}(\mathbf{n} \otimes \boldsymbol{\rho}_i)(\mathbf{v}_s - \mathbf{v}) + (\delta\mathbf{r}_s - \delta\boldsymbol{\rho}) \cdot a^{ij}(\boldsymbol{\rho}_j \otimes \mathbf{n})\mathbf{v}_{,i}\right) ds,}_{\text{leads to } \mathbf{K}_N^r} \quad (10.13b)$$

where a^{ij} are contravariant components of the metric tensor for the master surface and $\boldsymbol{\rho}_j = \frac{\partial\boldsymbol{\rho}}{\partial\xi^j}$, $j = 1,2$ are covariant basis (tangent) vectors. Here, the first term (resp. the second term) after approximation of the geometry leads to the main part of the contact matrix $\mathbf{K}_N^m$ (resp. the rotational part of the contact matrix $\mathbf{K}_N^r$).

All parts can be algorithmically implemented for any order of approximation. In order to do this only the operator $\mathbf{A}$ for the approximation of the surface geometry has to be introduced. The derivatives with respect to convective coordinates $\mathbf{A}_\xi$ are also necessary. Thus, the approximation can be written as

$$\mathbf{r}_s - \boldsymbol{\rho} = \mathbf{A}\{\mathbf{x}\}, \quad \boldsymbol{\rho}_\xi = \mathbf{A}_\xi\{\mathbf{x}\} \quad (10.14)$$

where $\mathbf{x}$ is a nodal vector for the standard FE implementation, or a control points (knots) vector for a CAD approximation. This leads e.g. to the following structure of the contact matrix $\mathbf{K}_N^m$

$$\mathbf{K}_N^m = -\int_s \epsilon_N\mathbf{A}^T \cdot (\mathbf{n} \otimes \mathbf{n})\mathbf{A}ds, \quad (10.15)$$

where the integral is computed via the set of Gauss points defined on the slave segment and penetrating into the master surface (so-called penalty based Mortar method, see Fischer and Wriggers [40]).

10.3.2 Linearization of the tangential part δW_c^T, sticking case

The sticking case is understood as a case where the tangential traction remains in the elastic region and, therefore, is computed via the evolution equations *ii)*. Linearization leads to

$$D(\delta W_c^T) = D(\int_s T_i^{tr} \delta\xi^i ds) = \tag{10.16}$$

$$= \underbrace{\int_s (\delta\mathbf{r}_s - \delta\boldsymbol{\rho}) \cdot \mathbf{B}(\mathbf{v}_s - \mathbf{v})ds-}_{\text{leads to } \kappa_{T,st}^m}$$

$$\underbrace{- \int_s T_i \left((\delta\mathbf{r}_s - \delta\boldsymbol{\rho}) \cdot a^{il}a^{jk}\,\boldsymbol{\rho}_k \otimes \boldsymbol{\rho}_l\,\mathbf{v}_j + \delta\boldsymbol{\rho}_{,j} \cdot a^{ik}a^{jl}\,\boldsymbol{\rho}_k \otimes \boldsymbol{\rho}_l\,(\mathbf{v}_s - \mathbf{v}) \right) ds.}_{\text{leads to } \kappa_{T,st}^r}$$

10.3.3 Linearization of the tangential part δW_c^T, sliding case

The sliding case is understood as a case where the yield condition *iii)* is fulfilled and, therefore, the tangential traction is computed as a sliding traction given in eqn. (10.12). The corresponding linearization leads to

$$D(\delta W_c^T) = D(\int_s T_i^{sl} \delta\xi^i ds) = \tag{10.17}$$

$$= \underbrace{\int_s \left((\delta\mathbf{r}_s - \delta\boldsymbol{\rho}) \cdot \frac{\epsilon_N \mathbf{BFT}^{tr} \otimes \mathbf{n}}{\sqrt{\mathbf{BFT}^{tr} \cdot \mathbf{FBFT}^{tr}}}(\mathbf{v}_s - \mathbf{v}) \right) ds}_{\text{leads to } \kappa_{T,sl}^{m,1}}$$

$$\underbrace{- \int_s \left((\delta\mathbf{r}_s - \delta\boldsymbol{\rho}) \cdot \frac{|N|\,\mathbf{BFB}}{\sqrt{\mathbf{BFT}^{tr} \cdot \mathbf{FBFT}^{tr}}}(\mathbf{v}_s - \mathbf{v}) \right) ds}_{\text{leads to } \kappa_{T,sl}^{m,2}}$$

$$\underbrace{+ \int_s \left((\delta\mathbf{r}_s - \delta\boldsymbol{\rho}) \cdot \frac{|N|\,\mathbf{BFT}^{tr} \otimes (\mathbf{BFB})^T \mathbf{FBFT}^{tr}}{\sqrt{(\mathbf{BFT}^{tr} \cdot \mathbf{FBFT}^{tr})^3}}(\mathbf{v}_s - \mathbf{v}) \right) ds}_{\text{leads to } \kappa_{T,sl}^{m,3}}$$

$$- \underbrace{\int_s T_i^{sl} \left[(\delta\mathbf{r}_s - \delta\boldsymbol{\rho}) \cdot a^{il} a^{jk} \, \boldsymbol{\rho}_k \otimes \boldsymbol{\rho}_l \, \mathbf{v}_j + \delta\boldsymbol{\rho}_{,j} \cdot a^{ik} a^{jl} \, \boldsymbol{\rho}_k \otimes \boldsymbol{\rho}_l \, (\mathbf{v}_s - \mathbf{v}) \right] ds.}_{\text{leads to } \kappa^r_{T,sl}}$$

Now we can summarize the results concerning the symmetry of the necessary tangent matrices. As expected, all parts concerning non-frictional and sticking contact, namely $\mathbf{K}^m_N$, $\mathbf{K}^r_N$, $\mathbf{K}^m_{T,st}$ and $\mathbf{K}^r_{T,st}$ are symmetric. Attention should be paid to the sliding tangent matrix, because it contains both, symmetric and nonsymmetric parts:

$\mathbf{K}^{m,1}_{T,sl}$ is the first main part due to the coupling of the normal and the sliding tractions. It appears due to linearization of the normal traction N and it is nonsymmetric.

$\mathbf{K}^{m,2}_{T,sl}$ is the second main part. It appears due to linearization of the normal trial tangential traction $\mathbf{T}^{tr}$ and preserves symmetry.

$\mathbf{K}^{m,3}_{T,sl}$ is the third main part and nonsymmetric. It appears due to linearization of the complex term $\dfrac{1}{\sqrt{\mathbf{BFT}^{tr} \cdot \mathbf{FBFT}^{tr}}}$ reflecting the coupling of anisotropy for adhesion and friction.

Finally, $\mathbf{K}^r_{T,sl}$ is the rotational, symmetric part reflecting rotation of the master segment, for more details see [92].

10.4 Augmented Lagrangian method and symmetric Uzawa algorithm

The main advantage of the Augmented Lagrangian in comparison to the penalty method is the possibility to select a value of the penalty parameter leading to well conditioned tangent matrices together with the enforcement of the constraint conditions (non penetration and sticking) within a specified tolerance. The method is constructed as a nested algorithm known as Uzawa algorithm and possesses linear convergence. The algorithm using the full consistent tangent matrix is known as exact Uzawa algorithm, while the inexact algorithm is exploiting a somehow simplified matrix, see results on convergence in Stadler [166]. Laursen and Simo [109] developed a symmetrized Uzawa algorithm, where symmetric matrices for the sliding case have been obtained under the assumption that the normal traction remains constant for the solution of the equilibrium equations within an internal loop. This preserves the quadratic rate of convergence for the internal loop. The correct value is enforced then within the external loop where a linear rate of convergence is preserved.

10.4.1 Limitations of the Augmented Lagrangian approach for the coupled anisotropic model

The complexity for a generalization of the discussed algorithm into anisotropic friction especially including anisotropic adhesion is that the sliding tangent matrix is fully nonsymmetric, moreover, the assumption of isotropy together with the constant normal traction would lead only to the rotational part of the tangent matrix. Since, this part has no influence for small displacement problems as well as in the case when the master segment has no rotation, we obtain only zero matrices. This makes it impossible to obtain any solution. Fortunately, the covariant approach allows to estimate the influence of the matrices part by part. Thus, we can construct a symmetric algorithm and analyze it numerically.

Another problem for the coupled anisotropy is that the sticking case is defined when the slave contact point lays inside the elliptic adhesion domain, see Fig. 10.1. However, as is known from numerical results the ratio $\frac{a}{b} = \frac{\mu_1 \varepsilon_2}{\mu_2 \varepsilon_1}$ does not influence the convergence result in the penalty based approach and can lead to correct kinematics even for large sliding problems. An enforcement to put a slave contact point inside the adhesion ellipse via the Augmented Lagrangian approach necessarily leads to dis-convergence in cases as $\frac{a}{b} \to 0$, or $\frac{a}{b} \to \infty$. In computation due to the linear convergence of the Augmented Lagrangian method the global number of iteration is proportional to the ratio $\frac{a}{b}$ for $a > b$, e.g. a simply computable case with $\frac{a}{b} = 10$ would lead to a 10-times increase of the global number of iterations for the Augmented Lagrangian method. Thus, one should judge the coupled anisotropic model as an interface model for tangential traction, rather than a penalty based approach.

Summarizing the discussion, we can define *computable cases* for the Augmented Lagrangian method in application to the coupled anisotropic model for:

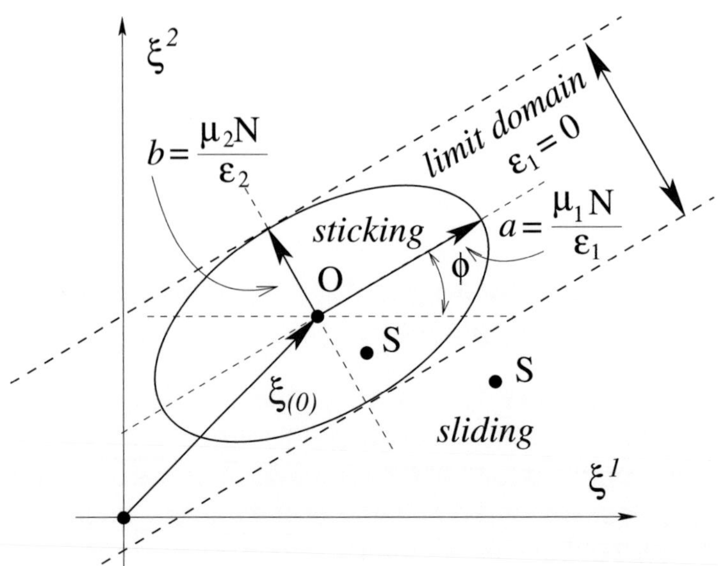

Figure 10.1: Allowable elastic region (adhesion domain).

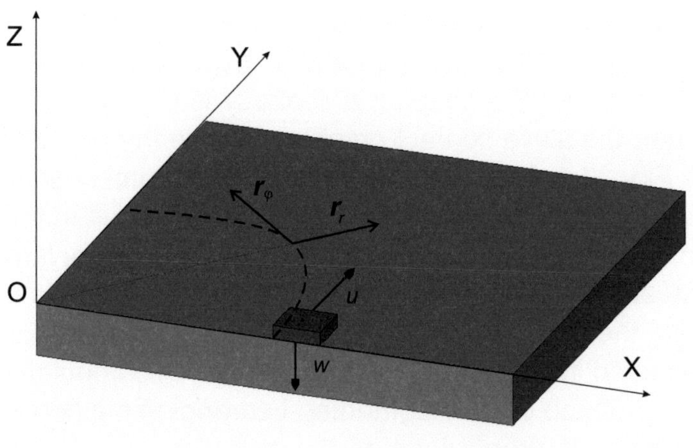

Figure 10.2: Geometry and loading for the case of polar orthotropy.

1. *Small sliding problems with low anisotropy.* Small sliding can be numerically defined as a case where it is important to compute the

distribution of the sticking-sliding zone. The low anisotropy is defined then as a case with $\frac{a}{b} \approx 1$.

In this case, Lagrangian multipliers for both normal N, and tangential traction $\mathbf{T}$ are augmented. The convergence for tangential displacements is checked by the proximity to the initial contact point:

$$\sqrt{a_{ij}(\xi^i_{(I)} - \xi^i_{(K-1)})(\xi^j_{(I)} - \xi^j_{(K-1)})} \leq \epsilon_T, \qquad (10.18)$$

where ϵ_T is a specified tolerance for tangential displacements. The convergence for normal displacements is checked as in the standard isotropic case by the normal penetration

$$|\xi^3| \leq \epsilon_N, \qquad (10.19)$$

where ϵ_N is a specified tolerance for the normal penetration.

2. *Large sliding problem with arbitrary anisotropy.* Large sliding can be numerically defined as a case where sliding is reached within a single load step. In this case, the global kinematical behavior of a contacting body is of interest rather than the distribution of the sticking-sliding zone.

   In this case, the Lagrangian multiplier only for the normal traction N needs to be updated because the tolerance of the sticking condition will normally not influence the tolerance of the computed trajectory. Thus, convergence is checked only for the normal displacement.

The symmetrized algorithm in accordance with the inexact Uzawa approach which appears to be numerically effective for the current anisotropic problem is constructed as follows. The internal loop, No. 3 in Table 10.1 and 10.2, serves for the solution of the equilibrium equations. The normal traction $N^{K,L,I}$ is computed via the augmented scheme and updated inside loop 2. The trial tangential traction $\mathbf{T}^{K,L,I}$ for *small sliding problems* is computed via the augmented scheme. The update according to the scheme $\Delta\lambda_T^L = \Delta\lambda_T^{L-1} + \mathbf{B}(\boldsymbol{\xi}_{(I)} - \boldsymbol{\xi}_{(K-1)})$ inside loop 2 allows to enforce sticking conditions similar to the normal penetration. The modification for *large sliding problems* is as follows: the sticking condition is satisfied in the penalty form, then the trial tangential traction $\mathbf{T}^{K,L,I}$ is computed as $\mathbf{T}^{K,L,I} = \mathbf{T}^{K-1,L,I} + \mathbf{B}(\boldsymbol{\xi}_{(I)} - \boldsymbol{\xi}_{(K-1)})$. The Lagrange multiplier

Table 10.1: **Update scheme for normal traction** N

1. Loop over applied in load incremental steps: $K = K + 1$
     initialization of Lagrange multiplier $\lambda_N^0 = 0$

> 2. Loop over augmented multipliers: $L = L + 1$
>
> > 3. Iterative solution of
> >   global equilibrium equations:
> >   with normal force
> >
> > $$N^{K,L,I} = \lambda_N^{L-1} + \varepsilon_N \xi_I^3$$
> >
> > penetration ξ_I^3 is computed in each iteration I.
>
> Update multipliers
> $$\lambda_N^L = N^{K,L,I}$$
>
> Convergence is checked by the non-penetration condition:
>
> $$|\xi_I^3| \le \epsilon_N, \ for \ N^{K,L,I} < 0,$$
>
> where ϵ_N - specified tolerance for normal displacements

for the tangential traction vector $\Delta\lambda_T^{L-1}$ is not introduced and, therefore, an update loop 2 does not exist.

An important modification for symmetrization should be done for the return-mapping scheme, see Table 10.3, where the sliding force instead of eqn. (10.12) is computed via the augmented multiplier for the normal traction λ_N^{L-1} as

$$\mathbf{T}^{sl} = -\frac{\mathbf{BFT}^{tr}}{\sqrt{\mathbf{BFT}^{tr} \cdot \mathbf{FBFT}^{tr}}}|\lambda_N^{L-1}|. \tag{10.20}$$

The yield function (10.6) is respectively modified as

$$\Phi_\lambda = \sqrt{\mathbf{T}_{tr}^{K,L,I} \cdot \mathbf{FT}_{tr}^{K,L,I}} - |\lambda_N^{L-1}|. \tag{10.21}$$

This leads to a constant sliding force for the internal loop and, therefore, the first main part of the sliding tangent matrix is zero $\mathbf{K}_{T,sl}^{m,1} = 0$.

Table 10.2: **Update scheme for trial tangential traction** $\mathbf{T}$

> 1. Loop over load applied in incremental steps: $K = K + 1$
>   initial condition for tangential traction $\mathbf{T}^{0,0,0} = 0$
>   initialization of Lagrange multiplier $\Delta\lambda_T^0 = 0$
>
> > 2. Loop over augmented multipliers: $L = L + 1$
> >
> > > 3. Iterative solution of
> > >   global equilibrium equations:
> > >   with tangential traction
> > >
> > > $$\mathbf{T}^{K,L,I} = \mathbf{T}^{K-1,L,I} + \Delta\lambda_T^{L-1} + \mathbf{B}(\boldsymbol{\xi}_{(I)} - \boldsymbol{\xi}_{(K-1)})$$
> > >
> > > $\boldsymbol{\xi}_{(I)}$ is a projection point in each iteration I,
> > > $\boldsymbol{\xi}_{(K-1)}$ is a projection point in the previous load step.
> >
> > Update multipliers
> >
> > $$\Delta\lambda_T^L = \Delta\lambda_T^{L-1} + \mathbf{B}(\boldsymbol{\xi}_{(I)} - \boldsymbol{\xi}_{(K-1)})$$
> >
> > Convergence is checked by proximity
> > to the initial sticking point:
> >
> > $$\sqrt{a_{ij}(\xi_{(I)}^i - \xi_{(K-1)}^i)(\xi_{(I)}^j - \xi_{(K-1)}^j)} \leq \epsilon_T,$$
> >
> > where ϵ_T - specified tolerance for tangential displacements

Then, the full tangent matrix for the sliding case becomes:

$$\mathbf{K}_{sl}^{full} = \mathbf{K}_N^m + \mathbf{K}_N^r + \mathbf{K}_{T,sl}^{m,2} + \mathbf{K}_{T,sl}^{m,3} + \mathbf{K}_{T,sl}^r. \tag{10.22}$$

The part $\mathbf{K}_{T,sl}^{m,3}$ is still nonsymmetric due to anisotropy. The matrix is fully symmetric only for isotropic friction. As it was found in numerical computations we can exclude this part with only a small loss of efficiency. This finally leads to the following matrix in the sliding case according to the inexact Uzawa algorithm:

$$\mathbf{K}_{sl}^{full} = \mathbf{K}_N^m + \mathbf{K}_N^r + \mathbf{K}_{T,sl}^{m,2} + \mathbf{K}_{T,sl}^r. \tag{10.23}$$

375

The tangent matrix for sticking remains symmetric:

$$\mathbf{K}_{st}^{full} = \mathbf{K}_N^m + \mathbf{K}_N^r + \mathbf{K}_{T,st}^m + \mathbf{K}_{T,st}^r.$$

(10.24)

Table 10.3: **Return-mapping scheme for
the symmetric Augmented Lagrangian method**

<table>
<tr><td>

1. Loop over load applied in incremental steps: $K = K + 1$
   initial condition for tangential traction $\mathbf{T}^{0,0,0} = 0$
   initialization of Lagrange multiplier $\Delta\lambda_T^0 = 0$
   initialization of Lagrange multiplier $\Delta\lambda_N^0 = 0$

<table>
<tr><td>

2. Loop over load augmented multipliers: $L = L + 1$

<table>
<tr><td>

3. Iterative solution of
   global equilibrium equations:
   a) compute trial tangential traction $\mathbf{T}_{tr}^{K,L,I}$ (Table 10.2)

   b) compute trial yield function Φ_λ

$$\Phi_\lambda = \sqrt{\mathbf{T}_{tr}^{K,L,I} \cdot \mathbf{F}\mathbf{T}_{tr}^{K,L,I}} - |\lambda_N^{L-1}|.$$

   c) return-mapping: real tangential traction $\mathbf{T}^{K,L,I}$

$$\mathbf{T}^{K,L,I} = \begin{cases} \mathbf{T}_{tr}^{K,L,I} & if\ \Phi_\lambda < 0 \\[2em] \mathbf{T}^{sl} = -\dfrac{\hat{\mathbf{T}}}{\sqrt{\hat{\mathbf{T}}\cdot\mathbf{F}\hat{\mathbf{T}}}}|\lambda_N^{L-1}| & if\ \Phi_\lambda \geq 0 \end{cases}$$

where $\hat{\mathbf{T}} = \mathbf{B}\mathbf{F}\mathbf{T}_{tr}^{K,L,I}$

</td></tr>
</table>

</td></tr>
</table>

</td></tr>
</table>

10.5 Numerical examples

Two cases with different anisotropy have been selected to illustrate the convergence for the proposed approach: constant orthotropy and polar orthotropy on the plane.

10.5.1 Small sliding problem. Constant orthotropy

We present here an example, which has been analyzed for the penalty based approach in [91]. The rectangular block, see Fig. 10.3, is considered on an orthotropic plane. The dimensions of the block are $10 \times 10 \times 4$ with linear elastic properties: Young's modulus $E = 2.10 \cdot 10^4$ and Poisson ratio $\nu = 0.3$, assumed within a consistent dimension system.

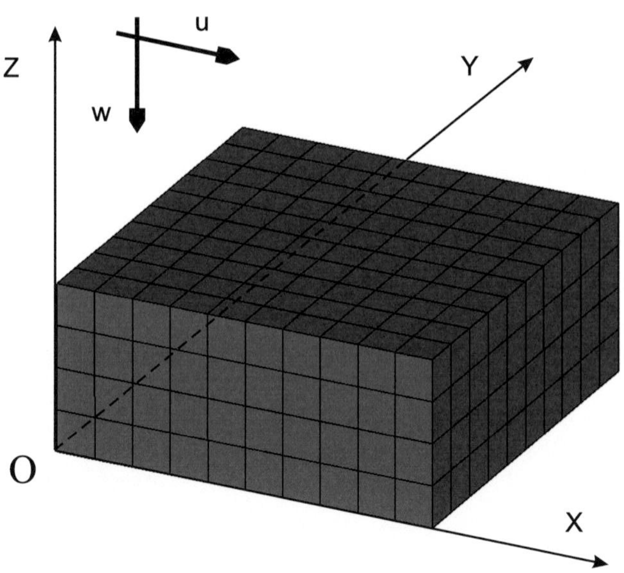

Figure 10.3: Geometry and loading of the rectangular in plane block.

The case of constant orthotropy is defined by the spectral representation of both, the adhesion tensor $\mathbf{B}$, and the friction tensor $\mathbf{F}$ as follows:

$$\mathbf{B} = [b^i_j] = - \begin{bmatrix} \varepsilon_1 \cos^2 \alpha + \varepsilon_2 \sin^2 \alpha & (\varepsilon_1 - \varepsilon_2) \sin \alpha \cos \alpha \\ (\varepsilon_1 - \varepsilon_2) \sin \alpha \cos \alpha & \varepsilon_1 \sin^2 \alpha + \varepsilon_2 \cos^2 \alpha \end{bmatrix}. \quad (10.25)$$

$$\mathbf{F} = [f_k^i] = \begin{bmatrix} \dfrac{1}{\mu_1^2}\cos^2\beta + \dfrac{1}{\mu_2^2}\sin^2\beta & (\dfrac{1}{\mu_1^2} - \dfrac{1}{\mu_2^2})\sin\beta\cos\beta \\[2em] (\dfrac{1}{\mu_1^2} - \dfrac{1}{\mu_2^2})\sin\beta\cos\beta & \dfrac{1}{\mu_1^2}\sin^2\beta + \dfrac{1}{\mu_2^2}\cos^2\beta \end{bmatrix}. \qquad (10.26)$$

The following parameters are taken for the computations: normal penalty parameter: $\varepsilon_N = 2.1 \cdot 10^4$; adhesion parameters: $\varepsilon_1 = 3.0 \cdot 10^4$, $\varepsilon_1 = 2.0 \cdot 10^4$; friction coefficients: $\mu_1 = 3.0$, $\mu_2 = 2.0$; tolerance for penetration: $\epsilon_N = 1.0 \cdot 10^{-5}$; tolerance for tangential displacement: $\epsilon_T = 1.0 \cdot 10^{-5}$; orthotropic angles: $\alpha = 45^o, \beta = 45^o$. This leads to a, so-called, geometrical isotropic case because of the ratio $\frac{\varepsilon_1\mu_2}{\varepsilon_2\mu_1} = 1.0$.

The block (see Fig. 10.3) is located on the XOY plane and loaded by vertically prescribed displacements on the upper surface $w = 1.0 \cdot 10^{-2}$. The penalty based approach gives convergence in 4 iterations, but both normal and tangential displacements inside the sticking region do not satisfy the prescribed tolerance. The results for the Augmented scheme are presented in Table 10.4. The method shows linear convergence as can be seen from the *Tolerance* column.

10.5.2 Large sliding problem. Polar orthotropy

For the case with large sliding, we consider the block on a plane with polar orthotropy see [91]. An elastic block with dimensions $1 \times 1 \times 0.25$ and mesh $4 \times 4 \times 1$ is positioned on a rigid block, see Fig. 10.2. Linear elastic material is assumed within a consistent dimension system: Young's modulus $E = 2.10 \cdot 10^4$; Poisson ratio $\nu = 0.3$. The loading is applied sequentially by prescribing displacements on the upper surface in $(1 + n)$ steps: 1) vertical loading with $w = 1.0 \cdot 10^{-2}$, 2) n steps with horizontal displacement increments $\Delta u = 1.0 \cdot 10^{-2}$ along the X axis. The frictional tensor is isotropic with $\mu_1 = \mu_2 = 0.2$, but the adhesion tensor has the following structure in the Cartesian coordinate system:

$$\mathbf{B} = -\frac{1}{x^2 + y^2}\begin{bmatrix} \varepsilon_r x^2 + \varepsilon_\varphi y^2 & (\varepsilon_r - \varepsilon_\varphi)xy \\ (\varepsilon_r - \varepsilon_\varphi)xy & \varepsilon_r y^2 + \varepsilon_\varphi x^2 \end{bmatrix}. \qquad (10.27)$$

Parameters for the adhesion tensor are chosen as $\varepsilon_r = 1000$, $\varepsilon_\varphi = 0.0$, in order to obtain the circular trajectory of the sliding block. Now only the normal tractions N are updated and a tolerance for penetration $\epsilon_N = 1.0 \cdot 10^{-5}$ is chosen. The gap for the penalty based approach is computed as $\xi^3 = 7.26 \cdot 10^{-3}$ if the normal penalty $\varepsilon_N = 2.1 \cdot 10^4$ is chosen. The total number of iterations (4556) compared with the penalty approach (1834) is influenced only by the normal penetration. The final coordinates of the nodal point $X = 0.292$, $Y = 4.996$ are compared with the results focused with the penalty approach $X = 0.306$, $Y = 4.994$ in order to show that the main kinematical effect of the anisotropic surfaces is preserved for the Augmented Lagrangian method. A comparison of both trajectories leads to a slight difference comparable to the value of the initial penetration and the initial sticking displacements.

Table 10.4: Constant orthotropy on a plane. Convergence results for the symmetric Augmented Lagrangian method.

Aug. No.	Number of eq. iter.	Tolerance
1	7	$1.8742 \cdot 10^{-3}$
2	14	$6.0419 \cdot 10^{-4}$
3	12	$1.7536 \cdot 10^{-4}$
4	12	$4.0497 \cdot 10^{-5}$
5	12	$3.5972 \cdot 10^{-6}$
	$\sum 56$	

10.6 Conclusions

In this contribution a symmetrization of the stiffness matrix has been developed within the Augmented Lagrangian method for anisotropic contact surfaces including both, anisotropy for adhesion and anisotropy for friction domains. It is shown that in general a fully coupled model necessarily leads to a fully nonsymmetric matrix in the case of sliding, but the covariant approach allows to estimate the structure of the tangent matrix part by part and, therefore, allows to construct a symmetric matrix used in accordance with the inexact Uzawa algorithm. However, as is shown, the Augmented Lagrangian method can not be directly applied to arbitrary anisotropic surfaces due to convergence problems. Thus, to create a robust algorithm, contact problems have to be subdivided into *small sliding problems* with low anisotropy where the distribution of the sticking-sliding zone is of interest, and into *large sliding problems* where the trajectory of the sliding body is of interest.

For *the small sliding problems*, both the normal, and the tangential contact tractions are augmented within the nested Uzawa algorithm. This makes it possible to enforce both normal and tangential sticking displacements to satisfy a prescribed tolerance. For *the large sliding problems* only the normal traction is augmented leading to the enforcement of only the normal displacements to satisfy a prescribe tolerance. In both cases the return-mapping scheme is exploited to obtain the real sliding tractions. Numerical examples including constant as well as non-linear orthotropy e.g. a polar orthotropy showed the effectiveness of the proposed approach.

11

On coupled models of anisotropic contact surfaces and their experimental validation[*]

Abstract

The necessity to apply a coupled contact interface model including anisotropy for both adhesion and friction is shown via a set of experiments for a rubber surface possessing a periodical waviness, and therefore, an obvious anisotropic structure. The focus of experimental investigations is placed upon the measurements of the global macro characteristics such as global forces and trajectories of a sliding block in order to validate the proposed computational model.

Keywords

anisotropy adhesion friction contact coupled model experimental verification

11.1   Introduction

Smoothness and isotropy of contacting body surfaces can vary considerably for different contact problems. Classifying the surfaces roughness two types can be distinguished: a) surfaces with randomly distributed asperities, and b) asperities with algorithmic structure, e.g. the considered surface shows different macro properties in different directions.

Mechanical characteristics for the associated contact problems of the

---

[*]The chapter is published in [99]: A. Konyukhov, P. Vielsack, K. Schweizerhof *On coupled models of anisotropic contact surfaces and their experimental validation*, Wear, **264**(7-8):579–588, 2008.

first type a) are obtained via statistically distributed asperities. Statistical analysis of a real rough surface and experimental aspects of its measurements have been developed in a series of publications: Longuet-Higgins [116], Greenwood and Williamson [50], Whitehouse and Archard [179] and more recently Whitehouse and Phillips [181] and Greenwood [48]. A comparative analysis of these surface models is presented in McCool [122]. These experimentally proved models later have been incorporated into finite element models, see e.g. Wriggers and Zavarise [196], [195], Buczkowski and Kleiber [24]. More advanced numerical analyses including homogenization methods and multi-scaled modeling are presented in Bandeira et. al. [9], [8]. Carbone and Mangialardi [26] derived contact tractions analytically for a particular example with a rigid wavy surface with a sinusoidal profile, assuming the presence of an adhesion hysteresis for 2D plane strain elasticity problem.

Constitutive modeling is applied for problems of the second type b). Such models are based on the generalization of Coulomb's friction law into the anisotropic domain. One of the first models has been proposed by Michalowski and Mroz [127] considered the sliding of a rigid block on an inclined surface. A model of orthotropic friction has been analyzed and consistently developed in Zmitrowicz [210], Curnier [32]. Various cases of anisotropy were presented in He and Curnier [62] based on the theory of tensor function representations and in Zmitrowicz [211] based on consideration of a relative sliding velocity. In the latter contribution, a classification of anisotropic surfaces based on the number of eigenvalues of the friction tensor has been proposed.

When looking at practical problems concerning contact interactions with friction between bodies made of soft rubber-like materials there are some situations in which the tangential elasticity of the contact surfaces should be taken into account. **In these cases anisotropy for elastic forces (adhesion) and frictional forces might be coupled.** Such a model including coupling of anisotropy for both friction and adhesion has been developed and analyzed numerically in Konyukhov and Schweizerhof [90], [91]. In the current contribution we discuss the validation of this model with a particular experimental test. The contact surfaces are chosen to possess visual orthotropic properties, thus a corrugated rubber mat is taken. **The results of the experiments show the necessity to**

use the coupled model including anisotropy for both friction and adhesion. Some originally surprising experimental phenomena, such as geometrically isotropic observed behavior of a sliding block despite obvious physical anisotropies can be explained only within the proposed model.

11.2 Experimental investigation

A series of experimental tests are performed in order to investigate the global characteristics of the system "block on a rough surface". The rough surface possesses visually a clear periodical structure and, therefore, the mechanical constitutive model for an observable orthotropic structure can be applied. Since we are trying to verify the average interface model, the measured values in experiments are intentionally chosen to be global, namely we measure global forces leading to the macro friction coefficients and trajectory of a block instead of micro friction coefficients and corresponding stiffnesses of asperities. The focus of the discussion is placed upon the kinematical behavior of the block driven by a constant force together with the measurement of force components leading to this motion. Therefore, the main measurable characteristics during these experiments are global forces and trajectories of the block, which create a main basis for further calibration of models for orthotropic friction. For the judgment of the results Coulomb like models are assumed a-priori to be valid for the global behavior, i.e. that the tangential driving global force F is proportional to the normal reaction N: $F = f(x, y)N$, where a function $f(x, y)$ describes the orthotropic properties of a surface.

11.2.1 Experimental setup

A massive block positioned on a plane is moved with constant velocity by a sliding carriage guided by rods on both sides, see Fig. 11.1. The block made from steel has dimensions $110 \times 110\,mm$ in plane and $20\,mm$ height. The mass is $m = 1.875kg$. The contact surface between the sliding carriage and the block is covered with a $Teflon^{(R)}$ strip to minimize the friction between them due to relative sliding. In contrast to this, the contact surface of the steel block is covered by a suede-like material with square $90 \times 90\,mm$ to increase the interaction between the block and the basement. A constant driving velocity is achieved by a step motor acting on a rack, which allows a straight displacement of $500\,mm$. The contact force between the rack and the sliding carriage is measured by a force sensor. The displacement of the block during sliding is captured by an optoelectronic device which is installed on a tripod above the surface. The corresponding LED (light-emitting diode) is fixed on the block.

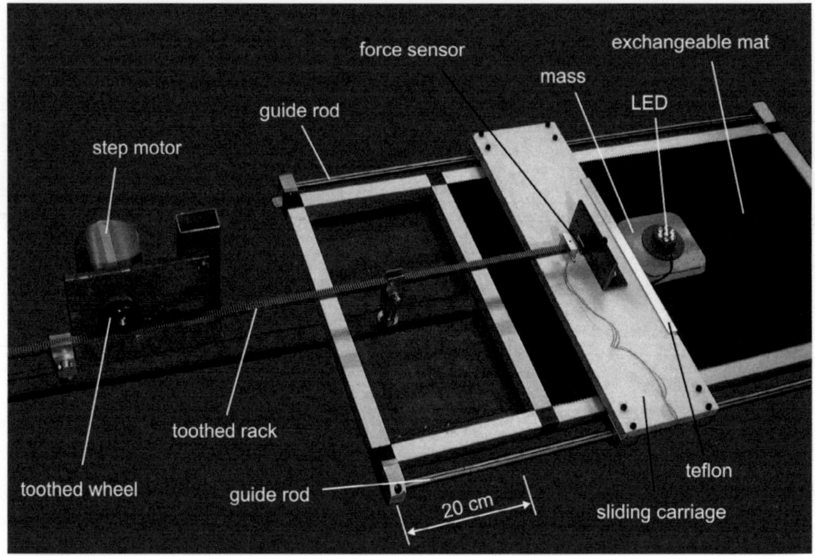

Figure 11.1: View on experimental setup.

For the first set of experiments a rubber mat with rather stiff ripples has been taken. Naturally it was represented by an aged corrugated rubber map. The frictional orthotropy is given by the wavy profile of the contact rubber surface with parallel ripples possessing in the cross section a periodical structure, see the CAD model in Fig. 11.2. The distance between the ripples is rather small in comparison to the dimension of the contact area of the block allowing approximately 30 ripples in contact area depending on orthotropy angle in experiments. The orientation of the ripples with respect to the fixed driving direction can be varied from 0° up to 90° by repositioning the mat, see Fig. 11.3. Again we should mention that the aim of the experiments is to show the necessity of a coupled anisotropic model for adhesion and friction, therefore, we intentionally skip any measurement of ripple stiffnesses concentrating on finding the global macro characteristics of the coupled behavior.

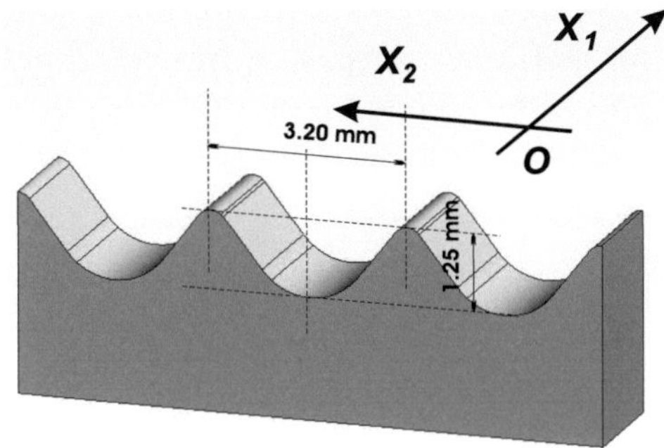

Figure 11.2: Geometrical structure of the corrugated rubber mat. Wavy periodical profile, CAD model.

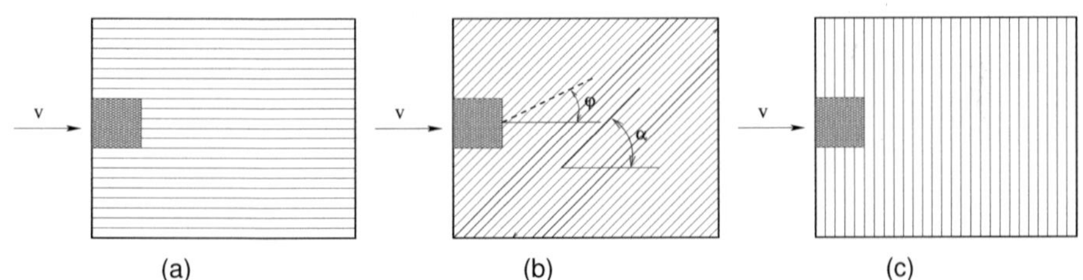

Figure 11.3: Orientation of the orthotropy with respect to the fixed direction of the velocity: a) $\alpha = 0^o$, b) $\alpha = 0^o < \alpha < 90^o$, c) $\alpha = 90^o$. The trajectory of the block is a straight line declined at angle φ.

11.2.2 Experimental results

At the beginning some experiments are performed to find out the global macro mechanical properties of the system. All experiments were reproduced with a driving velocity of the carriage $v = 24.4\,cm/sec$. First of all, the sliding carriage was moved without the block in order to define the internal resistant force F_{int}. Then experiments have been performed with the sliding block to define the resulting driving force in the case $\alpha = 0^o$, corresponding to the X_1-axis along the ripples, see CAD model in Fig. 11.2, and $\alpha = 90^o$, corresponding to the X_2-axis across the ripples, see CAD model in Fig. 11.2, respectively. These measurements together with subtracting the internal force lead to the definition of macro friction coefficients μ_1 and μ_2 corresponding to angles $\alpha = 0^o$ and $\alpha = 90^o$. Assuming a Coulomb friction law the friction coefficients μ_i were computed as

$$\begin{aligned}
\mu_1 &= \frac{F_{\alpha=0^o} - F_{int}}{N} = \frac{12.50 - 5.00}{1.875 \cdot 9.806} = 0.408 \approx 0.41 \\
\mu_2 &= \frac{F_{\alpha=90^o} - F_{int}}{N} = \frac{16.50 - 5.00}{1.875 \cdot 9.806} = 0.625 \approx 0.63.
\end{aligned} \qquad (11.1)$$

We note for further references that the macro friction coefficient across the ripples is found to be higher than the macro friction coefficient along the ripples.

The second set of experiments to find macro parameters of the interface model is made by setting the angle α varying from $\alpha = 0^o$ up to $\alpha = 90^o$ with a step $\Delta\alpha = 5^o$. The focus lies on the definition of the trajectory of the sliding block. **In all cases, the trajectory was observed as a straight line inclined with the angle $\varphi = \varphi(\alpha)$, see the sketch in Fig. 11.3(b).** The mean value $\bar\varphi$ of sequences of 10 experiments for each angle α is taken for the later analysis. The standard deviation has been found as $s \approx 0.3^o$ (defined by $s = \sqrt{\frac{1}{N-1}\sum_{i=1}^{N}(\varphi_i - \bar\varphi)^2}$). Combining all results leads to the diagram in Fig. 11.4 showing the dependency of the inclination angle φ on the orientation of the orthotropy given by the angle α of the ripples. The maximum of the inclination angle φ is located in the range of small angles α (the maximum is on the left side on the graph).

11.2.2.1 Geometrically isotropic observed behavior of a sliding block

As a fairly surprising result detected in the experiments a large sensitivity to the elastic properties of the rubber ripples was obtained. Thus, if the rubber mat with highly elastic rubber ripples (e.g. a new mat) has been taken for the second set of experiments, then the inclination angle φ was only varying in a very small range about $0^o - 2^o$ with a standard

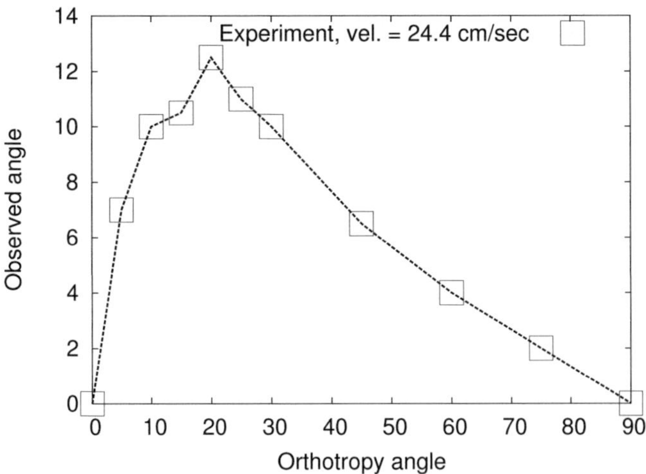

Figure 11.4: Observed mean value of the inclination angle φ vs. orthotropy angle α. Experimental results for different velocities of the block.

deviation $s \approx 0.5^o$ independently from the angle α. Since an interval $3s$ of the mean value defined as $\overline{\varphi} - 3s \leq \varphi \leq \overline{\varphi} + 3s$ covers the zero value $\varphi = 0$, and assuming a Gaussian error distribution, we can conclude that the inclination angle φ does not depend on the orthotropy angle α. However, the measurement of forces still showed the difference between the macro coefficients of friction μ_1 and μ_2. We observe that macro friction orthotropy is still present, but the kinematical effect of the orthotropy disappears.

As we show later, the orthotropic friction model is not capable to describe this effect from the macro-model point of view, but the coupled orthotropic adhesion − orthotropic friction interface model allows to qualitatively describe the observed phenomena.

11.3 Analysis of various models for anisotropic friction and applicability to the observed phenomenon

In this section, the range of applicability of a classical model of orthotropic friction, based only on the orthotropic friction tensor and its generalization including orthotropy for both friction (inelastic region) and adhesion (elastic region) is discussed. The necessity to assume in addition elastic properties for the surface will be shown.

As a first simple model which can be investigated analytically a material point on a plane is considered. According to the experimental tests we assume a quasi-statical motion of the material point $\mathbf{A}$ with weight P loaded by the force $\mathbf{F}$ acting along the X^1-axis, see Fig. 11.5. The orthotropic properties of the surface are defined in the coordinate system ξ^1, ξ^2 inclined with an angle α to the original coordinate system. During quasi-statical loading, point $\mathbf{A}$ is moving along a line with velocity vector $\mathbf{v}$ inclined with an angle φ. The reaction force $\mathbf{T}$ with Cartesian components T_1, T_2 is acting on the point. The values of components depend on the hypothesis concerning the orthotropic friction law. Here two variants of the orthotropic law are considered: the well known orthotropic friction Coulomb law and a contact interface model including orthotropy for both friction and adhesion, see Konyukhov and Schweizerhof [90], [91].

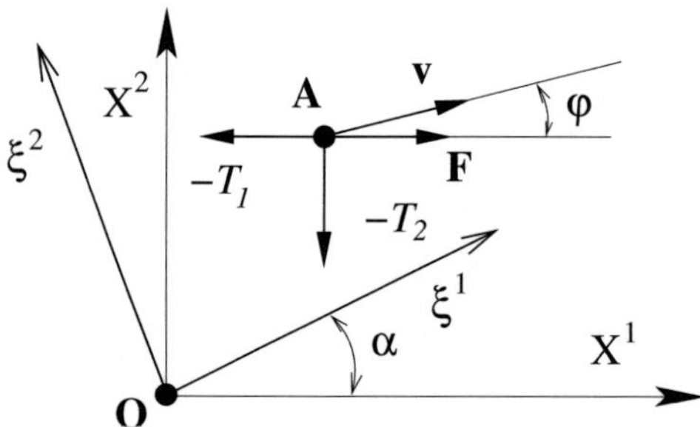

Figure 11.5: Motion of material point $\mathbf{A}$ on an orthotropic plane loaded by force $\mathbf{F}$.

The equilibrium equations for the system in Fig. 11.5 are given as:

$$\begin{cases} X^1 : & F + T_1 = 0; \\ X^2 : & T_2 = 0; \\ X^3 : & -P + N = 0. \end{cases} \qquad (11.2)$$

where N is the reaction force along the X^3 axis.

The principle of maximum dissipation is applied to obtain relation between the sliding force $\mathbf{T}$ and sliding displacements $\Delta \mathbf{r}^{sl}$. This principle requires that the dissipation function Ψ reaches its maximum

$$\Psi := \Delta \mathbf{r}^{sl} \cdot \mathbf{T} = \Delta x^i_{sl} T_i \longrightarrow \max, \tag{11.3}$$

where $\Delta \mathbf{r}^{sl}$ is an increment of the sliding vector. The dissipation function Ψ must also satisfy the sliding condition, formulated via inequalities, reflecting the assumed friction law, e.g. Coulomb's law.

11.3.1 Orthotropic Coulomb friction law

First, we recall the standard case known in literature, see e.g. [127], [32], [129], [211], where orthotropy is defined only for the sliding forces. The model is formulated according to the generalization of the sliding criteria. The yield function for the Coulomb friction law is then written as

$$\Phi := \sqrt{\mathbf{T} \cdot \mathbf{FT}} - |N| = \sqrt{T_i T_j f^{ij}} - |N|. \tag{11.4}$$

The sticking and the sliding conditions are defined by the rule:

$$\Phi < 0 \rightarrow sticking; \quad \Phi \geq 0 \rightarrow sliding. \tag{11.5}$$

According to equations (11.4-11.5) the material point is not moving during sticking (no adhesion) and the motion starts when $\Phi = 0$. The components of the friction tensor f^{ij} are defined for the orthotropy on the plane via e.g. the spectral representation plane as follows:

$$\mathbf{F} = \mathbf{Q}_\alpha \mathbf{\Lambda}_F \mathbf{Q}_\alpha^T = \tag{11.6}$$

$$= \begin{bmatrix} \cos\alpha & -\sin\alpha \\ \sin\alpha & \cos\alpha \end{bmatrix} \cdot \begin{bmatrix} \dfrac{1}{\mu_1^2} & 0 \\ 0 & \dfrac{1}{\mu_2^2} \end{bmatrix} \cdot \begin{bmatrix} \cos\alpha & -\sin\alpha \\ \sin\alpha & \cos\alpha \end{bmatrix}^T$$

$$= \begin{bmatrix} \dfrac{1}{\mu_1^2}\cos^2\alpha + \dfrac{1}{\mu_2^2}\sin^2\alpha & \left(\dfrac{1}{\mu_1^2} - \dfrac{1}{\mu_2^2}\right)\sin\alpha\cos\alpha \\[3mm] \left(\dfrac{1}{\mu_1^2} - \dfrac{1}{\mu_2^2}\right)\sin\alpha\cos\alpha & \dfrac{1}{\mu_1^2}\sin^2\alpha + \dfrac{1}{\mu_2^2}\cos^2\alpha \end{bmatrix},$$

where $\mu_i > 0$ are friction coefficients along the axis ξ^i inclined at angle α.

The standard method of the convex analysis is applied to obtain the sliding forces with regard to the principle of maximum dissipation (11.3). Thus, the Lagrange function with the multiplier λ is specified as

$$\mathcal{L} := -\Psi + \lambda\Phi = -\Delta x_{sl}^i T_i + \lambda \left(\sqrt{T_i T_j f^{ij}} - |N| \right) \tag{11.7}$$

together with the complementary Kuhn-Tucker conditions:

$$\lambda \geq 0, \quad \lambda\Phi = 0. \tag{11.8}$$

The optimality conditions $\dfrac{\partial \mathcal{L}}{\partial T^i} = 0$ lead to the following sliding displacement components:

$$\Delta x_{sl}^i = \lambda \frac{T_j f^{ij}}{\sqrt{T_k T_l f^{kl}}}. \tag{11.9}$$

These equations recover the trajectory of a block as a straight line declined by angle φ, which is confirmed by experiments.

Now, taking into account the second equilibrium equation (11.2) $\tan\varphi$ can be determined:

$$\left. \begin{aligned} \Delta x^1 &= \lambda \frac{T_1 f^{11}}{\sqrt{T_k T_l f^{kl}}}, \\ \Delta x^2 &= \lambda \frac{T_1 f^{12}}{\sqrt{T_k T_l f^{kl}}}, \end{aligned} \right\} \implies \tan\varphi = \frac{\Delta x^2}{\Delta x^1} = \frac{f^{12}}{f^{11}}, \tag{11.10}$$

and, after transformations taking into account the values determined in eqn. (11.6), we finally obtain:

$$\tan\varphi = \frac{(\mu_2^2 - \mu_1^2)}{\mu_2^2 + \mu_1^2 \tan^2\alpha} \tan\alpha. \tag{11.11}$$

11.3.2 Model for orthotropic contact interfaces including both adhesion and friction

An alternative model involving coupling orthotropy for both adhesion and friction can be proposed including the elastic-plastic analogy and the return-mapping scheme. This model is investigated theoretically and developed into the computational model by Konyukhov and Schweizerhof [90], [91]. Then the problem is formulated in **continuous form** as follows

a) The relative velocity vector of the contact point is decomposed additively into an elastic part $\mathbf{v}^{el}$ and a sliding part $\mathbf{v}^{sl}$

$$\mathbf{v}^r = \mathbf{v}^{el} + \mathbf{v}^{sl}. \tag{11.12}$$

b) The elastic part $\mathbf{v}^{el}$ is responsible for reversible deformations (adhesion) and satisfies the evolution equations

$$\frac{d\mathbf{T}}{dt} = \mathbf{B}\mathbf{v}^{el}. \tag{11.13}$$

At this point an adhesion tensor $\mathbf{B}$ describing orthotropic properties for the elastic region is introduced.

c) The tangential force $\mathbf{T}$ must satisfy the following inequalities defined via the yield function, which in tensor form can be written as:

$$\Phi := \sqrt{f^{ij}T_i T_j} - |N| = \sqrt{\mathbf{T} \cdot \mathbf{FT}} - |N| : \tag{11.14}$$

• if $\Phi < 0$ then the contact point is inside the elastic domain and $\mathbf{T} = \mathbf{T}^{el}$ is an elastic force,
• if $\Phi = 0$ then the contact point is sliding and $\mathbf{T} = \mathbf{T}^{sl}$ is a sliding force.

d) The power of the sliding forces, described by the energy dissipation function D achieves its maximum:

$$D := \dot{x}^i_{sl}T^{sl}_i = \mathbf{v}^{sl} \cdot \mathbf{T}^{sl}, \quad D \longrightarrow \max. \tag{11.15}$$

The principle of maximum dissipation requires that the plastic dissipation function D subjected to the inequality conditions (11.14) achieves a maximum. For **the computational treatment**, the model is reformulated in **incremental form** and then the return-mapping scheme is applied. The incremental analog is given as *i)* The full incremental displacement vector $\Delta x^i = \Delta x^i_{(n+1)} - \Delta x^i_{(n)}$ is decomposed additively into an elastic increment Δx^i_{el} and into a sliding increment Δx^i_{sl}:

$$\Delta x^i = \Delta x^i_{el} + \Delta x^i_{sl}. \tag{11.16}$$

ii) The elastic increment Δx^i_{el} is computed via the incremental evolution equations, for which the tensor $\mathbf{B}$ is assumed to be constant:

$$T^{tr}_{i\,(n+1)} = b_{ij}\Delta x^{i\,(n+1)}_{el} = b_{ij}(x^{i\,(n+1)}_{el} - x^{i\,(0)}). \tag{11.17}$$

iii) In order to decide whether the trial force $\mathbf{T}^{tr}$ is a sliding force $\mathbf{T}^{sl}$ or a sticking force $\mathbf{T}^{st}$ the yield condition is checked in each load step:

$$\Phi^{tr} := \sqrt{f^{ij}T^{tr}_{i\,(n+1)}T^{tr}_{j\,(n+1)}} - N_{(n+1)} \tag{11.18}$$

- If $\Phi^{tr} < 0$ then the trial force is a real sticking force $\mathbf{T} = \mathbf{T}^{tr}$.

- If $\Phi^{tr} \geq 0$ then the sliding force must be obtained via the maximum of the energy dissipation function given in incremental form.

iv) The incremental analog of the continuous formulation eqn. (11.15) is then:

$$D^{(n+1)}_{min} := -\Delta\mathbf{r}^{sl}\cdot\mathbf{T}^{sl}_{(n+1)} = -\Delta x^i_{sl}T^{sl}_{i\,(n+1)}, \quad D^{(n+1)}_{min} \longrightarrow \min. \tag{11.19}$$

We recall the results obtained in [90], [91]. There the sliding force $\mathbf{T}^{sl}$ can be defined after the necessary transformations as

$$\mathbf{T}^{sl} = -\frac{\mathbf{BFT}^{tr}}{\sqrt{\mathbf{BFT}^{tr}\cdot\mathbf{FBFT}^{tr}}}|N|. \tag{11.20}$$

Now, we must follow the return-mapping scheme in order to define the inclination angle φ. The problem is considered as a displacement driven one, therefore the incremental displacement $\Delta\mathbf{r} = \{\Delta x^1, \Delta x^2\}$

is applied. Thus, in each load step the sliding force in eqn. (11.20) is computed as:

$$\mathbf{T}^{sl} = -\frac{\mathbf{BFB}\Delta\mathbf{r}}{\sqrt{\mathbf{BFT}^{tr}\cdot\mathbf{FBFT}^{tr}}}|N| = \mathbf{A}\Delta\mathbf{r}. \qquad (11.21)$$

Now, if sliding is assumed, the second component of the sliding force T_2 in the formulation depicted in Fig. 11.5 becomes zero, see equilibrium eqn. (11.2). Thus, the displacement vector components $\Delta x^1, \Delta x^2$ are coupled via the equation:

$$T_2 = 0 \implies a_{21}\Delta x^1 + a_{22}\Delta x^2 = 0, \qquad (11.22)$$

leading to the equation for the angle φ:

$$\tan\varphi = \frac{\Delta x^2}{\Delta x^1} = -\frac{a_{21}}{a_{22}}. \qquad (11.23)$$

11.3.2.1 Analysis of the model by general spectral representation

In order to calibrate later a theoretical curve $\varphi(\alpha)$ from the experimental tests presented in Fig. 11.4, we consider a spectral decomposition of the matrix $\mathbf{A}$ given in eqn. (11.21) as

$$\mathbf{A} = [a_{ij}] = \begin{bmatrix} \lambda_1^2\cos^2\alpha + \lambda_2^2\sin^2\alpha & (\lambda_1^2 - \lambda_2^2)\sin\alpha\cos\alpha \\ (\lambda_1^2 - \lambda_2^2)\sin\alpha\cos\alpha & \lambda_1^2\sin^2\alpha + \lambda_2^2\cos^2\alpha \end{bmatrix}, \qquad (11.24)$$

leading together with the condition (11.23) to the observed sliding angle φ defined as

$$\tan\varphi = -\frac{a_{21}}{a_{22}} = -\frac{(\lambda_1^2 - \lambda_2^2)\sin\alpha\cos\alpha}{\lambda_1^2\sin^2\alpha + \lambda_2^2\cos^2\alpha} = -\frac{(\lambda_1^2 - \lambda_2^2)\tan\alpha}{\lambda_1^2\tan^2\alpha + \lambda_2^2}. \qquad (11.25)$$

An analysis for extremal values gives us

$$\frac{d\tan\varphi}{d\tan\alpha} = 0, \implies (\lambda_1^2 - \lambda_2^2)(\lambda_2^2 - \lambda_1^2\tan^2\alpha) = 0. \qquad (11.26)$$

The first bracket leads to the isotropic case, whereas from the second one the following critical value is obtained:

$$\tan \alpha_{ext} = \frac{\lambda_2}{\lambda_1},\tag{11.27}$$

leading to the extremum of the observed inclination angle φ_{ext} for the motion of the point

$$\tan \varphi_{ext} = \frac{\lambda_2^2 - \lambda_1^2}{2\lambda_1\lambda_2}.\tag{11.28}$$

Considering the last equation (11.28) we can obtain a critical ratio of the eigenvalues

$$ratio_{ext} = \frac{\lambda_1}{\lambda_2} = -\tan \varphi_{ext} \pm \sqrt{\tan \varphi_{ext}^2 + 1}.\tag{11.29}$$

This value will be used during the validation procedure.

For further considerations we adopt the spectral decomposition also for the adhesion tensor

$$
\begin{aligned}
\mathbf{B} \;=\; & \mathbf{Q}_\alpha \mathbf{\Lambda}_B \mathbf{Q}_\alpha^T = \tag{11.30}\\[4pt]
=\; & \begin{bmatrix} \cos\alpha & -\sin\alpha \\ \sin\alpha & \cos\alpha \end{bmatrix} \cdot \begin{bmatrix} -\varepsilon_1 & 0 \\ 0 & -\varepsilon_2 \end{bmatrix} \cdot \begin{bmatrix} \cos\alpha & -\sin\alpha \\ \sin\alpha & \cos\alpha \end{bmatrix}^T = \\[4pt]
=\; & \begin{bmatrix} \varepsilon_1 \cos^2\alpha + \varepsilon_2 \sin^2\alpha & (\varepsilon_1 - \varepsilon_2)\sin\alpha\cos\alpha \\ (\varepsilon_1 - \varepsilon_2)\sin\alpha\cos\alpha & \varepsilon_1 \sin^2\alpha + \varepsilon_2 \cos^2\alpha \end{bmatrix},
\end{aligned}
$$

where $\varepsilon_i > 0$ are stiffnesses along the axis ξ^i inclined at angle α.

Remark:

The trajectory of a block is a straight line inclined by the angle φ. Moreover, the observed inclination angle φ does not depend on particular values of λ_1, λ_2, but only on its ratio λ_1/λ_2.

11.3.2.2 Mechanical interpretation of the model

As is known, the mechanical interpretation of the regularized friction model assuming elastic deformations is a spring-slider system, see Simo and Hughes [160]. As generalization of this; according to our model, we consider a material point with two spring-slider systems, see

Fig. 11.6. The properties of these systems are the following: ε_i – stiffness of ith spring, μ_i – coefficient of friction for i^{th} sliding device. Each system i is constrained to move parallel along the axis X_i respectively. The constant force $\mathbf{F}$ inclined with angle α to the coordinate axis X_1 is applied to the point. Then the trajectory of the point lies either above the force line or below the force line depending on the ratio of eigenvalues λ_1 and λ_2 as discussed later, see also computational analysis in [91].

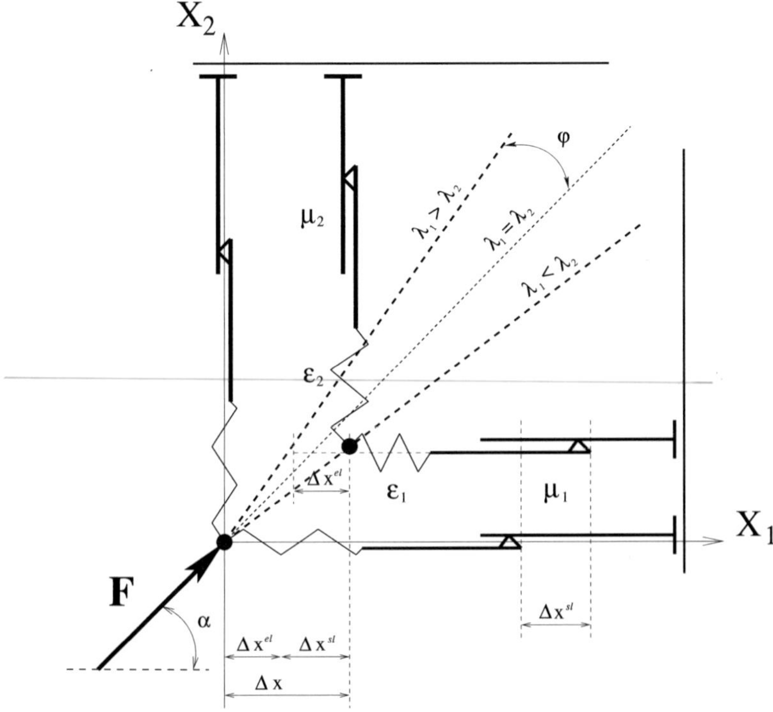

Figure 11.6: Mechanical interpretation of the orthotropic adhesion – orthotropic friction model. A material point on a plane with a two spring – two slider system loaded by the force **F** in plane.

11.4 Calibration of parameters for different models

As a representative parameter we take the curve $\varphi(\alpha)$ known from the experiment, see Fig. 11.4. In addition, we distinguish two orthotropy angles: α – orthotropy angle for the adhesion tensor $\mathbf{B}$ and β – orthotropy angle for the friction tensor $\mathbf{F}$. In order to unify the computations we chose the orthotropic values as $\mu_1 < \mu_2$ and $\varepsilon_1 < \varepsilon_2$. The following test computations are performed for calibration purposes:

1. The orthotropic friction model as discussed in Section 11.3.1.

2. The interface model including coupled orthotropy for both adhesion and friction as discussed in Section 11.3.2 with the specific case of isotropic adhesion $\mathbf{B} = -\varepsilon\mathbf{E}$.

3. The interface model including coupled orthotropy for both adhesion and friction with the specific case of isotropic friction $\mathbf{F} = \dfrac{1}{\mu^2}\mathbf{E}$.

4. The interface model including orthotropy for both adhesion and friction with the specific case of coinciding orthotropy angle $\alpha = \beta$.

5. The interface model including orthotropy for both adhesion and friction with the specification of the friction orthotropy angle β by 90° degrees as $\beta = \alpha + \pi/2$.

11.4.1 Case 1

We start the validation from the simple model including only orthotropic friction as discussed in Sect. 11.3.1. Our aim is to find out a case describing qualitatively the experimental results. Therefore, we perform a test computation with the following friction coefficients $\mu_1 = 0.1, \mu_1 = 0.5$. In Fig. 11.7 the results are depicted. In addition, at an extremal angle $\alpha = \arctan\frac{\mu_2}{\mu_1} = (78.69^o)$ the maximum value $\varphi_{\max} = \arctan\dfrac{\mu_2^2 - \mu_1^2}{2\mu_1\mu_2} = (67.38^o)$ is computed by analyzing the shape of a curve. It can be seen, that the point is moving into the direction with a smaller friction coefficient, which contradicts the experimental curve, see Fig. 11.4.

11.4.2 Case 2

As a next step, we choose the orthotropic adhesion – orthotropic friction interface model, but including orthotropy only for the friction tensor and

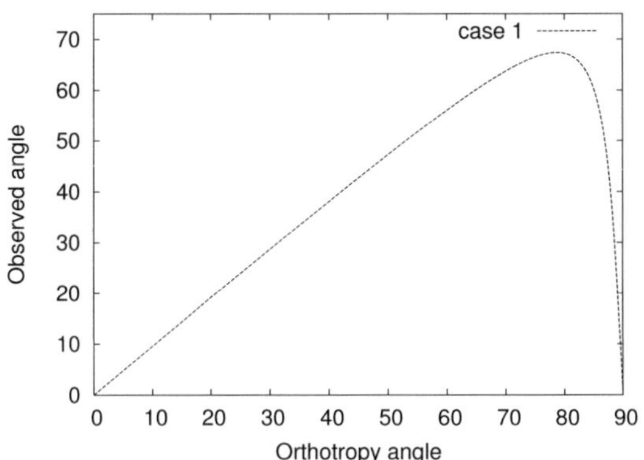

Figure 11.7: Computed inclination angle vs. orthotropy angle.
Case 1: Purely orthotropic friction model.

keeping the adhesion tensor to be isotropic, $\mathbf{B} = -\varepsilon_T \mathbf{E}$. In this case, a structure of the matrix in eqns. (11.21) and (11.24) is given as $\mathbf{A} = -\varepsilon_T^2 \mathbf{F}$ leading to the inclination angle in eqn. (11.25) to be defined as

$$\tan \varphi = \frac{(\mu_1^2 - \mu_2^2) \tan \beta}{\mu_1^2 + \mu_2^2 \tan^2 \beta}. \tag{11.31}$$

The analysis with the values $\mu_1 = 0.1$ and $\mu_2 = 0.5$ gives the curve presented in Fig. 11.8 with the extremal parameters $\beta_{ext} = 11.31^o$ and $\varphi_{\min} = -67.38^o$. In this case, the point tends to move into the direction with the larger friction coefficient. This is also a contradiction to the experimental results.

11.4.3 Case 3

As a next step, within the orthotropic adhesion − orthotropic friction interface model orthotropy is assumed only for adhesion with parameters $\varepsilon_1 = 1 \cdot 10^4$ and $\varepsilon_2 = 5 \cdot 10^4$. The following structure of the matrix $\mathbf{A}$ is obtained

$$\mathbf{A} = \mathbf{BFB} = \frac{1}{\mu^2} \mathbf{B}, \tag{11.32}$$

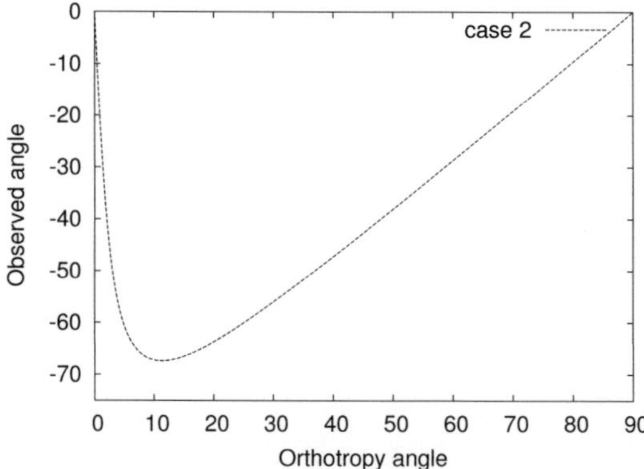

Figure 11.8: Orthotropic adhesion – orthotropic friction interface model. Computed inclination angle vs. orthotropy angle. Case 2: isotropic adhesion – orthotropic friction.

leading to an inclination angle φ defined as

$$\tan \varphi = \frac{\left(\varepsilon_2^2 - \varepsilon_1^2\right) \tan \alpha}{\varepsilon_2^2 + \varepsilon_1^2 \tan^2 \alpha}. \tag{11.33}$$

The result is depicted in Fig. 11.9 with the extremal values as $\alpha_{ext} = 78.69^o$ and $\varphi_{\max} = 67.38^o$. The curve shows a motion into the direction with the smaller stiffness ε_1. This contradicts the experimental results as well.

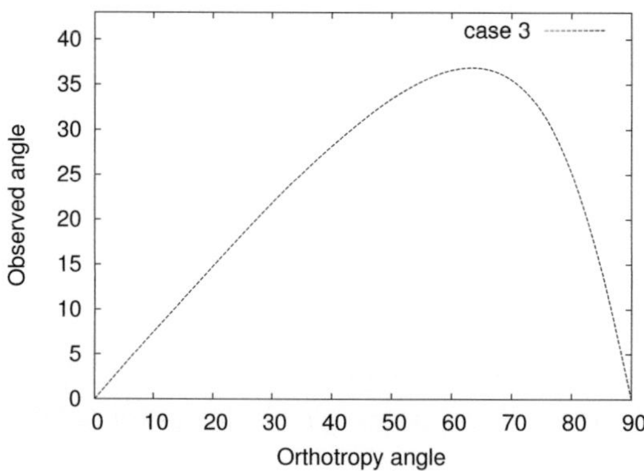

Figure 11.9: Orthotropic adhesion – orthotropic friction interface model. Computed inclination angle vs. orthotropy angle. Case 3: orthotropic adhesion – isotropic friction.

Summary
We observed that neither the purely orthotropic friction model, nor the coupled model with separately included orthotropy either for friction, or for adhesion is unable to capture the phenomena even qualitatively.

11.4.4 Case 4

Now we keep the orthotropy for both the adhesion tensor and the friction tensor with the same angle, namely $\alpha = \beta$. The structure of the tensor $\mathbf{A}$ is found from the spectral decomposition in eqns. (11.6) and (11.30) with $\mathbf{Q}_\alpha \equiv \mathbf{Q}_\beta$

$$\mathbf{A} = \mathbf{BFB} = \mathbf{Q}_\alpha \mathbf{\Lambda}_B \underbrace{\mathbf{Q}_\alpha^T \mathbf{Q}_\beta}_{\mathbf{E}} \mathbf{\Lambda}_F \underbrace{\mathbf{Q}_\beta^T \mathbf{Q}_\alpha}_{\mathbf{E}} \mathbf{\Lambda}_B \mathbf{Q}_\alpha^T = \mathbf{Q}_\alpha \mathbf{\Lambda}_B^2 \mathbf{\Lambda}_F \mathbf{Q}_\alpha^T, \quad (11.34)$$

leading to the eigenvalues $\lambda_i = \varepsilon_i/\mu_i$. This case gives a more comprehensive information for the analysis of the physical experiments. The most important issue is that by the combination of two orthotropic tensors an isotropic case can be recovered. This case appears if we take the value in proportion $\frac{\varepsilon_1}{\mu_1} = \frac{\varepsilon_2}{\mu_2}$ leading to $\lambda_1 = \lambda_2$, e.g. with the combination of two previous cases with $\varepsilon_1 = 10^4$ and $\varepsilon_2 = 5 \cdot 10^4$ and $\mu_1 = 0.1, \mu_2 = 0.5$ isotropy of the motion is recovered. For $\lambda_1 < \lambda_2$ we obtain a behavior similar to case 3; for $\lambda_1 > \lambda_2$ we find a behavior similar to case 2. E.g. the computation with the parameters $\varepsilon_1 = 0.5 \cdot 10^4$, $\varepsilon_2 = 5 \cdot 10^4$ and $\mu_1 = 0.1, \mu_2 = 0.5$ shows similarity to case 3 as shown in Fig. 11.10. In this computational case, we assumed that the ripples are softer in direction X_1, see Fig. 11.2, which is obviously not the case even without measurements of the ripple stiffnesses.

11.4.5 Case 5

Finally, we can choose the last possible modification for the observed geometrical orthotropy for surfaces as taken in the experiment. We define a new angle $\hat{\beta}$ as a main angle of surface asperities in the experiment, see Fig. 11.11. The orthotropy angle β for the friction tensor is shifted by 90° degrees to $\beta = \alpha + \pi/2$ with respect to the orthotropy angle α for the adhesion tensor. The structure of the tensor $\mathbf{A}$ is given

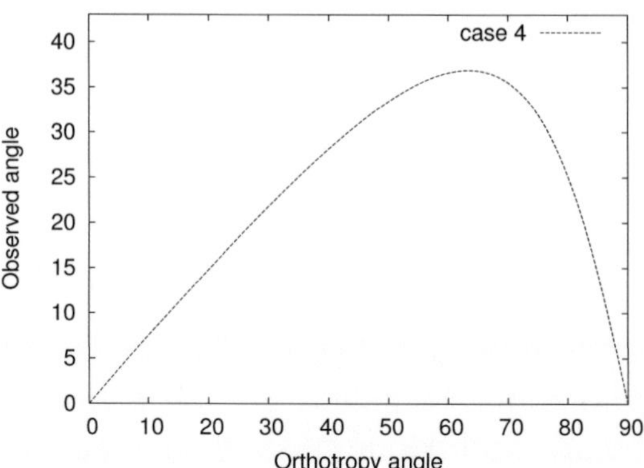

Figure 11.10: Orthotropic adhesion – orthotropic friction interface model. Computed inclination angle vs. orthotropy angle. Case 4: orthotropic adhesion – orthotropic friction with eigenvalues $\lambda_1 < \lambda_2$.

according to eqn. (11.34), but now the composition of the two orthogonal matrices $\mathbf{Q}_\alpha^T \mathbf{Q}_\beta$ leads to:

$$
\mathbf{Q}_\alpha^T \mathbf{Q}_\beta \;=\; \tag{11.35}
$$

$$
= \begin{bmatrix} \cos\alpha & -\sin\alpha \\ \sin\alpha & \cos\alpha \end{bmatrix}^T \cdot \begin{bmatrix} \cos\beta & -\sin\beta \\ \sin\beta & \cos\beta \end{bmatrix} =
$$

$$
= \begin{bmatrix} \cos\alpha & -\sin\alpha \\ \sin\alpha & \cos\alpha \end{bmatrix}^T \cdot \begin{bmatrix} -\sin\alpha & -\cos\alpha \\ \cos\alpha & -\sin\alpha \end{bmatrix} =
$$

$$
= \begin{bmatrix} 0 & -1 \\ 1 & 0 \end{bmatrix}.
$$

Then, the matrix $\mathbf{A}$ in eqn. (11.34) is derived as

$$
\mathbf{A} = \mathbf{B}\mathbf{F}\mathbf{B} = \; \mathbf{Q}_\alpha \Lambda_B \underbrace{\mathbf{Q}_\alpha^T \mathbf{Q}_\beta}\, \Lambda_F \underbrace{\mathbf{Q}_\beta^T \mathbf{Q}_\alpha}\, \Lambda_B \mathbf{Q}_\alpha^T \tag{11.36}
$$

$$
= \; \mathbf{Q}_\alpha \begin{bmatrix} \dfrac{\varepsilon_1^2}{\mu_2^2} & 0 \\[2ex] 0 & \dfrac{\varepsilon_2^2}{\mu_1^2} \end{bmatrix} \mathbf{Q}_\alpha^T ,
$$

leading to the following eigenvalues $\lambda_1 = \varepsilon_1/\mu_2$ and $\lambda_2 = \varepsilon_2/\mu_1$ in eqn. (11.24). The computation with $\varepsilon_1 = 10^4$ and $\varepsilon_2 = 10^3$ and $\mu_1 = 0.1, \mu_2 = 0.5$ gives the curve φ vs. $\hat{\beta}$ depicted in Fig. 11.12, which quantitively has a shape similar to the experimental one (maximum from the left side). The extremal values are found as $\hat{\beta}_{ext} = 26.56^o$ and $\varphi_{\max} = 36.87^o$.

Thus, summarizing the numerical investigations and focusing on the comparison to the experiments it becomes obvious that for the surface as given in the experiment it is necessary to apply the orthotropic adhesion – orthotropic friction interface model.

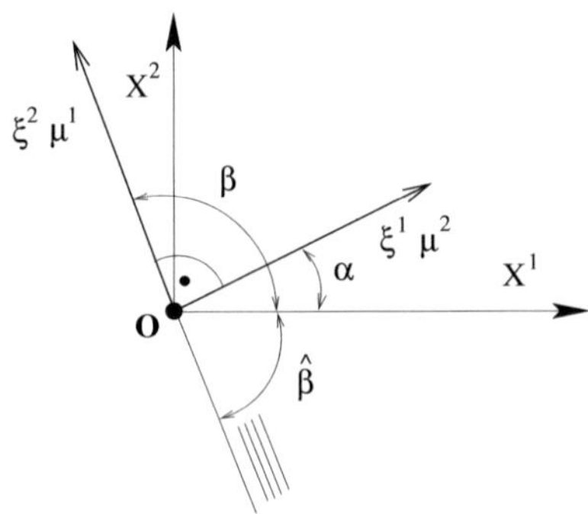

Figure 11.11:   Definition of the experimentally observed angle $\hat{\beta}$, by an orthotropy angle α for the adhesion tensor and an orthotropy angle β for the friction tensor.

In this case, the force T_2 measured in direction X_2 is larger, while the ripples are softer in the same direction, which is visually observed in experiments even without measurement of ripple stiffnesses, see Fig. 11.2. Moreover, assuming $\frac{\varepsilon_1}{\varepsilon_2} = \frac{\mu_1}{\mu_2}$ we obtain $\varphi(\alpha) = 0$, i.e. the block is no longer inclined, and we recover the geometrically isotropic behavior already observed in experiments, see Sect. 11.2.2.1. In this case, orthotropy for friction is compensated by orthotropy for adhesion leading to the observed isotropic behavior of the block, though the reason for this remains uncovered.

11.4.6   Calibration of the theoretical curve by extremal values

As found from the proposed model, the inclination angle φ depends only on the ratio of eigenvalues λ_1/λ_2, see eqn. (11.25). This ratio contains

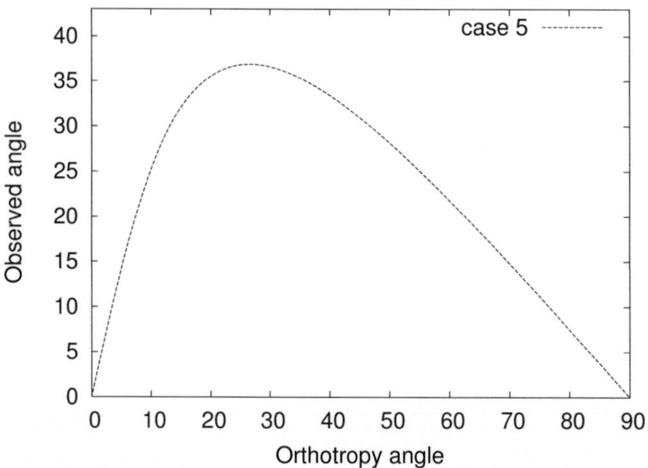

Figure 11.12: Orthotropic adhesion – orthotropic friction interface model. Computed inclination angle φ vs. redefined orthotropy angle $\hat{\beta}$. Case 5: orthotropic adhesion – orthotropic friction with angles $\beta = \alpha + \frac{\pi}{2}$.

information also about the ratio of adhesion parameters $\varepsilon_1/\varepsilon_2$:

$$ratio_{ext} = \frac{\lambda_1}{\lambda_2} = \frac{\varepsilon_1}{\varepsilon_2} \cdot \frac{\mu_1}{\mu_2}. \tag{11.37}$$

Since, the friction coefficients μ_1, μ_2 are defined via the measurement of forces in the experiment, the inclination angle φ depends only on the ratio of eigenvalues $\varepsilon_1/\varepsilon_2$, see also the Remark in Section 11.3.2.1. This fact gives the possibility to judge macro properties for ripple stiffnesses without measurements. Thus, we will use eqn. (11.25) for a calibration of the model. Calibration is provided according to following rules: 1) a maximum rule – both theoretical and experimental curves must achieve the same maximum; 2) a least square fit method. The friction coefficients have been determined previously to $\mu_1 = 0.408$, $\mu_2 = 0.625$.

According to the maximum rule the value for the angle φ_{ext} defined in eqn. (11.29) is used for calibration purposes. Taking e.g. the maximum angle $\varphi_{\max} = 11.5^o$ measured for the velocity $12.5\,cm/sec$ the ratio of the eigenvalues given in eqn. (11.29) becomes

$$ratio_{ext} = \frac{\lambda_1}{\lambda_2} = -0.20345 \pm 1.02048 = 0.817, \tag{11.38}$$

where only the positive solution is taken. The ratio of the stiffness coefficient is then obtained as

$$\frac{\varepsilon_1}{\varepsilon_2} = \frac{\mu_2}{\mu_1} \cdot ratio_{ext} = \frac{0.625}{0.408} \cdot 0.817 = 1.251 \tag{11.39}$$

A more mathematically precise least square fit method leads to the statement derived from eqn. (11.25). The following sum must be minimized:

$$\sum_{k=1}^{N} \left\{ \tan \varphi^{(k)} + \frac{(\lambda_1^2 - \lambda_2^2) \tan \beta^{(k)}}{\lambda_1^2 \tan^2 \beta^{(k)} + \lambda_2^2} \right\}^2 =$$

$$= \sum_{k=1}^{N} \left\{ \tan \varphi^{(k)} + \frac{(r_\lambda^2 - 1) \tan \beta^{(k)}}{r_\lambda^2 \tan^2 \beta^{(k)} + 1} \right\}^2 \longrightarrow \min, \tag{11.40}$$

where $\varphi^{(k)}$ are measured declination angles vs. applied orthotropy angles $\alpha^{(k)}$ and $\beta^{(k)} = \pi/2 - \alpha^{(k)}$. Minimization with regard to the variable r_λ^2 leads to the following expression:

$$r_\lambda^2 = \frac{\sum_{k=1}^{N} \left\{ \tan \beta^{(k)} (1 + \tan^2 \beta^{(k)})(\tan \beta^{(k)} - \tan \varphi^{(k)}) \right\}}{\sum_{k=1}^{N} \left\{ \tan \beta^{(k)} (1 + \tan^2 \beta^{(k)})(\tan \beta^{(k)} + \tan \varphi^{(k)}) \right\}}. \tag{11.41}$$

The cases including the computed parameters are shown in Fig. 11.13 and do not exhibit a quantitively good correlation, though we reached the aim to show the necessity of coupling both orthotropy for friction and orthotropy for adhesion in addition to the general applicability of the proposed model. Among the possible reasons for the bad correlations might be the following:
a) a too simple linear elastic model for adhesion region;
b) the simple Coulomb friction model for the friction region without e.g. hysteresis etc.
Nevertheless, it seems to be important to consider fairly complex models including the coupled orthotropy for the adhesion and the friction in

order to describe the observed phenomena more precisely.

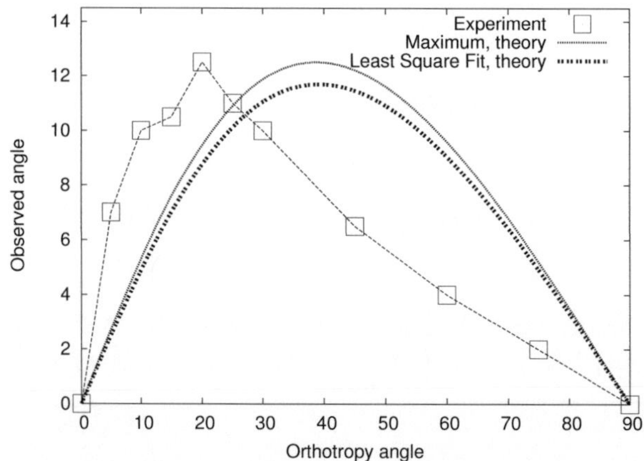

Figure 11.13: Observed inclination angle φ vs. redefined orthotropy angle $\hat{\beta}$. Calibration: (1) by the absolute maximum value for different velocities of the block; (2) by the least square fit method.

11.5 Acknowledgments

We thank mechanics master W. Wendler for the careful preparation and performing of the experiments. His actions were very creative and are greatfully acknowledged. We also thank the DFG for the support given by grant SCHW 307/18-2.

11.6 Conclusions

The necessity to apply a coupled model including anisotropy for both adhesion and friction is shown for particular soft anisotropic surfaces, such as a periodic wavy rubber profile. The macro characteristics of the proposed model such as macro friction coefficients μ_1, μ_2 can be defined via the measurement of global forces, while the trajectory, namely, the observed sliding angle $\varphi(\alpha)$ gives information about the ratio of orthotropic stiffnesses $\varepsilon_1/\varepsilon_2$. The model does not show the dependence of particular values of stiffnesses ε_1 and ε_2 on the trajectory, which was also confirmed by experiments. The model allows to skip measurements of surface microstiffnesses. Moreover, particular phenomena when a sliding block shows geometrically isotropic behavior can be described correctly with the proposed model.

The considered model contains only the coupling of linear models for adhesion and friction, which is a possible explanation for the rather poor quantitative correlation with experiments. The key to achieve better correlation – from our point of view – can be the coupling of a more general law for adhesion as well as for friction. Thus, a more complex elastic law (e.g. Ogden material law for 2D case) can be taken for the adhesion region together with a more complex friction law for the friction region (e.g. see the proposals of He and Curnier [62] and recently Zmitrowicz [214]). The calibration process, in due course, can be provided by experimental investigations as well as by numerical tests involving homogenization processes and multi-scale techniques, for which the methodology for the isotropic case is shown in [9].

12

Geometrically exact covariant approach for contact between curves representing beam and cable type structures*

Abstract

For curve-to-curve contact situation a geometrically exact description in a covariant form is developed. The contact kinematics, the variational formulation and the constitutive relations for contact tractions are described in a specially defined curvilinear coordinate system based on the closest point projection (CPP) procedure. The fundamental problems about the solvability of the CPP procedure for the curve-to-curve case are investigated in detail as well. The introduced coordinate system is independent of the choice of a contacting curve. This allows to describe all possible relative motions of both curves including not only normal and tangential interactions, but also rotational interaction between curves representing e.g. circular cross sections of beams. All necessary derivations for the iterative solution method such as linearization of weak forms and the return-mapping schemes are fulfilled in a covariant form for the arbitrary distance between the curves. This allows to apply the developed theory to both contact cases: when bodies are contacting by their edges and to contact between beams. Another advantage is the complete independence concerning the order of approxi-

---

*The chapter is published in [96] A. Konyukhov, K. Schweizerhof *Geometrically exact covariant approach for contact between curves*, Computer Methods in Applied Mechanics and Engineering, **199**(3):2510–2531, 2010.

mation involved in finite element construction. The numerical examples are chosen to verify the proposed theory and to compare different cases for both, the beam-to-beam and the edge-to-edge contact cases as well as to illustrate the possibilities of the theory to describe relative contact kinematics.

Keywords
contact curves, edges, beams geometrically exact covariant

12.1 Introduction

Contact between beams as a branch of computational contact considerations appears to be a "stand alone specific problem" due to its special geometrical tasks and following difficulties. Thus only fairly recently, a first computational algorithm for beam-to-beam contact was proposed by Wriggers and Zavarise [197] (1997) for non-frictional contact between straight beams with circular cross sections. The measure of contact has been introduced as the shortest distance between the central lines. Then the specific task is that two projection points on both contact elements have to be found. Further, in order to extend this into the frictional case in Zavarise and Wriggers [200] (2000) the return-mapping algorithm is exploited. The solution algorithm is based on an iterative solution and involves the linearization for the particular linear contact elements. A generalization of this algorithm for the case of rectangular cross-sections is given in Litewka and Wriggers [113] (2002) and [114] (2002) for both frictional and non-frictional cases respectively. While the mentioned publications were based on a penalty method to enforce the contact constraints, in Litewka and Wriggers [115] (2003) the Lagrange multiplier method has been involved. One of the obvious – and up-to now not resolved – difficulties is the linearization procedure for the case of arbitrarily curved geometry of beams. As a first step thus, Litewka in [111] (2007) and in [112] (2007) considered a smooth Hermite type element for contact formulations, however, the linearization was provided as the direct derivation of the obtained functionals via mathematical software (Mathematica in the current case).

Another approach to contact between beams as an application for cable contact is described in Zhou et.al.[208] (2004) in which, a so-called "sliding cable" element leading to an element with a "broken line" is derived. This model, however, became a standard for a model of "seat belt connections" since many years in commercial codes, such as LS-DYNA, see [52]. A special problem of mooring cables has been considered in Souza de Cursi [165] (1992) as a 2D contact problem of Signorini type.

These references reveal the main difficulties and limitations of the current state of beam-to-beam contact models:

1. Existing finite elements models are yet not serving for arbitrary order of approximations due to problems with the linearization, though, specific applications of high order finite elements are known e.g. in [112], [111].

2. kinematics of contact considered in the existing models are rather limited; by far not all possible variances of the relative motion and corresponding relative constitutive equations are fully involved.

Thus, the main goal of the current contribution is to develop a general theory for curve-to-curve contact from a geometrical point of view considering a particular case of contact between bodies when the contacting bodies are interacting along two curves. It can be viewed as either contact between two sharp edges for full 3D bodies, or between two beams as a mechanical model in the latter, if center lines of beams or edges of bodies are considered in contact. Our aim is to describe all contact parameters such as relative displacements and velocities, contact forces in the specific coordinate system which is most suited to describe the various geometrical diversities of both two curves. This concept includes internal geometrical characteristics of curves such as curvature and torsion as well as their relative position in 3D space. The basics of such a concept have been developed by the authors previously for contact between surfaces as a covariant approach for the nonfrictional case in [86] (2004), for the frictional case in [89] (2005), specially reconsidered for 2D geometries in [92] (2006) and for high order finite elements in [97] (2009), and generalized for the case of surfaces possessing coupled anisotropies for tangential adhesion and friction in [90] and [91] (2006).

The important feature for any curve-to-curve contact is a suitable definition of the coordinate system which will allow to distinguish all kinematical details of the mutual interaction. Thus all contact measures for relative motions should be directly obtained in this coordinate system. Obviously a distance between two curves provides a definition of a measure of normal contact interaction. This leads to the well known idea of the closest point projection (CPP) procedure. How the closest point is representing the contact appropriately will be discussed later. Further, the specific CPP procedure allows to set up two local curvilinear coordinate systems attached to the each curve. These coordinate systems are given in terms of Serret-Frenet formulas for 3D curved lines. This allows to consider any motion of one curve relative to another. A good observation of the motion gives us then a choice of proper measures for such an interaction between two curves. Afterwards, all possible parameters such as sliding distances, variations, forces as well as the necessary operations such as derivations, formulation of the return-mapping algorithm are considered in a covariant form using the curvilinear metrics of the specific two coordinate systems attached to the curves. As a result, the complete diversity of contact interactions including even a possible moment interaction are described. Finally, a compact form of the computational algorithm – a-priori independent on the type and order of approximation of the curves – is established.

The outline of the contribution is then given as follows:

First, the closest point procedure in the local Serret-Frenet coordinate system attached to the curves is presented leading to a special local 3D coordinate system. A geometrical analysis of solvability for the CPP procedure is then given in this local coordinate system illustrated by some examples showing the geometrical properties of the CPP procedure including restrictions on curvature and position of an osculating circle in space. Then geometrical characteristics of the curvilinear coordinate system – structure of the metric tensor and covariant derivatives – are studied. The kinematics of the curve-to-curve interaction can be then considered in the local coordinate system and corresponding measures of contact interaction are derived. The components of a contact force vector are hereby obtained as energy conjugated pairs for the rate of defined contact measures as a contravariant vector in the corresponding basis. A rotational moment between curves is appearing in addition to normal and tangential forces. By taking into account equilibrium between two interacting curves a weak formulation of contact, morever, a symmetric form is obtained. Since a curve-to-curve interaction is resulting from various geometrical situations constitutive equations for contact traction have to be considered separately for the cases of beams or cables and for edges of solid bodies. A separate section follows devoted to the linearization of the nonlinear expressions exploiting the covariant differentiations of the weak form together with the return-mapping scheme. The description of the finite element discretization shows the implementation details for the simplest case of a linear approximation of the edges of the contacting bodies. The full formulation is derived for arbitrary large motions and an arbitrary structure of the curves. The numerical examples are specifically selected to show the range of the applicability of the developed computational scheme.

12.2 Closest point projection and definition of local coordinate systems

The closest distance between two curves is a natural measure of any contact interaction between two curves. This distance can be defined in any convenient coordinate system. The standard coordinate system for the final discretization and thus for the numerical computation is a global Cartesian coordinate system, but in order to consider some local geometrical properties it is far more convenient to consider the kinematics of deformation and to describe the differential properties of the curves in coordinate systems geometrically attached to the curves. Known in differential geometry such a system is the Serret-Frenet coordinate system to be defined in any point of the curve e.g. in a point which gives rise to a contact measure between curves. This allows to exploit standard notations known in differential geometry ([103], [34]) throughout the article.

Let us consider, see Fig. 12.1, two curves in 3D Cartesian space arbitrarily parameterized with parameters ξ^I, $I = 1, 2$, where I is a number assigned to the curve. For further developments – except special cases

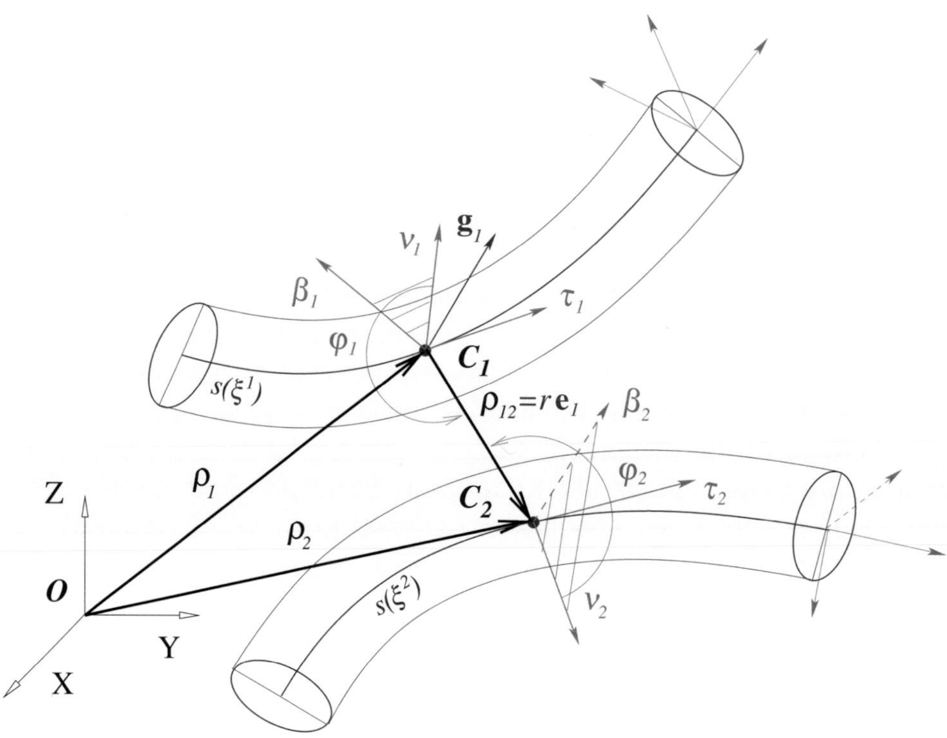

Figure 12.1: Closest point projection (CPP) procedure. Definition of the Serret-Frenet coordinate system $(\boldsymbol{\tau}_I, \boldsymbol{\nu}_I, \boldsymbol{\beta}_I)$ and a local coordinate system $(\boldsymbol{\tau}_I, \mathbf{e}_I, \mathbf{g}_I)$ based on CPP procedure.

– the curves are assumed to be continuously differentiable, namely C^1-continuous with resp. to the parameter ξ^I. The parameterization in vector notation can be written as

$$\boldsymbol{\rho}_I = \boldsymbol{\rho}(\xi_I), \quad I = 1, 2 \text{ number of curve} \tag{12.1}$$

or via coordinates x_I^1, x_I^2, x_I^3 in Cartesian space as

$$x_I^j = x_I^j(\xi_I), \quad I = 1, 2, \quad j = 1, 2, 3. \tag{12.2}$$

An arc-length s_I for the I-th curve is then used in the differential geometry considerations for the curves as a natural coordinate to describe all their geometrical characteristics. It is defined via the differential of a vector ρ_I

$$ds_I = ds(\xi_I) = \sqrt{\rho_{I,\xi_I} \cdot \rho_{I,\xi_I}}\, d\xi_I = J_{(I)} d\xi_I \quad \text{(no summation)} \quad I = 1,2. \quad (12.3)$$

Here the following short notations are introduced for derivatives:

- derivative with respect to a convective coordinate ξ_I

$$\dot{\rho}_I = \rho_{I,\xi_I} \equiv \frac{\partial \rho_I}{\partial \xi^I}$$

- derivative with respect to the arc-length (natural coordinate) s_I

$$\rho_I' \equiv \frac{\partial \rho_I}{\partial s};$$

- Jacobian of the transformation $s_I \rightarrow \xi_I$

$$J_{(I)} = \frac{ds_I}{d\xi_I} = \sqrt{\rho_{I,\xi_I} \cdot \rho_{I,\xi_I}} = \sqrt{\sum_{j=1}^{3}\left(\frac{\partial x^j}{\partial \xi_I}\frac{\partial x^j}{\partial \xi_I}\right)}$$

The coordinates s_1, s_2 will be used then to define measures of the tangential interaction.

The closest distance $|\overrightarrow{C_1 C_2}| = |\rho_{12}|$ between two curves can be chosen as a natural measure of the contact interaction (see Fig. 12.1). However, the definition of this distance requires the knowledge about the points C_1 and C_2, and, therefore, its arc-lengths s_1 and s_2 as local coordinates on both curves respectively. It leads to the auxiliary minimization problem for the following distance function $\mathbf{F}(s_1, s_2)$:

$$\mathbf{F}(s_1, s_2) = \frac{1}{2}\|\rho_1(s_1) - \rho_2(s_2)\|^2 \longrightarrow \min \qquad (12.4)$$

This mathematical problem is called **closest point projection (CPP) procedure**. The solution gives us two coordinates s_1^p and s_2^p. Then the closest distance is defined as

$$r = |\rho_{12}| = |\rho_1(s_1^p) - \rho_2(s_2^p)|. \qquad (12.5)$$

The necessary conditions minimizing $\mathbf{F}(s_1, s_2)$ in eqn. (12.4) lead to derivatives with respect to the arc-lengths s_1 and s_2 as local coordinates being zero:

$$\mathbf{F'} = \begin{cases} (\boldsymbol{\rho}_1 - \boldsymbol{\rho}_2) \cdot \boldsymbol{\rho}_1' &= 0 \\ -(\boldsymbol{\rho}_1 - \boldsymbol{\rho}_2) \cdot \boldsymbol{\rho}_2' &= 0 \end{cases}$$

$$= \begin{cases} (\boldsymbol{\rho}_1 - \boldsymbol{\rho}_2) \cdot \boldsymbol{\tau}_1 &= 0 \\ -(\boldsymbol{\rho}_1 - \boldsymbol{\rho}_2) \cdot \boldsymbol{\tau}_2 &= 0 \end{cases}, \tag{12.6}$$

with unit tangent vectors introduced as derivatives with respect to the arc-length $\boldsymbol{\tau}_I = \boldsymbol{\rho}_I'$.

In general, equation system (12.6) is non-linear and an iterative method should be applied for its solution in a numerical realization. For the nonlinear system the question of solvability becomes then important. Such an analysis is possible in a special coordinate system defined in the following section.

12.2.1 Definition of a local coordinate system

First, we start with a standard Serret-Frenet basis connected to the I-th curve. As is known, at each sufficiently smooth point of a curve one can define three unit vectors: a unit tangent vector $\boldsymbol{\tau}$, a unit normal vector $\boldsymbol{\nu}$ pointing to a center of curvature of the curve and a unit binormal vector $\boldsymbol{\beta}$ defined as a cross product

$$\boldsymbol{\beta} = \boldsymbol{\tau} \times \boldsymbol{\nu}. \tag{12.7}$$

These three unit vectors are connected via the derivatives of the curve ρ with respect to the arc-length s. The formulas are known in the standard literature on differential geometry, see e.g. in [103], [34], as Serret-Frenet formulas:

$$\begin{cases} \dfrac{d\boldsymbol{\tau}}{ds} &= & k\boldsymbol{\nu} \\[2mm] \dfrac{d\boldsymbol{\nu}}{ds} &= -k\boldsymbol{\tau} &+ & \varkappa\boldsymbol{\beta} \\[2mm] \dfrac{d\boldsymbol{\beta}}{ds} &= & -\varkappa\boldsymbol{\nu} \end{cases}. \tag{12.8}$$

Here, k is a curvature, and $\varkappa$ is a torsion of a spatial curve. Assuming an arbitrary parameterization in eqn. (12.1) and in the Serret-Frenet formulas (12.8) the curvature k and the torsion $\varkappa$ of a spatial curve can be computed as

$$k\boldsymbol{\beta} = \frac{\dot{\boldsymbol{\rho}} \times \ddot{\boldsymbol{\rho}}}{|\dot{\boldsymbol{\rho}}|^3} \longrightarrow k = \frac{|\dot{\boldsymbol{\rho}} \times \ddot{\boldsymbol{\rho}}|}{|\dot{\boldsymbol{\rho}}|^3}, \tag{12.9}$$

$$\varkappa = \frac{\det(\dot{\boldsymbol{\rho}}, \ddot{\boldsymbol{\rho}}, \partial^3\boldsymbol{\rho}/\partial\xi^3)}{|\dot{\boldsymbol{\rho}} \times \ddot{\boldsymbol{\rho}}|^2} = \frac{(\dot{\boldsymbol{\rho}} \times \ddot{\boldsymbol{\rho}}) \cdot \partial^3\boldsymbol{\rho}/\partial\xi^3}{|\dot{\boldsymbol{\rho}} \times \ddot{\boldsymbol{\rho}}|^2}. \tag{12.10}$$

Now, we attach these Serret-Frenet frames $\boldsymbol{\tau}_I(s_I), \boldsymbol{\nu}_I(s_I), \boldsymbol{\beta}_I(s_I)$ to the projection points (C_1 and C_2), see Fig. 12.1.

The closest distance between the curves denoted as absolute value of the vector $|\overrightarrow{C_1 C_2}|$ becomes a second coordinate r in our specially defined coordinate system, while the arc-length will be the first coordinate s_I. A vector corresponding to the closest distance $\overrightarrow{C_1 C_2}$ and defined as $\boldsymbol{\rho}_{12}$ is orthogonal to both vectors $\boldsymbol{\tau}_1$ and $\boldsymbol{\tau}_2$ and, therefore, is representing an intersection line of two planes $\boldsymbol{\nu}_1 C_1 \boldsymbol{\beta}_1$ and $\boldsymbol{\nu}_2 C_2 \boldsymbol{\beta}_2$. The unit vector of $\boldsymbol{\rho}_{12}$ is denoted as $\mathbf{e}_1$. Introducing the parameterization in the plane $\boldsymbol{\nu}_1 C_1 \boldsymbol{\beta}_1$ for the vector $\mathbf{e}_1$ via an angle φ_1, see again Fig. 12.1 we obtain

$$\boldsymbol{\rho}_{12} = \boldsymbol{\rho}_2 - \boldsymbol{\rho}_1 = r\mathbf{e}_1(\varphi_1) \in \boldsymbol{\nu}_I C_I \boldsymbol{\beta}_I - \text{ plane}. \tag{12.11}$$

Here, the angle φ_1 between the unit normal $\boldsymbol{\nu}_I$ and the closest distance vector $\boldsymbol{\rho}_{12}$ becomes then a third coordinate φ_1 in our new coordinate system. Thus, we can redefine a new coordinate system based on

a) a unit tangent vector $\boldsymbol{\tau}_I$,

b) a unit vector of the closest distance vector $\mathbf{e}_I$,

c) a unit vector $\mathbf{g}_I$ orthogonal to both $\boldsymbol{\tau}_I$ and $\mathbf{e}_I$.

The last two vectors are defined via the basic unit vectors $\boldsymbol{\nu}_I, \boldsymbol{\beta}_I$ as follows:

$$\mathbf{e}_1(\varphi_1) \equiv \boldsymbol{\nu}_1 \cos \varphi_1 + \boldsymbol{\beta}_1 \sin \varphi_1 \tag{12.12a}$$

$$\mathbf{g}_1(\varphi_1) = \frac{\partial \mathbf{e}_1}{\partial \varphi_1} = -\boldsymbol{\nu}_1 \sin \varphi_1 + \boldsymbol{\beta}_1 \cos \varphi_1. \tag{12.12b}$$

In fact $\mathbf{e}_1$ and $\mathbf{g}_1$ are defining a polar coordinate system in the plane orthogonal to $\boldsymbol{\tau}_I$. The new introduced coordinate system defined by the unit vectors τ_I, e_I, g_I is orthogonal, but changing in space with regard to a given curve. Summarizing, a relative motion of a point C_2 and thus the motion of a second curve can now be considered in the following coordinate system:

$$\boldsymbol{\rho}_2(s_1, r, \varphi_1) = \boldsymbol{\rho}_1(s_1) + r e_1(\varphi_1), \quad 1 \leftrightarrow 2. \tag{12.13}$$

Thus, a 3D-local curvilinear coordinate system with convective coordinates s_1, r, φ_1 is defined. These coordinates are representing a set of convective coordinates for the curve in analogy to convective surface coordinates ξ^1, ξ^2 and the penetration ξ^3. Extensive usage of convective coordinates for surface-to-surface contact is described in Laursen [106].

Remark.
the description is symmetric with regard to any choice of a curve $1 \leftrightarrow 2$.

Remark.
the coordinate r is always positive and identical for both curves.

12.2.2 Analysis of uniqueness and existence of solutions for the CPP: Definition of a projection domain

In this section we discuss a choice in which solutions of the CPP procedure exist and are unique a-priori. A similar analysis of the CPP procedure for surfaces is given in Konyukhov and Schweizerhof [95]. As is known, the convexity of the function $\mathbf{F}$ in eqn. (12.4) would lead to the fulfillment of uniqueness and existence. Then the positivity of a second derivative $\mathbf{F}'' > 0$ follows. The second derivative exploiting the Serret-Frenet formulas (12.8) can be written as

$$
\mathbf{F}'' = \begin{bmatrix} \boldsymbol{\rho}_1' \cdot \boldsymbol{\tau}_1 + (\boldsymbol{\rho}_1 - \boldsymbol{\rho}_2) \cdot \boldsymbol{\tau}_1' & -\boldsymbol{\rho}_2' \cdot \boldsymbol{\tau}_1 \\ -\boldsymbol{\rho}_1' \cdot \boldsymbol{\tau}_2 & \boldsymbol{\rho}_2' \cdot \boldsymbol{\tau}_2 - (\boldsymbol{\rho}_1 - \boldsymbol{\rho}_2) \cdot \boldsymbol{\tau}_2' \end{bmatrix}
$$

$$
= \begin{bmatrix} \boldsymbol{\tau}_1 \cdot \boldsymbol{\tau}_1 + k_1(\boldsymbol{\rho}_1 - \boldsymbol{\rho}_2) \cdot \boldsymbol{\nu}_1 & -\boldsymbol{\tau}_1 \cdot \boldsymbol{\tau}_2 \\ -\boldsymbol{\tau}_1 \cdot \boldsymbol{\tau}_2 & \boldsymbol{\tau}_2 \cdot \boldsymbol{\tau}_2 - k_2(\boldsymbol{\rho}_1 - \boldsymbol{\rho}_2) \cdot \boldsymbol{\nu}_2 \end{bmatrix}
$$

$$
= \begin{bmatrix} 1 + k_1(\boldsymbol{\rho}_1 - \boldsymbol{\rho}_2) \cdot \boldsymbol{\nu}_1 & -\boldsymbol{\tau}_1 \cdot \boldsymbol{\tau}_2 \\ -\boldsymbol{\tau}_1 \cdot \boldsymbol{\tau}_2 & 1 + k_2(\boldsymbol{\rho}_2 - \boldsymbol{\rho}_1) \cdot \boldsymbol{\nu}_2 \end{bmatrix}.
$$

$$\tag{12.14}$$

Defining then ψ as an angle between two tangent lines $\boldsymbol{\tau}_1$ and $\boldsymbol{\tau}_2$

$$\boldsymbol{\tau}_1 \cdot \boldsymbol{\tau}_2 = \cos \psi \tag{12.15}$$

and taking into account the coordinate system in eqn. (12.13) we obtain

$$\mathbf{F}'' = \begin{bmatrix} 1 - k_1 r \cos \varphi_1 & -\cos \psi \\ -\cos \psi & 1 - k_2 r \cos \varphi_2 \end{bmatrix} \tag{12.16}$$

with the determinant as

$$\det \mathbf{F}'' = (1 - k_1 r \cos \varphi_1)(1 - k_2 r \cos \varphi_2) - \cos^2 \psi. \tag{12.17}$$

A Sylvester criterion can be now applied to enforce the positivity of the matrix $\mathbf{F}''$. This leads to both, a positive first term of the matrix $F''_{11} > 0$ and a positive determinant $\det \mathbf{F}'' > 0$:

$$\begin{cases} 1 - k_I r \cos \varphi_I & > & 0 \\ \sin^2 \psi + r(r k_1 k_2 \cos \varphi_1 \cos \varphi_2 - (k_1 \cos \varphi_1 + k_2 \cos \varphi_2)) & > & 0. \end{cases} \tag{12.18}$$

We shall now study step-by-step several particular cases in order to understand the geometrical properties of the inequality conditions in eqn. (12.18).

12.2.2.1   Case 1: both lines are straight lines

In this – simplest – case both curvatures are zero $k_1 \equiv 0$, $k_2 \equiv 0$ and eqn. (12.17), resp. the Sylvester criterion in eqn. (12.18), is transformed as

$$\sin^2 \psi \neq 0 \longrightarrow \psi \neq \pi n, \ n = 0, 1, 2, ... \tag{12.19}$$

This has a simple geometrical interpretation – **for the existence and uniqueness of the CPP procedure for straight lines the angle defined between tangent vectors $\boldsymbol{\tau}_1$ and $\boldsymbol{\tau}_2$ in eqn. (12.15) should not be zero and, therefore, the lines should not be parallel.**

12.2.2.2 Case 2: two circles in a plane

As a next step for the complexification we consider first two overlapping circles laying in a plane, see Figs. 12.2, 12.3 and 12.4. In this case an angle between the tangent vectors τ_1 and τ_2 is zero $\psi = 0$ (the critical case for the straight lines). The situation is also rather difficult for overlapping circles with the angle between the normal ν_1 and the vector e_1 being zero $\varphi_1 = \varphi_2 = 0$. Denoting the radii of curvature as as $R_I = \dfrac{1}{k_I}$, the first equation in the Sylvester criterion (12.18) leads to $r < R_I$, however the determinant gives us the requirement $r > R_1 + R_2$. This inconsistency for an enforcement of positivity means that the second derivative $\mathbf{F}''$ is neither positive, nor negative, making it impossible to use the second derivative $\mathbf{F}''$ only. One can see the difficulties in Fig. 12.2 where the multiplicity of the solution is depicted: the shortest distance can be defined as $C_1 S_2$, while the initial procedure assumes $C_1 C_2$. We note that even in this case the determinant is not zero which makes it possible to exploit the Newton scheme for the iterative solution.

It is possible to simplify the analysis of the uniqueness if we **consider projection domains for both curves** separately. The projection domain for a curve, see Konyukhov and Schweizerhof [95], is defined as a 3D domain from which any point can be projected uniquely onto the current curve. This domain is obtained as follows.

Projection domain for a spatial curve

1. a half-space domain with $\varphi_I \in [\pi/2, 3\pi/2]$:

$$\Omega(x,y,z) = \left\{ \mathbf{r} \in \mathbf{R}^3 \,|\, \mathbf{r} = \boldsymbol{\rho}_I(s_I) + r\mathbf{e}_I(\varphi_I),\ \varphi_I \in (\frac{3\pi}{2}, 2\pi),\ r \in (0, \inf) \right\}$$

2. a layer with $\varphi \in (-\pi/2, \pi/2)$. The projection of the vector $r\mathbf{e}_I$ onto the normal ν_I must be smaller than the radius of curvature $R_I = \dfrac{1}{k_I}$ (shadowed domains in Fig. 12.7):

$$\Omega(x,y,z) = \left\{ \mathbf{r} \in \mathbf{R}^3 \,|\, \mathbf{r} = \boldsymbol{\rho}_I(s_I) + r\mathbf{e}_I(\varphi_I),\ \varphi_I \in (-\frac{\pi}{2}, \frac{\pi}{2}),\ 0 < r\cos\varphi_I < R_I. \right\}$$

In a case of overlapping circles we are falling into case (2), see Fig. 12.3.

Summarizing, we can derive the domain of existence and uniqueness of the CPP procedure as an overlapping of two projection domains for the corresponding curves. Thus **for the existence and uniqueness of the CPP procedure for two overlapping circles the distance between two points should be less than the minimal radius of curvature**.

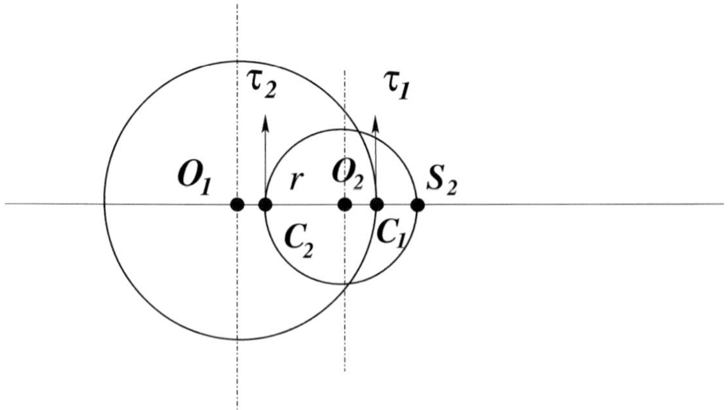

Figure 12.2: Two overlapping circles in a single plane. The CPP procedure can have a multiple solution: which choice C_1C_2 or C_1S_2?

For non-overlapping circles, see Fig. 12.4, the angle is $\varphi_1 = \pi$, and the Sylvester criterion (12.18) is always fulfilled leading to the conclusion that **the closest distance between two non-overlapping circles is uniquely defined**. This situation is also falling into case (1) as the overlapping of projection domains for two separated circles.

Remark.
The conclusions of the current case are still valid if we consider arbitrary curves with the osculating circles laying in parallel planes.

12.2.2.3 Case 3: osculating circles in orthogonal planes

Let us consider now two arbitrary curved lines ($k_1 \neq 0$, $k_2 \neq 0$) which have two osculating circles laying in orthogonal planes, but still possessing parallel tangent vectors, e.g. $\psi = \pi$, for the case shown in Fig. 12.5.

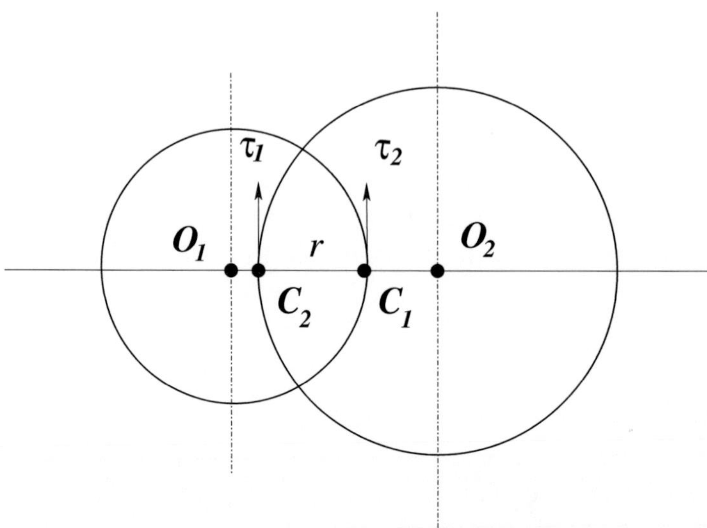

Figure 12.3: Two overlapping circles in a single plane. The CPP procedure has a unique solution in the overlapping region of two projection domains $\{r \in (0, R_1)\} \cap \{r \in (0, R_2)\}$ and, therefore, if $r < \min\{R_1, R_2\}$

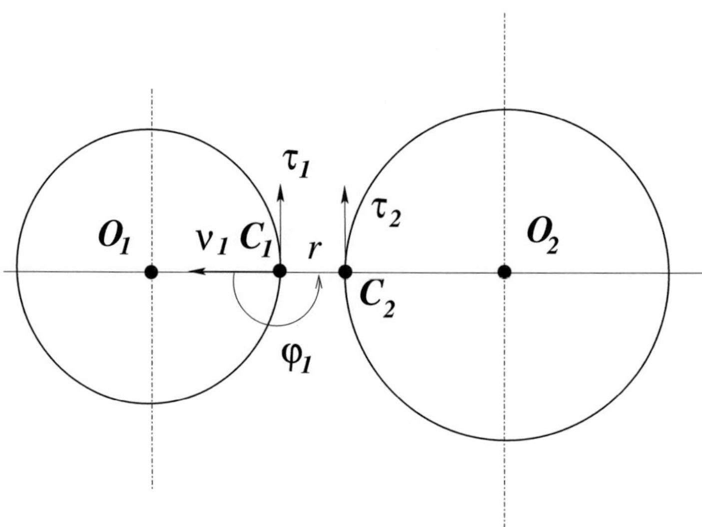

Figure 12.4: Two non-overlapping circles in a single plane. The CPP procedure has always a unique solution (Closest distance is uniquely defined).

One can see that for the exterior projection with $\varphi_1, \varphi_2 \in [\pi/2, 3\pi/2]$. the Sylvester criterion is fulfilled.

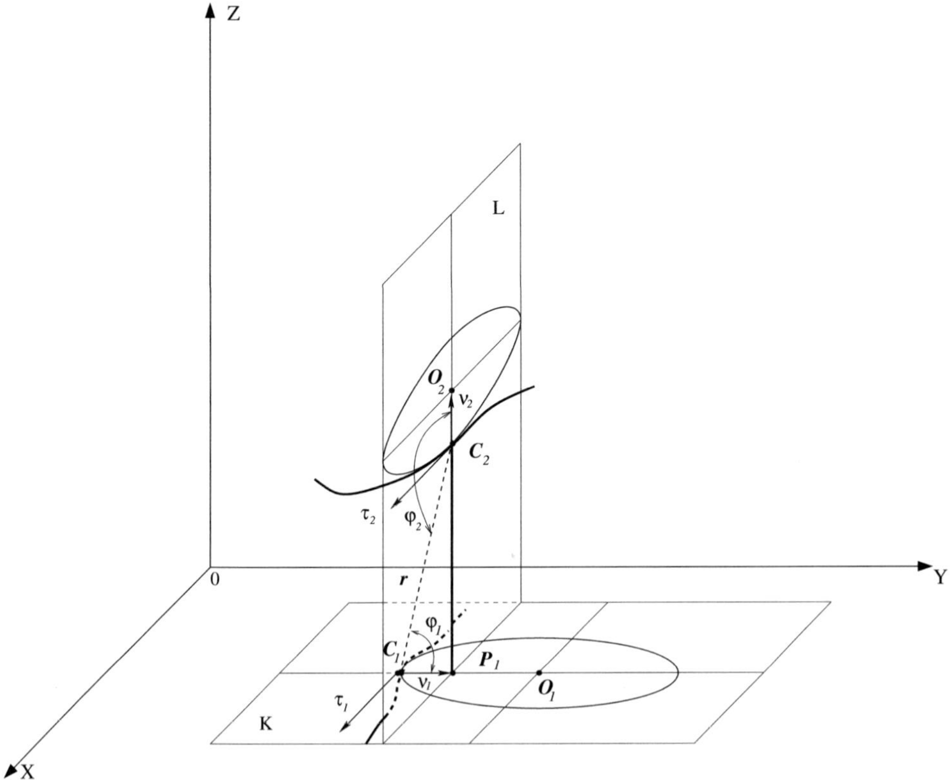

Figure 12.5: Two curves with osculating circles laying in orthogonal planes. The closest distance is uniquely defined if two points are situated in the overlapping projection domains. $C_1 \in \{(r, \varphi_2) \,|\, 0 < r + \infty, \; \varphi_2 \in [\pi/2, 3\pi/2]\}$ $C_2 \in \{(r, \varphi_1) \,|\, r \cos \varphi_1 < R_1, \; \varphi_1 \in [-\pi/2, \pi/2]\}$.

12.2.2.4 Case 4: a singular determinant for curved lines

The example of an extremal situation with singular determinant $\mathbf{F}''$ in eqn. (12.17) is illustrated in Fig. 12.6. In this case the tangent vectors τ_1 and τ_2 are orthogonal leading to $\psi = \pi/2$, and therefore, $\cos \psi = 0$. Moreover, the closest point C_2 is projected exactly into a center of curvature O_1 leading to the fulfillment of $R_1 - r \cos \varphi_1 = 0$ and finally to $\mathbf{F}'' = 0$.

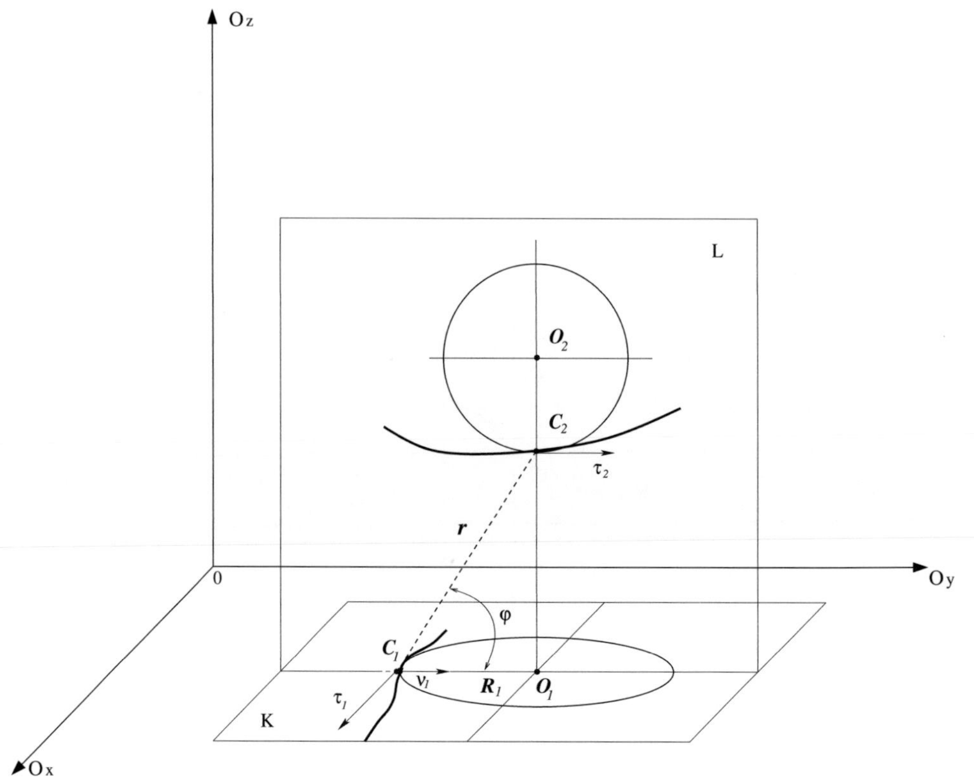

Figure 12.6: Singular determinant $\det \mathbf{F}'' = 0$ for arbitrary curves. The tangent vectors are orthogonal and the closest point C_2 is projected into a center of curvature.

12.2.2.5 Case 5: arbitrary curves in space

For arbitrary situations the direct analysis of eqns. (12.18) is rather complicated, however, the concept of two overlapping projection domains described in paragraph 12.2.2.2 can be effectively applied. Fig. 12.7 illustrates the analysis. A point C_2 lays in the projection domain of type (1) in paragraph 12.2.2.2 (a half-space domain) of the first curve. A point C_1 lays in the projection domain of type (2) in paragraph 12.2.2.2 (a layer) of the first curve because the projection onto the osculating circle P_2C_2 is smaller than the radius of curvature $R_2 = O_2C_2$. The domain of existence and uniqueness of the CPP procedure is constructed as the overlapping of projection domains for two curves.

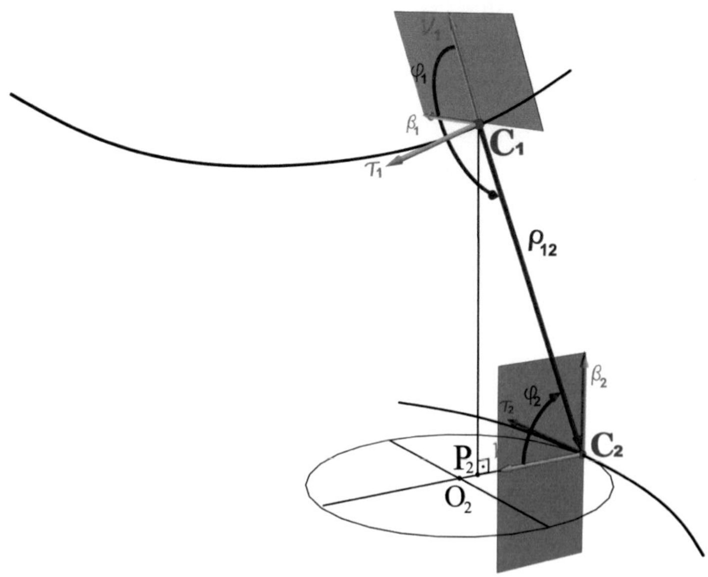

Figure 12.7: Two arbitrary curves. The solution depends on the positions of osculating circles which give rise to corresponding projection domains. The domain for uniqueness of the solution of a CPP procedure for both curves is constructed as the overlapping of projection domains for two curves.

12.2.3 Computational issues of the CPP procedure

The CPP procedure written in terms of the arc-length parameter in eqn. (12.4) is efficiently solved numerically via a Newton-type solver. The procedure can be outlined as follows. With arbitrary parameterizations via ξ_1 and ξ_2, the second derivative $\mathbf{F}''$ in eqn. (12.14) becomes then a matrix

$$\frac{\partial^2 \mathbf{F}}{\partial \xi^2} = \ddot{\mathbf{F}} = \frac{\partial \mathbf{\Phi}}{\partial \xi} = \begin{bmatrix} \dot{\rho}_1 \cdot \dot{\rho}_1 + (\rho_1 - \rho_2) \cdot \ddot{\rho}_1 & -\dot{\rho}_1 \cdot \dot{\rho}_2 \\ -\dot{\rho}_1 \cdot \dot{\rho}_2 & \dot{\rho}_2 \cdot \dot{\rho}_2 + (\rho_2 - \rho_1) \cdot \ddot{\rho}_2 \end{bmatrix} \quad (12.20)$$

while the first derivative $\mathbf{F}'$ is defined similar to eqn. (12.6)

$$\dot{\mathbf{F}} = \mathbf{\Phi} = \left\{ \begin{array}{c} (\rho_1 - \rho_2) \cdot \dot{\rho}_1 \\ -(\rho_1 - \rho_2) \cdot \dot{\rho}_2 \end{array} \right\} . \quad (12.21)$$

The last equation should be solved iteratively as $\dot{\mathbf{F}} = \mathbf{\Phi} = 0$. Here an additional variable $\mathbf{\Phi}$ is introduced to illustrate the classical structure of the Newton iterative scheme.

The tangential coordinates ξ_1^p and ξ_2^p along both lines are computed then via a Newton iteration procedure as

$$\boldsymbol{\xi}^{(n+1)} = \boldsymbol{\xi}^{(n)} + \Delta\boldsymbol{\xi} \text{ with an initial guess } \boldsymbol{\xi}^{(0)}$$

$$\text{and increments } \Delta\boldsymbol{\xi} = \left\{ \begin{array}{c} \Delta\xi_1 \\ \Delta\xi_2 \end{array} \right\} = - \left[\frac{\partial\mathbf{\Phi}}{\partial\boldsymbol{\xi}}\right]^{-1} \mathbf{\Phi} = -(\dot{\mathbf{F}})^{-1}\dot{\mathbf{F}}. \quad (12.22)$$

In the section devoted to the finite element implementation we specify this procedure for a particular contact element.

12.3 Kinematics of contact.
Measures of contact interaction

In this section we study differential properties of the coordinate system defined in eqn. (12.13). Since the coordinate system is arbitrarily curvilinear it is necessary to define the covariant operations. As a preparation, we compute partial derivatives of the unit vectors $\mathbf{e}_1, \mathbf{g}_1$ introduced in eqns. (12.12a, 12.12b).

The derivatives of the vector $\mathbf{e}_1$ with respect to s_1 and with respect to φ_1 are

$$\frac{\partial\mathbf{e}_1}{\partial s_1} = \cos\varphi_1(-k_1\boldsymbol{\tau}_1 + \varkappa_1\boldsymbol{\beta}_1) + \sin\varphi_1(-\varkappa_1)\boldsymbol{\nu}_1 = \quad (12.23a)$$

$$= -k_1\boldsymbol{\tau}_1\cos\varphi_1 + \varkappa_1\mathbf{g}_1$$

$$\frac{\partial\mathbf{e}_1}{\partial\varphi_1} = \mathbf{g}_1. \quad (12.23b)$$

The derivatives of the vector $\mathbf{g}_1$ with respect to s_1 and with respect to

φ_1 are

$$\frac{\partial \mathbf{g}_1}{\partial s_1} = -\sin\varphi_1[-k_1\boldsymbol{\tau}_1 + \varkappa_1\boldsymbol{\beta}_1] - \cos\varphi_1\varkappa_1\boldsymbol{\nu}_1 = \underline{k_1\boldsymbol{\tau}_1\sin\varphi_1 - \varkappa_1\mathbf{e}_1}$$

$$(12.24a)$$

$$\frac{\partial \mathbf{g}_1}{\partial \varphi_1} = -\mathbf{e}_1.$$

$$(12.24b)$$

Now, we can define coordinate vectors $\mathbf{r}_1, \mathbf{r}_2, \mathbf{r}_3$ (a subscript defining the first curve is omitted) of the curvilinear coordinate system as partial derivatives with respect to coordinates s_1, r, φ_1:

$$\mathbf{r}_1 = \frac{\partial \boldsymbol{\rho}_2}{\partial s_1} = \frac{\partial \boldsymbol{\rho}_1}{\partial s_1} + r\frac{\partial \mathbf{e}_1}{\partial s_1} = \boldsymbol{\tau}_1(1 - k_1 r\cos\varphi_1) + \varkappa r \mathbf{g}_1 \qquad (12.25a)$$

$$\mathbf{r}_2 = \frac{\partial \boldsymbol{\rho}_2}{\partial r} = \mathbf{e}_1 \qquad (12.25b)$$

$$\mathbf{r}_3 = \frac{\partial \boldsymbol{\rho}_2}{\partial \varphi} = r\mathbf{g}_1. \qquad (12.25c)$$

The corresponding covariant metric tensor is given by components (subscript $(.)_1$ is omitted)

$$g_{11} = (\mathbf{r}_1 \cdot \mathbf{r}_1) = (1 - kr\cos\varphi)^2 + \varkappa r^2$$
$$g_{12} = (\mathbf{r}_1 \cdot \mathbf{r}_2) = 0$$
$$g_{13} = (\mathbf{r}_1 \cdot \mathbf{r}_3) = \varkappa r^2$$
$$g_{22} = 1$$
$$g_{33} = r^2$$

leading to a matrix

$$[g_{ij}] = \begin{bmatrix} (1 - kr\cos\varphi)^2 + \varkappa^2 r^2 & 0 & \varkappa r^2 \\ 0 & 1 & 0 \\ \varkappa r^2 & 0 & r^2 \end{bmatrix}. \qquad (12.26)$$

The system is, in general, non-orthogonal (see $g_{13} \neq 0$) and is changing in space reflecting the geometry of a given curve. However, at the given point on a curve with $r = 0$ the system is orthogonal.

For the case of edge-to-edge contact the distance r between curves is a small value and the terms containing torsion $\varkappa_1$ and curvature k_1 are also usually small at least within a single finite element. It motivates us to consider all parameters in such a case in a coordinate system exactly attached to the curve at $r = 0$, removing all small terms containing combinations of either a curvature or a torsion together with the r-coodinate, see the analogy for a tangent plane for the surface-to-surface contact in [86], [89] and [92]. Thus, we consider also, an orthogonal basis as

$$\mathbf{r}_1 = \boldsymbol{\tau}_1, \quad \mathbf{r}_2 = \mathbf{e}_1, \quad \mathbf{r}_3 = r\mathbf{g}_1 \qquad (12.27)$$

with the corresponding metric tensor defined by a matrix

$$[g_{ij}] = \begin{bmatrix} 1 & 0 & 0 \\ 0 & 1 & 0 \\ 0 & 0 & r^2 \end{bmatrix}. \qquad (12.28)$$

Exactly for this situation we define the contravariant components of the metric tensor as inverse components of the matrix in eqn. (12.28). It leads to the standard definition of the contravariant basis $\mathbf{r}^i$, $i = 1, 2, 3$ for the curve with $r = 0$ as

$$\mathbf{r}^i = g^{ij}\mathbf{r}_j \qquad (12.29a)$$

$$\mathbf{r}^1 = \boldsymbol{\tau}_1, \quad \mathbf{r}^2 = \mathbf{e}_1, \quad \mathbf{r}^3 = \frac{1}{r}\mathbf{g}_1. \qquad (12.29b)$$

12.3.1 Rates and variations of measures for contact interaction

Now we can compute a full time derivative for a vector $\mathbf{r}_2(s_1, r, \varphi_1)$ describing a full velocity vector in our coordinate system

$$\frac{d\boldsymbol{\rho}_2}{dt} = \frac{\partial\boldsymbol{\rho}_1}{\partial t} + (1 - rk_1\cos\varphi_1)\dot{s}_1\boldsymbol{\tau}_1 + \dot{r}\mathbf{e}_1 + r(\varkappa_1\dot{s}_1 + \dot{\varphi}_1)\mathbf{g}_1. \qquad (12.30)$$

Redefining now the full velocity vector of the second point C_2 as

$$\mathbf{v}_2 = \frac{d\boldsymbol{\rho}_2}{dt} \qquad (12.31)$$

and the translational velocity of the first point C_1 as

$$\mathbf{v}_1^t = \frac{\partial \boldsymbol{\rho}_1^t}{\partial t} \tag{12.32}$$

we can rewrite the relative velocity vector of a point C_2 in the coordinate system of the first curve as

$$\mathbf{v}_2 - \mathbf{v}_1^t = (1 - rk_1 \cos \varphi_1)\dot{s}_1 \boldsymbol{\tau}_1 + \dot{r}\mathbf{e}_1 + r(\varkappa_1 \dot{s}_1 + \dot{\varphi}_1)\mathbf{g}_1. \tag{12.33}$$

Keeping in mind the edge-to-edge contact applications it is more convenient to consider the relative velocity vector in the coordinate system attached exactly to the contact point. Thus, we obtain the expression of the relative velocity vector (12.33) via the vectors $\boldsymbol{\tau}_1, \mathbf{e}_1, \mathbf{g}_1$ as

$$\mathbf{v}_2 - \mathbf{v}_1^t = \dot{s}_1 \boldsymbol{\tau}_1 + \dot{r}\mathbf{e}_1 + (\varkappa_1 \dot{s}_1 + \dot{\varphi}_1)r\mathbf{g}_1. \tag{12.34}$$

Here, the last term remains reflecting the geometry of a 3D spatial curve, because $r\mathbf{g}_1$ is a covariant basis vector $\mathbf{r}_3$, but not a small value, however a term $rk_1 \cos \varphi_1$ is assumed to be small.

Computing a scalar product with the coordinate vectors $\mathbf{r}_I$ at $r = 0$ and, therefore, with $\boldsymbol{\tau}_1, \mathbf{e}_1, \mathbf{g}_1$ we obtain the rate of convective coordinates as

$$\left\{ \begin{array}{rcl} \dot{s}_1 & = & (\mathbf{v}_2 - \mathbf{v}_1^t) \cdot \boldsymbol{\tau}_1 \\ \dot{r}_1 & = & (\mathbf{v}_2 - \mathbf{v}_1^t) \cdot \mathbf{e}_1 \\ \dot{\varphi}_1 + \varkappa_1 \dot{s}_1 & = & (\mathbf{v}_2 - \mathbf{v}_1^t) \cdot \dfrac{\mathbf{g}_1}{r} \end{array} \right. \qquad 1 \leftrightarrow 2. \tag{12.35}$$

This gives us the rate of measures for the contact interaction between curves

- $\dot{s}_1$ – for tangential interaction;

- $\dot{r}$ – for normal interaction;

- $\dot{\varphi}_1$ – for angular interaction.

The last coordinate is describing the rotation of the first curve (plane $\mathbf{e}_1, \mathbf{g}_1$) along the vector $\boldsymbol{\tau}_1$.

For arbitrary large motions we also derive the exact – not simplified –

expression for the rates of measures from eqn. (12.33):

$$\begin{cases} \dot{s}_1 &= \dfrac{(\mathbf{v}_2 - \mathbf{v}_1^t) \cdot \boldsymbol{\tau}_1}{(1 - rk_1 \cos \varphi_1)} \\[2mm] \dot{r}_1 &= (\mathbf{v}_2 - \mathbf{v}_1^t) \cdot \mathbf{e}_1 \\[2mm] \dot{\varphi}_1 &= \dfrac{(\mathbf{v}_2 - \mathbf{v}_1^t) \cdot \mathbf{g}_1}{r} - \varkappa_1 \dot{s}_1 = \\[2mm] &= \dfrac{(\mathbf{v}_2 - \mathbf{v}_1^t) \cdot \mathbf{g}_1}{r} - \varkappa_1 \dfrac{(\mathbf{v}_2 - \mathbf{v}_1^t) \cdot \boldsymbol{\tau}_1}{(1 - rk_1 \cos \varphi_1)}. \end{cases} \qquad 1 \leftrightarrow 2 \qquad (12.36)$$

Remark.

Using the analogy between the kinematical values and the variations we can write the variations of the relative displacements in full form (see eqn. (12.33)) as

$$\delta\boldsymbol{\rho}_2 - \delta\boldsymbol{\rho}_1 = (1 - rk_1 \cos \varphi_1)\delta s_1 \boldsymbol{\tau}_1 + \delta r \mathbf{e}_1 + r(\varkappa_1 \delta s_1 + \delta\varphi_1)\mathbf{g}_1, \qquad (12.37)$$

or taking into account small values at $r = 0$ (see eqn. (12.34)) as

$$\delta\boldsymbol{\rho}_2 - \delta\boldsymbol{\rho}_1 = \delta s_1 \boldsymbol{\tau}_1 + \delta r \mathbf{e}_1 + r(\varkappa_1 \delta s_1 + \delta\varphi_1)\mathbf{g}_1 \qquad (12.38)$$

as well as the variations of convective coordinates at $r = 0$ as

$$\begin{cases} \delta s_1 &= (\delta\boldsymbol{\rho}_2 - \delta\boldsymbol{\rho}_1) \cdot \boldsymbol{\tau}_1 \\[1mm] \delta r_1 &= (\delta\boldsymbol{\rho}_2 - \delta\boldsymbol{\rho}_1) \cdot \mathbf{e}_1 \\[1mm] r(\varkappa_1 \delta s_1 + \delta\varphi_1) &= (\delta\boldsymbol{\rho}_2 - \delta\boldsymbol{\rho}_1) \cdot \mathbf{g}_1 \end{cases} \qquad 1 \leftrightarrow 2. \qquad (12.39)$$

12.3.2 Linearization in a covariant form of variations for contact measures

For the forthcoming linearization we need also the covariant derivatives of the variation of the relative displacement vector in eqn (12.37). The covariant derivation of vector components is the derivation operation taking into account the change of a basis vector in space. Following the standard derivation the Christoffel symbols Γ_{ij}^k, see [34], are required

$$\frac{\partial \boldsymbol{\rho}_i}{\partial \xi^j} = \boldsymbol{\rho}_{ij} = \Gamma_{ij}^k \boldsymbol{\rho}_k, \qquad (12.40)$$

but instead of computing the Christoffel symbols, we will directly compute derivatives of coordinate vectors using their definitions in eqns. (12.12a-12.12b) and the Serret-Frenet formulas (12.8). For the derivation of the relative virtual displacements we can take the full values given in eqn. (12.37) without any simplification. The following derivatives are obtained

with respect to s_1

$$\frac{\partial}{\partial s_1}(\delta\boldsymbol{\rho}_2 - \delta\boldsymbol{\rho}_1) = \qquad (12.41a)$$

$$= k_1(1 - k_1 r \cos\varphi_1)\delta s_1 \boldsymbol{\nu}_1 + (-k_1 \cos\varphi_1 \boldsymbol{\tau}_1 + \varkappa_1 \mathbf{g}_1)\delta r + r(\varkappa_1 \delta s_1 + \delta\varphi_1)(k_1 \sin\varphi_1 \boldsymbol{\tau}_1 - \varkappa_1 \mathbf{e}_1)$$

$$= [-k_1 \cos\varphi_1 \delta r + (\varkappa_1 \delta s_1 + \delta\varphi_1) r k_1 \sin\varphi_1]\boldsymbol{\tau}_1 +$$

$$k_1(1 - k_1 r \cos\varphi_1)\delta s_1 \boldsymbol{\nu}_1 - r\varkappa_1(\varkappa_1 \delta s_1 + \delta\varphi_1)\mathbf{e}_1 + \varkappa_1 \delta r \mathbf{g}_1;$$

with respect to r

$$\frac{\partial}{\partial r}(\delta\boldsymbol{\rho}_2 - \delta\boldsymbol{\rho}_1) = -k_1 \cos\varphi_1 \delta s_1 \boldsymbol{\tau}_1 + (\varkappa_1 \delta s_1 + \delta\varphi_1)\mathbf{g}_1 \qquad (12.42)$$

and with respect to φ_1

$$\frac{\partial}{\partial\varphi_1}(\delta\boldsymbol{\rho}_2 - \delta\boldsymbol{\rho}_1) = k_1 r \sin\varphi_1 \delta s_1 \boldsymbol{\tau}_1 + \delta r \mathbf{g}_1 - (\varkappa_1 \delta s_1 + \delta\varphi_1) r \mathbf{e}_1. \qquad (12.43)$$

The full covariant derivative operator $\mathcal{L}_{s,r,\varphi}[...]$ has the following form (taking into account the derivation of coordinate vectors)

$$\mathcal{L}_{s,r,\varphi}[...] \equiv \dot{s}\frac{\partial}{\partial s}[...] + \dot{r}\frac{\partial}{\partial r}[...] + \dot{\varphi}\frac{\partial}{\partial\varphi}[...]. \qquad (12.44)$$

Summarizing all partial derivatives in eqns. (12.41a), (12.42), (12.43) in one operator (12.44) and expressing a normal $\boldsymbol{\nu}$ via e and g (eqns. (12.12a, 12.12b)) as

$$\boldsymbol{\nu}_1 = \mathbf{e}_1 \cos\varphi_1 - \mathbf{g}_1 \sin\varphi_1 \qquad (12.45)$$

we obtain the derivative of the variation of the relative displacement vector, or in fact the second derivative of the relative displacement vector, as

$$\mathcal{L}_{s,r,\varphi}[(\delta\boldsymbol{\rho}_2 - \delta\boldsymbol{\rho}_1)] = [rk_1\varkappa_1\sin\varphi\,\dot{s}_1\delta s_1 - \tag{12.46a}$$

$$- k_1\cos\varphi_1(\dot{s}_1\delta r + \dot{r}\delta s_1) + rk_1\sin\varphi_1(\dot{s}_1\delta\varphi_1 + \dot{\varphi}_1\delta s_1)]\,\boldsymbol{\tau}_1 \tag{12.46b}$$

$$+ [k_1\cos\varphi_1(1 - rk_1\cos\varphi_1)\dot{s}_1\delta s_1 - r(\varkappa_1\dot{s}_1 + \dot{\varphi}_1)(\varkappa_1\delta s_1 + \delta\varphi_1)]\,\mathbf{e}_1 \tag{12.46c}$$

$$+ [-k_1\sin\varphi_1(1 - rk_1\cos\varphi_1)\dot{s}_1\delta s_1 + (\dot{r}\delta\varphi_1 + \dot{\varphi}_1\delta r) + \varkappa_1(\dot{s}_1\delta r + \dot{r}\delta s_1)]\,\mathbf{g}_1 \tag{12.46d}$$

Remark.

The long expressions in eqns. (12.46b), (12.46c), (12.46d) represent the exact linearization in a covariant form for the case of an arbitrary motion (even for a large distance r between curves). For simplification the derivatives of the curvature k and the torsion $\varkappa$ are not presented, however, due to the derivation with respect to the arc-length s, the expression is exact for a line of constant curvature and torsion, e.g. for circular and spiral lines.

We have to note that the linearization is given by *a symmetric bilinear form with respect to convective variables* s, r, φ.

12.4   Weak form

A weak form, or variational form, is representing the equilibrium conditions between two curves via the principle of virtual work. Let us consider $\mathbf{R}_I ds_I$ as a force vector distributed along a curve segment ds_I, $I = 1, 2$. For an infinitesimal curve segment ds_I a virtual displacement vector $\delta\boldsymbol{\rho}_I$ is defined. Then a virtual work δW for two contacting curves is computed as integral along the mutual contact area

$$\delta W = \int_{s_1} \delta\boldsymbol{\rho}_1 \cdot \mathbf{R}_1 ds_1 + \int_{s_2} \delta\boldsymbol{\rho}_2 \cdot \mathbf{R}_2 ds_2. \tag{12.47}$$

Considering that two curves are contacting along the mutual boundary the equilibrium conditions should be point-wisely fulfilled

$$\mathbf{R}_1 ds_1 + \mathbf{R}_2 ds_2 = 0. \tag{12.48}$$

Using eqn. (12.48) in eqn. (12.47) the virtual work can be equivalently written as an integral either along the first curve, or along the second curve

$$\delta W = \int_{s_1} \mathbf{R}_1 \cdot (\delta\boldsymbol{\rho}_1 - \delta\boldsymbol{\rho}_2)ds_1 = \int_{s_2} \mathbf{R}_2 \cdot (\delta\boldsymbol{\rho}_2 - \delta\boldsymbol{\rho}_1)ds_2. \tag{12.49}$$

For symmetry reasons we can use the following:

$$\delta W = \frac{1}{2}\left\{\int_{s_1} \mathbf{R}_1 \cdot (\delta\boldsymbol{\rho}_1 - \delta\boldsymbol{\rho}_2)ds_1 + \int_{s_2} \mathbf{R}_2 \cdot (\delta\boldsymbol{\rho}_2 - \delta\boldsymbol{\rho}_1)ds_2\right\}. \tag{12.50}$$

Now taking into account the developed contact kinematics between curves – *only point-wise contact between curves with regard to the CPP procedure* – we can define forces acting point-wisely via the Dirac delta function $D(s_I - s_I^p)$ along curve s_I at the contact (projection) point s_I^p

$$\mathbf{R}_I(s_I) = \mathbf{R}_I D(s_I - s_I^p). \tag{12.51}$$

Thus, the virtual work in eqn. (12.50) is immediately transformed into

$$\delta W = Sym\,\{\mathbf{R}_1 \cdot (\delta\boldsymbol{\rho}_1 - \delta\boldsymbol{\rho}_2)\} = \frac{1}{2}\{\mathbf{R}_1 \cdot (\delta\boldsymbol{\rho}_1 - \delta\boldsymbol{\rho}_2) + \mathbf{R}_2 \cdot (\delta\boldsymbol{\rho}_2 - \delta\boldsymbol{\rho}_1)\}. \tag{12.52}$$

The force vector $\mathbf{R}_1$ should be defined as the energy conjugate vector, i.e. force components and variation of coordinates (12.38) are conjugated pairs representing that the scalar product in eqn. (12.52) leads to the virtual work. For arbitrary large motions of curves the force vector should be expressed via the vectors $\mathbf{r}_i$ in eqns. (12.25a), (12.25b), (12.25c) exploiting then the full metic tensor in eqn. (12.26). Thus, we can define $\mathbf{R}_1$ in a contravariant form as

$$\mathbf{R}_1 = R^i \mathbf{r}_i = R^i_{phys} \frac{\mathbf{r}_i}{\sqrt{g_{ii}}}, \tag{12.53}$$

where the physical components R^i_{phys} are introduced as

$$
\begin{aligned}
R^1_{phys} &= T(s_1) - \text{ tangential force along } \boldsymbol{\tau}_1, \\
R^2_{phys} &= N(r) - \text{ normal force along } \mathbf{e}_1, \\
R^3_{phys} &= R_\varphi(\varphi) - \text{ rotational (torsional) force along } \mathbf{g}_1.
\end{aligned}
\tag{12.54}
$$

Thus, the full force in the basis of a first curve, see Fig. 12.8, is written as

$$
\begin{aligned}
\mathbf{R}_1 &= T_1 \left[\frac{(1 - rk_1 \cos\varphi_1)\,\boldsymbol{\tau}_1 + \varkappa_1 r\,\mathbf{g}_1}{\sqrt{g_{11}}} \right] + N\,\mathbf{e}_1 + R_{\varphi_1}\,\mathbf{g}_1 = \\
&= T_1 \frac{(1 - rk_1 \cos\varphi_1)}{\sqrt{g_{11}}}\,\boldsymbol{\tau}_1 + N\,\mathbf{e}_1 + \left(R_{\varphi_1} + T_1 \frac{\varkappa_1 r}{\sqrt{g_{11}}} \right)\,\mathbf{g}_1.
\end{aligned}
\tag{12.55}
$$

Since the coordinate vectors $\mathbf{r}_i$ are changing, the tangential component T on a curve has then a contribution projected on both $\boldsymbol{\tau}_1$ and $\mathbf{g}_1$ vectors at $r \neq 0$. However, the force at the projection point of the first curve with $r = 0$ is expressed as follows

$$
\mathbf{R}_1 = T_1\boldsymbol{\tau}_1 + N\mathbf{e}_1 + \frac{M_1}{r}\mathbf{g}_1 \quad 1 \leftrightarrow 2.
\tag{12.56}
$$

The force N is a mutual normal force between two curves acting along the common normal vector $\mathbf{e}_1 = -\mathbf{e}_2$ (thus, subscript is omitted) and $R_\varphi = \dfrac{M_1}{r}$ represents a force acting at a distance r. This force results in an equivalent rotational moment M_1 at the projection point acting along the $\boldsymbol{\tau}_1$ axis of the first curve (plane $\mathbf{e}_1, \mathbf{g}_1$), see Fig. 12.8. The moment is appearing due to the dimension reduction and can be interpreted as a moment in a $1D$-Cosserat continuum. Using symmetry the corresponding components with the index $(...)_2$ are acting on the second curve.

Now, the virtual work in eqn. (12.52) taking into account eqn. (12.37) for the variation of the relative displacement vector is transformed into

$$
\delta W = Sym\{\mathbf{R}_1 \cdot (\delta\boldsymbol{\rho}_1 - \delta\boldsymbol{\rho}_2)\} = -Sym\{R^i \delta\xi^j g_{ij}\} = -Sym\left\{ \frac{R^i_{phys}}{\sqrt{g_{ii}}} g_{ij}\,\delta\xi^j \right\},
\tag{12.57}
$$

where the scalar product is expressed via the full metric tensor g_{ij} in eqn. (12.26) and, for a while, convective coordinates denoted as $s = \xi^1, r = \xi^2, \varphi = \xi^3$. Expressing eqn. (12.57) in the basis of a curve via the physical components we obtain:

$$\delta W = -Sym\left\{ T_1\sqrt{(1 - rk_1\cos\varphi_1)^2 + (r\varkappa_1)^2}\,\delta s_1 + N\,\delta r + M_1\,\delta\varphi_1 \right. \tag{12.58a}$$

$$\left. +T_1\frac{\varkappa_1 r^2}{\sqrt{(1 - rk_1\cos\varphi_1)^2 + (r\varkappa_1)^2}}\,\delta\varphi_1 + M_1\varkappa_1\,\delta s_1 \right\}. \tag{12.58b}$$

The minus sign appears as the expression in eqn. (12.37) is given for $(\delta\boldsymbol{\rho}_2 - \delta\boldsymbol{\rho}_1)$. The weak form expressed in the basis $\mathbf{r}_i$ with arbitrary r contains – due to the non-orthogonality of the coordinate vectors $\mathbf{r}_i$ – the last mixed terms in eqn. (12.58b). One can also obtain this result using eqn. (12.37) together with eqn. (12.55) directly. At the projection point on a curve ($r = 0$) the variational equation becomes

$$\delta W = Sym\left\{ \underbrace{T_1\delta s_1}_{\delta W_T} + \underbrace{N\delta r}_{\delta W_N} + \underbrace{M_1(\delta\varphi_1 + \varkappa\delta s_1)}_{\delta W_M} \right\}. \tag{12.59}$$

Exploiting the variations of convective coordinates at $r = 0$ in eqn. (12.39) the variational equation can be written in vector form as

$$\delta W = Sym\left\{ \underbrace{T_1(\delta\boldsymbol{\rho}_2 - \delta\boldsymbol{\rho}_1)\cdot\boldsymbol{\tau}_1}_{\delta W_T} + \underbrace{N(\delta\boldsymbol{\rho}_2 - \delta\boldsymbol{\rho}_1)\cdot\mathbf{e}_1}_{\delta W_N} + \underbrace{\frac{M_1}{r}(\delta\boldsymbol{\rho}_2 - \delta\boldsymbol{\rho}_1)\cdot\mathbf{g}_1}_{\delta W_M} \right\}. \tag{12.60}$$

We obtain the weak form allowing to describe the:

1. normal interaction between curves (part δW_N);

2. relative tangential interaction independently along each curve (part δW_T);

3. relative moment interaction independently for each curve (part δW_M).

The last two positions are the obvious achievement of the current covariant description, because

- *it is not possible within the previously known models, see the overview in Wriggers [188], to distinguish independently the full 3D relative motion including the relative normal, tangential and rotational motions along each curve;*

* *only the current covariant description allows to define the moment interaction between curves in an energy consistent form as the full relative motion is included.*

Remark.

Interaction in the circumferential direction is considered to be essential in the mechanics of wire ropes, see [30]. Thus, the developed model can be directly applied to the analysis of contact interaction inside the wire ropes and similar objects.

Remark.

The weak form presented in eqns. (12.59), (12.60) is obtained with the assumption of a small distance r and thus by construction is suitable for the case of edge-to-edge contact. For the case of beam-to-beam contact the value r includes the geometrical parameters of a beam and, therefore, is not small for arbitrary beams. In such a case as well as in a case with extremely curved geometry (e.g. for high-order FE with exact geometry) the full variational equation (12.58a), (12.58b) should be taken.

The developed theory produces an open question:

How to define additional constitutive equations for the moments M_1 **and** M_2. Some proposals considered in Section 12.5 are devoted to constitutive equations, however, it is clear that *a non-frictional problem can be characterized only by the functional* δW_N, while **the full sticking case will require all parts of the functional** δW in the corresponding variational equations (12.58a), (12.58b), or in the case of small r in (12.59).

12.5 Contact constraints and constitutive equations for contact tractions

Two major approaches are used in contact mechanics to enforce contact constraints and to define contact tractions:

- The Lagrange multiplier method allows to define the contact tractions as separate unknowns and the contact constraints are enforced via additional Lagrange multipliers. This leads to a formulation known in the optimization theory as *a saddle-point problem*. Then all constraints are exactly enforced.

- The penalty method allows to define the contact constraints via specifically defined and a-priori convex penalty functions. Thus, the method is enforcing the contact constraint only approximately.

However, since the penalty method can be mechanically interpreted as a specification of an additional interface law acting on the contact area and coupled with contact constraints then *the question, especially in experimental contact mechanics, still remains open whether to interpret this approach as an approximate mathematical method to enforce contact constraints, or to interpret it as a specification of an additional constitutive law for contact interfaces.* The following discussion is illustrating some difficulties. The normal contact constraints have a geometrical interpretation as non-penetration, and, so far, can be satisfied via a stable numerical algorithm with Lagrange multipliers. However, the tangent constraints including e.g. exact satisfaction for the Coulomb friction law have many numerical problems arising from the proved non-uniqueness of the solution (see the discussion in Eck and Jarušek [35] (1998)). Actually all stable numerical algorithms for Coulomb friction for large displacement problems are mainly based either on the penalty method, or on the Augmented Lagrangian method. Moreover, there are many experimental proofs for the existence of elastic tangent deformations and, therefore, are motivating to work with the penalty method interpreting it as an additional constitutive law for contact interfaces.

Thus, in the current contribution we exclusively will work with the second approach, namely, we are specifying the constitutive equations for contact tractions coupled with contact constraints.

12.5.1 Normal contact. Specification of constitutive laws for the traction N coupled with contact constraints for the variable r

The following situations with the curve-to-curve contact kinematics may arise for different mechanical models, namely,

- beam-to-beam contact,

- a special case of solid-to-solid contact when contact is appearing along sharp edges.

These two cases are influencing the specification of normal contact constraints for the variable r and, in due course, the specification of constitutive laws for a normal traction N. They are now considered separately.

12.5.1.1 Case of beam-to-beam contact

Beam models in mechanics are characterized as a line with corresponding cross sections. The line represents a reduction of a 3D solid to a 1D line, though in 3D space, according to the chosen kinematical hypothesis (e.g. Bernoulli, Timoshenko, Reissner). Thus we can always define for each I-th curve a function $R_I(\varphi_I)$ defining a cross section of the beam in our coordinate system, i.e. in the direction normal to a curve in the plane spanned by the vectors ν_I, β_I.

First, we define a mutual penetration of two beams as a part of intersection of two cross sections $R_1(\varphi_1)$ and $R_2(\varphi_2)$ along coordinate r as

$$p = r - (R_1(\varphi_1) + R_2(\varphi_2)). \qquad (12.61)$$

A constitutive relation coupled with the corresponding contact constraints for the normal force N mutual for both curves can be written as:

$$N = \begin{cases} 0 & if \quad p > 0 \quad \text{no contact} \\ \varepsilon_N(r - (R_1(\varphi_1) + R_2(\varphi_2)) & if \quad p \leq 0 \quad \text{contact.} \end{cases} \qquad (12.62)$$

Here, a parameter ε_N is defining a penalty parameter, or a stiffness for the normal interaction. If a penetration p is positive then the curves are out of contact. For the negative penetration the contact force N is defined as a linear elastic force.

However, even for the beam-to-beam contact some difficulties may arise for complex shapes of the cross-section used in practice (e.g. for I, C, T-shaped domains). In those cases taking the edge-to-edge contact instead of specifying $R_1(\varphi_1)$ and $R_2(\varphi_2)$ may lead to a simpler procedure.

12.5.1.2 Case of Edge-to-Edge contact

The case is representing a situation when 3D solid bodies are contacting along their sharp edges. Assuming that the parameterizations of both edge curves are separately given we can define the following constitutive equation for the normal traction:

$$N = \begin{cases} 0 & if \quad \varphi \notin [\varphi_0, \varphi_1] & \text{no contact} \\ -\varepsilon_N r & if \quad \varphi_0 \leq \varphi \leq \varphi_1 \text{ and } p = r \leq 0 & \text{contact.} \end{cases} \tag{12.63}$$

Here, the normal force is acting along the common normal e_1, see Fig. 12.8, and is defined in proportion to the distance r (representing the penetration into the edge) if the corresponding vector e_1 is laying inside the sector defined in plane $\nu_1 C_1 \beta_1$ via the angles φ_0 and φ_1. As in the previous case ε_N is defining a penalty parameter, or a stiffness for the normal interaction.

12.5.2 Tangential contact. Specification of constitutive laws for tractions T_I coupled with contact constraints for the variables s_I

The specification of constitutive laws for tangential variables s_I is unique for the curve-to-curve kinematics, i.e. is common for both beam-to-beam and edge-to-edge cases. One of the advantages of the current theory is the possibility to specify the contact law separately for each curve. For

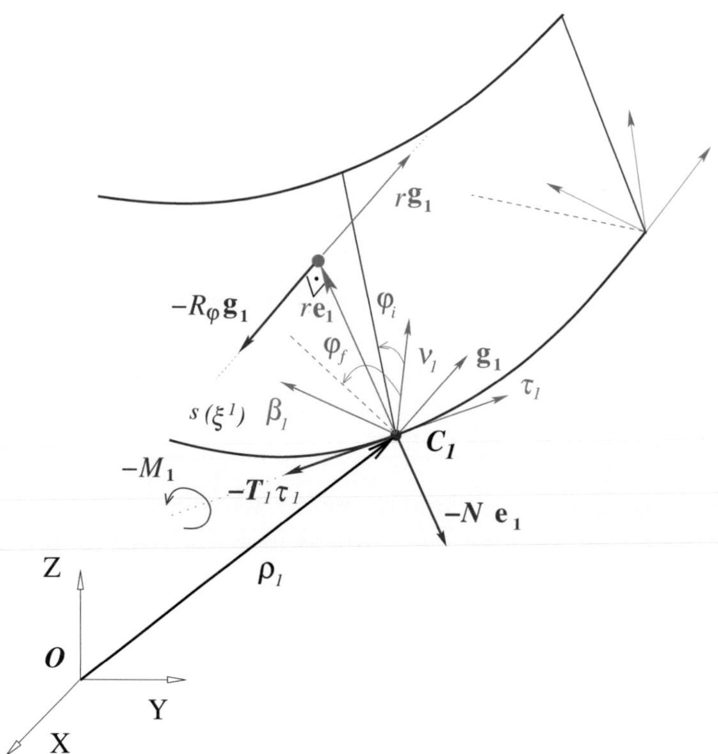

Figure 12.8: Contact tractions in the spatial coordinate system. Edge-to-Edge contact case.

tangential contact *the sticking case and the sliding case* should be differentiated. Thus we give a definition of **tangential sticking** as follows:

- if $\dot{s}_1 = 0$ then the first curve is **tangentially sticking** along the second curve;

- if $\dot{s}_1 \neq 0$ then the first curve is **tangentially sliding** along the second curve.

This definition is symmetric with regard to the choice of the curves, therefore, can be repeated in the sense *the first curve $\leftrightarrow$ the second curve*. Using the elasto-plastic analogy for the Coulomb friction law we can specify the constitutive law for tangential friction between curves as elasto-plastic one *(allowing elastic deformations for a sticking case)* in the following rate forms:

$$\dot{T}_1 = \begin{cases} -\varepsilon_1 \dot{s}_1 & if \quad |T_1| < \mu_1|N| \quad tangential\ sticking \text{ along } 2^{nd} \text{ curve.} \\ -\mu_1|N|\dfrac{\dot{s}_1}{|\dot{s}_1|} & if \quad |T_1| \geq \mu_1|N| \quad tangential\ sliding \text{ along } 2^{nd} \text{ curve.} \end{cases}$$

$$(12.64)$$

Here,

ε_1 – tangential stiffness along the first curve;

μ_1 – coefficient of friction along the first curve.

Applying then the return-mapping scheme to eqn. (12.64) we can get a trial tangential force as

$$\Delta T_1^{(trial)} = -\varepsilon_1 \Delta s_1 \implies T_1^{(trial,n+1)} = T_1^{(n)} - \varepsilon_1(s_1^{(n+1)} - s_1^{(n)}), \quad (12.65)$$

$T_1^{(n)},\ s_1^{(n)}$ – tangential force (resp. arc-length coordinate) for the previous converged load step;

$T_1^{(trial,n+1)},\ s_1^{(n+1)}$ – tangential force (resp. arc-length coordinate) in the current iteration.

After the return-mapping step we get the real tangential force as

$$T_1 = \begin{cases} T_1^{(trial,n+1)} & \text{if} \quad |T_1^{(trial,n+1)}| < \mu_1|N_1| \quad sticking. \\ T_1^{(sl,n+1)} = \mu_1|N_1|sign(T_1^{(trial,n+1)}) & \text{if} \quad |T_1^{(trial,n+1)}| \geq \mu_1|N_1| \quad sliding. \end{cases}$$

$$(12.66)$$

All equations (12.64), (12.65), (12.66) are, of course, symmetric concerning choice of the curves $1 \leftrightarrow 2$.

12.5.3   Rotational contact. Specification of a constitutive law for the rotational moment M_I coupled with contact constraints for the variables φ_I

The moment M_1 appearing in the current theory is energetically coupled with the variable φ_1, and therefore, is responsible for the rotational motion along the τ_1 axis (plane $e_1\,C_1\,g_1$). A specification of this moment allows e.g. to describe a full sticking case specifying *the rotational sticking* in addition to the tangential sticking to prevent the motion of the first curve relative to the second curve. **The rotational sticking** can be defined as follows:

- if $\dot{\varphi}_1 = 0$ then the first curve is **rotationally sticking** along the second curve, i.e. rotation along the τ_1-axis is not possible.

- if $\dot{\varphi}_1 \neq 0$ then the first curve is **rotationally sliding** along the second curve, i.e. can be rotated along the τ_1-axis.

This definition is again symmetric with regard to the choice of the curves $1 \leftrightarrow 2$. One of the simplest way to define the constitutive relation M_1 is by using an analogy to the Tresca friction law, but now defined for the angular variable φ_1. We can specify the following relation for rotational friction between curves in rate form:

$$\dot{M}_1 = \begin{cases} -\varepsilon_1^M \dot{\varphi}_1 & if \quad |M_1| < M^{cr} & rotational\ sticking \text{ along } 2^{nd} \text{ curve.} \\ -\mu_1^M \dfrac{\dot{\varphi}_1}{|\dot{\varphi}_1|} & if \quad |M_1| \geq \mu_1^M M^{cr} & rotational\ sliding \text{ along } 2^{nd} \text{ curve.} \end{cases}$$

$$(12.67)$$

Here,

ε_1^M – is a rotational stiffness for the first curve;

μ_1^M – is a coefficient of rotational friction for the first curve;

M^{cr} – is a critical given value for the rotational moment after which the curve begins to rotate. The simplicity of this law is emphasized by the independence of the critical moment M^{cr} on any other parameters, e.g. a normal contact force. Applying then the return-mapping scheme to eqn. (12.67) we can get a trial rotational moment as

$$\Delta M_1^{(trial)} = -\varepsilon_1^M \Delta \varphi_1 \implies M_1^{(trial,n+1)} = M_1^{(n)} - \varepsilon_1^M (\varphi_1^{(n+1)} - \varphi_1^{(n)}),$$

$$(12.68)$$

similar to the scheme (12.65). The return-mapping step is constructed fully similar to (12.66)

$$M_1 = \begin{cases} M_1^{(trial,n+1)} & if \quad |M_1^{(trial,n+1)}| < M^{cr} & sticking; \\ M_1^{(sl,n+1)} = \mu_1^M sign(M_1^{(trial,n+1)}) & if \quad |M_1^{(trial,n+1)}| \geq M^{cr} & sliding. \end{cases}$$

$$(12.69)$$

Remark.

Full sticking will require the enforcement of both tangential sticking and rotational sticking.

12.6 Rate of contact forces in a covariant form

For the further linearization it is necessary to obtain the frame-indifferent rate of the force vector $\mathbf{R}_1$. As is known, the rate in the form of covariant derivatives is a frame indifferent one. Following its formal definition the total time derivative of a vector can be written in a covariant form

$$\frac{d\mathbf{R}}{dt} = \frac{dR^i \mathbf{r}_i}{dt} = \nabla_j R^i \dot{\xi}^j \mathbf{r}_i, \tag{12.70}$$

where the components in the coordinate system $\mathbf{r}_1, \mathbf{r}_2, \mathbf{r}_3$ are defined as follows:

$$\nabla_j R^i = \frac{\partial R^i}{\partial \xi^j} + R^k \Gamma^i_{jk}, \tag{12.71}$$

where again the Christoffel symbols representing the derivation of coordinate vectors are necessary. All necessary derivations have been established in Section 12.3.2 via the direct derivation of coordinate vectors. Thus, *the main step now is to express all constitutive relations presented in Section 12.5 in the form of covariant derivatives* with respect to a metrics given by g_{ij} in eqn. (12.26).

12.6.1 Covariant form for sticking

All sticking cases, discussed in Section 12.5 are characterized by the proportionality of the force rate components $\dot{T}, \dot{N}, \dot{M}$ to the corresponding velocity components $\dot{s}, \dot{r}, \dot{\varphi}$. Summarizing, we can write in a vector form as

$$\frac{d\mathbf{R}}{dt} = -\mathbf{D}(\mathbf{v}_2 - \mathbf{v}_1^t) \quad \Longrightarrow \quad \nabla_j R^i \dot{\xi}^j \mathbf{r}_i = -d_i \dot{\xi}^i \mathbf{r}_i. \tag{12.72}$$

Here, the tensor expressed in the mixed metrics $\mathbf{D} = d_i \mathbf{r}_i \otimes \mathbf{r}^i$ is given by the diagonal matrix containing the stiffness parameters $\varepsilon_1, \varepsilon_N, \varepsilon_1^M$

$$\mathbf{D} = \begin{bmatrix} \varepsilon_1 & 0 & 0 \\ 0 & \varepsilon_N & 0 \\ 0 & 0 & \varepsilon_1^M \end{bmatrix}. \tag{12.73}$$

Remark.

Since the rate forms given in the definition of constitutive laws, see eqns. (12.64) and (12.67) are coinciding with derivatives of the corresponding trial forces in eqns. (12.65) and (12.68) for the return-mapping scheme in eqns. (12.66) and (12.69) we can directly use the vector form in eqn. (12.72) for the forthcoming linearization.

12.6.2 Covariant form for sliding

Similar to the sticking cases the rate for the sliding cases can be written in a proportional form, however, the form suitable for the linearization should contain the linearization of the real forces given in eqns. (12.66) and (12.69) of the return-mapping scheme. Thus, we will linearize the weak form directly for the sliding cases in the next section.

12.7 Linearization of the weak form

Numerical solution schemes for contact between bodies are realized via a Newton type iterative solver, thus a knowledge about derivatives of the functional δW is required. This procedure known as linearization will be taken in a covariant form, i.e. provided in a coordinate system exploiting covariant derivatives. We start with the full form in eqn. (12.57) without any simplification on small distance r and curvature. For the linearization the total time derivative in a form of the covariant derivative operator in eqn. (12.44) is used:

$$\mathcal{L}_{s,r,\varphi}[\delta W] = \mathcal{L}_{s,r,\varphi}[Sym\{\mathbf{R}_1 \cdot (\delta\boldsymbol{\rho}_1 - \delta\boldsymbol{\rho}_2)\}] =$$

$$= Sym\{\mathcal{L}_{s,r,\varphi}[\mathbf{R}_1] \cdot (\delta\boldsymbol{\rho}_1 - \delta\boldsymbol{\rho}_2)\} + Sym\{\mathbf{R}_1 \cdot \mathcal{L}_{s,r,\varphi}[(\delta\boldsymbol{\rho}_1 - \delta\boldsymbol{\rho}_2)]\}. \tag{12.74}$$

The second term contains derivatives of the variations and remains in the same form for the sticking and sliding cases for the real force vector $\mathbf{R}_1$. This part represents a tangent matrix due to the geometrical nonlinearity of contact interaction, while the first term in eqn. (12.74) represents a constitutive relation for the contact force.

12.7.1   First part, representing geometrical nonlinearity

The derivatives in eqn. (12.46a) and the full expression for the force in eqn. (12.55) are used to obtain the part representing the geometrical nonlinearity.

$$Sym\{\mathbf{R}_1 \cdot \mathcal{L}_{s,r,\varphi}[(\delta\boldsymbol{\rho}_1 - \delta\boldsymbol{\rho}_2)]\} = \qquad (12.75a)$$

$$Sym\left\{\left(T_1\left[\frac{(1 - rk_1\cos\varphi_1)\,\boldsymbol{\tau}_1 + \varkappa_1 r\,\mathbf{g}_1}{\sqrt{g_{11}}}\right] + N\,\mathbf{e}_1 + R_{\varphi_1}\,\mathbf{g}_1\right) \cdot \mathcal{L}_{s,r,\varphi}[(\delta\boldsymbol{\rho}_1 - \delta\boldsymbol{\rho}_2)]\right\} =$$

$$= -Sym\{\,T_1\,[rk_1\varkappa_1 \sin\varphi \dot{s}_1\delta s_1 - k_1\cos\varphi_1(\dot{s}_1\delta r + \dot{r}\delta s_1) \qquad (12.75b)$$

$$+ rk_1 \sin\varphi_1(\dot{s}_1\delta\varphi_1 + \dot{\varphi}_1\delta s_1)]\,\frac{(1 - rk_1\cos\varphi_1)}{\sqrt{(1 - rk_1\cos\varphi_1)^2 + (r\varkappa_1)^2}}$$

$$+N\,[k_1\cos\varphi_1(1 - rk_1\cos\varphi_1)\dot{s}_1\delta s_1 - r(\varkappa_1\dot{s}_1 + \dot{\varphi}_1)(\varkappa_1\delta s_1 + \delta\varphi_1)]$$
$$(12.75c)$$

$$+\left(R_{\varphi_1} + T\,\frac{r\varkappa_1}{\sqrt{(1 - rk_1\cos\varphi_1)^2 + (r\varkappa_1)^2}}\right)[-k_1\sin\varphi_1(1 - rk_1\cos\varphi_1)\dot{s}_1\delta s_1 \qquad (12.75d)$$

$$+ (\dot{r}\delta\varphi_1 + \dot{\varphi}_1\delta r) + \varkappa_1(\dot{s}_1\delta r + \dot{r}\delta s_1)]\,\}.$$

The long expression can be tremendously simplified if **we consider the linearized term at** $r = 0$:

$$Sym\{\mathbf{R}_1 \cdot \mathcal{L}_{s,r,\varphi}[(\delta\boldsymbol{\rho}_1 - \delta\boldsymbol{\rho}_2)]\} = \qquad (12.76a)$$

$$= -Sym\{-\,T_1\,k_1\cos\varphi_1\,(\dot{s}_1\delta r + \dot{r}\delta s_1) \qquad (12.76b)$$

$$+N\left[k_1\cos\varphi_1\,\dot{s}_1\delta s_1\right] \tag{12.76c}$$

$$+\frac{M_1}{r}\left[-k_1\sin\varphi_1\,\dot{s}_1\delta s_1+(\dot{r}\delta\varphi_1+\dot{\varphi}_1\delta r)+\varkappa_1(\dot{s}_1\delta r+\dot{r}\delta s_1)\right]\Big\}. \tag{12.76d}$$

For finite element implementations it is rather convenient to give a tensor form directly leading to the corresponding tangent matrices. In order to derive this form all pairs containing derivatives and variations of coordinates in eqns. (12.76b), (12.76c) and (12.76d) are transformed using eqns. (12.35) and (12.39). The following example for $\dot{r}\delta s_1$ is illustrating the simple transformation procedure for two scalar products into a tensor product:

$$\dot{r}\delta s_1=(\mathbf{v}_2-\mathbf{v}_1^t)\cdot\mathbf{e}_1(\delta\boldsymbol{\rho}_2-\delta\boldsymbol{\rho}_1)\cdot\boldsymbol{\tau}_1=(\delta\boldsymbol{\rho}_2-\delta\boldsymbol{\rho}_1)\cdot\boldsymbol{\tau}_1\otimes\mathbf{e}_1(\mathbf{v}_2-\mathbf{v}_1^t). \tag{12.77}$$

Transformation of the linearized equation (12.76a) according to this pattern leads to

$$Sym\{\mathbf{R}_1\cdot\mathcal{L}_{s,r,\varphi}[(\delta\boldsymbol{\rho}_1-\delta\boldsymbol{\rho}_2)]\}= \tag{12.78a}$$

$$=-Sym\{(\delta\boldsymbol{\rho}_2-\delta\boldsymbol{\rho}_1)\cdot[\,(-\,T_1\,k_1\cos\varphi_1\,(\boldsymbol{\tau}_1\otimes\mathbf{e}_1+\mathbf{e}_1\otimes\boldsymbol{\tau}_1)+ \tag{12.78b}$$

$$+N\,k_1\cos\varphi_1\,(\boldsymbol{\tau}_1\otimes\boldsymbol{\tau}_1)+ \tag{12.78c}$$

$$+\frac{M_1}{r}\left(-k_1\sin\varphi_1\,\boldsymbol{\tau}_1\otimes\boldsymbol{\tau}_1+\frac{\mathbf{e}_1\otimes\mathbf{g}_1+\mathbf{g}_1\otimes\mathbf{e}_1}{r}\right)]\,(\mathbf{v}_2-\mathbf{v}_1^t)\,\}. \tag{12.78d}$$

The term with $\varkappa_1$ is disappearing because of the following transformation

$$(\dot{r}\delta\varphi_1+\dot{\varphi}_1\delta r)+\varkappa_1(\dot{s}_1\delta r+\dot{r}\delta s_1)\longrightarrow$$

$$\frac{\mathbf{e}_1\otimes\mathbf{g}_1+\mathbf{g}_1\otimes\mathbf{e}_1}{r}-\varkappa_1(\boldsymbol{\tau}_1\otimes\mathbf{e}_1+\mathbf{e}_1\otimes\boldsymbol{\tau}_1)+\varkappa_1(\boldsymbol{\tau}_1\otimes\mathbf{e}_1+\mathbf{e}_1\otimes\boldsymbol{\tau}_1)=$$

$$=\frac{\mathbf{e}_1\otimes\mathbf{g}_1+\mathbf{g}_1\otimes\mathbf{e}_1}{r}.$$

12.7.2 Constitutive part for sticking

Due to the Remark 12.6.1 in Section 12.6.1 we can use the rate equation given in eqn. (12.72) keeping in mind the equivalence of operators $\dfrac{d}{dt} \sim \mathcal{L}_{s,r,\varphi}$. We are starting with a tensor form as

$$Sym\{\mathcal{L}_{s,r,\varphi}[\mathbf{R}_1] \cdot (\delta\boldsymbol{\rho}_1 - \delta\boldsymbol{\rho}_2)\} = -\mathbf{D}(\mathbf{v}_2 - \mathbf{v}_1^t) \cdot (\delta\boldsymbol{\rho}_1 - \delta\boldsymbol{\rho}_2) = \qquad (12.79)$$

$$= Sym\{d_i\dot{\xi}^i\mathbf{r}_i \cdot \delta\xi^j\mathbf{r}_j\} = Sym\{d_i g_{ij}\dot{\xi}^i\delta\xi^j\}$$

and then obtaining

$$= Sym\{\varepsilon_1\left((1 - rk_1\cos\varphi_1)^2 + (r\varkappa_1)^2\right)\dot{s}_1\delta s_1 + \varepsilon_r\,\dot{r}\delta r+ \qquad (12.80a)$$

$$+\varepsilon_1^M r^2\,\dot{\varphi}_1\delta\varphi_1 + r^2\varkappa_1\left(\varepsilon_1\,\dot{s}_1\delta\varphi_1 + \varepsilon_1^M\dot{\varphi}_1\delta s_1\right)\}. \qquad (12.80b)$$

Remark.
Obviously in a case of not equal stiffnesses for the spatial curve $\varepsilon_1 \neq \varepsilon_1^M$ the symmetry is destroyed. Due to the coupled kinematics for the spatial curve, see the velocity vector in eqn. (12.33), we are restricted to take $\varepsilon_1 = \varepsilon_1^M$, otherwise the sum of force rates along $\mathbf{r}_1$ and $\mathbf{r}_2$ will not be parallel to the relative velocity vector along the curve. For the latter we can take then

$$\varepsilon_1^M = \varepsilon_1. \qquad (12.81)$$

Taking into account this remark after grouping eqns. (12.80a) and (12.80b) the following form is obtained:

$$Sym\{\mathcal{L}_{s,r,\varphi}[\mathbf{R}_1] \cdot (\delta\boldsymbol{\rho}_1 - \delta\boldsymbol{\rho}_2)\} =$$

$$Sym\left\{\varepsilon_1(1 - rk_1\cos\varphi_1)^2\,\dot{s}_1\delta s_1 + \varepsilon_r\,\dot{r}\delta r + \varepsilon_1 r^2\,(\dot{\varphi}_1 + \varkappa_1\dot{s}_1)(\delta\varphi_1 + \varkappa_1\delta s_1)\right\}. \qquad (12.82)$$

Now for the tensor form we will use the non-reduced equations for the convective velocities in eqn. (12.36) and their variational analog. The final result has a simple form as

$$Sym\{\mathcal{L}_{s,r,\varphi}[\mathbf{R}_1] \cdot (\delta\boldsymbol{\rho}_1 - \delta\boldsymbol{\rho}_2)\} =$$

$$= Sym\left\{ (\delta\boldsymbol{\rho}_2 - \delta\boldsymbol{\rho}_1) \cdot (\varepsilon_1\,\boldsymbol{\tau}_1 \otimes \boldsymbol{\tau}_1 + \varepsilon_r\,\mathbf{e}_1 \otimes \mathbf{e}_1 + \varepsilon_1\,\mathbf{g}_1 \otimes \mathbf{g}_1)\,(\mathbf{v}_2 - \mathbf{v}_1^t)\right\}, \qquad (12.83)$$

leading to a constitutive part of the tangent matrix in the sticking case.

12.7.3 Constitutive part for tangential sliding

The tangential sliding force obtained in the computation via the return-mapping scheme in eqn. (12.66) can be written then in vector form as:

$$\mathbf{R} = -\mu|N|\frac{\mathbf{r}_1}{\sqrt{g_{11}}}. \qquad (12.84)$$

Inserting this into the first term of the linearized weak form in eqn. (12.74) we obtain:

$$Sym\left\{\mathcal{L}_{s,r,\varphi}\left[-\mu|N|\frac{\mathbf{r}_1}{\sqrt{g_{11}}}\right] \cdot (\delta\boldsymbol{\rho}_1 - \delta\boldsymbol{\rho}_2)\right\} = \qquad (12.85)$$

$$= Sym\left\{\mathcal{L}_{s,r,\varphi}\left[\frac{\mu|N|}{\sqrt{g_{11}}}\right]\mathbf{r}_1 \cdot (\delta\boldsymbol{\rho}_2 - \delta\boldsymbol{\rho}_1)\right\} + \qquad (12.86a)$$

$$+ Sym\left\{\frac{\mu|N|}{\sqrt{g_{11}}}\mathcal{L}_{s,r,\varphi}[\mathbf{r}_1] \cdot (\delta\boldsymbol{\rho}_2 - \delta\boldsymbol{\rho}_1)\right\}. \qquad (12.86b)$$

The first term (eqn. (12.86a)) is transformed as

$$(1st) = Sym\left\{\mu_1\left(\frac{-\varepsilon_r\dot{r}}{\sqrt{g_{11}}} - |N|\frac{(-k_1\cos\varphi_1(1 - rk_1\cos\varphi_1)\dot{r}}{(\sqrt{g_{11}})^3}\right.\right.+$$

$$\left.\left. + |N|\frac{rk_1\sin\varphi_1(1 - rk_1\cos\varphi_1)\dot{\varphi}_1 - \varkappa_1^2 r\dot{r}}{(\sqrt{g_{11}})^3}\right)\right.$$

$$\left.[(1 - rk_1\cos\varphi_1)^2\delta s_1 + r^2\varkappa_1(\varkappa_1\delta s_1 + \delta\varphi_1)] \right\} \qquad (12.87a)$$

and after taking at $r = 0$

$$= Sym\{\mu_1(-\varepsilon_r + |N|k_1\cos\varphi_1)\dot{r}\delta s_1\}. \qquad (12.87b)$$

One can prove that the second term (eqn. (12.86b)) taking its value at $r = 0$ is transformed then as

$$(2nd) = Sym\left\{-\mu_1|N|k_1\cos\varphi_1\dot{r}\delta s_1\right\}. \tag{12.88}$$

Summing eqns. (12.87b) and (12.88) we obtain the simple value for **the linearized constitutive part in the case of tangential sliding**

$$Sym\left\{\mathcal{L}_{s,r,\varphi}\left[-\mu|N|\frac{\mathbf{r}_1}{\sqrt{g_{11}}}\right]\cdot(\delta\boldsymbol{\rho}_1 - \delta\boldsymbol{\rho}_2)\right\} = -Sym\{\mu_1\varepsilon_r\dot{r}\delta s_1\} = \tag{12.89a}$$

or in tensor form

$$= -Sym\{(\delta\boldsymbol{\rho}_2 - \delta\boldsymbol{\rho}_1)\cdot\varepsilon_r\mu_1\boldsymbol{\tau}_1\otimes\mathbf{e}_1(\mathbf{v}_2 - \mathbf{v}_1^t)\} \tag{12.89b}$$

12.7.4 Linearized part for rotational sliding

The rotational sliding force representing the moment obtained in the computation via the return-mapping scheme in eqn. (12.69) can be written then in vector form as:

$$\mathbf{R} = R_{\varphi_1}\mathbf{g}_1 = \frac{M_1}{r}\mathbf{g}_1 = -\mu_1^M sign(\dot{\varphi}_1)\frac{1}{r}\mathbf{g}_1. \tag{12.90}$$

Inserting this into the linearized weak form we obtain:

$$\mathcal{L}_{s,r,\varphi}[\delta W_M] =$$

$$= Sym\left\{\mathcal{L}_{s,r,\varphi}\left[-\mu_1^M sign(\dot{\varphi}_1)\frac{1}{r}\mathbf{g}_1\right]\cdot(\delta\boldsymbol{\rho}_1 - \delta\boldsymbol{\rho}_2)\right. \tag{12.91a}$$

$$\left. -\mu_1^M sign(\dot{\varphi}_1)\frac{1}{r}\mathbf{g}_1\cdot\mathcal{L}_{s,r,\varphi}[(\delta\boldsymbol{\rho}_1 - \delta\boldsymbol{\rho}_2)]\right\} \tag{12.91b}$$

The first term is transformed as

$$(1st) = Sym\{\frac{\mu_1^M sign(\dot{\varphi}_1)}{r}[k_1\sin\varphi_1(1 - rk_1\cos\varphi_1)\dot{s}_1\delta s_1$$

$$-(\varkappa_1\dot{s}_1 + \dot{\varphi}_1)\delta r - \dot{r}(\varkappa_1\delta s_1 + \delta\varphi_1)]\}. \tag{12.92}$$

The second term is representing the geometrically nonlinear part and can be taken directly from eqn. (12.75d) for the sliding value of M_1

$$(2nd) = Sym\{\frac{\mu_1^M sign(\dot{\varphi}_1)}{r}[-k_1 \sin\varphi_1(1 - rk_1\cos\varphi_1)\,\dot{s}_1\delta s_1$$
$$+(\dot{r}\delta\varphi_1 + \dot{\varphi}_1\delta r) + \varkappa_1(\dot{s}_1\delta r + \dot{r}\delta s_1)]\,\}. \quad (12.93)$$

The sum of the two terms is zero. Thus, **the tangent matrix for a rotational sliding due to the Tresca model is zero.**

12.8   Finite element implementation

In this section we present the structure of the iterative solution scheme for a Newton type method and describe the implementation of the developed theory based on the simplest finite element with linear shape functions.

12.8.1   Linear element for edge-to-edge contact

Let us consider two edges from contacting bodies covered with a finite element mesh possessing linear approximations. We construct a possible contact pair taking a linear approximation through nodes $\mathbf{r}_1^{(1)}$ and $\mathbf{r}_1^{(2)}$ for the edge $\rho_1(\xi^1)$ from one body and a linear approximation through nodes $\mathbf{r}_2^{(1)}$ and $\mathbf{r}_2^{(2)}$ for the edge $\rho_2(\xi^2)$ from another body. A pair of $\rho_1(\xi^1)$ and $\rho_2(\xi^2)$ defined by four nodes, see Fig. 12.9 gives us a contact element:

$$\rho_I(\xi_I) = \frac{1 - \xi_I}{2}\mathbf{r}_I^{(1)} + \frac{1 + \xi_I}{2}\mathbf{r}_I^{(2)}, \quad I = 1, 2, \quad -1 \le \xi^I \le 1. \quad (12.94)$$

A nodal vector for the contact element is defined as follows:

$$\{\mathbf{x}\}^T = \left\{x_1^{(1)}, y_1^{(1)}, z_1^{(1)}, x_1^{(2)}, y_1^{(2)}, z_1^{(2)}, x_2^{(1)}, y_2^{(1)}, z_2^{(1)}, x_2^{(2)}, y_2^{(2)}, z_2^{(2)}\right\}^T \quad (12.95)$$

In almost all parameters the relative displacement vector $\rho_2 - \rho_1$ is required, therefore, a matrix approximation operator $\mathbf{A}(\xi_1, \xi_2)$ introduced

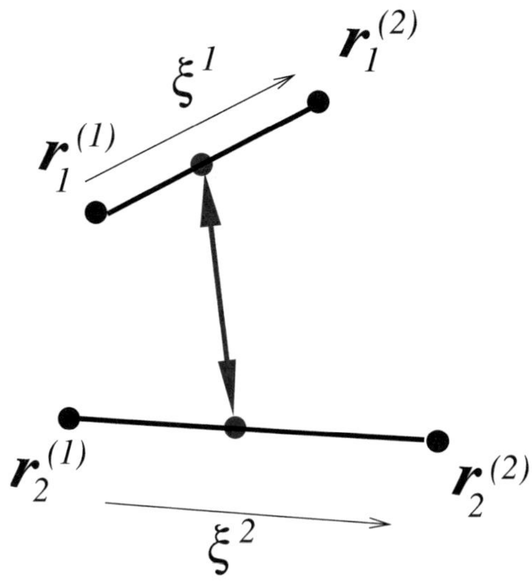

Figure 12.9: Two linear edges define a contact pair – a contact finite element with linear approximations.

as

$$
\mathbf{A}(\xi_1, \xi_2) =
\left[
\begin{array}{cccccc}
-\dfrac{1-\xi_1}{2} & 0 & 0 & -\dfrac{1+\xi_1}{2} & 0 & 0 \\[2ex]
0 & -\dfrac{1-\xi_1}{2} & 0 & 0 & -\dfrac{1+\xi_1}{2} & 0 \\[2ex]
0 & 0 & -\dfrac{1-\xi_1}{2} & 0 & 0 & -\dfrac{1+\xi_1}{2}
\end{array}
\right.
$$

$$
\left.
\begin{array}{cccccc}
\dfrac{1-\xi_2}{2} & 0 & 0 & \dfrac{1+\xi_2}{2} & 0 & 0 \\[2ex]
0 & \dfrac{1-\xi_2}{2} & 0 & 0 & \dfrac{1+\xi_2}{2} & 0 \\[2ex]
0 & 0 & \dfrac{1-\xi_2}{2} & 0 & 0 & \dfrac{1+\xi_2}{2}
\end{array}
\right] \tag{12.96}
$$

gives us the approximation of the aforementioned vector as

$$
\boldsymbol{\rho}_2(\xi_2) - \boldsymbol{\rho}_1(\xi_1) = \mathbf{A}\,\mathbf{x}. \tag{12.97}
$$

For arbitrarily curved lines and arbitrarily large displacements it is necessary to compute the following geometrical parameters of the contact:

1. The closest point projection (CPP) procedure for convective vari-

ables, see in eqn. (12.22). The procedure gives as results the ξ_p^1, ξ_p^2 as projection points and the first local coordinates. The corresponding arc-lengths s_1, s_2 can be computed using the Jacobian $J_{(I)}$.

2. The Jacobian for the corresponding transformation becomes $\xi_I \rightarrow s_I$

$$J_{(I)} = \frac{ds_I}{d\xi_I} = \sqrt{\boldsymbol{\rho}_{I,\xi_I} \cdot \boldsymbol{\rho}_{I,\xi_I}} = \sqrt{\sum_{j=1}^{3} \left(\frac{\partial x^j}{\partial \xi_I} \frac{\partial x^j}{\partial \xi_I} \right)}.$$

3. Local coordinate vectors of the Serret-Frenet frame:

- a unit tangent vector $\boldsymbol{\tau}_I$

$$\boldsymbol{\tau}_I = \frac{\partial \boldsymbol{\rho}_I}{\partial s_I} = \frac{\partial \boldsymbol{\rho}_I}{\partial \xi_I} \frac{1}{J_{(I)}}$$

Then a curvature k_I in eqn. (12.9) and a torsion $\varkappa_I$ in eqn. (12.10) can be computed.

- a binormal vector from eqn. (12.9) as

$$\boldsymbol{\beta}_I = \frac{\dot{\boldsymbol{\rho}}_I \times \ddot{\boldsymbol{\rho}}_I}{|\dot{\boldsymbol{\rho}}_I \times \ddot{\boldsymbol{\rho}}_I|}$$

- a normal vector as

$$\boldsymbol{\nu}_I = \boldsymbol{\beta}_I \times \boldsymbol{\tau}_I$$

4. The distance between curves at projection points as a second co-ordinate r:

$$r = |\boldsymbol{\rho}_2(\xi_2^p) - \boldsymbol{\rho}_1(\xi_1^p)|$$

5. Coordinate vectors for the local coordinate system:

$$\mathbf{e}_1 = \frac{\boldsymbol{\rho}_2(\xi_2^p) - \boldsymbol{\rho}_1(\xi_1^p)}{r}, \quad \mathbf{g}_1 = \boldsymbol{\tau}_1 \times \mathbf{e}_1$$

6. The angle φ_I as the third coordinate defined via

$$\cos \varphi_I = (\mathbf{e}_I \cdot \boldsymbol{\nu}_I), \quad \sin \varphi_I = (\mathbf{e}_I \cdot \boldsymbol{\beta}_I).$$

For the element with linear approximations (both, curvature k and torsion $\varkappa$ are zero) the procedure is simplified.

1. Tangent vectors for the corresponding linear segments

$$\dot{\boldsymbol{\rho}}_1(\xi_1) = \frac{\mathbf{r}_1^{(2)} - \mathbf{r}_1^{(1)}}{2} \quad \longrightarrow \quad \boldsymbol{\tau}_1 = \frac{\boldsymbol{\rho}_1}{|\boldsymbol{\rho}_1|}$$

$$\dot{\boldsymbol{\rho}}_2(\xi_2) = \frac{\mathbf{r}_2^{(2)} - \mathbf{r}_2^{(1)}}{2} \quad \longrightarrow \quad \boldsymbol{\tau}_2 = \frac{\boldsymbol{\rho}_2}{|\boldsymbol{\rho}_2|}$$

2. The coordinate vector $\mathbf{e}_1$ is derived by its definition to be orthogonal to both tangent vectors, therefore, via the cross product

$$\mathbf{e}_1 = \boldsymbol{\tau}_1 \times \boldsymbol{\tau}_2.$$

3. The coordinate vector $\mathbf{g}_1$ is derived via a cross product

$$\mathbf{g}_1 = \boldsymbol{\tau}_1 \times \mathbf{e}_1.$$

4. The closest point projection procedure is resolved exactly. The projection points ξ_1^p, ξ_2^p (first coordinates) are defined via the following linear system:

$$\begin{cases} (\dot{\boldsymbol{\rho}}_1 \cdot \dot{\boldsymbol{\rho}}_1)\xi_1 \;-\; (\dot{\boldsymbol{\rho}}_1 \cdot \dot{\boldsymbol{\rho}}_2)\xi_2 \;=\; \frac{1}{2}(\mathbf{r}_2^{(1)} + \mathbf{r}_2^{(2)} - \mathbf{r}_1^{(1)} - \mathbf{r}_1^{(2)}) \cdot \dot{\boldsymbol{\rho}}_1 \\[2ex] -(\dot{\boldsymbol{\rho}}_2 \cdot \dot{\boldsymbol{\rho}}_1)\xi_1 \;+\; (\dot{\boldsymbol{\rho}}_2 \cdot \dot{\boldsymbol{\rho}}_2)\xi_2 \;=\; \frac{1}{2}(\mathbf{r}_2^{(1)} + \mathbf{r}_2^{(2)} - \mathbf{r}_1^{(1)} - \mathbf{r}_1^{(2)}) \cdot \dot{\boldsymbol{\rho}}_2 \end{cases}$$

The determinant of the system can be transformed as

$$\det = (\dot{\boldsymbol{\rho}}_1 \cdot \dot{\boldsymbol{\rho}}_1)(\dot{\boldsymbol{\rho}}_2 \cdot \dot{\boldsymbol{\rho}}_2) - (\dot{\boldsymbol{\rho}}_1 \cdot \dot{\boldsymbol{\rho}}_2)^2 = |\dot{\boldsymbol{\rho}}_1 \times \dot{\boldsymbol{\rho}}_2| \neq 0$$

leading to the known condition **"tangent vectors should not be parallel"**, see Sect. 12.2.2.1.

5. **Local searching procedure.**
   If any of projection points is laying outside the corresponding segment i.e.

$$if\ \xi_1^p \notin [-1,1],\ or\ \xi_2^p \notin [-1,1]\ then$$

   a contact element is excluded from the computation at the current iteration.

6. The closest distance between lines (the second coordinate r) is defined as

$$r = \frac{1}{2}(\mathbf{r}_2^{(1)} + \mathbf{r}_2^{(2)} - \mathbf{r}_1^{(1)} - \mathbf{r}_1^{(2)}) \cdot \mathbf{e}_1$$

7. Definition of the coordinate increments. Since the straight line has an infinite number of normal vectors, the third coordinate φ_I can be defined arbitrarily, e.g. via *"the orientation node"* similar to the definition of linear finite element for beams. However, all coordinate increments are defined via displacement increment vectors.

   Defining a displacement increment vector for the first curve as $\Delta\mathbf{u}_1$ and resp. $\Delta\mathbf{u}_2$ for the second curve all increments for convective coordinates are computed from the incremental analog of eqn. (12.35)

$$\begin{cases} \Delta s_1 &= (\Delta\mathbf{u}_2 - \Delta\mathbf{u}_1) \cdot \boldsymbol{\tau}_1 \\ \Delta r_1 &= (\Delta\mathbf{u}_2 - \Delta\mathbf{u}_1) \cdot \mathbf{e}_1 \\ \Delta\varphi_1 &= (\Delta\mathbf{u}_2 - \Delta\mathbf{u}_1) \cdot \dfrac{\mathbf{g}_1}{r} \end{cases} \qquad 1 \leftrightarrow 2.$$

   They are directly used for the corresponding return-mapping schemes in eqns. (12.65), (12.66), (12.68), (12.69).

12.8.2 Structure of tangent matrices

Several situations should be distinguished for the curve-to-curve contact element. The first situation is based on the definition of the penetration p for the specific case either for edges, or for beams, see Section 12.5. Thus, for the case *no contact* with positive penetration $p > 0$ the contact element is excluded from the computation in the current iteration. The situations for tangential as well as for rotational sticking and sliding should be distinguished for each line separately leading to the algorithmization of as many as four different situations based on the corresponding return-mapping algorithm. For the numerical examples with linear solid-shell elements and edge-to-edge contact elements, the matrices are simplified due to the linear shape functions and small penetration, i.e. all curvatures $k_I = 0$ and torsions $\varkappa_I = 0$ as well as penetration $r = 0$ are zero. For the numerical example with curvilinear beam-to-beam contact elements we are concentrating on the non-frictional case and show the full tangent matrix for the normal part only, all parameters such as curvature, torsions are computed as well and the coordinate r is also taken into account because it contains implicitly the size of the cross sections of the beams.

12.8.2.1 Tangent matrix for the normal force N

The tangent matrix for normal forces is constructed from one part with the normal force N representing the geometrical nonlinearity as presented in Sect. 12.7.1 and from another part containing ε_r from Section 12.7.2. The tangent matrix represents the linearization of the normal force which is always present for contact for both, sticking and sliding cases. Thus, eqn. (12.75c) together with the part in eqn. (12.83) containing a normal stiffness ε_r should be discretized, namely

$$(\delta\boldsymbol{\rho}_2 - \delta\boldsymbol{\rho}_1) \cdot (\varepsilon_r\, \mathbf{e}_1 \otimes \mathbf{e}_1)\, (\mathbf{v}_2 - \mathbf{v}_1^t) \tag{12.98a}$$

$$+N\, Sym\{\, (\delta\boldsymbol{\rho}_2 - \delta\boldsymbol{\rho}_1) \cdot [k_1 \cos\varphi_1(1 - rk_1\cos\varphi_1)\dot{s}_1\delta s_1$$

$$-r(\varkappa_1\dot{s}_1 + \dot{\varphi}_1)(\varkappa_1\delta s_1 + \delta\varphi_1)]\, (\mathbf{v}_2 - \mathbf{v}_1^t)\} \tag{12.98b}$$

Rates and variations of coordinates are transformed using eqns. (12.36) for arbitrary geometry. Following the transformation pattern given in eqn. (12.77) and using the approximation operator $\mathbf{A}$ in eqn. (12.97) we can obtain the following matrix:

$$\mathbf{K}^N = \qquad\qquad \varepsilon_r\,\mathbf{A}^T\left[\mathbf{e}_1 \otimes \mathbf{e}_1\right]\mathbf{A} \tag{12.99}$$

$$+\;\; N\,\mathbf{A}^T\!\left[Sym\left\{\frac{k_1\cos\varphi_1}{(1-rk_1\cos\varphi_1)}\boldsymbol{\tau}_1\otimes\boldsymbol{\tau}_1 - \tfrac{1}{r}\mathbf{g}_1\otimes\mathbf{g}_1\right\}\right]\mathbf{A}. \tag{12.100}$$

This matrix consists of a main part eqn. (12.99) together with a curvature and a rotational part given in eqn. (12.100). In the case of a linear approximation and edge-to-edge contact approach the matrix is simplified only to the main part as

$$\mathbf{K}^N = \varepsilon_r\,\mathbf{A}^T\left[\mathbf{e}_1 \otimes \mathbf{e}_1\right]\mathbf{A}. \tag{12.101}$$

The matrix represents both segments.

12.8.2.2   Tangent matrices for tangential sticking

This part is computed and then added to the global matrix if the tangential sticking case is determined within the corresponding return-mapping scheme for the certain I-th segment:

$$\mathbf{K}_I^{T,st} = \varepsilon_I\,\mathbf{A}^T\left[\boldsymbol{\tau}_I \otimes \boldsymbol{\tau}_I\right]\mathbf{A}. \tag{12.102}$$

The matrix is taken only for the I-th tangentially sticking segment.

12.8.2.3   Tangent matrices for tangential sliding

This part, see eqn. (12.89b), is computed and then added to the global matrix if the tangential sliding is determined within the corresponding return-mapping scheme for the I-th segment:

$$\mathbf{K}_I^{T,sl} = -\varepsilon_r\,\mu_1\,\mathbf{A}^T\left[\boldsymbol{\tau}_1 \otimes \mathbf{e}_1\right]\mathbf{A}. \tag{12.103}$$

The matrix is then taken only for the I-th tangentially sliding segment.

12.8.2.4   Tangent matrices for rotational sticking

This matrix includes a part representing the geometrical nonlinearity from eqn. (12.76d) for the linear segment ($k_I \equiv 0,\; \varkappa_1 \equiv 0$) and a constitutive part for the rotational sticking from eqn. (12.83). This part is computed and then added to the global matrix if the rotational sticking is determined within the corresponding return-mapping scheme for the I-th segment:

$$\mathbf{K}_I^{M,st} = \mathbf{A}^T \left[\varepsilon_1 \, \mathbf{g}_1 \otimes \mathbf{g}_1 - \frac{M_1}{r^2} \left(\mathbf{e}_1 \otimes \mathbf{g}_1 + \mathbf{g}_1 \otimes \mathbf{e}_1 \right) \right] \mathbf{A} \qquad (12.104)$$

The matrix is then only for the I-th rotationally sticking segment.

Here, ε_1 represents a rotational stiffness, see Remark in Sect. 12.7.2.

12.8.2.5 Tangent matrices for rotational sliding

These matrices are zero, because the sliding contribution is computed according to the Tresca model independent of other parameters and remains constant, see proof in Section 12.7.4.

Remark.
The tangent matrices are representing only the contact kinematics according to the developed theory and, therefore, must be added to the global tangent matrices. Thus, finite element models can be enriched with a contact possibility. The full finite element model must contain then finite elements describing the behavior of a continuum according to the selected model, e.g. shell, solid-shell and beam elements.

12.8.3 Residual vector

The residual vector represents discretization of the weak form. For the considered linear finite element the weak form in eqn. (12.60) should be discretized with special attention to its symmetrical structure. We consider this procedure separately for each part for normal, tangential and moment interactions.

12.8.4 Part for normal interaction

The normal part δW_N according to the symmetry $1 \leftrightarrow 2$ is fully written as

$$\begin{aligned}
\delta W_N &= Sym\left\{N(\delta\boldsymbol{\rho}_2 - \delta\boldsymbol{\rho}_1) \cdot \mathbf{e}_1\right\} = \\
&= \frac{1}{2}\{N(\delta\boldsymbol{\rho}_2 - \delta\boldsymbol{\rho}_1) \cdot \mathbf{e}_1 + N(\delta\boldsymbol{\rho}_1 - \delta\boldsymbol{\rho}_2) \cdot \mathbf{e}_2\}. \quad (12.105)
\end{aligned}$$

We note that the normal force N is mutual for both curves and vectors defining the shortest distance are satisfying $\mathbf{e}_2 = -\mathbf{e}_1$, see definition in eqn. (12.11). Introducing the approximation operator $\mathbf{A}$ we obtain:

$$\begin{aligned}
\delta W_N &= \frac{1}{2}\{N(\mathbf{A}\delta\mathbf{x}) \cdot \mathbf{e}_1 - N(\mathbf{A}\delta\mathbf{x}) \cdot (-\mathbf{e}_1)\} \\
&= N(\mathbf{A}\delta\mathbf{x}) \cdot \mathbf{e}_1 \\
&= \{\delta\mathbf{x}\}^T N[\mathbf{A}]^T\{\mathbf{e}_1\}. \quad (12.106)
\end{aligned}$$

The last line in eqn. (12.106) is written using the square and figure brackets to emphasize the vector-matrix notation after the discretization. Thus, this part of the residual for the normal interaction becomes then

$$\{\mathbf{R}_N\} = N[\mathbf{A}]^T\{\mathbf{e}_1\} \quad (12.107)$$

12.8.5 Part for tangential interaction

The tangential part is written as

$$\begin{aligned}
\delta W_T &= Sym\{T_1(\delta\boldsymbol{\rho}_2 - \delta\boldsymbol{\rho}_1) \cdot \boldsymbol{\tau}_1\} \\
&= \frac{1}{2}\{T_1(\delta\boldsymbol{\rho}_2 - \delta\boldsymbol{\rho}_1) \cdot \boldsymbol{\tau}_1 + T_2(\delta\boldsymbol{\rho}_1 - \delta\boldsymbol{\rho}_2) \cdot \boldsymbol{\tau}_2\}. \quad (12.108)
\end{aligned}$$

Now the tangential forces T_1, T_2 are defined independently according to the corresponding return-mapping algorithms. The tangent vectors $\boldsymbol{\tau}_1$ and $\boldsymbol{\tau}_2$ are individual for each segment as well. Thus, the discretization leads to

$$\delta W_T = \frac{1}{2}\{T_1(\delta\boldsymbol{\rho}_2 - \delta\boldsymbol{\rho}_1)\cdot\boldsymbol{\tau}_1 + T_2(\delta\boldsymbol{\rho}_1 - \delta\boldsymbol{\rho}_2)\cdot\boldsymbol{\tau}_2\}$$

$$= \frac{1}{2}\{T_1(\mathbf{A}\delta\mathbf{x})\cdot\boldsymbol{\tau}_1 - T_2(\mathbf{A}\delta\mathbf{x})\cdot\boldsymbol{\tau}_2\}$$

$$= \frac{1}{2}\{\delta\mathbf{x}\}^T[\mathbf{A}]^T(T_1\{\boldsymbol{\tau}_1\} - T_2\{\boldsymbol{\tau}_2\}). \tag{12.109}$$

The residual for the tangtential part is written then as

$$\{\mathbf{R}_T\} = \frac{1}{2}[\mathbf{A}]^T(T_1\{\boldsymbol{\tau}_1\} - T_2\{\boldsymbol{\tau}_2\}). \tag{12.110}$$

12.8.6 Part for moment (rotational) interaction

The part for the moment interaction starting from the equation

$$\delta W_M = Sym\left\{\frac{M_1}{r}(\delta\boldsymbol{\rho}_2 - \delta\boldsymbol{\rho}_1)\cdot\mathbf{g}_1\right\} \tag{12.111}$$

is discretized completely similar to the tangential part. The residual for the rotational part is written then as

$$\{\mathbf{R}_M\} = \frac{1}{2\,r}[\mathbf{A}]^T(M_1\{\mathbf{g}_1\} - M_2\{\mathbf{g}_2\}). \tag{12.112}$$

The full residual vector is finally constructed as the sum of all parts:

$$\{\mathbf{R}\} = \{\mathbf{R}_N\} + \{\mathbf{R}_T\} + \{\mathbf{R}_M\} \tag{12.113}$$

12.9 Numerical examples

In this section, first, a comparison of the non-frictional edge-to-edge contact formulation, see Section 12.5.1.2, with a special case of *Equilibrium of elastica problem* is presented for the verification. This case is also compared with the beam-to-beam contact formulation, see Section 12.5.1.1. Afterwards, an example with edge-to-edge contact for two intersecting beams is analyzed in details.

A "solid-shell" bilinear finite element, see [59], is chosen to model the behavior of the beams, thus modeling beams with rectangular cross section directly. A contact element with two linear segments, see implementation details in Section 12.8, is used to enrich the model with edge-to-edge contact.

In order to study beam-to-beam contact in detail, it is necessary, first to choose a proper finite element model for the beam itself and discuss then the consistency with our model concerning the rotational moment M_1, i.e. it should be consistently transfered to the corresponding rotational degrees of freedom. The consistent application of the developed contact theory for the curvilinear beam contact requires the application of high order finite element techniques, see e.g. [152], or as an alternative the iso-geometrical technique, see e.g. [72]. However, the development of such a curvilinear finite beam element together with a curvilinear beam-to-beam contact element especially for the arbitrary frictional case as well as further numerical analysis are outside of the scope of the current publication. This is planned to be discussed in detail in a following article. Some details are available in [44], [43] and can be summarized as follows:

1. The finite beam element model is based on Reissner theory. The finite element then contains six degrees of freedom per node: 3 displacements and 3 rotations. The stiffnesses of the mid-line $\mathbf{C}_n$ with regard to a load vector containing axial and two shear forces and the stiffnesses $\mathbf{C}_m$ with regard to a moment vector containing torsional and two bending moments represent the mechanical properties of the beam. In a beam coordinate system they have the following form:

$$\mathbf{C}_n = \begin{bmatrix} EA & 0 & 0 \\ 0 & GA_{s_1} & 0 \\ 0 & 0 & GA_{s_2} \end{bmatrix}, \quad \mathbf{C}_m = \begin{bmatrix} GI_t & 0 & 0 \\ 0 & EI_2 & 0 \\ 0 & 0 & EI_3 \end{bmatrix}, \quad (12.114)$$

where EA is the axial stiffness, GA_{s_1}, GA_{s_2} are shear stiffnesses, GI_t is the torsional stiffness, EI_2, EI_3 are bending stiffnesses.

2. Finite rotations are taken into account within the finite element formulation according to Ibrahimbegovic [73].

3. The iso-geometrical technique, see [72], has been exploited to obtain $C1$-continuous approximations for the curvilinear beams.

4. Beam-to-beam contact elements are obtained using the same approximation functions for the contact pairs in consistent fashion.

12.9.1   Bending of a flexible beam by a rigid beam. Non frictional case

The first flexible beam, see Fig. 13.4 modeled with 50 solid-shell finite elements is positioned in the XOY-plane and clamped at the left end. Material parameters: Linear Hooke's material with Young's modulus $E = 2.1 \cdot 10^4$ and Poisson ratio $\nu = 0.3$. Geometrical parameters: length 1.0 and square cross section 0.02×0.02 – units are consistent. The second beam with the same cross section is rigid (only one solid-shell element) and is positioned parallel to the OY-axis under the right end of the first beam. The first flexible beam is turned at 45° along the OX-axis such that during further loading it is contacting with the rigid beam only along the lower edge. Thus, only edge-to-edge contact is realized during the deformation process. In the case of modeling with bi-linear solid-shell elements, the displacement vector $\mathbf{u} = \{-1.0000, 0.0000, 0.6366\}^T$ is applied with 1000 equal increments (load steps) to all nodes of the rigid beam. The contact between beams is assumed to be non-frictional with a penalty parameter $\varepsilon_N = 2.1 \cdot 10^5$. All nodes of the flexible beam are constrained along the OY-axis to prevent bouncing along the rigid beam. During the loading process the first flexible beam is sliding along the rigid beam. Thus, a quasi 2D-deformation is realized, see the diagram of deformation in Fig. 12.11.

12.9.1.1   Verification – Equilibrium of the elastica problem

The analytical solution of the current evolution contact problem is hardly possible, however, the final deformed state (for non-frictional problem) can be formulated as *an equilibrium of the elastica problem* which was a subject of interest since centuries for many researchers starting with Leonhard Euler. The equilibrium of the elastica problem in 2D can be formulated via the famous equation:

$$EJ\frac{d\varphi}{ds} = M, \qquad (12.115)$$

where φ is angle between a tangent line and the X-axis, s is the standard arc-length parameter, and M is a bending moment. Eqn. (12.115) shows the well known proportionality of the curvature $\varkappa = \dfrac{d\varphi}{ds}$ to the bending moment M with bending stiffness. EJ as a proportionality coefficient. The definition of tangent angle φ and the differential dependences of a shear normal force Q and and a tangential force N on the moment M should be taken into account in order to obtain then the deformed line. Though, several analytical approaches to solve *the equilibrium of the elastica problem* are known in literature, e.g. solutions via elliptic integrals see e.g. in Popov [145], we are going to solve the problem directly via a finite difference scheme. Our special problem can be formulated as a Boundary Value Problem (BVP) (we adapt the stiffness to our "solid-shell" formulation by setting $EJ/(1-\nu^2)$ instead of EJ) for the following system of Ordinary Differential Equation (ODE):

$$\begin{cases} \dfrac{EJ}{1-\nu^2}\dfrac{d\varphi}{ds} = M + Nx\cos\alpha - Ny\sin\alpha \\[2mm] \dfrac{dx}{ds} = \cos\varphi \\[2mm] \dfrac{dy}{ds} = \sin\varphi \end{cases} \qquad (12.116)$$

with boundary conditions

$$\begin{cases} \varphi(0) = 0;\, \varphi(s_c) = \alpha \\[2mm] x(0) = 0;\, x(s_c) = x_c \\[2mm] y(0) = 0;\, y(s_c) = y_c, \quad 0 \le s \le 1. \end{cases} \qquad (12.117)$$

where coordinate system is chosen at the clamped end; N is a contact force between beams, $x(s_c)$ and $y(s_c)$ are Cartesian coordinate of the contact point appeared at length s_c; α is the angle between the beam tangent vector and the X-axis. In order to solve the system (12.116) the aforementioned differential dependences between forces and bending moment M should be taken account. However, in order to avoid the direct solution of BVP we will solve it approximately by reformulation BVP into the Initial Value Problem (IVP) Moreover, some values we will use from the finite element solution. Namely, the following steps of verification are fulfilled for comparison of the solutions:

1. From the finite element solution a final force N and an angle α at the contact point (x_c, y_c) are defined. For the current problem $N = 2.8092$ and $\alpha = 35.5346^o$ are obtained at $x_c = 0$ $y_c = 0.6366$.

2. The system (12.116) is solved with given N and α as the Initial Value Problem with the following initial conditions

$$\begin{cases} \varphi(0) = 0; \\ x(0) = 0; \\ y(0) = 0, \quad 0 \leq s \leq 1. \end{cases} \tag{12.118}$$

3. The obtained deformed line is then compared with deformed line from FE solution.

Remark.
The current approach of solution for BVP as IVP allows to satisfy the boundary conditions at the contact point $x(s_c) = x_c$ and $y(s_c) = y_c$ only approximately.

The simple explicit Euler finite difference scheme is applied then to solve the problem with $\Delta s = 0.001$. Comparison of the deformed central-lines is presented in Fig. 13.5. As mentioned in Remark 12.9.1.1, the deformed line is not passing through the contact point x_c, y_c.

12.9.1.2 Computations with curvilinear beam-to-beam contact

For a further comparison of the final deformed configuration the beam finite element model shortly described at the beginning of the Section 13.2 is used. **Only four** C^1-spline continuous finite elements are used to model the elastic beam. The same material data is taken. In order to obtain a model mechanically equivalent to the solid-shell one, the same stiffness characteristics as given in eqn. (12.114) have been taken. Namely, the cross section A and all area moments of inertia in eqn. (12.114) are taken to be the same as for the square section of the solid-shell finite element model. However, the cross section for the beam is taken to be circular for the contact algorithm as shown in eqn. (12.62) with constant radius for both beams $R = R_1 = R_2$. The radius R is computed from the equivalency of the cross section area for the solid-shell and the circular section. Now, the displacement vector **u** is applied only with 100 load steps to all nodes of the rigid beam, see Fig. 12.12. The verification with the analytical solution is also presented in Fig. 13.5. The problem is analyzed with both tangent matrices for normal interaction and presented in Table 13.2 for the following cases:

- Solid-shell FE model with "edge-to-edge" contact elements;

- Curvilinear beam FE model with "beam-to-beam" contact. Normal contact matrix for arbitrary large distance r and curvature with parts in eqns. (12.99) and (12.100).

- Curvilinear beam FE model with "beam-to-beam" contact for linear contact elements with only one part as given in eqn. (12.101).

FE model	No. of elem.	No. of load steps	Global No. of iterations
1	50	1000	3986
2	4	100	396
3	4	100	428

Table 12.1: Comparison of solution for 1) linear solid-shell elements; 2) curvilinear beam elements with full contact matrix; 3) curvilinear beam elements with "linear" contact matrix.

During the loading process the flexible beam is undergoing large deformations (from a straight beam into a curved beam) as well as large sliding, especially at the end of the loading process, see diagram in Fig. 12.11. This leads, on one side, to the necessity of a relatively large number of bi-linear solid-shell elements (50 elements) and, on the other side, to a relatively small load step size (resp. a large number of load steps) in order to describe the sliding over segments correctly. Steps $7 - 10$ presented in Fig. 12.11 are crucial during the loading process because the beam undergoes large deformations and large sliding. Namely this part of loading requires small load steps for linear elements. For the beam-to-beam element the number of equilibrium iterations per load step is increasing from 4 to 11 during the last load steps. A relatively small advantage using the full matrix compared with the reduced one is observed in the global number of iterations – 396 against 428 iterations, see Table 13.2. This can be explained by the fact that the beam possesses a small curvature at the contact point together with a small radius for the circular cross section. A crucial example can be constructed by taking "a soft" beam, scaling all shear moduli in eqn. (12.114) with a factor 10^{-4}. Then displacements are now applied in only 75 load steps of the same size as in the previous example. The final configuration is presented in Fig. 12.13 (*visual non-smoothness is just the result of incomplete visualization – only element-wise tube surfaces are constructed during postprocessing via Matlab which is used here*). In the example shear softening leads to high curvatures at contact points, thus the influence of the curvature part should be more pronounced. That is essentially true!!! – no convergence after the 70th load step has been found with using the linear matrix given in eqn. (12.101) and 297 global iterations are necessary taking all parts into account in eqns. (12.99) and (12.100).

The advantage of C^1-spline continuous beam finite elements compared to the linear solid-shell finite elements is obvious – only four elements are sufficient to describe efficiently the highly deformed configuration.

The small differences between deformed lines obtained for the solutions using "solid-shell" and beam finite elements as well as using the a finite difference scheme for *the Equilibrium of the elastica problem* are explained by the different kinematical hypotheses which are the basis of the corresponding theories for the three compared solutions and by the application of different numerical approaches.

12.9.2 Analysis of contact for intersecting beams

Two clamped beams of unit length intersecting each other are chosen to study various combinations of parameters within the developed theory such as the tangential and the rotational sticking including various sticking and sliding cases of two lines. The *"first"* lower beam is parallel to the OY-axis and the *"second"* upper beam is parallel to the OX-axis at the initial configuration, see the position in Fig. 12.15. Both beams are turned at 45° along their central-lines in order to provide initial contact only along the edges at a distance 0.75 from both clamped ends. Now two beams are flexible, made of the same material: linear Hooke's material with Young's modulus $E = 2.1 \cdot 10^4$ and Poisson ratio $\nu = 0.3$; and possessing the same geometrical parameters: length $L = 1.0$ and square cross section $A = 0.02 \times 0.02$. The beams are modeled with 50 solid shell elements and edge-to-edge contact elements are supplied.

The prescribed displacement $u_z = 0.5$ in vertical OZ-direction is prescribed at the upper node of the lower beam, see Fig. 12.15. The vertical displacement is applied in 100 increments. In the following always the top view along the OZ-axis for both initial and deformed final configurations are presented for different combinations of parameters.

12.9.2.1 Non-frictional contact

For non-frictional contact only the normal tangent matrix in eqn. (12.101) is taken into account together with the corresponding normal force N in eqn. (12.63). The penalty parameter is taken as $\varepsilon_N = 2.1 \cdot 10^6$. Convergence is reached with 391 global iterations for 100 load steps. Both, relative sliding along the "first" lower beam increasing ξ^1 and along the "second" upper beam increasing ξ^2 are observed, see Fig. 12.17.

12.9.2.2 Full sticking (tied) contact

Full sticking, or tied contact, is enforced using all tangential and rotational sticking conditions (corresponding penalty for both lines $\varepsilon_1 = \varepsilon_2 = 2.1 \cdot 10^6$) without the return-mapping scheme. Convergence is reached with 375 global iterations for 100 load steps. The result is presented in Fig. 12.18. To study the influence of various parts for the enforcement of full sticking (tied) condition the following cases are computed:

1. Tied contact enforced by tangential and rotational sticking.

2. Tied contact enforced only by tangential sticking.

3. Tied contact enforced only by rotational sticking.

It is interesting to note that the same final configuration presented in Fig. 12.18 is reached for all three cases. To emphasize this effect, an angle between the vectors $\vec{a}$ and $\vec{b}$ defining the edge directions for elements as depicted in Fig. 12.15 is studied. This angle is equal to 45° at the initial undeformed state and is then changing during the deformation, see Fig. 12.16. However, one can see that the angle remains the same for all kinds of enforcement. The explanations for this effect appear, however, obvious: full sticking (tied contact) can be enforced only by rotational sticking because the corresponding tangent matrix $g \otimes g$ responsible for the g direction is orthogonal to the vector e direction, and therefore, is leading to the sticking of the intersecting lines. The reason for the constant angle lays in the application of solid-shell finite elements which are by construction not capable to transmit the moment applied to the edge. However, the large opportunities for coupling this moment between curves either with classical beams or with classical shell finite elements possessing both rotational degrees of freedom and, in due course, moments remain open.

12.9.2.3 Partial tied contact – sticking only along one line

Now, partial tied contact is modeled along only one line. First, partial tangential sticking is enforced along the upper edge of the "first" lower beam by setting the tangential penalty parameter $\varepsilon_1 = 2.1 \cdot 10^5$ without applying the return-mapping scheme, while the non-frictional contact is assumed for the lower edge of the "second" upper beam by setting $\varepsilon_1 = 0.0$, see Fig. 12.19. Convergence is reached with 386 global iterations for 100 load steps. Then, partial tangential contact is enforced along the second line with the same values only changing the beam $1 \leftrightarrow 2$, see Fig. 12.20. Convergence is reached with 379 global iterations for 100 load steps in the latter case.

One can notice that during the loading for the non-frictional case, see Fig. 12.17, the "first" lower beam is undergoing large relative sliding along the "second" beam increasing ξ^2, while the "second" beam is undergoing only a small relative sliding – expressed in tangential coordinates – $\xi^2 >> \xi^1$. Thus, partial tied contact with a locked variable ξ^1 supplying $\varepsilon_1 = 2.1 \cdot 10^5$ looks similar to the non-frictional case – compare Fig. 12.17 and Fig. 12.19. In due course, partial tied contact with a locked variable ξ^2 supplying $\varepsilon_2 = 2.1 \cdot 10^5$ looks similar to the full sticking case – compare Fig. 12.18 and Fig. 12.20.

12.9.2.4 Sliding contact

Finally, the four considered cases *non-frictional, full sticking and two partial sticking cases* lead to *"an envelope"* in which frictional cases with all possible coefficients of friction μ_1 and μ_2 are contained. In Fig. 12.21 a top view of this envelope (shadowed) and the "first" lower beam positioned inside this envelope are given. The computation is provided for tangential sliding with coefficients of friction: $\mu_1 = 0.001$ and $\mu_2 = 0.05$. Convergence is reached with 419 global iterations for 100 load steps. The choice of rather small coefficients of friction is found to show that two beams are visually positioned inside the envelope constructed by the deformed configurations for four extremal cases. We have to note that in the sliding case especially for large deformations considerable convergence difficulties appear for large coefficients of friction and the application of the Augmented Lagrangian scheme has shown to be more efficient in this case.

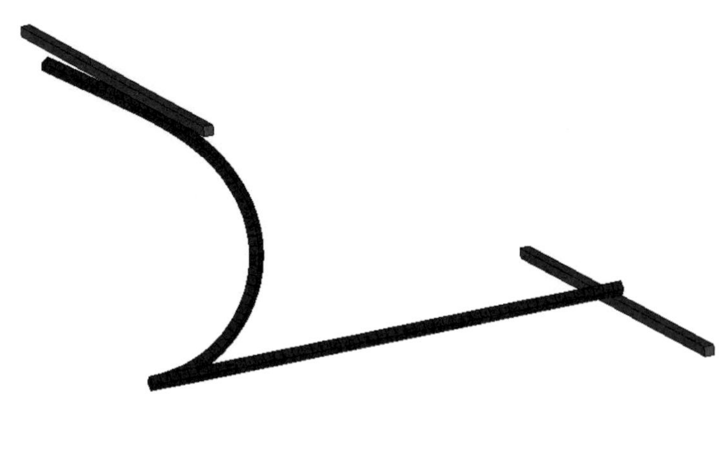

Figure 12.10: Bending of a flexible beam by a rigid beam – initial and final configurations applying solid-shell finite elements and **edge-to-edge contact elements**

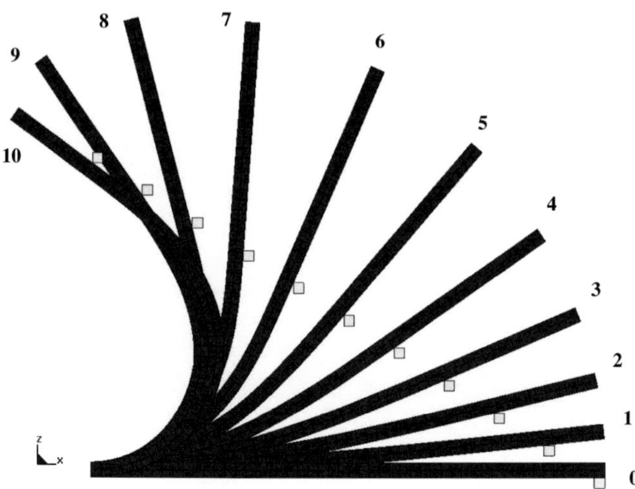

Figure 12.11: Diagram of deformation for equal load steps. Displacements of the rigid beam are prescribed.

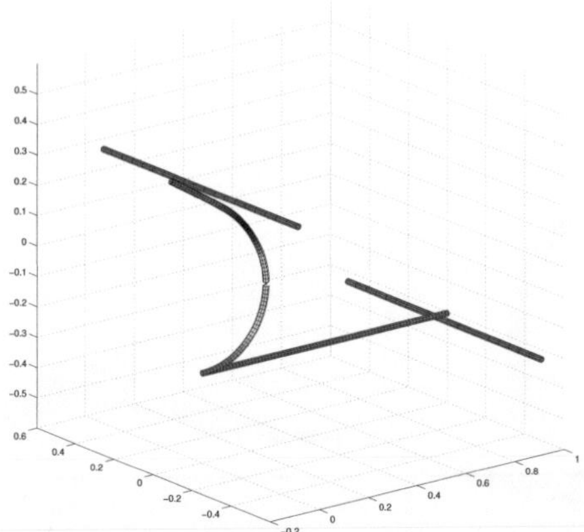

Figure 12.12: Bending of a flexible beam by a rigid beam applying beam finite elements and **beam-to-beam contact elements**

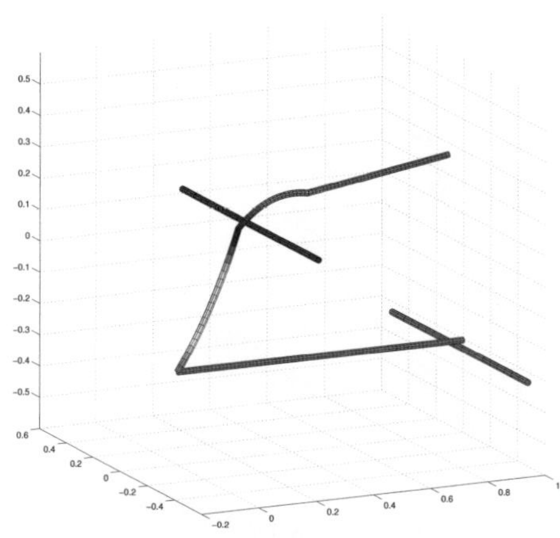

Figure 12.13: Bending of a flexible "soft" beam by a rigid beam applying beam finite elements and **beam-to-beam contact elements**

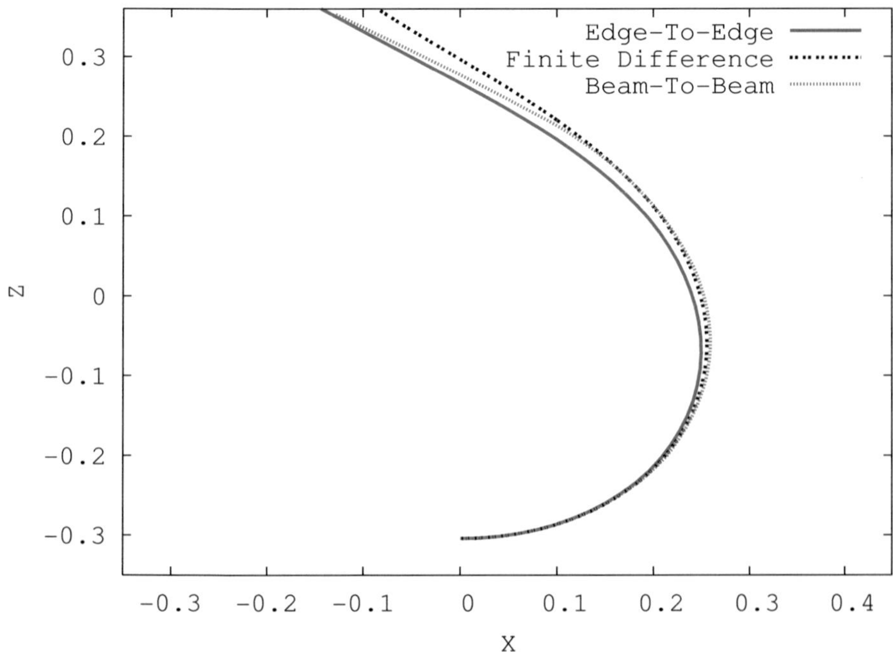

Figure 12.14: Comparison of the deformed central-lines for a) 50 bilinear solid-shell finite elements together with edge-to-edge contact; b) finite difference solution for *"the equilibrium of the elastica problem"*; c) 4 C^1-continuous curvilinear Reissner beam finite elements and beam-to-beam contact

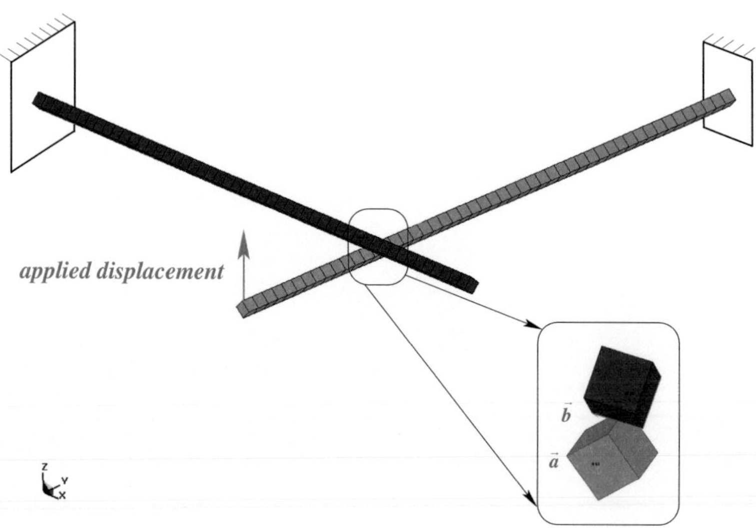

Figure 12.15: Intersecting deformable beams. Initial configuration and loading condition.

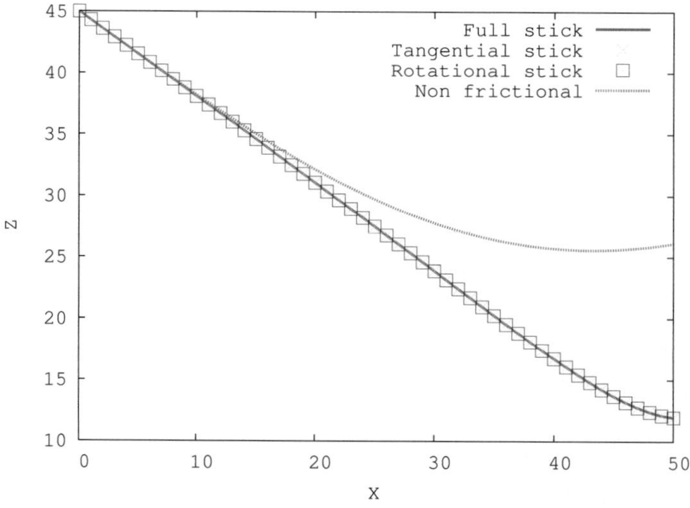

Figure 12.16: Evolution of the angle between edges with vectors $\vec{a}$ and $\vec{b}$ for contacting elements during deformation. Tied contact enforced: a) by tangential and rotational sticking (Full stick); b) only by tangential sticking (Tangential stick); c) only by rotational sticking (Rotational stick). d) Non-frictional contact.

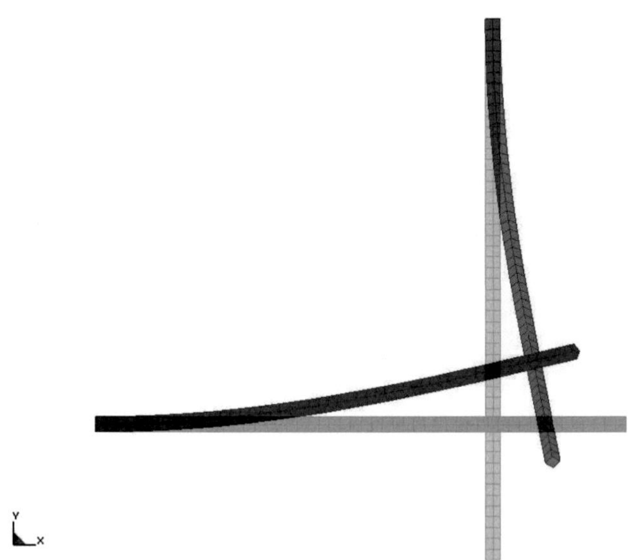

Figure 12.17: Non-frictional contact. Undeformed and deformed configurations.

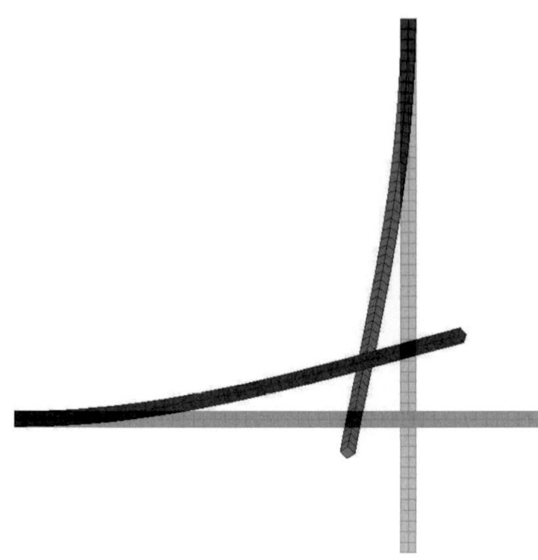

Figure 12.18: Full sticking (tied) contact. Undeformed and deformed configurations.

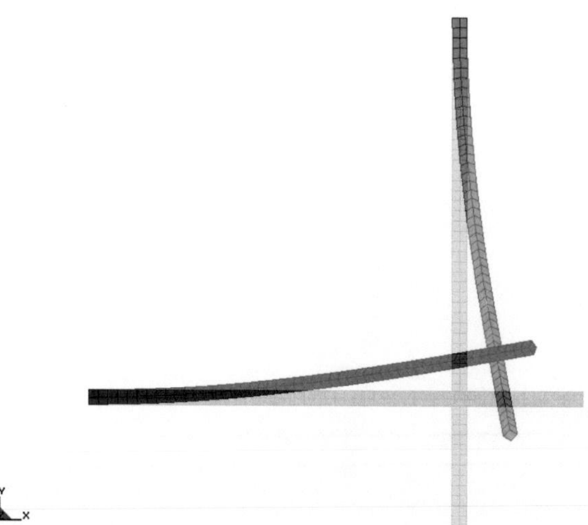

Figure 12.19: Partial tied contact. Tangential sticking is enforced along the upper beam with a penalty of $\varepsilon_1 = 2.1 \cdot 10^5$, however, non-frictional contact is supplied for the lower beam. Undeformed and deformed configurations.

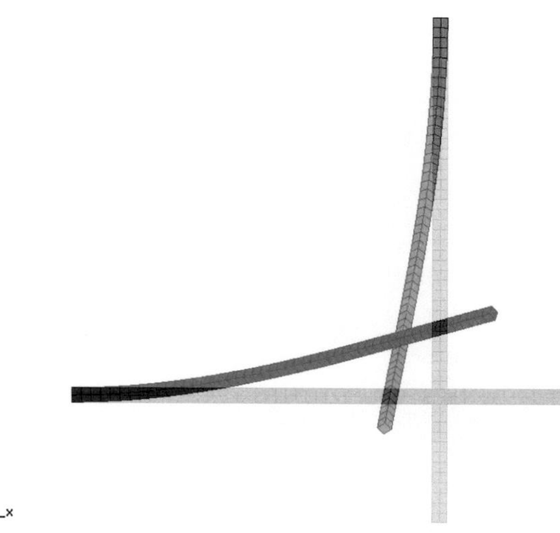

Figure 12.20: Partial tied contact. Tangential sticking is enforced along the lower beam with a penalty of $\varepsilon_2 = 2.1 \cdot 10^5$, however, non-frictional contact is supplied for the upper beam. Undeformed and deformed configurations.

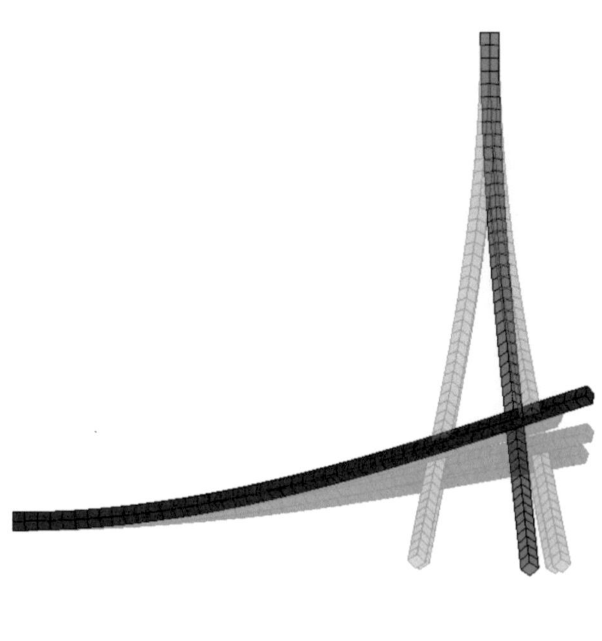

Figure 12.21: Sliding contact—tangential sliding with coefficients of friction: $\mu_1 = 0.001$ and $\mu_2 = 0.05$ is considered. The "lower" beam lays inside the envelope (shadowed) constructed by four extremal cases: 1) non-frictional contact, 2) full sticking (tied) contact, 3) partial tied contact for the "lower" beam and 4) partial tied contact for the "upper" beam.

Remark.

The considered effect with partial sticking and sliding can be considered as a result appearing during interaction between two beams possessing anisotropic surfaces with different properties (e.g. one beam has a very rough surface while the other has a very smooth surface).

Another possibility to apply the developed curve-to-curve contact approach is to model **a hinge slider joint** between two beams. In this joint the partial sticking as well as different elastic and friction properties can be prescribed by the construction.

12.10 Conclusion

A geometrically exact theory for curve-to-curve contact situations is proposed in the current contribution. The development begins consistently with the Closest Point Projection (CPP) procedure providing a shortest distance between curves as a natural measure of normal contact interaction. The CPP procedure leads to a special local coordinate system in which convective coordinates are used directly as measures of contact interaction between curves: normal, tangential and rotational. The existence and uniqueness of the CPP procedure is studied in detail – projection domains with a-priori unique solution are constructed in this coordinate system for curves with varying geometry.

Several achievements appear to be novel for the line-to-line contact description:

1. consideration of any relative motion including normal, tangential and rotational components separately for each curve is possible;
2. rotational interactions including corresponding rotational moments between curves can be considered consistently.

The Coulomb friction law for tangential interaction and the Tresca friction law for rotational interaction are considered as examples for constitutive relations between curves. All necessary linearizations for the iterative solution scheme are provided as covariant derivation in the introduced coordinate system for arbitrary large distances between curves. This leads to a closed form of tangent matrices independent of the approximation used for the finite elements.

The verification section contains the comparison between beam-to-beam and edge-to-edge finite element models as well as verification with a famous "Equilibrium of Euler elastica problem" computed via finite difference scheme. The further numerical examples are illustrating the ability to describe various kinematics for curve-to-curve contact situations e.g. partial sticking of a single curve. Though, "solid-shell" finite elements are predominantly used for the beam contact in this paper, the current theory can be efficiently applied for the beam finite element models with rotational degrees of freedom. The necessity to use the full matrices for the cases with high curvature at contact points has been shown in the case of the "soft" beam.

13

Geometrically exact theory for contact interactions of 1D manifolds. Algorithmic implementation with various finite element models[*]

Abstract

The intuitive understanding of contact between bodies is based on the geometry of adjoining bodies. A more sophisticated approach of an advanced analysis including the application of various numerical methods is to take advantage of the geometry of an analyzed object and describe the problem in the best coordinate system. The best coordinate system to describe contact interaction in all its geometrical details is a coordinate system attached to the geometrical features of contacting bodies. This leads to a systematical analysis of geometrical situations leading to contact pairs – surface-to-surface, line-to-surface, point-to-surface, line-to-line, point-to-line. Each contact pair is inherited with a special coordinate system based on its geometrical properties. The current contribution is concentrating on contact between 1D manifolds in 3D space – this is the majority of edge-to-edge, beam-to-beam, cable-to-edge etc. contact cases. The geometrically exact curve-to-curve contact approach is then systematically combined together with various finite element approaches – classical finite elements, isogeometric beam finite elements and also with a new developed solid-beam approach. Examples illus-

---

[*]The chapter is published in [96] A. Konyukhov, K. Schweizerhof *Geometrically exact theory for contact interactions of 1D manifolds. Algorithmic implementation with various finite element models*, Computer Methods in Applied Mechanics and Engineering, doi:10.1016/j.cma.2011.03.013, *Available online* 2 April 2011.

trating the diversity of various finite element combinations e.g. within knot mechanics are shown.

Keywords

curve-to-curve contact geometrically exact contact knot mechanics solid-beam

13.1 Combination with various finite element models of the continuum

There are several questions in further applications of the curve-to-curve approach to contact between bodies:

1. combination of classical 3D finite element discretization with the edge-to-edge approach;

2. combination of finite beam elements with the beam-to-beam algorithm;

3. development of special "solid-beam" elements with beam-to-beam algorithm which allow to take into account 3D effects also for beam type elements.

Combination of 3D finite elements with an edge-to-edge contact approach is rather straightforward. The continuum is discretized with solid or solid-shell elements and the chosen **edge-to-edge contact algorithm** is applied. Application with solid-shell elements, however with bilinear shape functions, is considered in detail in [96]. The consistent application with an isogeometric approach for the $3D$-continuum would require the iso-geometrical technique, see Hughes et.al. [72], or high order finite element techniques for shells, see Rank et.al. [152].

We consider here isogeometric techniques for beams and special cable like structures.

13.1.1 Combination of finite beam elements with the beam-to-beam contact algorithm

For a beam-to-beam algorithm we have to select a special finite beam element formulation allowing both finite rotations and enrichment with arbitrary curvilinear geometry. An extensive development is given e.g. in a complete series of articles by Ibrahimbegovic [73], Ibrahimbegovic [74], Ibrahimbegovic and Taylor [75]. The kinematics of deformation is subjected to the Reissner or the Timoshenko hypothesis; finite rotations are represented via quaternions. A short overview of the necessary equations is given in the following. The variational equation is given by

$$W_{int} = \frac{1}{2} \int_L \boldsymbol{\epsilon}^T \mathbf{n} + \boldsymbol{\kappa}^T \mathbf{m} \, ds_0 = \frac{1}{2} \int_L \boldsymbol{\epsilon}^T \Lambda \mathbf{C}_n \, \Lambda^T \boldsymbol{\epsilon} + \boldsymbol{\kappa}^T \Lambda \mathbf{C}_m \Lambda^T \boldsymbol{\kappa} \, ds_0, \quad (13.1)$$

where $\boldsymbol{\epsilon}$ is a strain measure for the mid-line of the curvilinear beam, $\boldsymbol{\kappa}$ is a bending strain for the curvilinear beam. Finite rotations are represented by the matrix Λ given by the Rodrigues formula. In the current approach, the matrix Λ, the update of Λ as well as $\boldsymbol{\kappa}$ within the iteration scheme are computed via quaternions operating with an incremental rotation vector $\Delta \mathbf{w}$, see details in [73]. The matrix $\mathbf{C}_n$ represents the stiffnesses of the mid-line with regard to a load vector $\mathbf{n}$ containing axial and two shear forces. The matrix $\mathbf{C}_m$ represents the bending stiffnesses with regard to a moment vector $\mathbf{m}$ containing torsional and two bending moments. In a beam coordinate system they have the following form:

$$\mathbf{C}_n = \begin{bmatrix} EA & 0 & 0 \\ 0 & GA_{s_1} & 0 \\ 0 & 0 & GA_{s_2} \end{bmatrix}, \quad \mathbf{C}_m = \begin{bmatrix} GI_t & 0 & 0 \\ 0 & EI_2 & 0 \\ 0 & 0 & EI_3 \end{bmatrix}, \quad (13.2)$$

where EA is the axial stiffness, GA_{s_1}, GA_{s_2} are shear stiffnesses, GI_t is the torsional stiffness, EI_2, EI_3 are bending stiffnesses. The finite element then contains six degrees of freedom: 3 displacements and 3 rotations. Thus, for the numerical algorithm, both a displacement vector $\mathbf{u}$ and an incremental rotation $\Delta \mathbf{w}$ have to be approximated.

13.1.1.1 C^1-continuous isogeometric approximation of the curve

For further implementation, we are using two types of approximation:

- C^1-continuous cubic spline in Hermite form;

- a special NURB spline representing the circular geometry exactly.

	$H_k(t)$	$\dfrac{\partial H_k}{\partial t}$	$\dfrac{\partial^2 H_k}{\partial t^2}$
$H_0^3(t)$	$1 - 3t^2 + 2t^3$	$-6t + 6t^2$	$-6 + 12t$
$H_1^3(t)$	$t - 2t^2 + t^3$	$1 - 4t + 3t^2$	$-4 + 6t$
$H_2^3(t)$	$t^3 - t^2$	$3t^2 - 2t$	$6t - 2$
$H_3^3(t)$	$3t^2 - 2t^3$	$6t - 6t^2$	$6 - 12t$

Table 13.1: **Hermite functions** $H_i(t)$**,** $t \in [0, 1]$ **and their derivatives** are forming a shape function space for smooth contact elements.

The Hermite type interpolation leads to the following finite element:

$$\mathbf{x}(t) = \mathbf{x}^{(1)} H_0^3(t) + \mathbf{m}^{(1)} H_1^3(t) + \mathbf{m}^{(2)} H_2^3(t) + \mathbf{x}^{(2)} H_3^3(t), \qquad (13.3)$$

where $H_i^3(t)$, $0 \le t \le 1$ are Hermite polynomials, see Table 13.1. $\mathbf{x}^{(1)}$, $\mathbf{x}_{(2)}$ are two nodes and $\mathbf{m}^{(1)}$, $\mathbf{m}^{(2)}$ are tangent vectors at corresponding nodes. Defining then the tangent vectors via a chord interpolation from nodes of the neighboring elements we obtain the following structure of the finite element, see Fig. 13.1:

$$\mathbf{x}(t) = \mathbf{x}^{(1)} \left(H_0^3(t) - \frac{H_2^3(t)}{2} \right) + \mathbf{x}_{(2)} \left(H_3^3(t) + \frac{H_1^3(t)}{2} \right) - \mathbf{x}_{(3)} \frac{H_1^3(t)}{2} + \mathbf{x}_{(4)} \frac{H_3^3(t)}{2} \quad (13.4)$$

Finally, the finite element is defined by nodes $\mathbf{x}^{(1)}$ and $\mathbf{x}^{(2)}$ i.e. all parameters such as forces, moments etc. are computed for the parameter $0 \le t \le 1$, see Fig. 13.1. Nodes $\mathbf{x}^{(3)}$ and $\mathbf{x}^{(4)}$ are taken from the neighboring element to keep C^1-continuity on element boundaries. The displacement vector $\mathbf{u}(t)$ and the vector of infinitesimal rotation $\Delta\mathbf{w}$ are approximated in the same fashion.

Equation (13.4) gives an approximation which is valid only for inner points of the curve. For boundary nodes the corresponding tangent vector $\mathbf{m}^{(i)}$ should be given according to the geometry.

This type of approximation is falling into a simplest type of NURBS approximation written in a very general form via Bernstein's polynomials

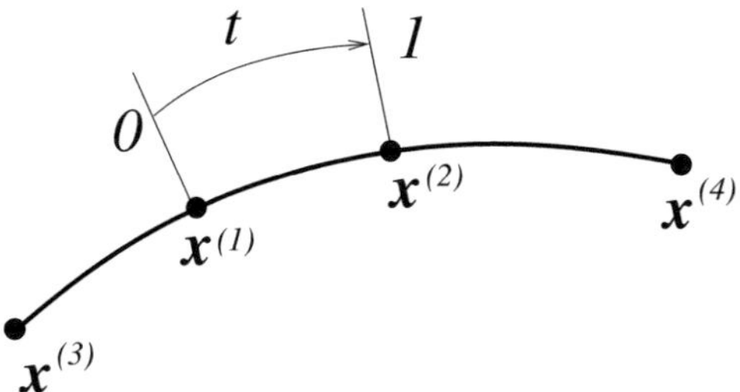

Figure 13.1: Approximation for the spline beam element. The beam element is given by nodes $x^{(1)}$ and $x^{(2)}$. Nodes $x^{(3)}$ and $x^{(4)}$ are taken from the neighboring element to **keep C^1-continuity on element boundaries**.

of order n

$$\mathbf{x}(u) = \frac{w_0 \mathbf{b}_0 B_0^n(t) + w_1 \mathbf{b}_1 B_1^n(t) + ... w_n \mathbf{b}_n B_n^n(t)}{w_0 B_0^n(t) + w_1 B_1^n(t) + ... w_n B_n^n(t)}. \tag{13.5}$$

This description of a spline has many advantages due to the additional degrees of freedom because of additional weights w_i. It leads to multiple possibilities to control the geometry, e.g. to represent a certain geometry exactly. A large number of special literature, however, is available containing all NURBS properties in detail, see e.g. Farin [38], Piegl and Tiller [143]. An example of a special NURB spline of the second order representing exact geometry of a circle is used in further implementation. For the quarter of a circle it has the following weights $w_0 = 1$, $w_1 = 1$, $w_2 = 2$, see Fig. 13.2.

A NURBS type contact is performed as two spline segments from the potentially contacting curves defining a beam-to-beam contact pair leading to a contact element possessing twice as many nodes, e.g. a contact element for the 4-nodes spline element is the 8-nodes curve-to-curve contact finite element.

13.1.2 Development of special "solid-beam" elements for the beam-to-beam algorithm

Beam finite element models are representing the dimension reduction from the 3D continuum into a 1D manifold. The costs of this are addi-

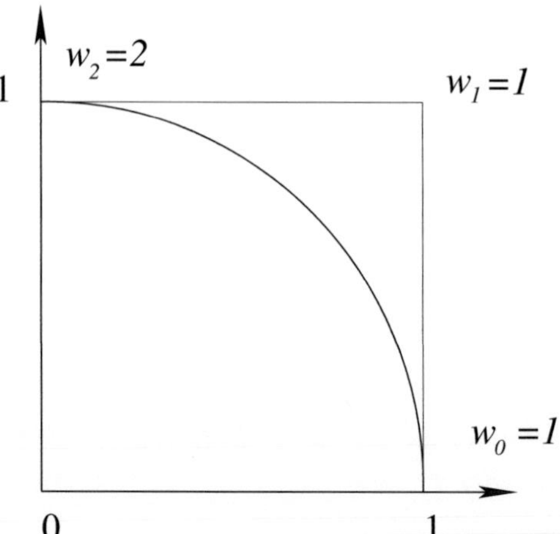

Figure 13.2: Quadratic NURBS approximation for the quarter of a circle. Weights are $w_0 = 1$, $w_1 = 1$, $w_2 = 2$.

tional rotational degrees of freedom and as a consequence for large rotations a costly update procedure for finite rotation parameters as well as for bending strains. For the cases of special beam-to-beam contact like cables possessing a particular elliptic geometry for their cross-sections it is rather attractive to construct a special 3D finite beam element. This element can be constructed as follows, see Fig. 13.3:

1. the mid-line of the beam is taken, first, with e.g. linear approximation. It is defined then by nodes 1 and 2;

2. the left cross-section is defined to be elliptic with the reference main axes defined by nodes 1-3 and 1-4;

3. the right cross-section is defined to be elliptic with the reference main axes defined by nodes 2-5 and 2-6;

Thus, an approximation of the 6-nodes "solid-beam" element is defined by the following shape functions:

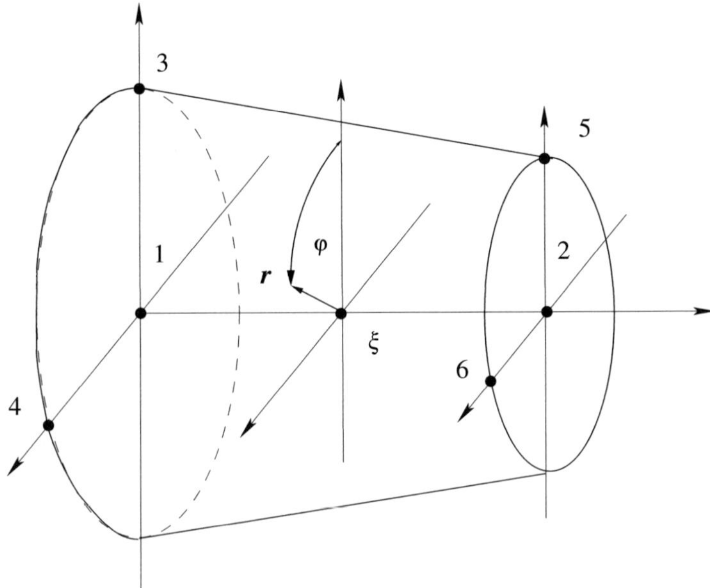

Figure 13.3: "Solid-Beam" with elliptic cross-sections defined by 6 nodes. Definition of local variables.

$$\mathbf{x} = \sum_{i=1}^{6} N(\xi, r, \varphi)\mathbf{x}_i =$$

$$\begin{aligned} = \; & \mathbf{x}_1(1-\xi)[1 - r(\cos\varphi + \sin\varphi)] \\ + \; & \mathbf{x}_2\xi[1 - r(\cos\varphi + \sin\varphi)] \\ + \; & \mathbf{x}_3(1-\xi)r\cos\varphi \\ + \; & \mathbf{x}_4(1-\xi)r\sin\varphi \\ + \; & \mathbf{x}_5\xi r\cos\varphi + \mathbf{x}_6\xi r\sin\varphi \end{aligned}$$

$$\text{with } 0 \leq \xi \leq 1; \;\; 0 \leq r \leq 1; \;\; 0 \leq \varphi \leq 2\pi \tag{13.7}$$

The nodal locations $\mathbf{x}_i = (x_i, y_i, z_i)$ are given in a global Cartesian coordinate system.

Contact for the solid-beam element is performed as follows: two mid-line segments from the potentially contacting solid-beam element are defining a beam-to-beam contact pair leading to the 4-nodes beam-to-beam contact finite element.

Remark.

Extension of the solid-beam approximation into a NURBS type approximation can be obtained, if the linear shape functions defining the approximation of a mid-line and depending on the variable ξ in eqn. (13.6) will be changed into corresponding shape functions of NURBS type, e.g. into those defined via Hermite functions in eqn. (13.4).

13.2 Numerical examples

13.2.1 Bending of a flexible beam by a rigid beam

Here the three finite element approaches are analyzed with the example "Bending of a flexible beam by a rigid beam" proposed in Konyukhov and Schweizerhof [96], see Fig. 13.4. This example is constructed as follows:

The first flexible beam with unit length 1.0 is positioned in the XOY-plane and clamped at the left end. The material parameters are: Linear Hooke's material with Young's modulus $E = 2.1 \cdot 10^4$ and Poisson ratio $\nu = 0.3$. The second beam with the same cross section is rigid (only one solid-shell element) and is positioned parallel to the OY-axis under the right end of the first beam. The total displacement vector $\mathbf{u} = \{-1.0000, 0.0000, 0.6366\}^T$ is applied with equal increments (load steps) to all nodes of the rigid beam. The contact between beams is assumed to be non-frictional with a penalty parameter $\varepsilon_N = 2.1 \cdot 10^5$. All nodes of the flexible beam are constrained along the OY-axis to prevent bouncing along the rigid beam. During the loading process the first flexible beam is sliding along the rigid beam. Thus, a quasi 2D-deformation is realized.

The following cases are computed now representing various combinations discussed in Section 13.1:

1. 50 solid-shell finite elements and edge-to-edge contact;

2. 4 C^1-continuous isogeometric beam elements and beam-to-beam contact;

3. 20 linear "solid-beam" element and beam-to-beam contact.

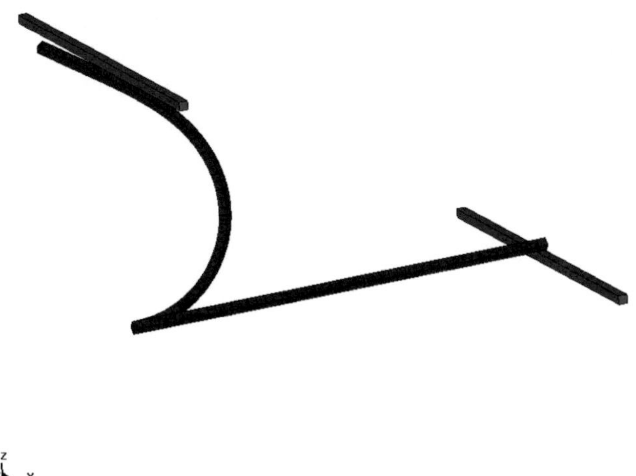

Figure 13.4: Bending of a flexible beam by a rigid beam – initial and final configurations applying solid-shell finite elements and **edge-to-edge contact elements**

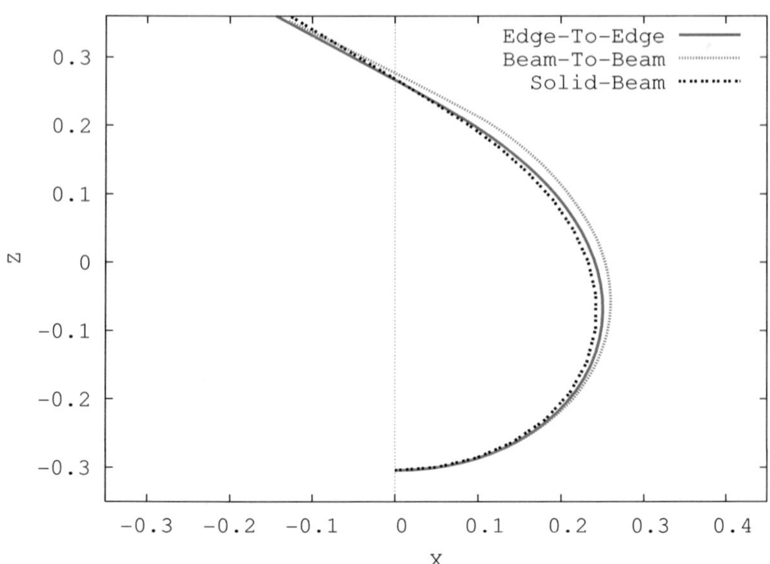

Figure 13.5: Comparison of the deformed central-lines for a) 50 bilinear solid-shell finite elements together with edge-to-edge contact; b) 4 C^1-continuous curvilinear beam finite elements and beam-to-beam contact; c) 20 linear solid-beam finite elements together with beam-to-beam contact.

For the first case, analyzed in detail in [96], the flexible beam (with a square cross section 0.02×0.02) is turned at 45° along the OX-axis such that during further loading it is contacting with the rigid beam only along the lower edge. Thus, only edge-to-edge contact is realized during the deformation process.

For the second case, also analyzed in detail in [96], a mechanically equivalent beam model with the same stiffness characteristics is taken. Namely, the cross section A and all area moments of inertia in eqn. (13.2) are taken to be the same as for the square section of the linear solid-shell finite element model. However, the cross section for the beam is taken to be circular for the contact algorithm as shown in eqn. (12.62) with constant radius for both beams $R = R_1 = R_2$. The radius R is computed from the equivalency of the cross section area for the solid-shell and the circular section.

The third case is geometrically equivalent to the second one, i.e. the cross section is circular, however, the stiffness properties will be computed directly for the solid-beam element with its stiffness matrices.

FE model	No. of elem.	No. of load steps	Global No. of iterations
1	50	1000	3986
2	4	100	396
3	20	100	752

Table 13.2: Comparison of solution for 1) linear solid-shell elements and edge-to-edge algorithm; 2) curvilinear beam elements and beam-to-beam algorithm; 3) linear solid-beam elements and beam-to-beam algorithm.

Remark.

An energy norm together with the relative tolerance $\varepsilon_{conv} = 10^{-12}$ has been used as a convergence criterion in all computations.

During the loading process the flexible beam is subjected to large deformations, especially at the end of the loading process. This leads, on one side, to the necessity of a relatively large number of bilinear solid-shell elements (50 elements) and, on the other side, to a relatively small load step size (resp. a large number of load steps) in order to describe the sliding over segments correctly. For the solid-beam case the local contact searching is performed according to the beam-to-beam case (larger area than edge-to-edge) – this allows to work with less elements

(20 FE) than for the solid-shell elements (50 FE). However, 20 *linear* solid-beams elements are resulting in a stiffer behavior than the first two models, see Fig. 13.5. This leads to a fairly large number of global iterations – 752. The number of iterations per load step is increasing towards the end of the deformation process to 18 per load step. We have to note again that in Fig. 13.5 the beam-to-beam case is shown for the mid-line; but both edge-to-edge and solid-beam cases are shown for contact lines (the point of intersection of both lines is at $x = 0$).

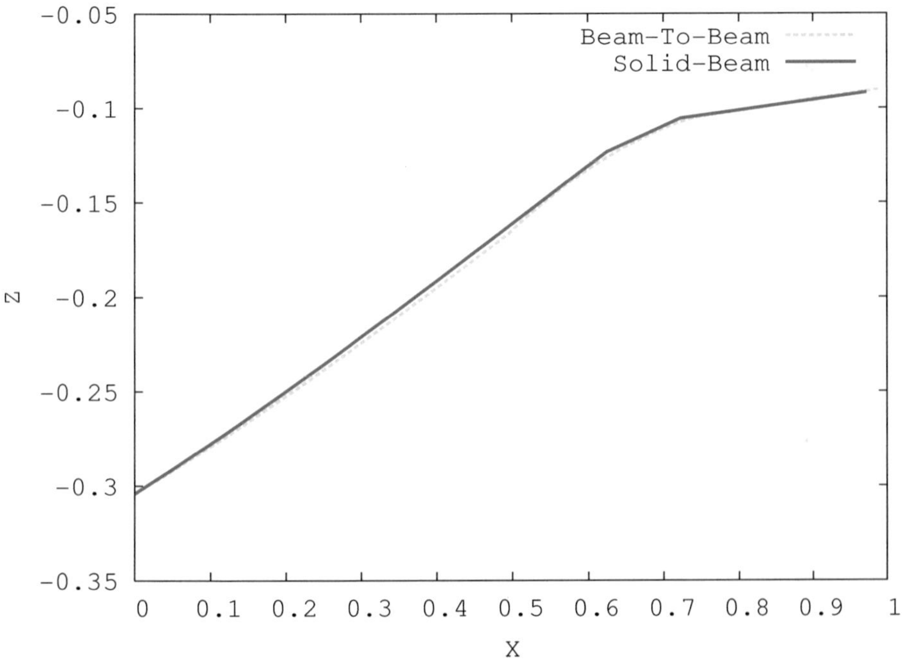

Figure 13.6: "Soft" cable. Comparison of the deformed lines for a) 4 C^1-continuous curvilinear beam finite elements and beam-to-beam contact; b) 20 linear solid-beam finite elements together with beam-to-beam contact.

13.2.1.1 Bending of a flexible beam (smooth "soft" cable) by a rigid beam

Convergence for a "soft" cable depending on the type of approximation has been studied in [96], where shear softening has been defined by setting the shear stiffnesses kGA with a scale coefficient $k = 10^{-4}$. This leads to "the softening" of the shear forces. For the solid-beam approach this effect can be defined as orthotropy inherited with the curvilinear coordinate system ξ, r, φ. In this case both orthotropic shear moduli $G_{\xi r}$, $G_{\xi\varphi}$ are scaled with the same factor $k = 10^{-4}$. The result is presented in Fig. 13.6 for the displacement vector $\mathbf{u} = \{-0.3000, 0.0000, 0.1910\}^T$ (only 30% compared to the previous case). The figures shows a good correlation, however, linear solid-beam elements are requiring (again due to stiffer behavior) a larger number of iterations, see Table 13.3.

FE model	No. of elem.	No. of load steps	Global No. of iterations
1	4	30	93
2	20	30	260

Table 13.3: Comparison of solutions for "soft" cable case with curvilinear beam elements together with beam-to-beam algorithm (FE model 1) and linear solid-beam elements together with beam-to-beam algorithm (FE model 2).

13.2.2 Contact between rings

Two equal rings represented by two intersecting circles in orthogonal planes are contacting each other, see Fig. 13.7. The distance between both is selected so that each circle is passing through the center of curvature of the other one. The radius of the midlines for rings is $R_0 = 1.0$, the radius of a circular cross-section is $r = 0, 2$. The material is linear elastic with $E = 200$ and $\nu = 0.3$. The cross section of the left ring is fixed – all displacements and rotations of the last left node resp. cross section are fixed. Both rings are meshed with 5 NURBS-elements each, the corresponding linear skeleton is shown in Fig. 13.7. The situation is more general than the simplest node-to-node contact – the spline elements are positioned along the ring so that in the reference configuration the shortest distance between rings is found between the middles of the spline elements. The assumption of non-frictional contact leads further to oscillations at the contact point, therefore, frictional sticking contact is assumed with high coefficients of friction $\mu_1 = \mu_2 = 1.0$ enforcing sticking behavior at the contact point.
The displacement $u = 1.273$ in the last right node of the right ring is applied with increments $\Delta u_1 = 0, 001$. The rings are contacting with

$u = 1.05 - 0.2 - 0.2 = 0.6$. The penalty parameter for contact is taken as 10^8. The final configuration before disconvergence is presented in Fig. 13.7. Computation with such small load steps is motivated to reach the highly deformed final configuration, however, even in this definitely not yet strongly deformed case severe convergence problems are observed with total number of iterations 8713. The possible reason for this is the rather small number of elements and that a linear material law taken into account. The ill-conditioning of the tangent matrix due to the large penalty in comparison with the elastic module is also influencing convergence.

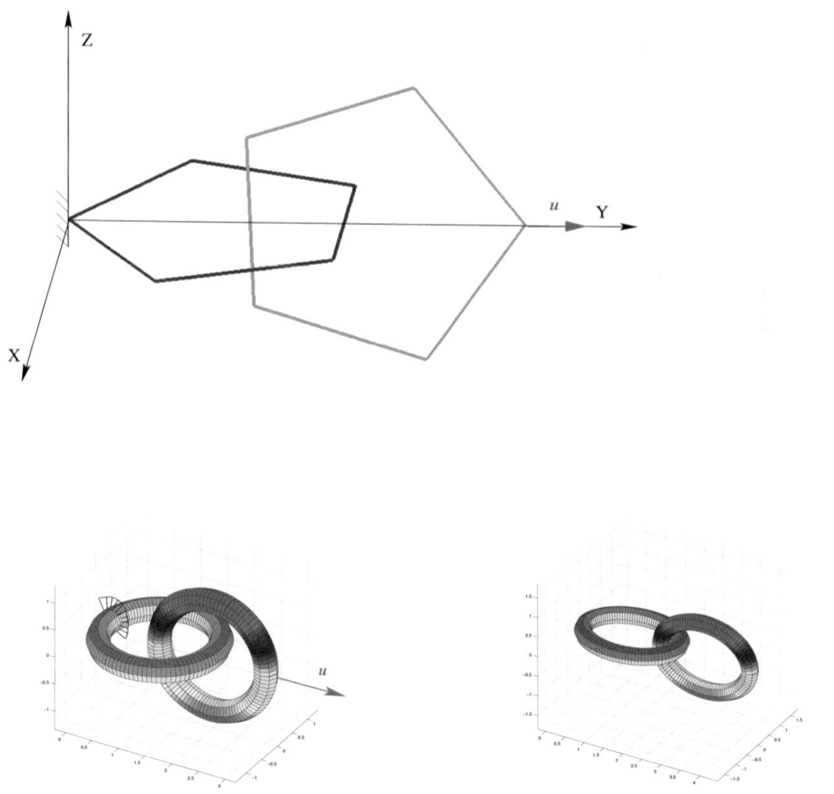

Figure 13.7: Contact between rings. a) reference linear skeleton; b) reference smooth (real) configuration after visualization; c) deformed configuration after visualization.

13.2.3 Tying of a knot

A knot is a method for fastening or securing linear material such as rope by tying or interweaving. **Wikipedia**

A knot is a perfect example requiring both a robust smooth cable element, and a robust curve-to-curve contact algorithm. Since many centuries knots have been in professional usage for climbers, fishermen, sailors and others. Among the huge number of knots, see *the Ashley book of knots* [7], there are **binding knots** and **bends**. The binding knots are used to constrict and hold objects together, and the bends serve to unite or join two ropes. In order to fulfill the security conditions such type of knots must not show relative sliding of ropes and must not spill or capsize (destroying the shape of a knot) under tension.

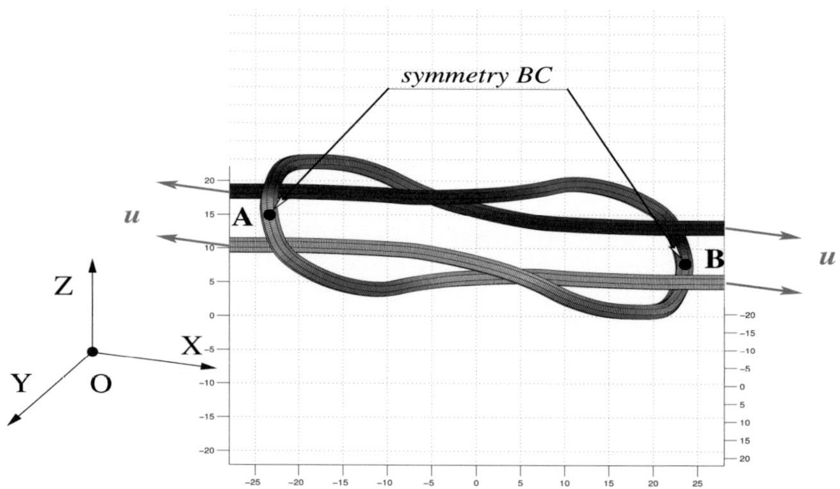

Figure 13.8: Geometry of a knot. Boundary conditions of symmetry and loading by prescribed displacement u.

From our knowledge there is no investigation yet of such a problem in the computational mechanics community, therefore the current example is a trial to begin to study this problem. We start with the most common **Reef Knot**, or **Square Knot**, see Fig. 13.8. This knot however is reported to be dangerous – *There have probably been more lives lost* *as a result of using a **Square Knot** as a bend to tie two ropes together than from the failure of any other half dozen knots combined* [7, p. 258]. The knot is modeled as follows, see the visualization via Matlab in Fig. 13.8:

- A spline forming a loop is passing through 33 characteristic nodes. Thus, each cable is modeled with 32 $C1$-smooth spline beam elements.

- Two loops are positioned initially without contact to form an opened **Reef Knot**, see Fig. 13.8.

- Material is linear elastic with $E = 200$ and $\nu = 0.3$.

- Cross section of both cables is circular with the radius $r = 1$.

- Dirichlet boundary conditions are applied at points $\mathbf{A}$ and $\mathbf{B}$ in order to supply the symmetry boundary conditions in plane $\mathbf{XOZ}$.

- The displacement vector along the $\mathbf{OX}$-axis is applied incrementally at both ends of both cables.

Different loading situations are studied:

1. beams are modeled as "soft" cables with a coefficient $k = 10^{-4}$ in eqn. (13.2), see Sect. 13.2.1.1. and displacements are free along the OY-axis, see Fig. 13.8;

2. beams are modeled as a "soft" cables, displacements are prescribed along the OY-axis in order to tie the knot, see Fig. 13.8.

Remark.
During modeling it has been found that only $C1$-smooth spline finite elements are capable to represent the result. Higher order finite elements (quadratic, cubic), but without $C1$-continuity are showing disconvergence during the first crossing of element boundaries.

If both ends are free to move in the XOZ plane as in case 1 then the stretching of them leads to opening the knot, namely both ends of the knots are moving outwards, see Fig. 13.9. A proper tying of the knot is thus enforced by additional application of the tying displacement along the OY-axis in case 2, see Fig. 13.10.

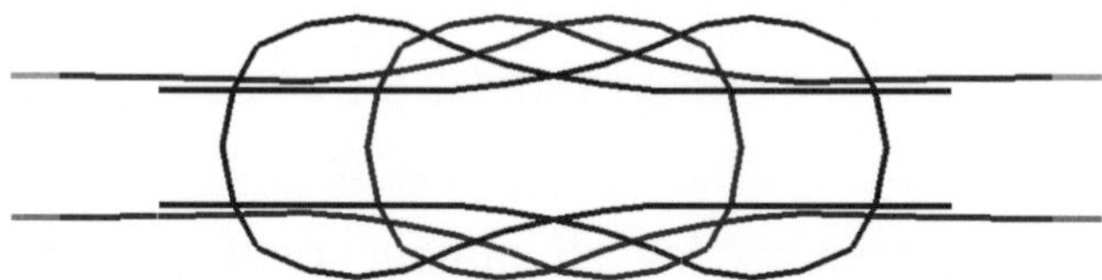

Figure 13.9: Both ends are free in the XOZ plane. Opening of the knot is observed though both ends are pulled.

The result is rather showing *"an unusual behavior"*. A more careful analysis recovers that the effect of untying the knot is caused by the CPP procedure. In the final position shown in the picture the tangents of the cables at contacting points are becoming almost parallel. In this position the CPP procedure becomes unstable because the corresponding Jacobian is close to zero. This leads to the necessity to develop an improved contact algorithm for *the problem of parallel tangent vectors*.

13.3   Conclusion

The contribution proposes a geometric view on contact appearing between bodies possessing different geometry. The presented geometrically exact theory is based on description of all kinematical and mechanical relations up to the solution methods in a proper coordinate system in a covariant form before any discretization. Selection of such a coordinate system is based on the Closest Point Projection procedure for which the solvability criterion allows a classification of all geometrical situations as well as a way to select such a coordinate system. Special attention is given to the case of $1D$-manifolds namely to the curve-to-

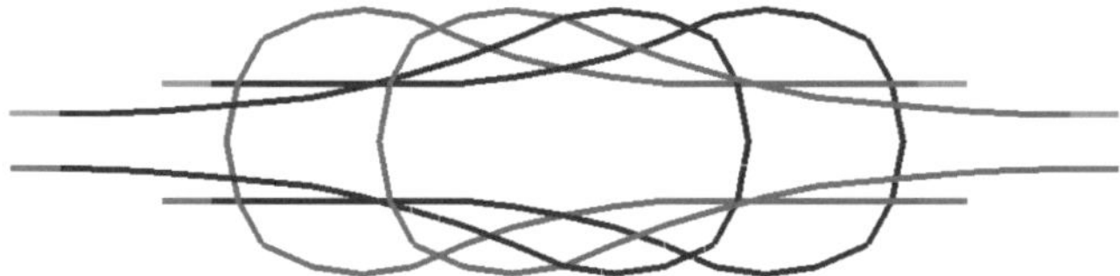

Figure 13.10: Additional displacements along the Y-axis are applied to tie the knot.

curve contact case. This case is considered together with several finite element models such as 3D solid-shell finite elements and finite beam elements. *A special solid-beam finite element* for elliptic cross-sections is developed to consider 3D continuum behavior in beam-to-beam contact. As a specific example a soft cable behavior is described within the orthotropy inherited with a curvilinear coordinate system. Various finite element techniques including isogeometric techniques have been employed. The numerical examples are selected to illustrate the possibility of the geometrically exact curve-to-curve contact formulation to work together with various combinations of finite element formulations. Interesting examples are knots for which the C^1 continuous isogeometric elements proved to be favorable.

14

Conclusions and outlook

The thesis contains a geometrical view on contemporary methods in computational contact mechanics. A unified systematic approach to deal with a certain geometrical situation and enrich it with mechanical properties such as anisotropy for adhesion and friction for surfaces as well as for curves is developed. The main goal of the development is the description of all kinematical and mechanical relations up to the solution methods in a proper coordinate system in a covariant form before any the discretization. Selection of such a coordinate system is based on the Closest Point Projection procedure for which the solvability criterion allows a classification of all geometrical situations as well as a way of the selection of such a coordinate system. Many numerical features for computational contact mechanics such as the solution of a patch test, smoothing of contact surfaces, contact algorithms for rigid surfaces, an algorithm for the transfer of history variables through element boundaries, an algorithm for updating of history variables for isotropic and anisotropic cases are developed in the thesis in combination with a covariant approach.

Several chapters are devoted to the coupled anisotropic adhesion friction model as a generalization of the classical Coulomb friction law. It is shown that the geometrical microstructure of surfaces can be defined by the adhesion tensor. The computational result has been validated in experiments. This opens the possibility to formulate many contact interface laws and construct computational algorithms in a covariant form.

Another important finding of the thesis is a novel approach for curve-to-curve contact interaction which is straightforwardly applied to both edge-to-edge and to beam-to-beam contact. This further opens the possibility together with the iso-geometrical technique to study systematically the behavior of structures including ropes and cables with multiple contact such as nets, fabrics etc.

Not all existing numerical methods to enforce contact conditions have been considered, however, the thesis is aimed to show the elegancy to formulate these methods in the most suitable geometrical way in a covariant form considering the discretization.

References

[1] *Abaqus. theory manual*, 2007. 38

[2] M. Al-Dojayli and S.A. Meguid, *Accurate modeling of contact using cubic splines*, Finite Elements in Analysis and Design **38** (2002), 337–352. 90, 104, 176

[3] P. Alart and A. Curnier, *A mixed formulation for frictional contact problems prone to Newton like solution methods*, Computer Methods in Applied Mechanics and Engineering **92** (1991), 353–375. 37

[4] P. Alart and A. Heege, *Consistent tangent matrices of curved contact operators involving anisotropic friction*, Revue Europeenne des Elements Finis **4** (1995), 183–207. 8, 16, 328

[5] Y. Araki and K. D. Hjelmstad, *Rate-dependent projection operators for frictional contact constraints*, International Journal for Numerical Methods in Engineering **57** (2003), 923–954. 7, 9, 110, 298

[6] F. Armero and E. Petoecz, *Formulation and analysis of conserving algorithms for frictionless dynamic contact-impact problems*, Computer Methods in Applied Mechanics and Engineering **158** (1998), 269–300. 14

[7] C. W. Ashley, *The ashley book of knots*, 12 ed., Farber and Farber Limited, London, Boston, 1993, Reprinted with amendments by Geoffrey Budworth. 488, 489

[8] A. A. Bandeira, P. M. Pimenta, and P. Wriggers, *A study of contact surface microstructure using a homogenization procedure considering elastoplastic behaviour of the asperities*, 4th Contact Mechanics International Symposium, CMIS 2005, 2005. 7, 382

[9] A. A. Bandeira, P. Wriggers, and P. M. Pimenta, *Numerical derivation of contact mechanics interface laws using a finite element approach for large 3d deformation*, International Journal for Numerical Methods in Engineering **59** (2004), 173–195. 7, 288, 382, 405

[10] T. Belytschko and O. Neal, *Contact-impact by the pinball algorithm with penalty and Lagrangian methods*, International Journal for Numerical Methods in Engineering **31** (1991), 547–572. 4, 15

[11] D. J. Benson and J. Q. Hallquist, *A single surface contact algorithm for the post-buckling analysis of shell structures*, Computer Methods in Applied Mechanics and Engineering **78** (1990), 141–163. 15, 267

[12] D.P. Bertsekas, *Constrained optimization and Lagrange multiplier methods.*, New York: Acad. Pr, 1982. 364

[13] ______ , *Convex analysis and optimization*, Athena Scientific, 2003. 11, 39

[14] P. Betsch, *The discrete null space method for the energy consistent integration of constrained mechanical systems. part I: Holonomic constraints*, Computer Methods in Applied Mechanics and Engineering **194** (2005), 5159–5190. 14

[15] P. Betsch and C. Hesch, *Energy-momentum conserving schemes for frictionless dynamic contact problems. part I: NTS method*, IUTAM Symposium on Computational Contact Mechanics (P. Wriggers and U. Nackenhorst, eds.), IUTAM Bookseries, IUTAM Symposium on Computational Contact Mechanics, Springer, 2007, pp. 77–96. 14

[16] J. Bonet and R.D. Wood, *Nonlinear continuum mechanics for finite element analysis*, Cambridge Univ. Press., 2000. 69, 101

[17] F. M. Borodich and B. A. Galanov, *Non-direct estimations of adhesive and elastic properties of materials by depth-sensing indentation*, Proc. of the Royal Society of London. Series A, Mathematical and Physical Sciences **464** (2008), 2759–2776. 9

[18] F. M. Borodich and L. M. Keer, *Evaluation of elastic modulus of materials by adhesive (no–slip) nano–indentatio*, Proc. of the Royal Society of London. Series A, Mathematical and Physical Sciences **460** (2004), 507–514. 9

[19] J. M. Borwein and A. S. Lewis, *Convex analysis and nonlinear optimization. theory and examples*, Springer: New York, Heidelberg, 2000. 11, 314, 316, 319

[20] F.P. Bowden and G.W. Rowe, *The adhesion of clean metals*, Proc. of the Royal Society of London. Series A, Mathematical and Physical Sciences **233** (1956), 429–442. 8

[21] H. C. J Bruneel and I. D. Rycke, *Quicktrace: a fast algorithm to detect contact*, International Journal for Numerical Methods in Engineering **54** (2002), 299–316. 15

[22] R. Buczkowski and M. Kleiber, *Elasto-plastic interface model for 3d-frictional orthotropic contact problems*, International Journal for Numerical Methods in Engineering **40** (1997), 599–619. 8, 289, 340

[23] ______ , *A stochastic model of rough surfaces for finite element contact analysis*, Computer Methods in Applied Mechanics and Engineering **199** (1999), 43–59. 7, 8, 288

[24] ______ , *Statistical model of strongly anisotropic rough surfaces for finite element contact analysis*, International Journal for Numerical Methods in Engineering **49** (2000), 1169–1189. 7, 8, 288, 289, 340, 382

[25] J. Campbell, L. D. Libersky, and R. Vignjevic, *A contact algorithm for smoothed particle hydrodynamics*, Computer Methods in Applied Mechanics and Engineering **184** (2000), 49–65. 4

[26] G. Carbone and L. Mangialardi, *Adhesion and friction of an elastic half-space in contact with a slightly wavy rigid surface*, Journal of the Mechanics and Physics of Solids **52** (2004), 1267–1287. 382

[27] S. Cescotto and R. Charlier, *Frictional contact finite elements based on mixed variational principles.*, International Journal for Numerical Methods in Engineering **36** (1993), 1681–1701. 11

[28] V. Chawla and T. A. Laursen, *Energy consistent algorithms for frictional contact problems*, International Journal for Numerical Methods in Engineering **42** (1998), 799–827. 14

[29] M. Cocou, M. Raous, and M. Schryve, *Analysis of a dynamic contact problem with adhesion and friction in viscoelasticity*, ECCM-III, Conference proceedings, 5-8 June 2006. 9

[30] G.A. Costello, *Theory of wire rope*, Springer-Verlag, New York, Berlin, Heidelberg, 1990. 434

[31] M. A. Crisfield, *Re-visiting the contact patch test*, International Journal for Numerical Methods in Engineering **48** (2000), 435–449. 11, 150, 190, 191

[32] A. Curnier, *A theory of friction*, International Journal of Solids and Structures **20** (1984), 637–647. 8, 9, 90, 288, 296, 299, 301, 382, 389

[33] P.J. Davis and P. Rabinowitz, *Methods of numerical integration*, third edition ed., Academic Press, 1984. 153, 172

[34] B.A. Dubrovin, T.A Fomenko, and S.P Novikov, *Modern geometry – methods and applications. Part 1*, Springer-Verlag, 1984. 411, 414, 428

[35] C. Eck and J. Jarusek, *Existence result for the static contact problem with Coulomb friction*, Mathematical Models and Methods in Applied Science **8** (1998), 445–468. 435

[36] N. El-Abbasi and K.J. Bathe, *Stability and patch test performance of contact discretization and new solution algorithm*, Computer and Structures **79** (2001), 1473–1486. 190

[37] N. El-Abbasi, S.A. Meguid, and A. Czekanski, *On the modeling of smooth contact surfaces using cubic splines*, International Journal for Numerical Methods in Engineering **50** (2001), 953–967. 90, 104, 176

[38] G.E. Farin, *NURBS: from projective geometry to practical use*, Peters, 1999. 158, 168, 177, 178, 179, 181, 183, 479

[39] Y. T. Feng and D. R. J. Owen, *An augmented spatial digital tree algorithm for contact detection in computational mechanics*, International Journal for Numerical Methods in Engineering **55** (2002), 159–176. 15

[40] K. A. Fischer and P. Wriggers, *Frictionless 2D contact formulations for finite deformations based on the Mortar method*, Computational Mechanics **36** (2005), 226–244. 364, 368

[41] ______, *Mortar based frictional contact formulation for higher order interpolations using the moving friction cone*, Computer Methods in Applied Mechanics and Engineering **195** (2006), 5020–5036. 4, 13, 151, 263, 273

[42] M. Fortin and R. Glowinski, *Augmented Lagrangian methods: applications to the numerical solutions of boundary-value problems*, Amsterdam: North-Holland, 1983. 364

[43] A. Francavilla and O. C. Zienkiewicz, *A note on numerical computation of elastic contact problems*, International Journal for Numerical Methods in Engineering **9** (1975), 913–924. 2

[44] K.N.G. Fuller and D. Tabor, *The effect of surface roughness on the adhesion of elastic solids*, Proc. of the Royal Society of London. Series A, Mathematical and Physical Sciences **345** (1975), 327–342. 9

[45] A. E. Giannakopoulos, *The return mapping method for the integration of friction constitutive relations*, Computers and Structures **32** (1989), 157–167. 36, 314

[46] R. Glowinski, J.-L. Lions, and R. Tremolieres, *Numerical analysis of variational inequalities*, rev. edition ed., Amsterdam: North-Holland, 1981. 314

[47] A. Gray, *Modern differential geometry of curves and surfaces*, Boca Raton: CRC Press, 1993. 71, 91, 94, 95, 106, 214

[48] J. A. Greenwood, *A unified theory of surface roughness*, Proc. of the Royal Society of London. Series A, Mathematical and Physical Sciences **393** (1984), 133–157. 7, 288, 382

[49] ______, *Adhesion of elastic spheres*, Proc. of the Royal Society of London. Series A, Mathematical and Physical Sciences **453** (1997), 1277–1297. 9

[50] J. A. Greenwood and J. B. P. Williamsom, *Contact of nominaly flat surfaces*, Proc. of the Royal Society of London. Series A, Mathematical and Physical Sciences **295** (1966), 300–319. 7, 288, 382

[51] M. E. Gurtin, J. Weissmueller, and F. Larche, *A general theory of curved deformable interfaces in solids at equilibrium*, Philosophical Magazine A **78** (1998), 1093–1109. 18

[52] J. Q. Hallquist, *Ls-dyna. theoretical manual*, http://www.lstc.com/, 2007. 38, 52, 64, 409

[53] J. Q. Hallquist, G. L. Goudreau, and D. J. Benson, *Sliding interface with contact-impact in large-scale lagrangian computations*, Computer Methods in Applied Mechanics and Engineering **51** (1985), 107–137. 3, 4, 12, 36, 38

[54] W. Han and M. Sofonea, *Quasistatic contact problems in viscoelasticity and viscoplasticity*, American Mathematical Society, International Press, 2002. 14

[55] W. Hardy and M Nottage, *Studies in adhesion. I*, Proc. of the Royal Society of London. Series A, Mathematical and Physical Sciences **112** (1926), 62–75. 8

[56] ______, *Studies in adhesion. II*, Proc. of the Royal Society of London. Series A, Mathematical and Physical Sciences **118** (1926), 209–229. 8

[57] M. Harnau, A. Konyukhov, and K. Schweizerhof, *Algorithmic aspects in large deformation contact analysis using 'solid-shell' elements*, Computers and Structures **83** (2005), 1804–1823. 4, 11, 140, 150, 196, 239, 249, 271, 273, 275, 279, 337, 364

[58] S. Hartmann, J. Oliver, R. Weyler, J.C. Cante, and J.A. Hernández, *A contact domain method for large deformation frictional contact problems. part 2: Numerical aspects*, Computer Methods in Applied Mechanics and Engineering **198** (2009), no. 33-36, 2607–2631. 12

[59] R. Hauptmann, S. Doll, M. Harnau, and K. Schweizerhof, *"Solid-shell" elements with linear and quadratic shape functions at large deformations with nearly incompressible materials.*, Computers and Structures **79** (2001), 1671–1685. 69, 85, 125, 248, 458

[60] R. Hauptmann and K. Schweizerhof, *A systematic development of "solid-shell" element formulations for linear and non-linear analysis employing only displacement degrees of freedom*, International Journal for Numerical Methods in Engineering **42** (1998), 49–69. 69, 125, 248

[61] R. Hauptmann, K. Schweizerhof, and S. Doll, *Extension of the solid-shell concept for large elastic and large elastoplastic deformations*, International Journal for Numerical Methods in Engineering **49** (2000), 1121–1141. 69, 83

[62] Q.C. He and A. Curnier, *Anisotropic dry friction between two orthotropic surfaces undergoing large displacements*, European Journal of Mechanics A-Solids **12** (1993), 631–666. 8, 289, 290, 296, 300, 313, 382, 405

[63] J. H. Heegaard and A. Curnier, *An augmented lagrangian method for discrete large-slip contact problems*, International Journal for Numerical Methods in Engineering **36** (1993), 569–593. 5, 37, 38, 52

[64] ———, *Geometric properties of 2d and 3d unilateral large slip contact operators*, Computer Methods in Applied Mechanics and Engineering **131** (1996), 263–286. 18, 38

[65] A. Heege and P. Alart, *A frictional contact element for strongly curved contact problems*, International Journal for Numerical Methods in Engineering **39** (1996), 165–184. 5, 16, 37, 64, 175, 259

[66] M. W. Heinstein and T. A. Laursen, *A three dimensional surface-to-surface projection algorithm for non-coincident domains*, Communications in Numerical Methods in Engineering **19** (2003), 421–432. 3, 191

[67] P. Heintz and P. Hansbo, *Stabilized lagrange multiplier methods for bilateral elastic contact with friction*, Computer Methods in Applied Mechanics and Engineering **195** (2006), 4323–4333. 12

[68] M. Hjiaj, Z.Q. Feng, G. de Saxce, and Z. Mroz, *Three-dimensional finite element computations for frictional contact problems with non-associated sliding rule*, International Journal for Numerical Methods in Engineering **60** (2004), 2045–2076. 8, 289, 340, 352

[69] S. Hueeber, G. Stadler, and B. I. Wohlmuth, *A primal-dual active set algorithm for three-dimensional contact problems with coulomb friction*, Preprint, Inst Angewandte Anal. und Num. Simulation, Stuttgart (2006). 13, 264, 364

[70] S. Hueeber and B. I. Wohlmuth, *Large deformation frictional contact mechanics: continuum formulation and augmented lagrangian treatment*, Computer Methods in Applied Mechanics and Engineering **194** (2005), 3147–3166. 12, 37

[71] T. J. R. Hughes, R. L. Taylor, J. L. Sackman, A. Curnier, and W. Kanoknukulchai, *A finite element method for a class of contact-impact problems*, Computer Methods in Applied Mechanics and Engineering **8** (1976), 249–276. 11

REFERENCES

[72] T.J.R. Hughes, J.A. Cottrell, and Y. Bazilevs, *Isogeometric analysis: Cad, finite elements, nurbs, exact geometry and mesh refinement*, Computer Methods in Applied Mechanics and Engineering **194** (2005), 4135–4195. 173, 258, 458, 459, 476

[73] A. Ibrahimbegovic, *On finite element implementation of geometrically nonlinear reissner's beam theory: three-dimensional curved beam elements*, Computer Methods in Applied Mechanics and Engineering **122** (1995), 11–26. 458, 477

[74] ______, *On the choice of finite rotation parameters*, Computer Methods in Applied Mechanics and Engineering **149** (1997), 49–71. 477

[75] A. Ibrahimbegovic and R. L. Taylor, *n the role of frame-invariance in structural mechanics models at finite rotations*, Computer Methods in Applied Mechanics and Engineering **191** (2002), 5159–5176. 477

[76] K. L. Johnson, *Adhesion and friction between a smooth elastic spherical asperity and a plane surface*, Proc. of the Royal Society of London. Series A, Mathematical and Physical Sciences **453** (1997), 163–179. 9

[77] ______, *Contact mechanics*, Cambridge: Cambridge Univ. Press, 2001. 276, 278, 345

[78] K. L. Johnson, K. Kendall, and A. D. Roberts, *Surface energy and the contact of elastic solids*, Proc. of the Royal Society of London. Series A, Mathematical and Physical Sciences **324** (1971), 301–313. 9

[79] R. E. Jones and P. Papadopoulos, *A yield-limited lagrange multiplier formulation for frictional contact*, International Journal for Numerical Methods in Engineering **48** (2000), 1127–1149. 11, 37, 264

[80] ______, *A novel three-dimensional contact finite element based on smooth pressure interpolations*, International Journal for Numerical Methods in Engineering **51** (2001), 791–811. 191

[81] ______, *A geometric interpretation of frictional contact mechanics*, Zeitschrift für Angewandte Mathematik und Physik (ZAMP) **57** (2006), no. 6, 1025–1041. 18

[82] ______, *Simulating anisotropic frictional response using smoothly interpolated traction fields*, Computer Methods in Applied Mechanics and Engineering **195** (2006), 588–613. 8, 289, 340

[83] C. Kane, E. A Reppetto, M. Ortiz, and J. E. Marsden, *Finite element analysis of nonsmooth contact. formulation and application to two-dimensional problems*, Computer Methods in Applied Mechanics and Engineering **180** (1999), 263–297. 4

[84] N. Kikuchi and J. T. Oden, *Contact problems in elasticity: a study of variational inequalities and finite element methods*, SIAM, 1988. 3, 12, 36, 68, 97, 98, 99, 109, 111, 212

[85] H. Kim, W.K. Kim, M.L. Falk, and D.A. Rigney, *MD simulations of microstructure evolution during high-velocity sliding between crystalline materials*, Tribology Letters **28** (2007), no. 3, 299–306. 10

500

[86] A. Konyukhov and K. Schweizerhof, *Contact formulation via a velocity description allowing efficiency improvements in frictionless contact analysis*, Computational Mechanics **33** (2004), 165–173. 5, 12, 16, 38, 41, 60, 67, 103, 116, 125, 212, 258, 260, 266, 276, 289, 291, 294, 329, 330, 331, 367, 409, 426

[87] ______, *Large deformation frictional contact formulation for low order "solid shell" elements*, ECCOMAS-2004, Jyv"askyl"a, Conference Proceedings eds. P. Neittanm"aki, T. Rossi, K. Majava, O. Pironneau, 2004, July 2004. 158, 201

[88] ______, *Application of a covariant description to the contact of shells with different approximation*, 5th IASS-IACM,Conference Proceedings, June 1-4 2005. 200

[89] ______, *Covariant description for frictional contact problems*, Computational Mechanics **35** (2005), 190–213. 5, 13, 16, 38, 41, 58, 60, 61, 89, 212, 213, 226, 240, 243, 258, 260, 266, 289, 291, 294, 329, 330, 367, 409, 426

[90] ______, *Covariant description of contact interfaces considering anisotropy for adhesion and friction: Part 1. formulation and analysis of the computational model*, Computer Methods in Applied Mechanics and Engineering **196** (2006), 103–117. 9, 11, 16, 41, 258, 265, 287, 329, 365, 382, 388, 391, 392, 409

[91] ______, *Covariant description of contact interfaces considering anisotropy for adhesion and friction: Part 2. linearization, finite element implementation and numerical analysis of the model*, Computer Methods in Applied Mechanics and Engineering **196** (2006), 289–303. 9, 11, 16, 41, 258, 265, 327, 365, 377, 378, 382, 388, 391, 392, 395, 409

[92] ______, *A special focus on 2d formulations for contact problems using a covariant description*, International Journal for Numerical Methods in Engineering **66** (2006), 1432–1465. 5, 16, 49, 61, 211, 258, 271, 276, 284, 370, 409, 426

[93] ______, *On a continuous transfer of history variables for frictional contact problems based on interpretations of covariant derivatives as a parallel translation*, IUTAM Symposium -"Multiscale problems in Multibody System Contacts", IUTAM Bookseries, 2007. 203

[94] ______, *Symmetrization of various friction models based on an augmented lagrangian approach*, IUTAM Bookseries, pp. 97–111, Springer, 2007. 13, 17, 264, 363

[95] ______, *On the solvability of closest point projection procedures in contact analysis: Analysis and solution strategy for surfaces of arbitrary geometry*, Computer Methods in Applied Mechanics and Engineering **197** (2008), no. 33-40, 3045–3056. 35, 260, 416, 418

[96] ______, *Geometrically exact covariant approach for contact between curves representing beam and cable type structures*, submitted (2009), 00–56. 17, 407, 475, 476, 482, 484, 485

[97] ______, *Incorporation of contact for high-order finite elements in covariant form*, Computer Methods in Applied Mechanics and Engineering **198** (2009), no. 13-14, 1213–1223, HOFEM07 - International Workshop on High-Order Finite Element Methods, 2007. 257, 409

REFERENCES

[98] A. Konyukhov, K. Schweizerhof, and P. Vielsack, *On models of contact surfaces including anisotropy for friction and adhesion and their experimental validations*, ECCM-III, European Conference on Computational Mechanics, Conference proceedings, III European Conference on Computational Mechanics. Solid, Structures and Coupled Problems in Engineering. ECCM-IIIConference Proceedings, June 5-9 2006. 10

[99] A. Konyukhov, P. Vielsack, and K. Schweizerhof, *On coupled models of anisotropic contact surfaces and their experimental validation*, Wear **264** (2008), no. 7-8, 579–588. 10, 381

[100] L. Krstulovic-Opara and P. Wriggers, *Convergence studies for 2d smooth contact elements*, ECCM. European Conference on Computational Mechanics, 2001. 16, 140

[101] ______, *A two-dimensional c1-continuous contact element based on the moving friction cone description*, WCCM V. Fifth World Congress on Computational Mechanics (2002). 16, 90, 114, 176, 295

[102] L. Krstulovic-Opara, P. Wriggers, and J. Korelc, *A c1-continuous formulation for 3d finite deformation frictional contact*, Computational Mechanics **29** (2002), 27–42. 16, 91, 176, 240, 259, 325

[103] S. Lang, *Fundamentals of differential geometry*, Springer, New York, 1999. 42, 56, 411, 414

[104] T. A. Laursen, *Formulation and treatment of frictional contact problems using finite elements*, Ph.D. thesis, Stanford University, 1992. 6, 13, 16, 68, 113, 122

[105] ______, *Convected description in large deformation frictional contact problems*, International Journal of Solids and Structures **31** (1994), 669–681. 5, 13, 18

[106] ______, *Computational contact and impact mechanics. fundamentals of modeling interfacial phenomena in nonlinear finite element analysis*, Berlin; Heidelberg; New York: Springer, 2002. 6, 11, 38, 61, 64, 68, 72, 76, 77, 78, 97, 98, 99, 109, 113, 122, 212, 225, 234, 235, 265, 291, 293, 364, 416

[107] T. A. Laursen and V. Chawla, *Design of energy conserving algorithms for frictionless dynamic contact problems*, International Journal for Numerical Methods in Engineering **40** (1997), 863–886. 14

[108] T. A. Laursen and M. W. Heinstein, *Consistent mesh tying methods for topologically distinct discretized surfaces in non-linear solid mechanics*, International Journal for Numerical Methods in Engineering **57** (2003), 1197–1242. 3

[109] T. A. Laursen and J. C. Simo, *A continuum-based finite element formulation for the implicit solution of multibody large deformation frictional contact problems.*, International Journal for Numerical Methods in Engineering **35** (1993), 3451–3485. 5, 6, 12, 16, 18, 37, 90, 113, 122, 212, 225, 226, 258, 264, 291, 364, 370

[110] L. D. Libersky and A. G. Petschek, *Smooth particle hydrodynamics with strength of materials*, Advances in the Free-Lagrange Method, pp. 248–257, Springer, New York, 1991. 4

[111] P. Litewka, *Hermite polynomial smoothing in beam-to-beam frictional contact*, Computational Mechanics **40** (2007), 815–826. 16, 408, 409

[112] ______, *Smooth frictional contact between beams in 3d*, IUTAM Symposium on Computational Contact Mechanics (U. Nackenhorst and P. Wriggers, eds.), IUTAM Bookseries, U.Nackenhorst, P. Wriggers eds., IUTAM Bookseries, Springer, November 5-8 2007, pp. 157–176. 16, 408, 409

[113] P. Litewka and P. Wriggers, *Contact between 3d beams with rectangular cross-sections*, International Journal for Numerical Methods in Engineering **53** (2002), 2019–2042. 5, 13, 16, 408

[114] ______, *Frictional contact between 3d beams.*, Computational Mechanics **28** (2002), 26–39. 5, 13, 16, 408

[115] ______, *Frictional beam-to-beam contact by the lagrange multipliers method*, VII International Conference on Computational Plasticity, COMPLAS 2003, April 7-10 2003, VII International Conference on Computational Plasticity. COMPLAS 2003. 13, 408

[116] M. S. Longuet-Higgins, *The statistical analysis of a random, moving surface.*, Philosophical Transaction of the Royal Society of London, Series A, Mathematical and Physical Sciences **249** (1957), 321–387. 7, 288, 382

[117] ______, *Statistical properties of an isotropic random surface*, Philosophical Transaction of the Royal Society of London, Mathematical and Physical Sciences **250** (1957), 157–174. 7

[118] S. Luding, *Contact models for very loose granular materials.*, IUTAM Bookseries, IUTAM Symposium on Multiscale Problems in Multibody System Contact, ed. by P. Eberhard, 2007. 8

[119] D.G. Luenberger, *Linear and nonlinear programming*, 2nd ed., Addison-Wesley,, 1984. 314, 316, 319

[120] B. N. Maker and T. A. Laursen, *A finite element formulation for rod-continuum interactions: the one-dimensional slideline*, International Journal for Numerical Methods in Engineering **37** (1994), 1–18. 13

[121] J. E. Marsden and T. J. R. Hughes, *Mathematical foundations of elasticity*, New York: Dover, 1994. 69, 73, 91, 96, 101, 104, 106, 111, 214, 255

[122] McCool and I. John, *Comparison of models for the contact of rough surfaces.*, Wear **107** (1986), 37–60. 7, 288, 382

[123] T. W. McDevitt and T. A. Laursen, *A mortar-finite element formulation for frictional contact problems*, International Journal for Numerical Methods in Engineering **48** (2000), 1525–1547. 5, 239

[124] J.S. McFarlane and D. Tabor, *Adhesion of solids and the effect of surface films*, Proc. of the Royal Society of London. Series A, Mathematical and Physical Sciences **202** (1950), 224–243. 8

[125] ______, *Relation between friction and adhesion*, Proc. of the Royal Society of London. Series A, Mathematical and Physical Sciences **202** (1950), 244–253. 8

[126] A.P. Merkle and L.D. Marks, *A predictive analytical friction model from basic theories of interfaces, contacts and dislocations*, Tribology Letters **26(1)** (2007), 73–84. 10

REFERENCES

[127] R. Michalowski and Z. Mroz, *Associated and non-associated sliding rules in contact friction problems*, Archives of Mechanics. (Archiwum mechaniki stosowanej), Polish Academy of Sciences **30** (1978), 259–276. 8, 300, 382, 389

[128] P. Montmitonnet and A. Hasquin, *Implementation of an anisotropic friction law in a 3d finite element model of hot rolling*, Simulation of Materials Processing: Theory, Methods and Applications (S.-F. Shen and Dowson P., eds.), Proc. of NUMIFORM'95, 1995, pp. 301–306. 8, 16, 328

[129] Z. Mroz and S. Stupkiewicz, *An anisotropic friction and wear model*, International Journal of Solids and Structures **31** (1994), 1113–1131. 8, 288, 289, 389

[130] A. Munjiza, *The combined finite-discrete element method*, Wiley, 2004. 15

[131] V.G. Oancea and T. A. Laursen, *On the constitutive modeling and finite element computation of rate-dependent frictional sliding in large deformations*, Computer Methods in Applied Mechanics and Engineering **143** (1997), 197–227. 298

[132] J. Oliver, S. Hartmann, J.C. Cante, R. Weyler, and J.A. Hernández, *A contact domain method for large deformation frictional contact problems. part 1: Theoretical basis*, Computer Methods in Applied Mechanics and Engineering **198** (2009), no. 33-36, 2591–2606. 12

[133] I. Paczelt, B.A. Szabo, and Szabo T., *Solution of contact problem using the hp-version of the finite element methods*, Computers and Mathematics with Applications **38** (1999), 49–69. 258, 280

[134] V. Padmanabhan and T. A. Laursen, *A framework for development of surface smoothing procedures in large deformation frictional contact analysis*, Finite Elements in Analysis and Design **37** (2001), 173–198. 90, 104, 176, 240, 259

[135] M. Paggi, A. Carpinteri, and G. Zavarise, *A unified interface constitutive law for the study of fracture and contact problems in heterogeneous materials*, Analysis and Simulation of Contact Problems (P. Wriggers and U. Nackenhorst, eds.), Springer, 2005, Lecture Notes in Applied and Computational Mechanics Vol. 27, pp. 297–304. 7

[136] A. Pandolfi, C. Kane, J. E. Marsden, and M. Ortiz, *Time-discreized variational formulation of non-smooth frictional contact*, International Journal for Numerical Methods in Engineering **53** (2002), 1801–1829. 4

[137] H. Parisch, *A consistent tangent stiffness matrix for three-dimensional nonlinear contact analysis*, International Journal for Numerical Methods in Engineering **28** (1989), 1803–1812. 12, 16, 36, 68, 212, 225

[138] H. Parisch and Ch. Luebbing, *A formulation of arbitrarily shaped surface elements for three-dimensional large deformation contact with friction*, International Journal for Numerical Methods in Engineering **40** (1997), 3359–3383. 5, 12, 16, 36, 90, 212, 291

[139] D. Peric and D. R. J. Owen, *Computational model for 3-d contact problems with friction based on the penalty method*, International Journal for Numerical Methods in Engineering **35** (1992), 1289–1309. 5, 12, 16, 36, 90, 125, 212, 291

[140] B. N. J Persson, *Sliding friction. physical principles and applications*, (2nd ed.), 2000. 11

[141] B. N. J Persson and E. Tosatti, *The effect of surface roughness on the adhesion of elastic solids*, Journal of Chemical Physics **115** (2001), 5597–5610. 9

[142] P. Persson, *Parallel numerical procedures for the solution of contact-impact problems*, Ph.D. thesis, Linkeoping University, 2000. 15

[143] L. Piegl and W. Tiller, *The nurbs book (monographs in visual communications)*, 2. ed. ed., Springer, 1997. 177, 181, 479

[144] G. Pietrzak and A. Curnier, *Large deformation frictional contact mechanics: continuum formulation and augmented lagrangian treatment*, Computer Methods in Applied Mechanics and Engineering **177** (1999), 351–381. 13, 37, 68, 176, 259, 264, 364

[145] V.L. Popov, A. Gerve, B. Kehrwald, and I.Yu. Smolin, *Simulation of wear in combustion engines*, Computational Materials Science **19** (2000), 285–291. 460

[146] V.L. Popov and S.G. Psakhie, *Numerical simulation methods in tribology*, Tribology International **40 (6)** (2007), 916–923. 10

[147] M. A. Puso, *A 3d mortar method for solid mechanics*, International Journal for Numerical Methods in Engineering **59** (2004), 315–336. 5

[148] M. A. Puso and T. A. Laursen, *A 3d contact smoothing method using Gregory patches*, International Journal for Numerical Methods in Engineering **54** (2002), 1161–1194. 91, 104, 176, 240, 259

[149] ______, *Mesh tying on curved interface in 3d.*, Engineering Computations **20** (2003), 305–319. 3

[150] ______, *A mortar segment-to-segment contact method for large deformation solid mechanics*, Computer Methods in Applied Mechanics and Engineering **193** (2004), 601–629. 4, 5, 13, 37, 239, 364

[151] ______, *A mortar segment-to-segment frictional contact method for large deformations*, Computer Methods in Applied Mechanics and Engineering **193** (2004), 4891–4913. 4, 5, 13, 37, 364

[152] E. Rank, A. Düster, V. Nübel, K. Preusch, and O.T. Bruhns, *High order finite elements for shells*, Computer Methods in Applied Mechanics and Engineering **194** (2005), 2494–2512. 258, 458, 476

[153] M. Raous, L. Cangemi, and A. Cocou, *A consistent model coupling adhesion, friction, and unilateral contact*, Computer Methods in Applied Mechanics and Engineering **177** (1999), 383–399. 9

[154] R. Schoen and S.T. Yau, *Lectures on differential geometry*, International Press, Boston, 1994. 94, 95, 96, 106

[155] K. Schweizerhof and J. Q. Hallquist, *Efficiency refinements of contact strategies and algorithms in explicit finite element programming*, Computational Plasticity (D. R. J. Owen and E. Onate, eds.), Pineridge Press, 1992, pp. 457–482. 15, 175, 259

[156] K. Schweizerhof and A. Konyukhov, *Contact with shells*, 7th World Congress on Computational Mechanics. WCCM-7, July 16-22 2006. 202

REFERENCES

[157] W. Sextro, *Dynamical contact problems with friction: models, methods, experiments, and applications*, Springer: Berlin, Heidelberg, 2002. 11

[158] M Shillor, M. Sofonea, and J. J. Telega, *Models and analysis of quasistatic contact: variational methods*, Springer, 2004. 14

[159] T. Shimizu and T. Sano, *An application of a penalty method contact and friction algorithm to a 3-dimensional tool surface expressed by a b-spline patch*, Journal of Material Processing Technology **48** (1995), 207–213. 175

[160] J. C. Simo and T. J. R. Hughes, *Elastoplasticity and viscoplasticity: computational aspects*, Springer-Verlag, Berlin, 1992. 37, 38, 112, 123, 313, 314, 325, 341, 394

[161] J. C. Simo and T. A. Laursen, *An augmented lagrangian treatment of contact problems involving friction*, Computers and Structures **42** (1992), 97–116. 13, 16, 68, 90, 111, 122

[162] J. C. Simo, P. Wriggers, K. Schweizerhof, and R. L. Taylor, *Finite deformation post-buckling analysis involving inelasticity and contact constraints*, International Journal for Numerical Methods in Engineering **23** (1986), 779–800. 3, 5

[163] J. M. Solberg and P. Papadopoulos, *An analysis of dual formulations for the finite element solution of two-body contact problems*, Computer Methods in Applied Mechanics and Engineering **194** (2005), 2734–2780. 11, 264, 275

[164] P. Solin, K. Segeth, and I. Dolezel, *Higher-order finite element methods*, Chapman Hall, 2004. 173, 266

[165] J. E. Souza de Cursi, *Stress unilateral analysis of mooring cables*, International Journal for Numerical Methods in Engineering **34** (1992), 279–302. 409

[166] G. Stadler, *Infinite-dimensional semi-smooth newton and augmented lagrangian methods for friction and contact problems in elasticity*, PhD Thesis, University of Graz, Graz., 2004. 370

[167] ______, *Path-following and augmented lagrangian methods for contact problems in linear elasticity*, Journal on Computational and Applied Mathematics **2** (2007), 533–547. 13

[168] M. Stadler and G. A. Holzapfel, *Subdivision schemes for smooth contact surfaces of arbitrary mesh topology in 3d*, International Journal for Numerical Methods in Engineering **60** (2004), 1161–1195. 176, 240

[169] M. Stadler, G. A. Holzapfel, and J. Korelc, *Cn continuous modeling of smooth contact surfaces using nurbs and application to 2d problems*, International Journal for Numerical Methods in Engineering **57** (2003), 2177–2203. 16, 104, 176, 240, 259

[170] O. D. L. Strack and P. A. Cundall, *A descrete numerical model for granular assemblies*, Geotechnique **29** (1979), 47–65. 4, 14

[171] D. Tan, *Mesh matching and contact patch test.*, Computational Mechanics **31** (2003), 135–152. 2

[172] R. L. Taylor, *FEAP. a finite element analysis program.*, FEAP - Mechanik - Karlsruhe, Institut fuer Mechanik, Universitaet Karlsruhe, 1987. 82, 125, 243, 358

[173] R. L. Taylor and P. Papadopoulos, *On a patch test for contact problems in two dimensions*, Computational Methods in Nonlinear Mechanics (P. Wriggers and W. Wagner, eds.), Springer, 1991, pp. 690–702. 11, 190

[174] J. Tomas, *Micromechanics of particle adhesion*, IUTAM Symposium on Multiscale Problems in Multibody System Contacts, ed. by P. Eberhard, IUTAM Bookseries. Springer, 2007. 8

[175] F. Wang, J. Cheng, and Y. Zhenhan, *FFS contact searching algorithm for dynamic finite element analysis*, International Journal for Numerical Methods in Engineering **52** (2001), 655–672. 15

[176] S. P. Wang and A. Makinouchi, *Contact search strategies for FEM simulation of the blow molding process*, International Journal for Numerical Methods in Engineering **48** (2000), 501–521. 15

[177] S. P. Wang and E. Nakamachi, *The inside-outside contact search algorithm for finite element analysis*, International Journal for Numerical Methods in Engineering **40** (1997), 3665–3685. 15

[178] C. Wellmann, C. Lillie, and P. Wriggers, *A contact detection algorithm for superellipsoids based on the common-normal concept*, Engineering Computations **25** (2008), no. 5, 432–442. 4

[179] D. J. Whitehouse and J. F. Archard, *The properties of random surfaces of significance in their contact*, Proc. of the Royal Society of London. Series A, Mathematical and Physical Sciences **316** (1970), 97–121. 7, 288, 382

[180] D. J. Whitehouse and M. J. Phillips, *Discrete properties of random surfaces*, Philosophical Transaction of the Royal Society of London, Mathematical and Physical Sciences **290** (1978), 267–298. 7

[181] ———, *Two-dimensional discrete properties of random surfaces*, Proc. of the Royal Society of London. Series A, Mathematical and Physical Sciences **305** (1982), 441–468. 7, 288, 382

[182] K. Willner, *Elasto-plastic contact of rough surfaces*, Proc. of International Conference, Contact Mechanics III, 1997. 7

[183] K. Willner and L. Gaul, *A penalty approach for contact description by fem based on interface physics*, Proc. of International Conference, Contact Mechanics II, 1995. 7

[184] ———, *Contact description by fem based on interface physics*, Proc. of International Conference, 1997. 7

[185] K. Willner and D. Goerke, *Contact of rough surfaces. a comparison of experimental and numerical results*, ECCM III. Solid, Structures and Coupled Problems, Proc. of International Conference, 2006. 7

[186] B. I. Wohlmuth, *A mortar finite element method using dual spaces for the lagrange multiplier*, SIAM J Numer Anal **38** (2000), 989–1012. 364

[187] P. Wriggers, *Finite element algorithms for contact problems*, Archives of Computational Methods in Engineerin **24** (1995), 1–49. 16, 75, 109, 225

[188] ———, *Computational contact mechanics*, John Wiley and Sons, 2002. 3, 5, 11, 68, 72, 77, 78, 90, 97, 98, 99, 109, 122, 212, 225, 226, 234, 235, 237, 240, 265, 288, 291, 293, 364, 433

REFERENCES

[189] ______ , *Computational contact mechanics*, Springer, 2006. 38, 61, 64, 65

[190] P. Wriggers, L. Krstulovic-Opara, and J. Korelc, *Smooth c1-interpolations for two-dimensional frictional contact problem*, International Journal for Numerical Methods in Engineering **51** (2001), 1469–1495. 90, 104, 176, 240, 325

[191] P. Wriggers and O. Scherf, *Different a posteriori error estimators and indicators for contact problems*, Mathematical and Computer Modelling **28** (1998), 437–447. 240, 276

[192] P. Wriggers and J. C. Simo, *A note on tangent stiffness for fully nonlinear contact problems*, Communications in Applied Numerical Methods **1** (1985), 199–203. 3, 4, 5, 12, 16, 36, 37, 68, 258

[193] P. Wriggers, J. C. Simo, and R. L. Taylor, *Penalty and augmented lagrangian formulations for contact problems*, Proceeding NUMETA Conference (editors J.Middleton and G.Pande, ed.), Proceedings of NUMETA Conference, 1985. 68

[194] P. Wriggers, Vu Van, and E. Stein, *Finite element formulation of large deformation impact-contact problem with friction*, Computers and Structures **37** (1990), 319–331. 5, 12, 16, 36, 37, 76, 90, 111, 150, 212, 291

[195] P. Wriggers and G. Zavarise, *On the application of augmented lagrangian techniques for nonlinear constitutive laws in contact interface*, Communications in Applied Numerical Methods **9** (1993), 815–824. 7, 288, 382

[196] ______ , *Thermomechanical contact – a rigorous but simple numerical approach*, Computer and Structures **46** (1993), 47–53. 7, 288, 382

[197] ______ , *On contact between three-dimensional beams undergoing large deflection*, Communications in Numerical Methods in Engineering **13** (1997), 429–438. 13, 16, 408

[198] ______ , *A formulation for frictionless contact problems using a weak form introduced by nitsche*, Computational Mechanics **41** (2008), 407–420. 12

[199] G. Zavarise and P. Wriggers, *A segment-to-segment contact strategy*, Mathematical and Computer Modelling **28** (1998), 497–515. 4, 190, 239

[200] ______ , *Contact with friction between beams in 3-d space*, International Journal for Numerical Methods in Engineering **49** (2000), 977–1006. 13, 16, 408

[201] G. Zavarise, P. Wriggers, E. Stein, and B. A. Schrefler, *A numerical model for thermomechanical contact based on microscopic interface laws*, Mechanics Research Communications **19** (1992), 173–182. 288

[202] ______ , *Real contact mechanisms and finite element formulation a coupled thermomechanical approach*, International Journal for Numerical Methods in Engineering **35** (1992), 767–785. 7

[203] H. W. Zhang, S. Y. He, X. S. Li, and P. Wriggers, *A new algorithm for numerical solution of 3d elastoplastic contact problems with orthotropic friction law*, Computational Mechanics **34** (2004), 1–14. 8, 289, 340

[204] Q.-S. Zheng, *Theory of representation for tensor function – a unified invariant approach to constitutive equations*, Applied Mechanics Reviews, ASME International **47** (1994), 545–587. 300

[205] Z. H. Zhong, *Finite element procedures for contact-impact problems*, Oxford Univ. Press, Oxford, 1993. 15, 37, 109

[206] Z. H. Zhong and L.A. Nilsson, *A contact searching algorithm for general contact problems*, Computer and Structures **33** (1989), 197–209. 15

[207] ______, *A unified contact algorithm based on the territory concept.*, Computer Methods in Applied Mechanics and Engineering **130** (1996), 1–16. 15, 38, 52

[208] B. Zhou, M. L. Accorsi, and J. W. Leonard, *Finite element formulation for modeling sliding cable elements*, Computers and Structures **82** (2004), 271–280. 409

[209] O. C. Zienkiewicz and R. L. Taylor, *Finite element method: The basis*, 5th edn ed., vol. 1, Butterworth-Heinemann, New York, 2000. 82, 173

[210] A. Zmitrowicz, *A theoretical model of anisotropic dry friction.*, Wear **73** (1981), 9–39. 8, 288, 296, 300, 382

[211] ______, *Mathematical description of anisotropic friction.*, International Journal of Solids and Structures **25** (1989), 837–862. 8, 288, 290, 296, 300, 382, 389

[212] ______, *An equation of anisotropic friction with sliding path curvature effects*, International Journal of Solids and Structures **36** (1999), 2825–2848. 340

[213] ______, *Illustrative examples of anisotropic friction with sliding path curvature effects*, International Journal of Solids and Structures **36** (1999), 2849–2863. 340

[214] ______, *Models of kinematics dependent anisotropic and heterogenous friction*, International Journal of Solids and Structures **43** (2006), no. 14-15, 4407–4451. 340, 405

Publications (after 2002)

[E1] Konyukhov A., Schweizerhof K. Description for smooth contact conditions based on the internal geometry of the contact surfaces. *PAMM*, 2003, **3**: 290–291.

[E2] Konyukhov A., Schweizerhof K. Contact formulation via a velocity description allowing efficiency improvements in frictionless contact analysis. *Computational Mechanics*, 2004, **33**: 165–173.

[E3] Konyukhov A., Schweizerhof K. Large deformation frictional contact formulation based on velocity description. *PAMM*, 2004, **4**: 334–335.

[E4] Harnau M., Konyukhov A., Schweizerhof K. Algorithmic aspects in large deformation contact analysis using 'Solid-Shell' elements. *Computers and Structures*, 2005, **83**: 1804–1823.

[E5] Konyukhov A., Schweizerhof K. Covariant description for frictional contact problems. *Computational Mechanics*, 2005, **35**: 190–213.

[E6] Müller I., Konyukhov A., Vielsack P., Schweizerhof K. Parameter estimation for finite element analyses of stationary oscillations of a vibro-impacting system. *Engineering Structures*, 2005, **27**: 191–201.

[E7] Konyukhov A., Schweizerhof K. Modeling of anisotropic surfaces within a covariant description. *PAMM*, 2005, **5**: 421–422.

[E8] Konyukhov A., Schweizerhof K. A special focus on 2D formulations for contact problems using a covariant description. *International Journal for Numerical Methods in Engineering*, 2006, **66**: 1432–1465.

[E9] Konyukhov A., Schweizerhof K. Covariant description of contact interfaces considering anisotropy for adhesion and friction. Part 1. Formulation and analysis of the computational model. *Computer Methods in Applied Mechanics and Engineering*, 2006, **196**(1-3): 103–117.

[E10] Konyukhov A., Schweizerhof K. Covariant description of contact interfaces considering anisotropy for adhesion and friction. Part 2. Linearization, finite element implementation and numerical analysis of the model. *Computer Methods in Applied Mechanics and Engineering*, 2006; **196**(1-3): 289–303 .

[E11] Konyukhov A., Schweizerhof K. On a geometrical approach in contact mechanics.In book *Analysis and Simulation of Contact Problems*, Lecture Notes in Applied and Computational Mechanics. Wriggers P. and Nackenhorst U. (eds.), pp. 23–30, 2006.

[E12] Konyukhov A., Schweizerhof K. High order FEM and covariant description for contact problem. *PAMM*, 2006, **6**:225–226.

[E13] Konyukhov A., Schweizerhof K. On a continuous transfer of history variables for frictional contact problems based on interpretations of covariant derivatives as a parallel translation. In *IUTAM Symposium -"Multiscale problems in Multibody System Contacts"*. P. Eberhard (ed.), pp. 95–101, IUTAM Bookseries, Springer, 2007.

[E14] Konyukhov A., Schweizerhof K. Symmetrization of various friction models based on an Augmented Lagrangian approach. In *IUTAM Symposium on "Computational Contact Mechanics"*, U.Nackenhorst, P. Wriggers eds., pp. 97–111, IUTAM Bookseries. Springer, 2007.

[E15] Konyukhov A., Schweizerhof K. Closest point projection in contact mechanics: existence and uniqueness for different type of surfaces. *PAMM*, 2007, **7**: 4040053-4040054.

[E16] Konyukhov A., Vielsack P., Schweizerhof K. On coupled models of anisotropic contact surfaces and their experimental validation. *Wear*, 2008, **7-8**: 579-588.

[E17] Konyukhov A., Schweizerhof K. On the solvability of closest point projection procedures in contact analysis: analysis and solution strategy for arbitrary surface approximation. *Computer Methods in Applied Mechanics and Engineering*, 2008, 197, (33-40), 3045-3056.

[E18] Konyukhov A., Schweizerhof K. Geometrical covariant approach for contact between curves representing beam and cable type structures. *PAMM*, 2008; **8**:10299-10300.

[E19] Konyukhov A., Schweizerhof K. Incorporation of contact for high-order finite elements in covariant form. *Computer Methods in Applied Mechanics and Engineering*, **198**(13-14), 2009, 1213-1223. HOFEM07 - International Workshop on High-Order Finite Element Methods, 2007

[E20] Konyukhov A., Schweizerhof K. Isogeometrical approach for cable type structures allowing large sliding contact. *PAMM*, 2009; to appear.

[E21] Izi R., Schweizerhof K., Konyukhov A. Stability of thin walled structures strongly coupled with contact. *PAMM*, 2009; to appear.

[E22] Konyukhov A., Schweizerhof K. On a geometrically exact theory for contact interactions. In book *Analysis and Simulation of Contact Problems*, Lecture Notes in Applied and Computational Mechanics. F.Pfeiffer, Wriggers P. (eds.), to appear, 2009.

[E23] Konyukhov A., Schweizerhof K. Geometrically exact covariant approach for contact between curves. *Computer Methods in Applied Mechanics and Engineering*, 2009, submitted.

Selected contributions before 2002.

1. Жигалко Ю.П., Конюхов А.В. Напряженное состояние тонкостенных элементов конструкций, нагреваемых локализованными потоками лучистой энергии. *Изв. ВУЗов «Авиационная техника»* 1997, **2**:19-25.
Zhigalko Yu.P., Konyukhov A.V. Stress-strain state of thinwalled structural elements heated by localized heat flow. *Izv. VUZ. Aviatsionnaya Tekhnika translated in Allerton Press, Inc., USA under the title "**Russian Aeronautics**",* 1997, **2**:19-25.

2. Конюхов А.В. Локальные эффекты в термоупругих пластинах и оболочках. Дисс. к.ф.м.н., Казань, КГУ, 1999.
Konyukhov A.V. Local effect in thermoelastic plates and shells, ***PhD thesis***, *Kazan State University, KSU,* 1999.

3. Конюхов А.В., Жигалко Ю.П. Термоупругие пластины и оболочки при локальных воздействиях. *Изв. ВУЗов «Авиационная техника»* 2000, **2**:21-32.
Konyukhov A.V., Zhigalko Yu.P. Thermoelastic plates and shells under various local interactions. *Izv. VUZ. Aviatsionnaya Tekhnika translated in Allerton Press, Inc., USA under the title "**Russian Aeronautics**",* 2000, **2**:21-32.

4. Конюхов А.В. Температурные поля и напряжения в пластинах и оболочках при локальном нагреве. *Изв. ВУЗов «Авиационная техника»* 2001, 2.
Konyukhov A.V. Thermal fields and stresses in plates and shells under local heatings: some analytical solutions. *Izv. VUZ. Aviatsionnaya Tekhnika translated in Allerton Press, Inc., USA under the title "**Russian Aeronautics**",* 2001, **2**.

5. Конюхов А.В. Основы анализа конструкций в ANSYS. Казань. Изд-во КГУ. 2001, 102 с.
Konyukhov A.V. *The Basics of Structural Analysis with ANSYS.* Kazan, Kazan State University Press, 2001.

6. GolovanovA.I., Mitryaykin V.I., Michaylov S.A., Konyukhov A.V. Numerical analysis and experimental investigation of a tail rotor blade composite torsion, *in Proceeding of the 27th European rotorcraft forum.* Moscow, 2001. 85. 12 P.

7. Конюхов А.В. Жигалко Ю.П. Влияние дефекта структуры на напряженное состояние термоупругих оболочек. Изв. вузов "Авиационная техника", 2002,1.
Konyukhov A.V. Influence of the structural deffects on stress-strain state of thermoelastic shells. *Izv. VUZ. Aviatsionnaya Tekhnika translated in Allerton Press, Inc., USA under the title "**Russian Aeronautics**",* 2002, **1**.

Presentations and other contributions
(after 2002)

[C1] Konyukhov A., Schweizerhof K. Thermomechanical contact problem compatible with "solid-shell" elements. *Dynamic Days Europe, International Conference, Heidelberg, July 15-19, 2002.*

[C2] Konyukhov A., Schweizerhof K. Description for smooth contact conditions based on the internal geometry of the contact surfaces. *75. GAMM Tagung, Abano Terme-Padua, Italy, March 24-28, 2003.*

[C3] Harnau M., Schweizerhof K., Konyukhov A. Solid-Shell Elements with Surface Contact Formulation for Large Deformation Contact Problems. *VII International Conference on Computational Plasticity. COMPLAS 2003. Barselona, Spain, April 7-10, 2003.*

[C4] Konyukhov A., Schweizerhof K. Kinematical approach for contact problems with arbitrary large deformations. *Conference Proceedings of 20th International Conference (in Russian and English)*, BEM-FEM, Sankt-Peterburg, Russia, Sept. 24-26, 2003.

[C5] Konyukhov A., Konoplev Yu.G. Ballon problems for hyperelastic thermomechanical shells. *Conference Proceedings of 20th International Conference (in Russian and English)*, BEM-FEM, Sankt-Peterburg, Russia, Sept. 24-26, 2003.

[C6] Konyukhov A., Schweizerhof K. Large Deformation Frictional Contact Formulation Based on Velocity Description. *75. GAMM Tagung, Dresden, Germany, March 21-27, 2004.*

[C7] Müller I., Schmieg H., Vielsack P., Konyukhov A. Experimentelle und numerische Untersuchung delaminierter Strukturen zur schwingungsbasierten Schadenidentifikation. *VDI-Schwingungstagung, 25-26 Mai, Wiesloch, Deutschland, VDI-Berichte, 1825: 157-176, 2004.*

[C8] Konyukhov A., Schweizerhof K., Harnau M. Large Deformation Frictional Contact Formulation for Low Order "Solid Shell" Elements. *ECCOMAS-2004, Jyv"askyl"a, Finland, July 24-28, 2004.* Conference Proceedings eds. P. Neittanm"aki, T. Rossi, K. Majava, O. Pironneau, 2004.

[C9] Konyukhov A.V. Covariant theory of the contact interaction. *Research Seminar. Moscow State University, Chair of "Mechanics of Composite Materials", Prof. Pobedrya, Nov. 22, 2004.*

[C10] Konyukhov A., Schweizerhof K.. Modeling of anisotropic surfaces within a covariant description. *76. GAMM Tagung, Luxemburg, 28 March-01 April 2005.*

[C11] Konyukhov A., Schweizerhof K. Application of a covariant description to the contact of shells with different approximation. *5th IASS-IACM, Salzburg, Austria, June 1-4, 2005. Conference Proceedings.*

[C12] Konyukhov A., Schweizerhof K. On a geometrical approach in contact mechanics. *4th Contact Mechanics International Symposium, Hannover/Loccum, Germany, July 4-6, 2005.*

[C13] Konyukhov A., Schweizerhof K. Covariant formulation of anisotropic contact interfaces. *8th US National Congress on Computational Mechanics, Austin, Texas, USA, July 25-27, 2005.*

[C14] Konyukhov A., Schweizerhof K. Covariant description of anisotropic contact surfaces. *VIII International Conference on Computational Plasticity. COMPLAS 2005. Barselona, Spain, September 5-8*, Conference Proceedings, Oñate, E. and Owen D.R.J. (eds.) 2005.

[C15] Konyukhov A., Schweizerhof K. On a continuous transfer of history variables for frictional contact problems based on interpretations of covariant derivatives as a parallel translation. *IUTAM-Symposium "Multiscale problems in Multibody System Contacts"*. Conference Proceedings, Stuttgart, Germany, February 20-23, 2006.

[C16] Konyukhov A., Schweizerhof K. High Order FE and Covariant Description for Contact Problems. *77. GAMM Tagung, Berlin, March 27-31, 2006.*

[C17] Konyukhov A. Covariant description for frictional contact problems and its generalization into anisotropy. *Research Seminar. Universität Siegen, Institut für Mechanik und Regelungstechnik, Prof. P. Betsch, April 11, 2006.*

[C18] Konyukhov A. Covariant description for frictional contact problems and its generalization into anisotropy. *LSTC, Livermore, California, USA, Research Seminar org. by Dr. J. Hallquist. April 20, 2006.*

[C19] Konyukhov A., Schweizerhof K., Vielsack P. On models of contact surfaces including anisotropy for friction and adhesion and their experimental validations. *III European Conference on Computational Mechanics. Solid, Structures and Coupled Problems in Engineering. ECCM-III*, Conference Proceedings, Lissabon, Portugal, June 5-9, 2006.

[C20] Schweizerhof K., Konyukhov A. Contact with shells. *7th World Congress on Computational Mechanics. WCCM-7, Los-Angeles, USA, July 16-22, 2006.*

[C21] Konyukhov A., Schweizerhof K. Development of a model for contact surfaces including friction and adhesion. *7th World Congress on Computational Mechanics. WCCM-7, Conference Proceedings, Los-Angeles, USA, July 16-22, 2006.*

[C22] Konyukhov A., Schweizerhof K. Symmetrization of various friction models based on an Augmented Lagrangian approach. *IUTAM-Symposium on Computational Contact Mechanics, Hannover, Germany, November 4-8, 2006.*

[C23] Konyukhov A. Covariant description for frictional contact problems and its generalization into anisotropy. *University of Linköping, Division of Solid Mechanics, Sweden, Research Seminar org. by Prof. Larsgunnar Nilsson, March 30, 2007.*

[C24] Konyukhov A.V. Geometrically exact theory for contact interaction: covariant approach for computations. *Research Seminar. Moscow State University, Chair of "Mechanics of Composite Materials", Prof. Pobedrya, Apr. 9, 2007.*

[C25] Konyukhov A., Schweizerhof K. Incorporation of contact for high order FEM in covariant form. *International Workshop on High-Order Finite Element Methods, 17-19 May, 2007, Herrsching am Ammersee (near Munich), Germany.*

[C26] Konyukhov A., Schweizerhof K. Closest point projection in contact mechanics: existence and uniqueness for different type of surfaces. *6th International Congress on Industrial and Applied Mathematics (GAMM is embedded), 16-20 July, 2007, Zürich, Switzerland.*

[C27] Konyukhov A., Schweizerhof K. Solvability of the closest point projection routines in contact analyses: Continuous projection domain for arbitrary surfaces. *9th US National Congress on Computational Mechanics, San-Francisco, USA, July 23-26, 2007.*

[C28] Schweizerhof K., Konyukhov A. On an Augmented Lagrangian method for anisotropic friction models. *9th US National Congress on Computational Mechanics, San-Francisco, USA, July 23-26, 2007.*

[C29] Konyukhov A., Schweizerhof K. Generalized closest point projection for contact analyses: on existence and uniqueness for arbitrary contact surfaces. *IX International Conference on Computational Plasticity. COMPLAS 2007. Barcelona, Spain, September 5-7*, Conference Proceedings, Oñate, E. and Owen D.R.J. (eds.) 2007.

[C30] Konyukhov A. On a geometrically exact (covariant) approach in contact mechanics. *FE im Schnee 2008, February 06-09, 2008, Söllerhaus, Austria.*

[C31] Konyukhov A., Schweizerhof K. Geometrical covariant approach for contact between curves representing beam and cable type structures. *79th Annual Meeting of Gesellschaft fï¿½r Angewandte Mathematik und Mechanik, 79th GAMM, Bremen, 31 March ? 4 April, 2008.*

[C32] Konyukhov A., Schweizerhof K. On a geometrically exact contact description for shells: from linear approximations to high-order FEM. *IASS-IACM 2008, 28-31 May, 2008, Ithaca, New York, USA.*

[C33] Schweizerhof K., Ewert E., Konyukhov A., R. Izi Stability and sensitivity analysis of imperfect shells involving contact. *IASS-IACM 2008, 28-31 May, 2008, Ithaca, New York, USA.*

[C34] Konyukhov A., Schweizerhof K. Covariant description for contact between arbitrary curves: general approach for beams, cables and surface edges. 8th.*World Congress on Computational Mechanics (WCCM8) 5th European Congress on Computational Methods in Applied Sciences and Engineering (ECCOMAS 2008), June 30 ?July 5, 2008, Venice, Italy.*

[C35] Schweizerhof K., Ewert E., Konyukhov A., Izi R. Stability and Sensitivity of Shell-Like Structures considering Imperfections and Contact. 8th. World Congress on Computational Mechanics (WCCM8) 5th European Congress on Computational Methods in Applied Sciences and Engineering (ECCOMAS 2008), June 30 –July 5, 2008, Venice, Italy.

[C36] Konyukhov A. On a geometrically exact theory in contact mechanics. *University of Linköping, Division of Solid Mechanics, Sweden, Research Seminar org. by Prof. Larsgunnar Nilsson, 12 December, 2008.*

PRESENTATIONS AND OTHER CONTRIBUTIONS (AFTER 2002)

[C37] Konyukhov A., Schweizerhof K. Isogeometrical approach for cable type structures allowing large sliding contact. *80th GAMM Gdansk-Poland, 9-13 February, 2009.*

[C38] Izi R., Schweizerhof K., Konyukhov A. Stability of thin walled structures strongly coupled with contact. *80th GAMM Gdansk-Poland, 9-13 February, 2009.*

[C39] Konyukhov A., Borodich F., Schweizerhof K. Mechanical Modeling of Adhesive Contact between Animal Hairy Feet and Rough Surfaces *28. WE-Heraeus-Seminar, Bad Honnef, 23-25 March, 2009*

[C40] Konyukhov A., Schweizerhof K. On a geometrically exact theory for contact interactions *5th Contact Mechanics International Symposium, CMIS2009 Chania, Crete, Greece, April 28-30, 2009.*

[C41] Schweizerhof K., Konyukhov A. Covariant description for contact between curves: application to edge-to-edge and beam-to-beam contact. *10th US National Congress on Computational Mechanics, Columbus, Ohio, USA, July 16-19, 2009.*

[C42] Konyukhov A., Schweizerhof K. Isogeometrical approach for curved beams allowing large sliding contact. *X International Conference on Computational Plasticity, COMPLAS X, Barcelona, Spain, 2-4 September, 2009.*

[C43] Schweizerhof K., Konyukhov A. Isogeometrical approach for curved cables – application to the tying of knots. *1st International Conference on Computational Contact Mechanics, ICCCM09, 16-18 September, 2009, Lecce, Italy.*

[C44] Konyukhov A., Schweizerhof K. Isogeometrical approach for curved beams allowing large sliding contact. *Conf. proceedings of COMPLAS X, Barcelona, Spain, 2-4 September, 2009.*

[C45] Konyukhov A., Schweizerhof K. On a geometrically exact theory for contact interactions. *1st International Conference on Computational Contact Mechanics, ICCCM09, 16-18 September, 2009, Lecce, Italy.*

[C46] Izi R., Schweizerhof K., Konyukhov A. FE Schemes for Stability Problems Strongly Coupled with Contact. *3rd GACM Colloquium on Computational Mechanics for Young Scientists from Academia and Industry, 21-23 September, 2009, Leibniz Universität Hannover.*